Multicore DSP

Multicore DSP

From Algorithms to Real-time Implementation
on the TMS320C66x SoC

Naim Dahnoun
University of Bristol
UK

Registered Office(s)
John Wiley & Sons, Inc., 111 River Street, Hoboken, NJ 07030, USA
John Wiley & Sons Ltd, The Atrium, Southern Gate, Chichester, West Sussex, PO19 8SQ, UK

Editorial Office
The Atrium, Southern Gate, Chichester, West Sussex, PO19 8SQ, UK

For details of our global editorial offices, customer services, and more information about Wiley products visit us at www.wiley.com.

Wiley also publishes its books in a variety of electronic formats and by print-on-demand. Some content that appears in standard print versions of this book may not be available in other formats.

Library of Congress Cataloging-in-Publication data applied for

ISBN: 9781119003823

Cover design by Wiley
Cover image: © matejmo/Gettyimages

Set in 10/12pt Warnock by SPi Global, Pondicherry, India

Printed in Singapore by C.O.S. Printers Pte Ltd

10 9 8 7 6 5 4 3 2 1

I dedicate this book to my children
Zahra, Yasmin and Riyad
and in memory of my parents

Contents

Preface

Today's many applications, such as medical, high-end imaging, high-performance computing and core networking, are facing increasing challenges in terms of data traffic, processing power and device-to-device communication. These put a high demand on the processor(s) and associated software and lead to processor manufacturers sustaining Moore's law by introducing multicore processors. Texas Instruments, with its leading-edge technology, introduced the multicore System-on-Chip (SoC) architecture family of processors to address these issues. As will be shown in this book, Texas Instruments introduced innovations at many levels, such as: powerful CPUs that support both fixed- and floating-point arithmetic (instruction by instruction) that can achieve more than 40G multiplications/core, a Navigator that enables direct communication between cores and memory access that removes data movement bottlenecks, a Hyperlink interface and advanced development tools.

The challenge is not only how many cores you can put on a piece of silicon, the processing power of each core and how fast they can communicate, but also in the programming model and ease of use. Unfortunately, programming models are not developed sufficiently to handle several cores. The improvement in performance gained by the use of a multicore processor depends very much on the application and software used. C and C++, which are commonly used in embedded systems, do not support partitioning and, therefore, porting sequential code to multicore is not trivial. In this book, it will be shown this complexity is alleviated by using: OpenMP, which is an Application Programming Interface (API) that supports multiplatform shared multiprocessing programming in C, C++ and Fortran; Open Computing Language (OpenCL); or the Inter-Processor Communication (IPC).

This book will help to innovate by making the reader understand the KeyStone SoC architectures, the development tools including debugging and various programming models with tested examples, and also help to broaden the knowledge by critically analysing each element (see Table of Contents) and understanding how these elements are working together. With the sheer number of practical examples and references provided, the reader will be able to quickly develop applications, take advantage of maximum performance and functionality of the processors, be able to easily use the tools to develop and debug applications and find the relevant references to pertinent material. Real-time multicore audio and video applications are provided. Applications will be based on TI's Multicore Software Development Kit (MCSDK), hand-optimised code, OpenMP, OpenCL and IPC.

Due to the sheer amount of documentation available, some information is either referred to or reproduced to avoid discontinuity and misinterpretation.

This book is divided into 20 chapters. Chapters 1 to 15 deal with the hardware and software issues, and Chapters 16 to 20 deal with applications. Most of the concepts are backed up with laboratory experiments and demos that have been thoroughly tested.

Chapter 1 Introduction: This introductory chapter provides the reader with general knowledge on multicore processors and their applications; gives a brief comparison between digital signal processor (DSP) SoCs, field-programmable gate arrays (FPGAs), graphic processors and CPUs; illustrates the challenges associated with multicore; and provides an up-to-date TMS320 roadmap showing the evolution of TI's DSP chips in terms of processing power.

Chapter 2 The TMS320C66x architecture overview: This chapter comprehensively describes the TMS320C66x architecture. This includes a detailed description of the DSP CorePacs and an overview of the peripherals, and it introduces some useful instructions and an overview of the memory organisation.

Chapter 3 Software development tools and the TMS320C6678 EVM: This chapter describes the software development tools that are required for testing the applications used in this book. It provides a step-by-step guide to the installation and use of the Code Composer Studio (CCS).

Chapter 4 Numerical issues: This chapter explains how fixed and floating points are represented and how to handle binary arithmetic. It provides examples showing how to display various data formats using the CCS.

Chapter 5 Software optimisation: This chapter discusses the different levels of optimisation for multicore and shows how code can be optimised for a DSP core. This chapter also shows how to use intrinsics and interface C language with intrinsics and assembly code. Multiple examples showing how to optimise code by hand and using the tools are provided.

Chapter 6 The TMS320C66x interrupts: This chapter shows how the interrupt controller events and the Chip-level Interrupt Controller work and how to program them to respond to events. The examples given use the general-purpose input–output (GPIO) pins to provide the interrupts.

Chapter 7 Real-time operating system: TI-RTOS: This chapter is divided into three main sections: (1) a real-time scheduler that is composed of the hardware and software interrupts, the task, the idle, clock and timer functions, synchronisation and events; (2) dynamic memory management; and (3) laboratory experiments.

Chapter 8 Enhanced Direct Memory Access (EDMA3) Controller: This chapter describes in detail the operation of the EDMA and provides examples with simple transfer, chaining transfer and linked transfer.

Chapter 9 Inter-Processor Communication (IPC): This chapter explains the need for IPC and describes the notify module, the messageQ, the ListMP module, the Multi-processor Memory Allocation, the transport mechanism and laboratory examples.

Chapter 10 Single and multicore debugging: This chapter introduces the need for debugging and describes the debug architecture that includes trace, Advanced Event Triggering and the Unified Breakpoint Manager. This chapter also describes the Unified Instrumentation Architecture, debugging with the System Analyzer tools, instrumentation with TI-RTOS and CCS and laboratory experiments.

Chapter 11 Bootloader for Keystone I and Keystone II: This chapter introduces the boot process for both the KeyStone I and KeyStone II, and provides laboratory experiments for both devices.

Chapter 12 Introduction to OpenMP: This chapter introduces the concept behind OpenMP and divides the content into three main sections: (1) work sharing, (2) data sharing and (3) synchronisation. Various examples with both KeyStone I and II are provided. For the KeyStone II, an example is implemented with OpenMP with the accelerator model.

Chapter 13 Introduction to OpenCL for the KeyStone II: In this chapter, another programming model called Open Computing Language (OpenCL) is introduced. This chapter will emphasise the OpenCL for the KeyStone rather than other devices. This chapter will show that OpenCL is easy to use since the programmer does not need to deal with details of communication between DSP cores or between the ARM and the DSP, which may be a daunting task.

Chapter 14 Multicore Navigator: This chapter shows how the Multicore Navigator can provide a high-speed packed data transfer to enhance CorePac to accelerator/peripheral data movements, core-to-core data movements, inter-core communication and synchronisation without loading the CorePacs. Examples are also provided.

Chapter 15 FIR filter implementation: The purpose of this chapter is twofold. Primarily, it shows how to design an FIR filter and implement it on the TMS320C66x processor; and, secondly, it shows how to optimise the code as discussed in Chapter 3. This chapter discusses the interface between C and assembly, how to use intrinsics, and how to put into practice material that has been covered in the previous chapters.

Chapter 16 IIR filter implementation: This chapter introduces the IIR filters and describes two popular design methods: the bilinear and the impulse invariant methods. Step by step, this chapter shows the procedures necessary to implement typical IIR filters specified by their transfer functions. Finally, this chapter provides complete implementation of an IIR filter in C language, assembly and linear assembly, and shows how to interface C with linear assembly.

Chapter 17 Adaptive filter implementation: This chapter starts by introducing the need for an adaptive filter in communications. It then shows how to calculate the filter coefficients using the mean squared error (MSE) criterion, exposes the least mean squares (LMS) algorithm and, finally, shows how the LMS algorithm is implemented in both C and assembly.

Chapter 18 FFT implementation: This chapter shows a derivation of an FFT algorithm and shows its implementation in C language. To improve the performance, the ping-pong EDMA has been used.

Chapter 19 Hough transform: This chapter shows the basic mathematics behind the Hough transform for detecting straight lines and how to implement it. This chapter also shows how to increase the performance by looking at the algorithm and minimising the number of operations required, and how to use the graphical display using the Code Composer Studio.

Chapter 20 Stereo vision implementation: This chapter shows the principle behind the stereo vision system and highlights different levels of optimisations for achieving real-time performance. Some techniques for reducing the processing time for calculating the disparity values for automotive applications are also introduced.

Acknowledgements

I didn't expect that writing another book would be challenging, considering that I have written previous material. This was mainly due to the fast-moving technology and because systems are getting more complex, but I had to keep up with it. Putting all knowledge gained in a simple form and sharing it give me great satisfaction.

My first thanks go to the Engineer to Engineer (e2e.ti.com) community that I found extremely generous and very helpful.

I am indebted to Cathy Wicks, Jocken Schyma, Jason Brand, Garry Clarkson and Rogerio Almeida, the key motivators and initiators for writing this book.

I would like to express my gratitude to the following reviewers for their insightful suggestions and comments: Pekka Varis, Dave Bell, Eric Stotzer, John Smrstik, Jennifer Stadelmann, Ran Katzur, Steve Preissig, Naga Chandrashekar, Greg Peake and Filip Moerman.

I owe my thanks to Professor John Rarity, Professor Andy Nix, Professor Dave Cliff, Professor David May, Dr Richard Nock, Dr Ross Xi, Dr Sergey Vityazev, Mr James Webley and all my colleagues at the Faculty of Engineering, University of Bristol, for their encouragement and support. I also thank all my students, and in particular Scott Tancock, Han Cui, Aleksandar Stanoev, Victor Prokhorov, Aliaksei Mikhailiuk, Akmal Ahmed, Ranger Fan, Elliott Worsey and Charles Khoury who were always eager to explore challenging issues, verify and test applications.

My thanks to my friends Professor Hamdani Abdelsalam, Judge Shamin and his wife Shabina, Hernandez Paul and his wife Khansa, Mr Baris Tanyeri, Mr Karaborek Yunus, Dr Tila Fai, Mr Ghoul Amar and Dr Seti Chenafa for their continuous support and friendship.

To my friend Gene Frantz, 'the father of DSP', who wrote the Foreword to my book and who has been a great inspiration to me.

Finally, many thanks to Ruth Thomas for her kindness in reading the whole manuscript and providing feedback, and Alex King and Preethi Belkese from Wiley who were very easy to deal with, encouraging and supportive.

Naim Dahnoun
Naim.Dahnoun@Bristol.ac.uk

Foreword

Having spent my professional career introducing the digital signal processing (DSP) technology and associated products to the industry, and now as a Professor in the practice at Rice University, I continue looking for the next use or user of DSP. One of the high points of my career has been working with professors and authors who are preparing the next generation of talented engineers.

I have known Naim for about 20 years, before he wrote his first and popular book entitled 'Digital Signal Processing Implementation: Using the TMS320C6000^{TM} Platform', which I reviewed. Since then, DSP processors have evolved into advanced heterogeneous multicore processors that are hard to program. To extract maximum performance, programmers need to master not only the applications that must be implemented but also the processor's hardware and supporting software. For instance, many programming models such as Message Passing Interface (MPI), Open Multi-Processing (OpenMP), Open Computing Language (OpenCL) and Inter-Processor Communication (IPC) have been introduced to ease development, in addition to the operating systems. Each of these programming languages is implemented differently by different device manufacturers, and each of these programming languages is covered in separate books. To make the best use of these programming models, one needs to compare and contrast them for a specific application. This book covers most of these programming models and gives the reader a good starting point.

This book is rich in its well-structured content and is worthy of deep and reflective reading. It starts by highlighting solutions of some problems on multicore processors, and then focusses on multicore DSPs. To gain maximum performance, this book provides details at the assembly and the linear assembly levels, and then shows how this could be achieved by using the appropriate compiler switches to save development time, increase portability and reduce maintenance. The book then tackles IPC, OpenMP, OpenCL and the Navigator to ease programming the Multicore DSP, and it provides a rich set of practical examples for both the KeyStone I and the KeyStone II platforms.

Debugging is as important as programming itself, especially in large and complex applications. With this in mind, silicon manufacturers have heavily invested in both hardware and software debugging, and in this book, Naim recognises the need to simplify its use.

In addition to hardware and development software, this book also shows how to implement main signal-processing algorithms such as FIR, IIR, adaptive filters, Hough transform, FFTs and disparity calculation for stereo vision applications.

There is no doubt that this book, with its comprehensive content, will provide the reader with knowledge and inspiration that will allow him or her to experiment and maybe push the boundaries even further.

Gene Frantz

About the Companion Website

Don't forget to visit the companion website for this book:

www.wiley.com/go/dahnoun/multicoredsp

There you will find valuable material designed to enhance your learning, including:

1) Appendix 1: Creating a Virtual Machine
2) Appendix 2: Software Directory
3) Appendix 3: Software updates
4) Exercises and Solutions
5) Source codes

Scan this QR code to visit the companion website:

1

Introduction to DSP

CHAPTER MENU

Learning how to master a system-on-chip (SoC) can be a long, daunting process, especially for the novice. However, keeping in mind the big picture and understanding why a specific piece of hardware or software is used will remove the complexity in the details.

The purpose of this chapter is give an overview for the need of multicore processors, list different types of multicore processors and introduce the KeyStone processors that are the subject of this book.

1.1   Introduction

Today's microprocessors are based on switching devices that provide alternation between two states, ON and OFF, that represent 1 s and 0 s. Up to now, the transistor is the only practical device to be used. Having small, fast, low-power transistors has always been the challenge for chip manufacturers. From the 1960s, as predicted by Gordon Moore (in Moore's law), the number of transistors that could be fitted in an integrated circuit doubled every 24 months [1]. That was possible due to new material, the development of chip process technology and especially the advances in photolithography that pushed the transistor size from 10 μm in the 1960s to about 10 nm currently. As the transistor scaled, industry not only took advantage of the transistor count but also increased the clock speed, using various architecture enhancements such as instruction-level parallelism (ILP) that can be achieved by superscaling (loading multiple instructions simultaneously and executing them simultaneously), pipelining (where different

Multicore DSP: From Algorithms to Real-time Implementation on the TMS320C66x SoC, First Edition. Naim Dahnoun.
© 2018 John Wiley & Sons Ltd. Published 2018 by John Wiley & Sons Ltd.
Companion website: www.wiley.com/go/dahnoun/multicoredsp

phases of instructions overlap), out-of-order execution (instructions are executed in any order, and the choice of the order is dynamic) and so on, and different power-efficient cache levels and power-aware software designs such as compilers for low-power consumption [2] and low-power instructions or instructions of variable length. However, the increase in clock frequency was not sustainable as power consumption became such a real constraint that it was not possible to produce a commercial device. In fact, chip manufacturers have abandoned the idea of continually increasing the clock frequency because it was technically challenging and costly and because power consumption was a real issue, especially for mobile computing devices such as smartphones and handheld computers and for high-performance computers. Recently, static power consumption has also become a concern as the transistor scales, and therefore both dynamic power and static powers are to be considered. It is also worth noting at this stage that increase in the operating frequency requires power consumption increase, that is not linear with the frequency, as one can assume.

This is due to the fact that an increase in frequency will require an increase in voltage. For instance, an increase of 50% of the frequency will also result in an increase of 35% of the voltage [2].

To overcome the problem of frequency plateau, processor manufacturers like Texas Instruments (TI), ARM and Intel find that by keeping the frequency at an acceptable level and increasing the number of cores, they will be able to support many application domains that require high performance and low power. Having multicore processors is not a new idea; for instance, TI introduced a 5-core processor in 1995 (TMS320C8x), a 2-core processor in 1998 (TMS320C54x) and the OMAP (Open Multimedia Application Platform) family in 2002 [3], and Lucent produced a 3-core processor in 2000. However, manufacturers and users were not that interested in multicore as the processors' frequency increase was sufficient to satisfy the market and multicore processors were complex and did not have real software support.

Ideally, a multicore processor should have the following features:

- Low power
- Low cost
- Low size (small form factor)
- High compute-performance
- Compute-performance that can scale through concurrency
- Software support (OpenMP, OpenCL etc.)
- Good development and debugging tools
- Efficient operating system(s)
- Good embedded debugging tools
- Good technical support
- Ease of use
- Chip availability.

It is important to stress that developing hardware alone is not enough; software plays a very important role. In fact, silicon manufacturers are now introducing software techniques to leverage inherent parallelism available on their devices and attract users. For instance, NVidia introduced CUDA and TI supports Open Event Machine (OpenEM), Open Multi-Processing (OpenMP) and Open Compute Language (OpenCL) to leverage the performance and reduce the time to market.

In the embedded computing market, the decision whether to select a digital signal processor (DSP), a CPU (such as an x86 or an ARM), a GPU or a field-programmable gate array (FPGA) has become very complex, and making the wrong decision can be very costly if not catastrophic if a large volume is involved; for instance, a one dollar difference for one million products will result in a total one million dollars difference. But, for low volume it is sometimes more interesting to select a costly device if development time and future upgrade are taken into account. However, factors like cost, performance per watt, ease of use, time to market, hardware and software support and chip availability can help in selecting the right device or a combination of devices for a specific application.

For embedded high-processing-power systems, the main competing types of devices are the DSPs, FPGAs and GPUs.

1.2 Multicore processors

The main features of a multicore device are high performance, scalability and low power consumption. There are two main types of multicore processors: homogeneous (also known as symmetric multiprocessing (SMP)) and heterogeneous multicore processors (also known as asymmetric multiprocessing (AMP)). A homogeneous processor, such as the KeyStone I family of processors [4], has a set of identical processors and a heterogeneous processor, such as the KeyStone II (second-generation KeyStone architecture) [5], has a set of different processors. From a hardware perspective, AMP offers more flexibility for tackling a wider range of complex applications at a lower power consumption. However, they may be more complex to program, as different cores may use different operating systems and different memory structures that need to be interfaced for data exchange and synchronisation. Saying that, it is not always the case when supporting tools are available. For instance, the KeyStone II, which is a heterogeneous processor, is preferred by programmers since the ARM cores provide a rich set of library functions provided by the Linux community and the user can dispatch tasks from the ARMs to the DSPs without dealing with the underlying memory when using OpenCL.

1.2.1 Can any algorithm benefit from a multicore processor?

To show the advantages and limitations of multicore processors, let's first explore Amdahl's law [6], which states that the performance improvement by parallelising a code will depend on the code that cannot be parallelised. If we refer to Figure 1.1, it shows an original code that is composed of a serial code and a code that can be parallelised. It also shows the code after being parallelised.

If we consider the ratio of the original code and the optimised code as shown in Equation (1.1), if $S(n)$ is the speed-up time and if $Ts = Tp$, then $S(n)$ will be equal to 1 and no speed-up will be obtained. However, if Ts is equal to $2 * Tp$, then the speed-up will be 2.

If N is large, then $\frac{T_{parallel}}{N} \approx 0$ and Equation (1.1) will be reduced to Equation (1.2), which show that the serial code will be dominant.

$$S(n) = \frac{Ts}{Tp} = \frac{T_{serial} + T_{parallel}}{T_{serial} + \frac{T_{parallel}}{N}} \tag{1.1}$$

$$S(n) = \frac{Ts}{Tp} \approx \frac{T_{serial} + T_{parallel}}{T_{serial}} = 1 + \frac{T_{parallel}}{T_{serial}} \tag{1.2}$$

Knowing the percentage p of code that is serial, one can derive Amdahl's law as shown in Equation (1.3) by replacing T_{serial} by $T * (1 - p)$ and $T_{parallel}$ by $T * p$ in Equation (1.2).

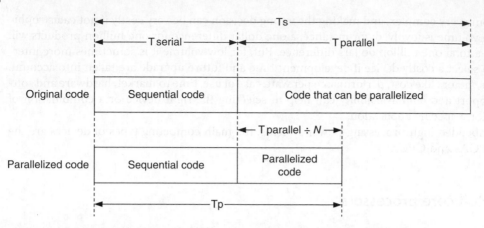

Figure 1.1 The impact of the serial code that cannot be parallelised on the performance.

$$S(n) = \frac{Ts}{Tp} = \frac{T_{serial} + T_{parallel}}{T_{serial}} = \frac{T*(1-p) + T*p}{T*(1-p) + \frac{T*p}{N}} \qquad (1.3)$$

Plotting $S(n)$, as shown in Figure 1.2, reveals that having a high number of cores for an application that has a low percentage of parallel code does not increase the speed. For instance, if the percentage of the parallel code is 50% (blue line), then having more cores will bring no real benefit if the number of cores is increased beyond 16 cores.

In Figure 1.1, the time it takes for cores to communicate is not shown. This is not the case in real applications, where communication and synchronisation times between cores comprise a real challenge, and the more cores that are used, the more time-consuming are the

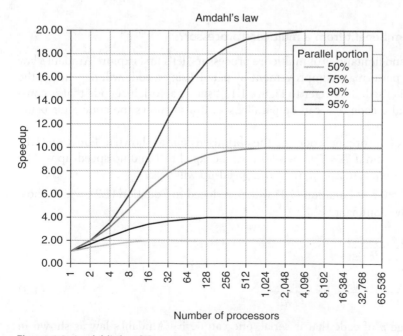

Figure 1.2 Amdahl's law [7].

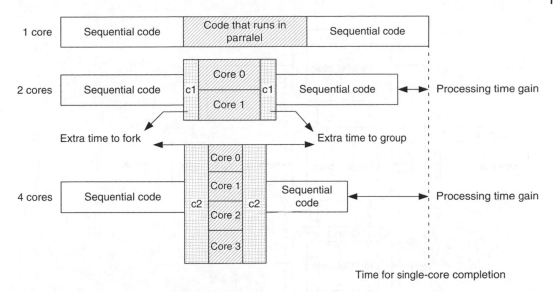

Figure 1.3 The inter-processor communication effect.

communication and synchronisation between cores; see illustration in Figure 1.3. It will be shown in this volume that increasing the number of cores does not necessarily increase the performance, and the drawback of parallelism can also introduce the potential for difficulty in debugging deadlocks and race conditions.

The second-generation KeyStone architecture (heterogeneous multicores) provides a better workload balance by distributing specific jobs to specific cores (the right core for the right job!).

1.2.2 How many cores do I need for my application?

Figure 1.2 showed that not all applications scale with the number of cores. There are three scenarios that need to be considered:

1) Scenario 1. This scenario has been discussed previously and is the case when an algorithm is composed of serial and parallel code. In this case, the number of cores to be used will depend on the parallel codes and the application. For instance, the example shown in Figure 1.4 can run on five cores or three cores, as core 0 can be reused to process part of the parallel code and one of core 1, core 2 or core 3 can be reused to run the final serial code.
2) Scenario 2. Some applications require different algorithms running sequentially. Consider the application shown in Figure 1.5. This application captures two videos of a road and performs a disparity calculation using the two videos, then performs a surface fitting to extract the surface of the road. The thresholding then removes the outliers (road surface), and the connected component labelling and detections are used to identify various objects below or above the road. In this application, each core can perform a function and therefore, six cores can perform six different jobs. More cores will not increase the performance if each core is not up to the task allocated to it.
3) Scenario 3. This scenario is a combination of scenarios 1 and 2. If we consider again the example shown in Figure 1.5 and the disparity calculation that requires more processing power (as in a practical situation), then more cores will be required. This is illustrated in Figure 1.6. In this application, eight cores will be required.

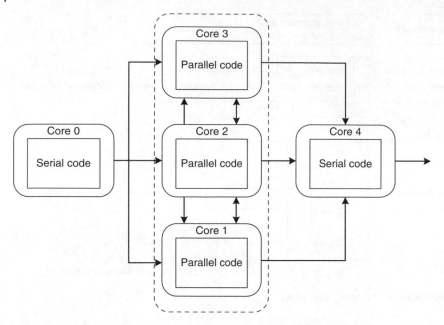

Figure 1.4 Example where three cores can perform the task required by the parallel code.

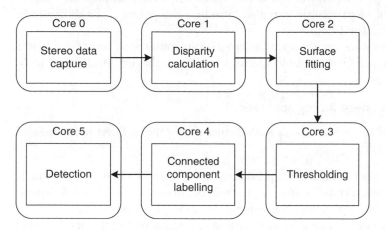

Figure 1.5 Example where cores are processing different algorithms.

1.3 Key applications of high-performance multicore devices

Reducing the operating clock frequency of the multiple processor cores and innovating in the inter-core communication on a single chip have led to a mirage of applications that are revealed every day and only limited by our own imagination. These applications range from scientific simulation, seismic wave imaging, avionics and defence, communications and telecommunications, consumer electronics, video and imaging, industrial, medical, security and space to high-performance computing (HPC). In turn, HPC is opening another window of scientific applications, such as advanced manufacturing, earth-system modelling and weather forecasting, life science, and big data analytics. Access to such machines is costly. However, the arrival of

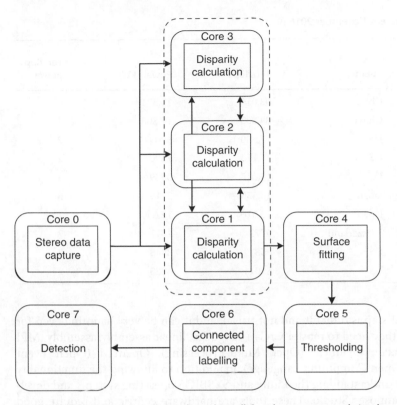

Figure 1.6 Example when serial code and parallel code are processed simultaneously.

low-cost, low-power, high-performance multicore processors is providing engineers and scientists with unprecedented low-cost tools.

HPC requires floating-point arithmetic that is essential for scientific applications, and therefore the performance is measured in floating-point operations per second (FLOPs). For instance, at the time of writing this book, the Sunway TaihuLight was number one according to TOP500. org [8, 9]. It was developed by the National Research Center of Parallel Computer Engineering & Technology (NRCPC), contained 10,649,600 cores with a peak performance of 125.4 petaflops (PFLOPs) and consumed 15.3 MW; see the list of the top ten supercomputers in Table 1.1. To put this in perspective, it has been reported that Google's data centres used 260 MW-hours, whereas a nuclear power station generates around 500–4000 MW [10]. Also at the time of writing this book, the Shoubu supercomputer from RIKEN was the most energy-efficient supercomputer and ranked as the top on the Green500 list [11]. The KeyStone SoC with its power efficiency and high performance is gaining momentum for use in green HPC; for instance, PayPal, a leader in online transaction processing, is using Hewlett-Packard's Moonshot system which is based on the KeyStone II SoC.

The development of an application for an SoC like the KeyStone can be a very long process: an idea is generated, algorithms are developed, selected algorithms are optimised and then they are normally evaluated in programming languages such MATLAB or Python depending on the application. Some algorithms are then developed in Visual Studio or a similar integrated development environment (IDE) to quickly test and debug the application, since the user can, for instance, use some libraries for getting real video or audio signals from a device using OpenCV, which is unlikely to be supported on an SoC. Then the code is translated to C/C++ language and

Table 1.1 Top 10 supercomputers, November 2016 [9]

Name	Country	Teraflops	Power (kW)	Teraflops power
Sunway TaihuLight	China	93,015	15,371	6
Tianhe-2	China	33,863	17,808	2
Titan	US	17,590	8209	2
Sequoia	US	17,173	7890	2
Cori	US	14,015	3939	4
Oakforest-PACS	Japan	13,555	2719	5
K Computer	Japan	10,510	12,660	1
Piz Daint	Switzerland	9779	1312	7
Mira	US	8587	3945	2
Trinity	US	8101	4233	2

ported to an SoC. The last step is difficult and not trivial, and it can be very daunting even for experienced engineers as they need to master C/C++ language, linear assembly/assembly, MPI (Message Passing Interface), OpenMP (Open Multi-Processing), OpenEM (Open Event Machine) and OpenCL (Open Computing Language), in addition to knowing the functionality of various peripherals and understanding the Linux and SYSBIOS operating systems and development tools such as Composer Studio. These tools are hardware-centric and require good understanding of the underlying hardware if maximum performance is to be achieved, especially when multicore programming is involved.

Increasing a multicore's performance is a twofold process: (1) to make the sequential part of the code run faster and (2) to exploit the parallelism offered by the multicore.

1.4 FPGAs, Multicore DSPs, GPUs and Multicore CPUs

In the past, FPGAs were the first choice for some applications that were not constrained by size or power consumption. For instance, there were not many commercial embedded devices that used FPGAs. However, recently, FPGA SoCs have integrated low-power software-programmable processing cores with the hardware programmability of an FPGA, like the Zynq-7000 from Xilinx, which targets embedded applications such as small cell base stations, multi-camera driver assistance systems and so on. However, critics may argue that power consumption and size are still not comparable to those of multicore DSPs. Despite further advantages of configurability, reconfigurability and programmability, an FPGA is still unattractive when time-to-market, maintenance and upgrades are issues. A comparison between FPGA multicore SoCs can be found in Ref. [12]; see Table 1.2. FPGAs also contribute to the development of SoCs that were traditionally designed using application-specific integrated circuits (ASICs) that have a substantial cost and time-to-market associated with them (they cost millions of dollars and take months to develop) [13].

On the other hand, graphic processors (e.g. GPUs) are gaining ground since they integrate low-power software-programmable processing cores and GPUs to form an SoC. These types of SoCs are referred to as *GPU-accelerated computing* by NVIDIA and finding applications

Table 1.2 Pros and cons of multicore SoCs and FPGA SoCs [12]

Feature/benefit	Multicore SoC	FPGA SoC
Futureproofing	Easily reprogrammed	Redesign required
Data flow	Very flexible	Unchangeable without a redesign
Processor diversity	Already integrated, highly programmable	General-purpose cores already integrated. Additional core types available as IP, but integration and licenses are required.
Power consumption	Low-power ARM cores, fine-grain power management strategies possible	Low-power ARM cores, no inherent power management
Footprint	Small, compact, stackable packages	Large footprint
System cost	High integration reduces system cost, and small footprint reduces PCB cost.	Costly IP integration required; larger footprint requires larger PCB space.
Cost of ownership	Shorter development cycle and faster time-to-market	More complex development
Time-to-market	Programmable resources shorten development cycles.	Complex development cycle lengthens time-to-market.

in HPC, deep learning, signal processing and so on. Theoretically, these GPUs can offer 100 to 1000 times faster speeds. However, in a practical situation, they may achieve around 2.5 times faster speed. To compare both multicores and GPUs, one must first select the application and perform code optimisation on both SoCs. Latency, power consumption, cost and size should also be considered. Graphic processors can perform well as all cores can run the same code.

More computing performance benchmarks among these emerging multicore SoCs, GPUs and FPGAs must be performed.

1.5 Challenges faced for programming a multicore processor

Chip designers did a very good job of packaging billions of transistors in a single chip package by making the transistors smaller to integrate on multiple cores, either homogeneous or heterogeneous. They also improved the communication between cores by providing coprocessors and fast buses, improved the memory hierarchy and even incorporated some hardware debugging tools. However, like any tool, a multicore processor will only be useful if one can use it. Therefore, new applications running on this new generation of processors can only run fast if programmers can write parallel code that takes advantage of the chips' features. Unfortunately, not all applications are embarrassingly parallel (meaning that no or little effort is required to make the code parallel).

The burden is now on the programmers as different types of applications require different approaches to parallelism. So, what do programmers need to have and know to make an application scale with the number of cores used? The first and most beneficial idea is to use an off-the-shelf software that transforms serial code to an efficient parallel code. Unfortunately, this has been attempted in the past but has not materialised yet to the point where it is fully automated, like for instance today's compilers which can achieve a better combination of performance and speed than code written by hand. Yes, there is software like OpenMP, OpenCL and OpenEM, but they still require good understanding of the underlying code to be parallelised.

The question now is 'What next?' Can chip makers continue to fit more cores on an SoC and make them faster, tools developers make serial-to-parallel code efficient and programmers learn more tricks to adapt an application to the hardware used? The answer depends not only on the application and cost involved but also on the time scale. For the near feature (in the next 10 years), then the answer is yes. For the long term, it is definitely no, because the data to process are ever increasing. For instance, if we consider the *Internet of things* (IoT), there will be around 20 billion 'things' connected by 2020, and these will generate large or big data that will need to be processed either locally or in a cloud server. In fact, even IoT gateways are based on multicores that can perform complex analytics and communication protocols for data normalisation and transmission. High-end applications that require high performance, high throughput and high capacity, such as genetic engineering, molecular dynamics, finance, cybersecurity, pharmaceuticals and weather forecasts that require hours, days or months to process, will definitely require a revolution in technology. Quantum, molecular, protein, DNA and optical computing are still in their infancy. Quantum computing may provide solutions considering the large investment in quantum technologies. For instance, the UK government alone is investing £270 million in quantum technologies [14].

1.6 Texas Instruments DSP roadmap

DSP has always been a front-runner in real-time embedded processing. However, with the emerging digital signal processing using GPUs and general-purpose processors (GPPs), and since only a handful of companies are still making DSP processors, engineers are wondering if this is the end of DSP processors. In fact, DSP manufacturers like TI now compete with different technologies. For instance, TI's DSPs now compete with Intel's CPUs and NVIDIA's graphic processors. Due to the continuous investment and innovation in DSP by Texas Instruments (which started in 1984 when they commercialised the first DSP chip, the TMS320C10) and due to the low-power and high-performance ARM processors, TI combined DSPs and ARMs to produce a processor with small form factor and low power per MHz/GMAC/GFLOP (gigaflop), which the industry is striving for.

Since this book deals with TI DSPs, it is worth summarising the TI DSP portfolio. Table 1.3 shows the three main embedded processors: the TMS320C6000, ARM and TMS320C5000 families. From TMS320C66x and the ARM processors sprang five SoCs: KeyStone I, KeyStone II, Sitara and two media processors, the DaVinci and the OMAP. As can be seen from Table 1.3, the KeyStone I is based on only the TMS320C66x. However, the KeyStone II is based on the TMS320C66x and ARM processors, and the Sitara is mainly based on the ARM but also can incorporate the TMS320C66x processors. The TMS320C66x is the most powerful in terms of performance among TI's processors, as shown in Figure 1.7, and also compared to the ARM cortex A15 and A9, as shown in Figure 1.8.

At the time of releasing this book, TI just upgraded its DSP roadmap by introducing the KeyStone III and the TMS320C7x DSP family that combines DSP and the Embedded Vision Engine (EVE), which is a flexible, programmable, low-latency, low-power-consumption and small-factor accelerator that performs vision-based analytics targeted at industries such as automotive, industrial machines and robotics. The DSP on the C7x is an upgrade of the TMS320C66x, and it is the first 64-bit DSP in the market and can achieve up to 16 times the performance of the TMS320C66x; see Figure 1.9. The KeyStone III combined the C7x processor and ARM cores.

Table 1.3 Main TI family of embedded processors

Core processor families	Processors	KeyStone I	KeyStone II	Sitara AMxxx [15]	Davinci DMxxx [16]	OMAP [17]
					Media processors	
TMS320C6000	TMS320C62x					
	TMS320C66x	✓	✓	✓		
	TMS320C64+					✓
	TMS320C67x				✓	✓
ARM	ARMA15		✓			
	ARM7					
	ARM8, 9, 15			✓		
	ARM8, 9					✓
TMS320C5000	TMS320C54					
	TMS320C55					

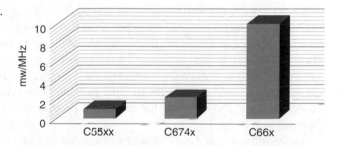

Figure 1.7 Texas Instruments DSPs.

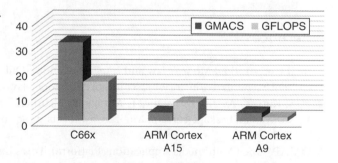

Figure 1.8 Performance comparison.

1.7 Conclusion

The days of increasing the performance by scaling the clock frequency are well over now. Multi-core processors are now the norm in a wide range of applications to the point that multicore is encompassing the complete spectrum of microprocessor applications from microcontrollers (multicore microcontrollers) to data centres (HPC).

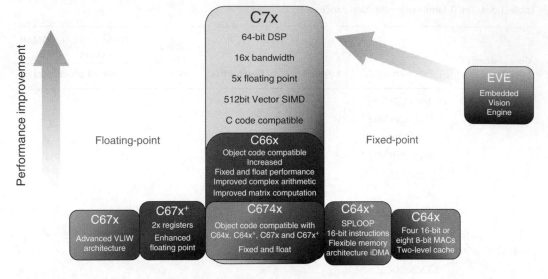

Figure 1.9 Texas Instruments DSP roadmap (courtesy Texas Instruments).

The KeyStone devices allow vectorisation (using single instruction multiple data (SIMD) instructions) and incorporate multiple cores (multicores), inter-processor communication, data transfer engines, memory management and debugging tools to increase system performance. The KeyStone II is seen as the green supercomputer as it can be used for HPC when performance per watt is a stringent requirement. To enter the low-power HPC arena, the KeyStone supports floating-point arithmetic and provides low-power heterogeneous multicores, optimised libraries and software for multicore programming (OpenMP, OpenEM, OpenCL and OpenMPI).

Finally, a multicore chip should be evaluated not just by its number of cores but also by its entire performance for a specific application.

References

1 Intel Corporation, Moore's law and Intel innovation, [Online]. Available: http://www.intel.co.uk/content/www/uk/en/history/museum-gordon-moore-law.html. [Accessed 2 December 2016].

2 N. S. Kim, T. Austin, D. Baauw, T. Mudge, K. Flautner, J. S. Hu, M. J. Irwin, M. Kandemir and V. Narayanan, Leakage current: Moore's law meets static power, *Computer*, vol. **36**, no. 12, pp. 68–75, 2003.

3 OMAP (Open Multimedia Applications Platform), Texas Instruments, [Online]. Available: https://en.wikipedia.org/wiki/OMAP#cite_note-1. [Accessed 6 December 206].

4 Texas Instruments, C66x Multicore DSP, [Online]. Available: http://www.ti.com/lsds/ti/processors/dsp/c6000_dsp/c66x/overview.page. [Accessed 2 December 2016].

5 Texas Instruments, C6000 Multicore DSP + ARM® SoC, [Online]. Available: http://www.ti.com/lsds/ti/processors/dsp/c6000_dsp-arm/overview.page. [Accessed 2 December 2016].

6 G. M. Amdahl, Validity of the single processor approach to achieving large scale computing capabilities, in *AFIPS*, Atlantic City, NJ, 1967.

7 Wikipedia, Amdahl's law, February 2017. [Online]. Available: https://en.wikipedia.org/wiki/Amdahl's_law. [Accessed January 2017].

8 TOP500, Home, TOP500.org, [Online]. Available: https://www.top500.org/. [Accessed 2 December 2016].

9 TOP500, June 2016, TOP500.org, June 2016. [Online]. Available: https://www.top500.org/lists/2016/06/. [Accessed 2 December 2016].

10 US Energy Infromation Administration, Frequently asked questions, 1 December 2015. [Online]. Available: http://www.eia.gov/tools/faqs/faq.cfm?id=104&t=21. [Accessed 2 December 2016].

11 TOP500, GREEN500 lists, TOP500.org, November 2016. [Online]. Available: https://www.top500.org/green500/lists/. [Accessed 2 December 2016].

12 P. Prakash, E. Blinka, S. Narnakaje, A. Friedmann, K. Garcia and R. Ferguson, Multicore SoCs stay a step ahead of SoC FPGAs, March 2016. [Online]. Available: http://www.ti.com/lit/wp/spry296/spry296.pdf. [Accessed 2 December 2016].

13 J. O. Hamblen and T. S. Hall, Using system-on-a-programmable-chip technology to design embedded systems, *International Journal of Computers and Their Applications*, vol. **13**, no. 3, pp. 142–152, 2006.

14 Engineering and Physical Sciences Research Council (EPSRC), Quantum technologies, [Online]. Available: https://www.epsrc.ac.uk/research/ourportfolio/themes/quantumtech/. [Accessed 2 December 2016].

15 Texas Instruments, Sitara processors, [Online]. Available: http://www.ti.com/lsds/ti/processors/sitara/overview.page. [Accessed January 2017].

16 Texas Instruments, DMxxx processor family overview, [Online]. Available: http://www.ti.com/general/docs/datasheetdiagram.tsp?genericPartNumber=TMS320DM8148&diagramId=63357. [Accessed January 2017].

17 Texas Instruments, OMAP processors, [Online]. Available: http://www.ti.com/lsds/ti/processors/dsp/media_processors/omap/products.page. [Accessed January 2017].

2

The TMS320C66x architecture overview

2.1   Overview

Building on a previous success with the first digital signal processor (DSP) generation based on the Texas Instruments (TI) VelociTI^{TM} architecture TMS320C6000, which used an enhancement of the VLIW (very long instruction word) architecture, TI has now pushed the frontiers a bit further by embracing the multicore system-on-chip (SoC) technology and adding

Multicore DSP: From Algorithms to Real-time Implementation on the TMS320C66x SoC, First Edition. Naim Dahnoun.
© 2018 John Wiley & Sons Ltd. Published 2018 by John Wiley & Sons Ltd.
Companion website: www.wiley.com/go/dahnoun/multicoredsp

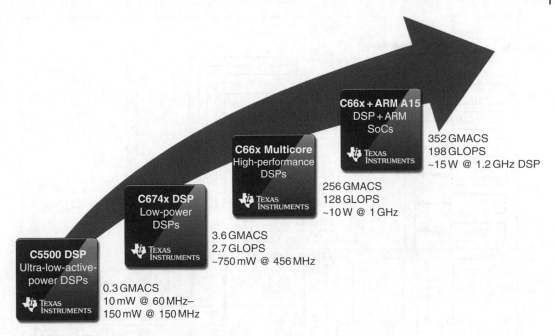

Figure 2.1 Texas Instruments (TI) digital signal processor (DSP) roadmap.

many features, such as: enhanced architecture, more configurable coprocessors, tiered memory architecture, high speed, a low-latency point-to-point communication interface known as the HyperLink, a TeraNet switch fabric which provides fast interconnection between the DSP CorePacs, the ARM CorePacs when available, memory, peripherals and a Multicore Navigator that can provide high-speed packed data movement without CPU loading, to create the new generation known as the TMS320C66x. See the TI DSP roadmap in Figure 2.1.

The TMS320C66x devices support both fixed- and floating-point arithmetic that can be mixed in order to combine low power and large dynamic range. The TMS320C66x is composed of four main parts: the CPUs, memories, peripherals and coprocessors, all connected by various buses as shown in Figure 2.2 and Figure 2.3.

At the time of writing this chapter, the TMS320C66x processors were divided into two families: the KeyStone I (see Table 2.1) and the KeyStone II (see Table 2.2). The KeyStone II family incorporates ARM cores in addition to DSP cores (known as CorePacs). A document on migration from KeyStone I to KeyStone II can be found in Ref. [1].

The KeyStone I (Figure 2.2) can be clocked from 600 MHz to 1.25 GHz depending on the device used; see Table 2.1. For the KeyStone II (Figure 2.3), both DSP and ARM cores can be clocked from 600 MHz to 1.4 GHz; see Table 2.2. The TMS320C66x CorePacs are an improved version of the C6000 CPUs covered in detail in Ref. [2].

2.2 The CPU

The TM320C66x CPUs are composed of two blocks known as data path 1 and data path 2, as shown in Figure 2.4. Each block has four execution units known as .L (logical unit), .M (multiplier unit), .S (shift unit) and .D (data unit) that can run in parallel; a register file containing 32 32-bit general-purpose registers; and multiple paths for (1) data communications between

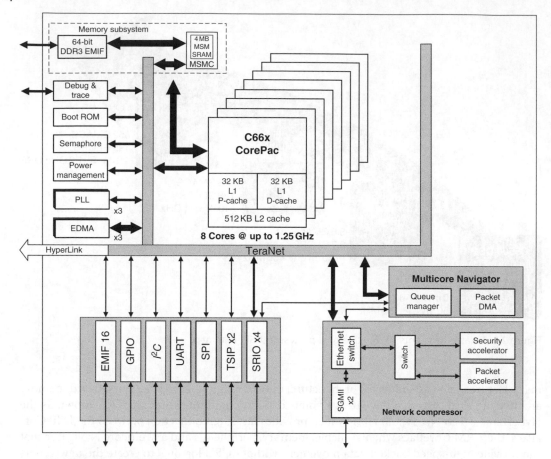

Figure 2.2 KeyStone I architecture [3].

each block and memory, (2) data communications within each block or (3) data communications between blocks.

From Figure 2.5, it can be seen that register file A can be written to or read from functional units .L1, .S1, .M1 and .D1 via the paths indicated by arrows. The same can be applied to register file B where all registers can be accessed by functional units .L2, .S2, .M2 and .D2. The CPU paths can be divided into two types: one is the data path, and the other is the address path. The data paths are used for data transfer between the register files and the units, or data transfer between the memory and the register files. However, the address path is used for sending the address from the data unit .D to the memory.

The challenge for optimising code on this processor is to make use of all units for every cycle. This is discussed in Chapter 5.

2.2.1 Cross paths

Cross paths enable linking of one side of the CPU (e.g. data path A) to the other (e.g. data path B). These are shown in bold arrows in Figure 2.5. Although the cross paths are useful in terms of the flexibility in using units with two or multiple operands from both sides of the CPU, there are restrictions which are discussed in this section.

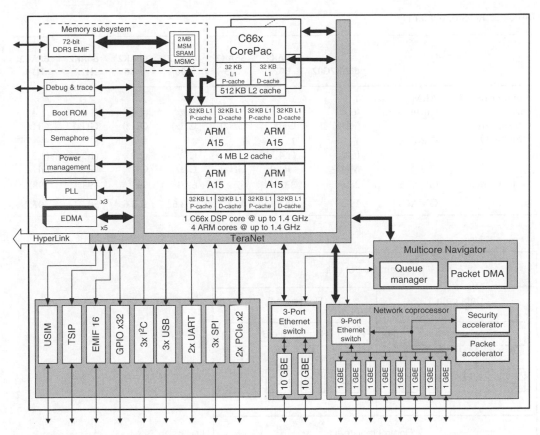

Figure 2.3 KeyStone II architecture [4].

Table 2.1 KeyStone I family

	C6678	C6674	C6657	C6655	C6654	C6652
MHz per core	1–1.25 GHz	1–1.25 GHz	1–1.25 GHz	1–1.25 GHz	750–850 MHz	600 MHz
Number of cores	8	4	2	1	1	1
Max GMACs	320 (@1.25 GHz)	160 (@1.25 GHz)	80 (@1.25 GHz)	40 (@1.25 GHz)	27.2 (@850 MHz)	19.2 (@600 MHz)
Max GFLOPs	160 @ 1.25 GHz	80 @ 1.25 GHz	40 @ 1.25 GHz	20 @ 1.25 GHz	13.6 @ 850 MHz	9.6@ 600 MHz

2.2.1.1 Data cross paths

The data cross paths are also referred to as the *register file cross paths*. These cross paths allow up to 64-bit operands from one side to cross to the other side. There are only two cross paths: one from side B to side A (1X), and one from side A to side B (2X). These limit the number of cross paths to two for each execute packet (instructions in parallel form an execute packet). The following points must be observed:

Table 2.2 KeyStone II family

		66AK2G02	66AK2 E02	66AK2 E05	66AK2 L06	66AK2 H06	66AK2 H12	66AK2 H14
Number of cores (maximum frequency)	ARM Cortex-A15	1 (600 MHz)	1 (1.4 GHz)	4 (1.4 GHz)	2 (1.2 GHz)	2 (1.4 GHz)	4 (1.4 GHz)	4 (1.4 GHz)
	C66x DSP	1 (600 MHz)	1 (1.4 GHz)	1 (1.2 GHz)	4 (1.2 GHz)	4 (1.2 GHz)	8 (1.2 GHz)	8 (1.2 GHz)
Performance	GFLOPs	28.8	33.6	67.2	69.0	99.2	198.4	198.4
	GMACs	19.2	44.8	44.8	153.6	153.6	307.2	307.2

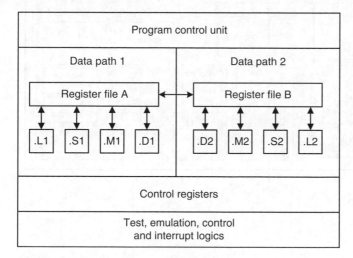

Figure 2.4 TMS320C66x CPU block diagram.

- Only one cross path per direction per execute packet is permitted.
- The destination register is always on the same side of the unit used.

2.2.1.2 Address cross paths

The addresses generated by the data unit .D1 and .D2 can be sent to either the data address path DA1 or the data address path DA2, as shown in bold arrows in Figure 2.5. The advantages of using an address cross path are to be able to generate the address using one register file, and to access the data from the other register file as illustrated in Figure 2.6. Here again, there are only two cross paths for each execute packet and the following points should be observed:

- Only one address cross path per direction per execute packet is allowed.
- When an address cross path is used, the destination register for the load (LD) instructions and the source register for the store (ST) instructions should come from the opposite side of the unit (see Figure 2.6), or simply the register pointers must come from the same side of the .D unit used.
- If both .D units are to be used, then either none or both of the address cross paths should be used.

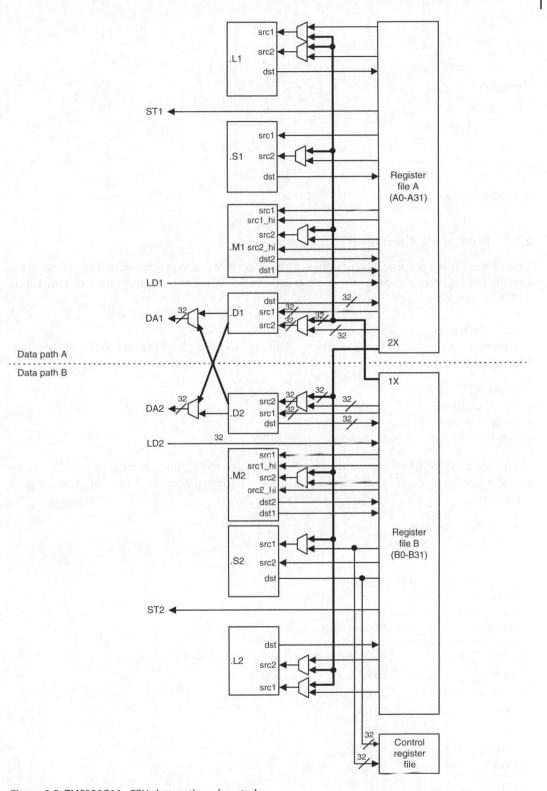

Figure 2.5 TMS320C66x CPU data path and control.

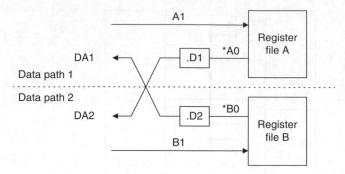

Figure 2.6 Address cross paths.

2.2.2 Register file A and file B

This processor is a reduced instruction set computer (RISC)-like processor, and all operands are specified in registers except for the *n*-bit constants. There are two register files each containing 32 32-bit registers.

2.2.2.1 Operands

An operand can be an *n*-bit constant or a 32-bit register, two 32-bit registers or four 32-bit registers, depending on the instruction:

- Constant
- 32-bit registers
- 64-bit registers
- 128-bit registers.

To create 40- or 64-bit operands, two registers have to be concatenated; see Table 2.3. To create a 128-bit operand, four registers have to be concatenated; see Table 2.4. The registers must be:

- From the same side
- Consecutively ordered

Table 2.3 Possible 40-/64-bit register pair combinations

Register file A	Register file B
A1:A0	B1:B0
A3:A2	B3:B2
A5:A4	B5:B4
A7:A6	B7:B6
A9:A8	B9:B8
A11:A10	B11:B10
A13:A12	B13:B12
A15:A14	B15:B14

Table 2.4 Possible 128-bit register pair combinations

Register file A	Register file B
A3:A2:A1:A0	B3:B2:B1:B0
A7:A6:A5:A4	B7:B6:B5:B4
A11:A10:A9:A8	B11:B10:B9:B8
A15:A14:A13:A12	B15:B14:B13:B12
A19:A18:A17:A16	B19:B18:B17:B16
A23:A22:A21:A20	B23:B22:B21:B20
A27:A26:A25:A24	B27:B26:B25:B24
A31:A30:A29:A28	B31:B30:B29:B28

- Ordered as even-odd from right to left for the 64-bit, as shown in Table 2.3, and ordered even-odd-even-odd from right to left for the 128-bit, as shown in Table 2.4.

2.2.3 Functional units

The four types of units (.M, .L, .S and .D) are designed to perform different operations. However, some operations can be performed with different units; for instance, the ADD instruction can be performed by the .L units, the .S units or the .D units. The TMS320C66x DSP CPU and Instruction Set Reference Guide [5] should be consulted before using an instruction.

The assembly syntax for this DSP core is as follows:

|condition|	instruction	.unit	operand 1,	operand 2,	destination	; comments

Example:

`|B0| ADD.S1 A0,A1,A2 ; comments`

where:

|B0|: If B0 is not equal to zero, then execute the instruction '**ADD .S1 A0,A1,A2**'.
ADD .S1 A0,A1,A2: Add A0 and A1, and store the result to register A2.
; comments: Used for comments and therefore not assembled.

2.2.3.1 Condition registers

1) The condition can be one of the following registers: A0, A1, A2, B0, B1 or B2.
2) Most instructions can be conditional.
3) The specified condition register is tested at the beginning of the E1 pipeline stage for all instructions. Refer to the user guide [5] for the pipeline operations.
4) Compact (16-bit) instructions on the DSP always execute unconditionally. See 'Compact instructions on the CPU' in Ref. [5].

The condition can be inverted by adding the exclamation symbol '!' as follows:

`|!B0| ADD.S1 A0,A1,A2`

where:

|!B0|: If B0 is equal to zero, then execute the instruction '**ADD .S1 A0,A1,A2**'.

2.2.3.2 .L units

The .L units support up to 64-bit operands. All instructions using these units complete in one cycle.

The .L unit can perform:

- Arithmetic operations (floating or fixed point)
- Logical operations
- Branch functions
- Data-packing operations
- Conversion to/from integer and single-precision values.

The .L unit has additional instructions for logical AND and OR instructions, as well as 90 degree or 270 degree rotation of complex numbers (up to two per cycle) [5].

Examples using the .L1 unit:

Example 1	AND	.L1	A1:A0,A3:A2,A9:A8	; AND 64-bit and 64-bit
Example 2	AND	.L1	A0,A1,A2	; AND 32-bit and 32-bit
Example 3	AND	.L1	0x9,A0,A2	; AND 5-bit constant (scst5) and 32-bit

2.2.3.3 .M units

There are two hardware multiplier units, .M1 (for data path 1) and .M2 (for data path 2), that can perform fixed-point or floating-point multiplications as shown in Table 2.5 and Table 2.6. The .M units support 128-bit.

Table 2.5 Fixed-point multiplications per unit

- Four 32 × 32 bit multiplies (e.g. QMPY32)
- Four 16 × 8 bit multiplies (e.g. DDOTP4)
- Two 16 × 16 bit multiplies (e.g. MPY2)
- 16 × 32 bit multiplies (e.g. MPYHI)
- Four 8 × 8 bit multiplies (e.g. MPYU4)
- Four 8 × 8 bit multiplies with add operations (e.g. DOTPU4)
- Four 16 × 16 multiplies with add/subtract capabilities (e.g. DOTP4H)
- One 16 × 16 bit complex multiply with or without rounding (e.g. CMPY/CMPYR)
- A 32 × 32 bit complex multiply with rounding (e.g. CMPY32R1)
- Complex multiply with rounding and conjugate, signed complex 16-bit (16-bit real/16-bit imaginary) (e.g. CCMPY32R1)
- Support for Galois field multiplication (e.g. GMPY)
- One multiplication of a [1 × 2] complex vector by a [2 × 2] complex matrix per cycle with or without rounding capability (e.g. CMATMPY)
- One multiplication of the conjugate of a [1 × 2] vector with a [2 × 2] complex matrix (e.g. CCMATMPY)

Table 2.6 Floating-point multiplications per unit

- One single-precision multiply each cycle
- One double-precision multiply every four cycles
- One double-precision multiply per cycle; also reduces the number of delay slots from 10 to 4
- One multiplication of two single-precision numbers, resulting in a double-precision number
- One, two, or four single-precision multiplies, or a complex single-precision multiply in one cycle

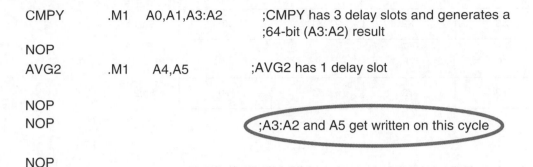

```
CMPY        .M1    A0,A1,A3:A2         ;CMPY has 3 delay slots and generates a
                                       ;64-bit (A3:A2) result
NOP
AVG2        .M1    A4,A5               ;AVG2 has 1 delay slot

NOP
NOP                                    ;A3:A2 and A5 get written on this cycle

NOP
```

Figure 2.7 Instructions completing in the same cycle.

As stated earlier, the instructions *load, store, multiply* and *branch* have different latencies and therefore complicate programming. All TMS320C66x instructions require only one cycle to execute (unit latency is one). However, some results are delayed (delay slots). When instructions are pipelined, the multiplier can issue one instruction per cycle.

Care should be taken when using the .M units to perform operations other than multiplications. Each .M unit has two 64-bit write ports to the register file, and therefore the results of a 4-cycle instruction and a 2-cycle instruction operating on the same .M unit can write their results on the same instruction cycle. This is not an issue as long as the programmer is aware of this; see the example in Figure 2.7.

2.2.3.4 .S units
These units (.S1 and .S2) contain 32-bit integer ALUs (arithmetic and logic units) and 40-bit shifters. They can be used for:

- 32-bit arithmetic, logic and bit field operations
- 32/40-bit shifts
- Branches
- Transfer to and from control registers (*.S2 only*)
- Constant generation.

Note: All instructions executing in the .L or .S are single-cycle instructions, except for the branch instructions.

2.2.3.5 .D units
The data units (.D1 and .D2) are the only units that can be used for accessing memory. They can be used for the following operations:

- Load and store with 5-bit constant offset
- Load and store with 15-bit constant offset (*.D2 only*)
- 32-bit additions and subtractions
- Linear and circular address calculations
- Logical operations
- Moving a constant or data from a register to another register.

X3	X2	X1	X0	Src1

α	α	α	α

Y3	Y2	Y1	Y0	Src2

=	=	=	=

Z3	Z2	Z1	Z0	Dst

Figure 2.8 Four-way SIMD operation.

2.3 Single instruction, multiple data (SIMD) instructions

To make maximum use of the units and therefore increase the performance, one should exploit the SIMD operations available with the TMS320C66x. Figure 2.8 shows an example of a 4-way SIMD with an instruction α operating on multiple data from Src1 and Src2 to produce multiple data on the Dst.

TMS320C66x supports 2-way, 4-way and 8-way SIMD operating on 8-bit, 16-bit, 32-bit, 64-bit or 128-bit, depending on the instruction used. Examples with different ways are shown in Table 2.7.

2.3.1 Control registers

The TMS320C66x devices have a number of registers for control purposes; see Table 2.8. Reading and writing to the control registers can be performed only via the .S2 unit. All control registers can be accessed by only the MVC (*move constant*) instruction.

Note: Only the .S2 unit and the MVC instruction can be used to access the control registers. However, some bit fields in some control registers can be modified by some instructions or events. For instance, when an interrupt occurs, a bit field in the Interrupt Flag Register (IFR) register will be modified.

2.4 The KeyStone memory

Memory is one of the predominant factors that establishes the final performance of any processor. In fact, the embedded memory system is one of the items that determines the system performance, efficiency, size and cost. The design of the memory (internal or external), the memory controller that manages the data flow and the buses that transport these data are very important for an efficient delivery of data at the bandwidth, latency and power required. In fact, the memory die takes more than 50% of the total area of a typical SoC. The TMS320C66x memory architecture is organised as shown in Figure 2.9. Each core has its own local level 1 memory (L1 Data Cache and L1 Program Cache) and its own local level 2 memory. Both local levels can be configured as memory-mapped SRAM, cache or a combination of SRAM and cache. Coherency is maintained between L1 and L2 for each core, as highlighted in Figure 2.9.

Table 2.7 SIMD examples

Table 2.8 TMS320C66x control registers [5]

Acronym	Register
AMR	Addressing mode register
CSR	Control status register
GFPGFR	Galois field multiply control register
ICR	Interrupt clear register
IER	Interrupt enable register
IFR	Interrupt flag register
IRP	Interrupt return pointer register
ISR	Interrupt set register
ISTP	Interrupt service table pointer register
NRP	Non-maskable interrupt (NMI) return pointer register
PCE1	Program counter, E1 phase
Control register file extensions	
DNUM	DSP core number register
ECR	Exception clear register
FR	Exception flag register
GPLYA	GMPY A-side polynomial register
GPLYB	GMPY B-side polynomial register
IERR	Internal exception report register
ILC	Inner loop count register
ITSR	Interrupt task state register
NTSR	NMI/exception task state register
REP	Restricted entry point address register
RILC	Reload inner loop count register
SSR	Saturation status register
TSCH	Time-stamp counter (high 32) register
TSCL	Time-stamp counter (low 32) register
TSR	Task state register
Control register file extensions for floating-point operations	
FADCR	Floating-point adder configuration register
FAUCR	Floating-point auxiliary configuration register
FMCR	Floating-point multiplier configuration register

The Multicore Shared Memory Controller (MSMC) allows all cores access to the shared memory (SL2) and the external memory. Note that the external memory is accessed via the external memory interface (EMIF) or the TeraNet. Multiple EMIFs may be available, depending on the device used.

The shared memory referred to in Figure 2.2 and Figure 2.3 as the SRAM is the MSMC that can be configured as shared memory SL2 or a shared memory level 3 (SL3), as shown in Figure 2.9.

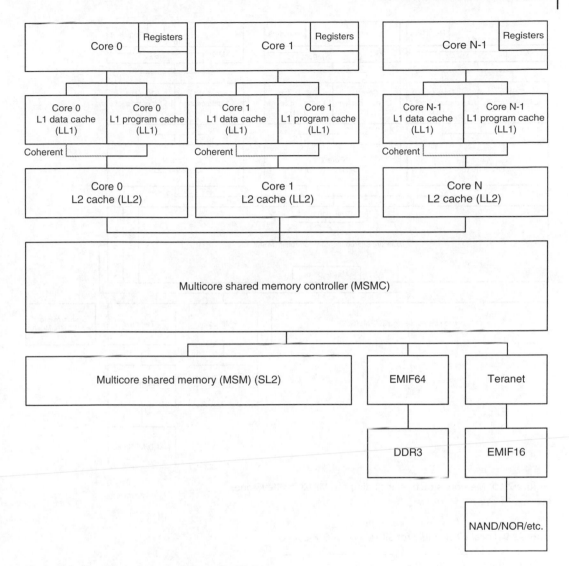

Figure 2.9 Simplified memory structure for KeyStone.

When the SRAM is configured as SL2, this memory will be cacheable with L1D and L1P memories.

When the SRAM is configured as SL3, this memory will be cacheable with both L1 and L2 memories.

Although the SL2 memory appears in Figure 2.9 as level 3, its performance is the same as that of the LL2 due to the optimal prefetching capability of the extended memory controllers (XMCs) that are placed within the cores (see Figure 2.10); hence it is called *level 2*.

2.4.1 Using the internal memory

When writing an application for a multicore processor, one tends to write code for one core and then run it in all cores. This simple task can be complicated to write as the local memories will have to have different addresses. For instance, the L2 SRAMs for the TMS320C6678 shown in

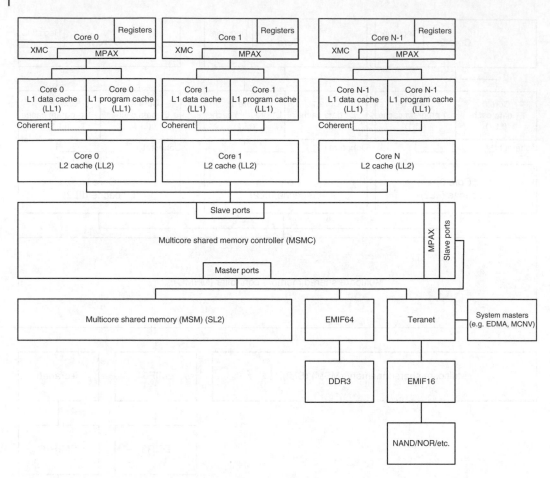

Figure 2.10 Memory structure, including the MPAX for KeyStone.

Table 2.9 Local L2 memory for all TMS320C6678 cores

Core 0: $0 \times 1080\ 0000$ to $0 \times 0087\ FFFF$
Core 1: $0 \times 1180\ 0000$ to $0 \times 1187\ FFFF$
Core 2: $0 \times 1280\ 0000$ to $0 \times 1287\ FFFF$
Core 3: $0 \times 1380\ 0000$ to $0 \times 1387\ FFFF$
Core 4: $0 \times 1480\ 0000$ to $0 \times 1487\ FFFF$
Core 5: $0 \times 1580\ 0000$ to $0 \times 1587\ FFFF$
Core 6: $0 \times 1680\ 0000$ to $0 \times 1687\ FFFF$
Core 7: $0 \times 1780\ 0000$ to $0 \times 1787\ FFFF$

Table 2.9 have different addresses. However, in a practical situation, each core sees its addresses differently. In the example shown in Figure 2.11, all cores use the same starting address $0 \times 0080\ 0000$ to access their local memories. For example, Core 0 can access the local memory of Core 1 by using the address $0 \times 1180\ 0000$, and Core 7 accessing the local memory L2 of Core 5 by using the address $0 \times 1580\ 0000$ and so on. In this way, a single code can be used by all cores without modifications.

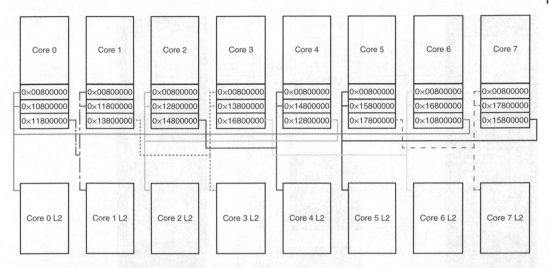

Figure 2.11 Example of cores accessing their local or other local memories.

2.4.2 Memory protection and extension

It has been shown that each core can use its own local memory (LL1, LL2, LS2 or LS3), and it has also been shown that cores can use the same code and the same addresses for accessing local variables. However, when data and/or code cannot fit in the internal memory, code and/or data will have to be located in the DDR memory. In this case, data and/or code located in the DDR will need to be accessed with different addresses unless they are shared. The Memory Protection and Address Extension (MPAX) unit can be used to make portions of the DDR look like local memories. For instance, if we consider the situation where Core 1 and Core 2 have the same code but different data (data 1 for Core 1 and data 2 for Core 2) and both code and data do not fit in the internal memory, one can use the MPAX registers to configure part of the DDR as a private memory to each core and part as shared memory, as shown in Figure 2.12. This has the advantage of increasing the performance, as no software is required to do the address translation and the same code is used by all cores. However, by doing so, the cache coherency must be maintained 'manually' by using the cache invalidate, cache writeback and cache writeback-invalidate, since there is no coherency between the external memory and the internal memory. In addition to address extension, the MPAX can also be used for internal and external memory protection. More details covering the MPAX can be found in Refs. [6] and [7].

2.4.3 Memory throughput

Knowing where to locate the program and data is very critical for performance. In this section, the maximum data throughput is highlighted.

Consider Figure 2.13, and be aware that the DSP cores for the TMS320C6678 can be clocked at 1.0 GHz, 1.25 GHz or 1.4 GHz. Let's assume that a core is clocked at 1.0 GHz and calculate the memory throughput.

L1D SRAM. This operates at the same frequency as the DSP core and can access a maximum of 128-bit data. Therefore, the throughput is 16 GB/s [(128) * (1.0)/8].

L1P SRAM. This operates at the DSP clock frequency, and the CPU can fetch up to 256-bit instructions. Therefore, the throughput is 32 GB/s [(256 * (1.0)/8)].

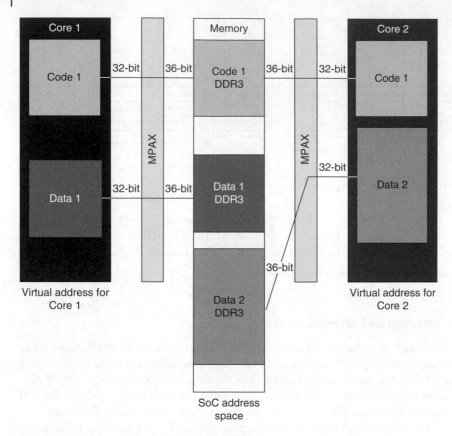

Figure 2.12 Example showing the use of MPAX.

L2 SRAM. This operates at half the frequency of the DSP core and can access a maximum of 256-bit data. Therefore, the throughput will be half of the L1D SRAM, that is, 16 GB/s [(256 * (0.5)/8].

MSMC SRAM. Operates at half the frequency of the DSP core but has four banks that can be accessed simultaneously. Therefore, the aggregate throughput will be four times that of the L2 SDRAM, which is 64 GB/s. Each KeyStone DSP core has a 256-bit path at half the DSP clock frequency for a throughput of 16 GB/s. The KeyStone II doubles the clock speed and throughput.

DDR3. The DDR3 has a 64-bit interface to the MSMC and can be clocked at a maximum frequency of 1.333 GHz; therefore, the throughput is 10.666 GB/s (64 * 1.333)/8).

It is also important to explore and contrast the data throughput using the CPU (as shown here) and the EDMA; see Ref. [8].

For the KeyStone II device throughput, refer to Ref. [9].

2.5 Peripherals

The KeyStone I and II have a rich set of peripherals that are shown in Figure 2.2 and Figure 2.3, respectively. Each peripheral is described in a user guide. The peripherals used in this book are summarised in this section.

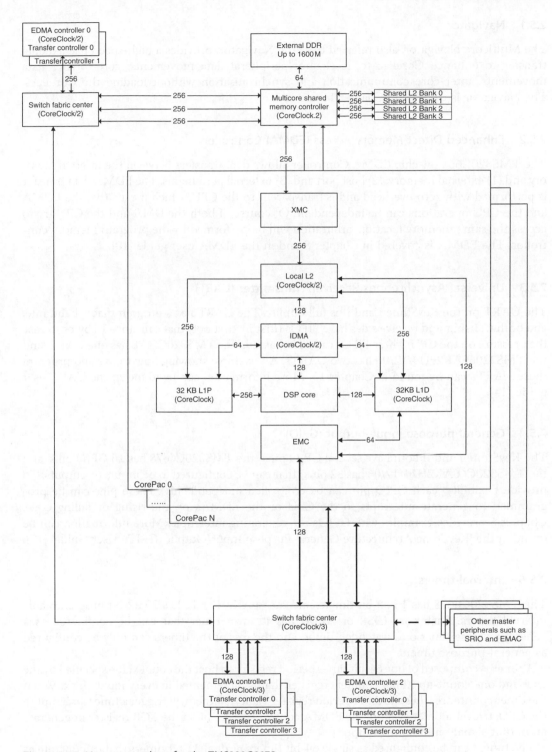

Figure 2.13 Memory topology for the TMS320C6678.

2.5.1 Navigator

The Multicore Navigator, also referred to as the Navigator, provides a high-speed packed data transfer to enhance CorePac to accelerator/peripheral data movements, core-to-core data movements, inter-core communication and synchronisation without loading the CorePacs. The Navigator is covered in Chapter 14.

2.5.2 Enhanced Direct Memory Access (EDMA) Controller

The TMS320C66x on-chip EDMA Controller allows data transfers between the internal memory and (1) external memory, (2) host port and (3) external peripherals. The EDMA data transfer is performed with zero overhead and is transparent to the CPU, which means that the EDMA and the CPU operations can be independent. Of course, if both the DMA and the CPU try to access the same memory location, arbitration will be performed by the program memory controller. The EDMA is covered in Chapter 8 and in the EDMA user guide [10].

2.5.3 Universal Asynchronous Receiver/Transmitter (UART)

The UART on the KeyStone I and II is full duplex. The UART has a programmable baud rate, and both transmit and receive sides have FIFOs (first in, first out) that can store 16 bytes to ease the pressure on the CPU. These FIFOs can be bypassed. The TMS320C6678 has one UART, and the TMS32066AK2H14/12/06 has two UARTs. An example showing how to use and program these UARTs can be found in Chapter 15. More information can be found in the UART user guide [11].

2.5.4 General purpose input–output (GPIO)

The KeyStone I and II both have several GPIO pins (the TMS320C6678 has 16 GPIO pins, and the TMS320C66AK2H14/12/06 has 32 pins) that can be configured to be inputs or outputs. To provide flexibility, each GPIO pin can be controlled independently. These pins can be programmed to generate interrupts to the CPU or the EDMAs on the rising or falling edge. Chapter 6 provides examples with GPIOs for generating interrupts. More information can be found in the 'KeyStone Architecture General Purpose Input/Output (GPIO) User Guide' [12].

2.5.5 Internal timers

The TMS320C6678 has 16 32-bit timers, and the 66AK2H14/12 has 20 32-bit programmable internal timers. Each core (DSP or ARM) has its own timer that can be configured as a general-purpose timer or a watchdog timer, and the rest of the timers can only be configured as general-purpose timers.

A timer is composed of one 64-bit timer period register to host the count value specified by the user and one count-up (timer counter) register that is incremented in every input clock. When the timer counter reaches the timer period register value, it either will trigger a timer interrupt to the CPU, trigger a timer event to the EDMA controller, set a bit in the TCR register or generate an output signal on the timer output pin.

The timers can be configured as single 64-bit timers or dual 32-bit timers that can operate as chained (chained mode), where one timer triggers the other which then generates the interrupt signals, or operate as unchained (unchained mode), where both timers can generate interrupts. The timers can also be configured to be used as 64-bit watchdog timers in order to provide a

Table 2.10 Timer modes

64-bit general-purpose timer (default)
Dual 32-bit timers (unchained)
Dual 32-bit timers (chained)
64-bit watchdog timer

control exit; see Table 2.10. More details can be found in Ref. [13]. See Chapter 7 for examples using the timers.

2.6 Conclusion

To get maximum performance from each DSP core, one should understand the architecture very well. It has been shown that each core has eight units, and the algorithm must make use of all these units as much as possible to extract maximum performance. To further exploit these units, SIMD operations should be used when feasible.

Understanding the operations of peripherals to use and the memory layout is important for developing applications with the required functionalities and performance.

References

1 Texas Instruments, KeyStone I-to-KeyStone II migration guide: SPRABW9A, July 2015. [Online]. Available: http://www.ti.com/lit/an/sprabw9a/sprabw9a.pdf. [Accessed 2 December 2016].

2 N. Dahnoun, *Digital Signal Processing Implementation Using the TMS320C6000 DSP Platform*, Reading, MA: Addison-Wesley Longman, 2000.

3 Texas Instruments, Multicore fixed and floating-point digital signal processor, March 2014. [Online]. Available: http://www.ti.com/lit/ds/symlink/tms320c6678.pdf. [Accessed 2 December 2016].

4 Texas Instruments, Multicore DSP + ARM KeyStone II System-on-Chip (SoC), November 2013. [Online]. Available: http://www.ti.com/lit/ds/symlink/66ak2h12.pdf. [Accessed 2 December 2016].

5 Texas Instruments, TMS320C66x DSP CPU and instruction set reference guide, November 2010. [Online]. Available: http://www.ti.com/lit/ug/sprugh7/sprugh7.pdf. [Accessed 2 December 2016].

6 Texas Instruments, TMS320C66x DSP CorePac user guide, July 2013. [Online]. Available: http://www.ti.com/lit/ug/sprugw0c/sprugw0c.pdf. [Accessed 2 December 2016].

7 Texas Instruments, KeyStone memory architecture, 2010. [Online]. Available: http://www.ti.com/lit/wp/spry150a/spry150a.pdf. [Accessed 2 December 2016].

8 Texas Instruments, TMS320C6678 memory access performance, April 2011. [Online]. Available: http://www.deyisupport.com/cfs-file.ashx/__key/telligent-evolution-components-attachments/00-53-00-00-00-02-19-24/TMS320C6678_5F00_Memory_5F00_Access_5F00_Performance.pdf. [Accessed 2 December 2016].

9 Throughput performance guide for KeyStone II devices, Texas Instruments, December 2015. [Online]. Available: http://www.ti.com/lit/an/sprabk5b/sprabk5b.pdf. [Accessed January 2017].

10 Texas Instruments, KeyStone Architecture Enhanced Direct Memory Access (EDMA3) Controller user's guide, May 2015. [Online]. Available: http://www.ti.com/lit/ug/sprugs5b/sprugs5b.pdf. [Accessed 2 December 2016].

11 Texas Instruments, KeyStone architecture Universal Asynchronous Receiver/Transmitter (UART) user guide, November 2010. [Online]. Available: http://www.ti.com/lit/ug/sprugp1/sprugp1.pdf. [Accessed 2 December 2016].

12 Texas Instruments, KeyStone architecture general purpose input/output (GPIO) user guide, November 2010. [Online]. Available: http://www.ti.com/lit/ug/sprugv1/sprugv1.pdf. [Accessed 2 December 2016].

13 Texas Instruments, KeyStone Architecture TIMER64P user guide, March 2012. [Online]. Available: http://www.ti.com/lit/ug/sprugv5a/sprugv5a.pdf. [Accessed 2 December 2016].

3

Software development tools and the TMS320C6678 EVM

3.1 Introduction

There has been massive growth in real-time applications demanding real-time processing power, and this has led DSP manufacturers to produce advanced chips and advanced development tools which not only allow engineers to develop complex algorithms with ease but also speed up time-to-market. Development tools can be divided into hardware and software

Multicore DSP: From Algorithms to Real-time Implementation on the TMS320C66x SoC, First Edition. Naim Dahnoun.
© 2018 John Wiley & Sons Ltd. Published 2018 by John Wiley & Sons Ltd.
Companion website: www.wiley.com/go/dahnoun/multicoredsp

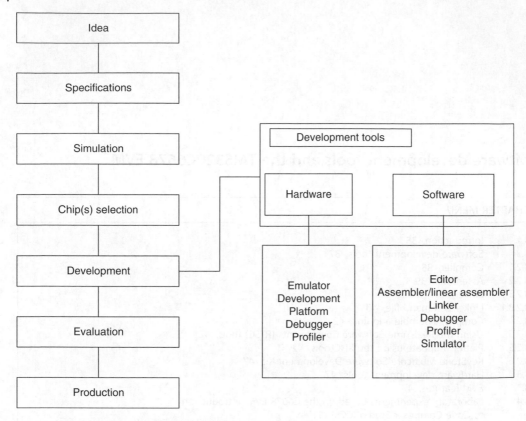

Figure 3.1 Hardware and software development tools.

development tools (see Figure 3.1). Development tools are very important and tend to shorten significantly the development time, which is itself the most time-consuming part of the production of a device.

Texas Instruments (TI) provides an advanced integrated software and hardware development toolset. On the software side, it provides the Multicore Software Development Kit (MCSDK) that provides foundational software for the KeyStone devices as shown in Figure 3.2. On the host side, the main part in the MCSDK is the Code Composer Studio™ (CCS) Integrated Development Environment (IDE) which is based on Eclipse, an open-source software framework used by many embedded software vendors. Included in CCS are a full suite of compilers (which support OpenMP) [1], a source code editor, a project building environment, a debugger, a profiler, an analyser suite, and many other code development capabilities (see video in Ref. [2]). On the target side, it includes mainly the instrument and trace, symmetric multiprocessing (SMP) Linux for ARM and other libraries.

CCS has been developed in such a way that its functionality can be extended by plug-ins. There are two types of plug-ins: ones written by TI and ones written by a third party. Some plug-ins are very useful as they provide graphical development tools which are intuitive and abstract device APIs (application programming interfaces) and hardware details. For instance, with a few clicks of the mouse, one can configure some interrupts, a peripheral or a debugging procedure.

It is essential for a developer to fully understand the use and capabilities of the tools. This chapter is divided into three main parts: the first part describes the development tools, the

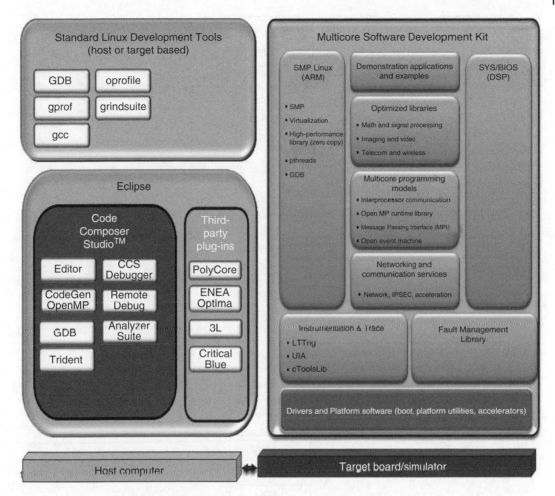

Figure 3.2 Texas Instruments' software ecosystem [3].

second part describes the evaluation module (EVM) and finally the third part describes provided laboratory exercises demonstrating the capabilities of the CCS and EVMs.

3.2 Software development tools

The software development tools consist of the following modules: the C-compiler, assembler, linker, simulator and code converter (see Figure 3.3).

If the source code is written in C language, the code should be compiled using the optimising C compiler provided by TI [4]. This compiler will translate the C source code into an assembly code. The assembly code generated by either the programmer, the compiler or the linear assembler (see Chapter 5) is then passed through the assembler that translates the code into object code. The resultant object files, any library code and a command file are all combined by the linker which produces a single executable file. The command file mainly provides the target hardware information to the linker and is described in Section 3.2.3.1.

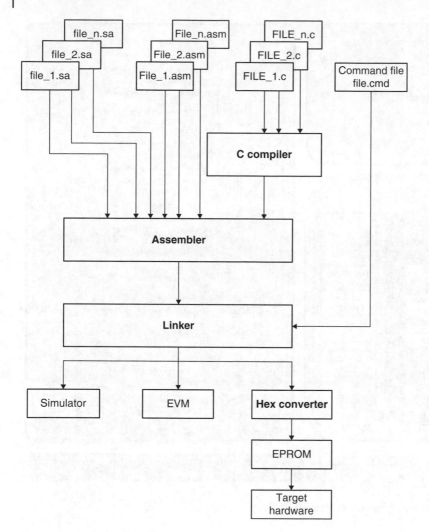

Figure 3.3 Basic development tools.

3.2.1 Compiler

The C code is not executable code and therefore needs to be translated to a language that the DSP understands. In general, programs are written in C language because of its portability, ease of use and popularity. Although for time-critical applications assembly is the most efficient language, the optimising C compiler for the TMS320C66x processors can achieve performances exceeding 70% compared with code written in assembly. This has the advantage of reducing the time-to-market and hence cost.

To evoke the compiler, use the **CL6x** command as shown here:

```
CL6x FIR1.c (This command line compiles the file called FIR1.c.)
```

Note: The **CL6x** command is not case sensitive for Windows.

The compiler uses options supplied by the user. These options provide information about the program and the system to the compiler. The most common options are described in Chapter 5.

Table 3.1 Common compiler options

Option	Description
-mv6600	Tells the compiler that the code is for the TMS320C66x processor
-k	Do not delete the assembly file (*.**asm**) created by the compiler.
-g	Symbolic debugging directives are generated to enable debugging.
-i	Specifies the directory where the **#include** files reside
-s	Interlists C and assembly source statements
-z	Adding the **-z** option to the command line will evoke the assembler and the linker.

However, for a complete description of the compiler options, the reader is referred to the optimising C compiler manual [4].

The options shown in Table 3.1 can be inserted between the **CL6x** command and the file name as shown here:

```
CL6x -gk FIR1.c
```

3.2.2 Assembler

The assembler translates the assembly code into an object code that the processor can execute. To evoke the assembler, type:

```
asm6x FIR1.asm FIR.obj
```

The above command line assembles the **FIR1.asm** file and generates the **FIR.obj** file. If '**FIR.obj**' is omitted from the command, the assembler automatically generates an object file with the same name as the input file but with the **.obj** extension, in this case **FIR1.obj**.

Note: The **asm6x** command is not case sensitive for Windows.

The assembler, as with the compiler, also has a number of 'switches' that the programmer can supply. The most common options are shown in Table 3.2.

The following command line assembles the **FIR1.asm** file and generates an object file called **fir1.obj** and a listing file called **fir1_lst.lst**.

Table 3.2 Common assembler options

Option	Description
-l	Generates an assembly listing file
-s	Puts labels in the symbolic table in order to be used by the debugger
-x	Generates a symbolic cross-reference table in the listing file (using the **-ax** option automatically evokes the **-l** option)

```
asm6x-g FIR1.asm fir1.obj -l fir1_lst.lst
```

Note: The file names are case sensitive.

3.2.3 Linker

The various object files which constitute an application are all combined by the linker to produce a single executable file. The linker also takes as inputs the library files and the command file that describes the hardware. To evoke the linker, type:

```
lnk6x FIR1.obj comd.cmd
```

The above command line links the **FIR1.obj** file with the file(s) contained in the command file **comd.cmd**. The linker options can also be contained in the command file.

The linker also has different options that are specified by the programmer. The most common options are shown in Table 3.3.

The following command line links the file **FIR1.obj** with the file(s) specified in the **comd.cmd** file and generates a map file (**FIR1.map**) and an output file (**FIR1.out**).

Note: The **-m FIR1.map** and the **-o FIR1.out** command could be included in the command file.

```
lnk6x FIR1.obj comd.cmd -m FIR1.map -o FIR1.out
```

Note: The **lnk6x** command is not case sensitive for Windows, and if **-o FIR1.out** is omitted, then an **A.out** file will be generated instead.

3.2.3.1 Linker command file

The command file serves three main objectives: the first objective is to describe to the linker the memory map of the system to be used, and this is specified by '**MEMORY** {...}'. The second objective is to tell the linker how to bind each section of the program to a specific section as defined by the **MEMORY** area; this is specified by '**SECTIONS** {...}'. The third objective is to supply the linker with the input and output files, and options of the linker. An excerpt of a command file for the TMS320C6678 EVM is shown in Figure 3.4.

As with all embedded systems, the command file is indispensable for real-time applications. The linker options specified in the **CL6x** command can be specified within the command file, as shown in Figure 3.4.

Table 3.3 Frequently used options for the linker

Options	Description
-o	Names an output file
-c	Uses auto-initialisation at runtime
-l	Specifies a library file
-m	Produces a map file

```
/***************************************************************************/
/*  C6678.cmd                                                            */
/*  Copyright (c) 2011 Texas Instruments Incorporated                    */
/*  Author: Rafael de Souza                                              */
/*                                                                       */
/*    Description: This file is a sample linker command file that can be */
/*                 used for linking programs built with the C compiler and */
/*                 running the resulting .out file on an C6678           */
/*                 device.  Use it as a guideline.  You will want to     */
/*                 change the memory layout to match your specific C6xxx */
/*                 target system.  You may want to change the allocation */
/*                 scheme according to the size of your program.         */
/*                                                                       */
/*    Usage:       The map below divides the external memory in segments */
/*                 Use the linker option --define=COREn=1                */
/*                 Where n is the core number.                           */
/*                                                                       */
/***************************************************************************/

MEMORY
{
    SHRAM:            o = 0x0C000000 l = 0x00400000   /* 4MB Multicore shared Memory */

    CORE0_L2_SRAM:    o = 0x10800000 l = 0x00080000   /* 512kB CORE0 L2/SRAM */
    CORE0_L1P_SRAM:   o = 0x10E00000 l = 0x00008000   /* 32kB CORE0 L1P/SRAM */
    CORE0_L1D_SRAM:   o = 0x10F00000 l = 0x00008000   /* 32kB CORE0 L1D/SRAM */

    CORE1_L2_SRAM:    o = 0x11800000 l = 0x00080000   /* 512kB CORE1 L2/SRAM */
    CORE1_L1P_SRAM:   o = 0x11E00000 l = 0x00008000   /* 32kB CORE1 L1P/SRAM */
    CORE1_L1D_SRAM:   o = 0x11F00000 l = 0x00008000   /* 32kB CORE1 L1D/SRAM */

    EMIF16_CS2:       o = 0x70000000 l = 0x04000000   /* 64MB EMIF16 CS2 Data Memory */
    EMIF16_CS3:       o = 0x74000000 l = 0x04000000   /* 64MB EMIF16 CS3 Data Memory */
    EMIF16_CS4:       o = 0x78000000 l = 0x04000000   /* 64MB EMIF16 CS4 Data Memory */
    EMIF16_CS5:       o = 0x7C000000 l = 0x04000000   /* 64MB EMIF16 CS5 Data Memory */

    CORE0_DDR3:       o = 0x80000000 l = 0x10000000   /* 256MB DDR3 SDRAM for CORE0 */
    CORE1_DDR3:       o = 0x90000000 l = 0x10000000   /* 256MB DDR3 SDRAM for CORE1 */
}

SECTIONS
{
#ifdef CORE0
    .text          >   CORE0_L2_SRAM
    .stack         >   CORE0_L2_SRAM
    .bss           >   CORE0_L2_SRAM
    .cio           >   CORE0_L2_SRAM
    .const         >   CORE0_L2_SRAM
    .data          >   CORE0_L2_SRAM
    .switch        >   CORE0_L2_SRAM
    .sysmem        >   CORE0_L2_SRAM
    .far           >   CORE0_L2_SRAM
    .args          >   CORE0_L2_SRAM
    .ppinfo        >   CORE0_L2_SRAM
    .ppdata        >   CORE0_L2_SRAM

    /* COFF sections */
    .pinit         >   CORE0_L2_SRAM
    .cinit         >   CORE0_L2_SRAM

    /* EABI sections */
    .binit         >   CORE0_L2_SRAM
    .init_array    >   CORE0_L2_SRAM
    .neardata      >   CORE0_L2_SRAM
    .fardata       >   CORE0_L2_SRAM
    .rodata        >   CORE0_L2_SRAM
    .c6xabi.exidx  >   CORE0_L2_SRAM
    .c6xabi.extab  >   CORE0_L2_SRAM
#endif

#ifdef CORE1
    .text          >   CORE1_L2_SRAM
    .stack         >   CORE1_L2_SRAM
```

Figure 3.4 Excerpt of command file for the TMS320C6678 EVM (**C6678.cmd**).

```
    .bss            >   CORE1_L2_SRAM
    .cio            >   CORE1_L2_SRAM
    .const          >   CORE1_L2_SRAM
    .data           >   CORE1_L2_SRAM
    .switch         >   CORE1_L2_SRAM
    .sysmem         >   CORE1_L2_SRAM
    .far            >   CORE1_L2_SRAM
    .args           >   CORE1_L2_SRAM
    .ppinfo         >   CORE1_L2_SRAM
    .ppdata         >   CORE1_L2_SRAM

    /* COFF sections */
    .pinit          >   CORE1_L2_SRAM
    .cinit          >   CORE1_L2_SRAM

    /* EABI sections */
    .binit          >   CORE1_L2_SRAM
    .init_array     >   CORE1_L2_SRAM
    .neardata       >   CORE1_L2_SRAM
    .fardata        >   CORE1_L2_SRAM
    .rodata         >   CORE1_L2_SRAM
    .c6xabi.exidx   >   CORE1_L2_SRAM
    .c6xabi.extab   >   CORE1_L2_SRAM
#endif
```

Figure 3.4 (Continued)

3.2.4 Compile, assemble and link

The command **CL6x**, combined with the **-z** linker option, can accomplish the compiling, assembling and linking stages all with a single command, as shown here:

```
CL6x -gs FIR1.c -z C6678.cmd
```

3.2.5 Using the Real-Time Software Components (RTSC) tools

By using the RTSC tools, one is able to take advantage of the higher levels of programming and performance that the RTSC can offer in addition to features that allow components to be added or modified/upgraded without the need to modify the source code. For a complete description of the RTSC, please refer to Ref. [5].

The XDCtools are the heart of the RTSC components; see Figure 3.5. These tools provide mainly efficient APIs for development and static configurations that can accelerate development to production time and ease maintenance.

3.2.5.1 Platform update using the XDCtools

In Section 3.2.3, a command file was written from scratch. With the use of the XDCtools, this can be imported as a package that is delivered by the developer (TI) and used by the consumer. This package can be in the form of a plug-in.

In the example shown in Figure 3.6, instead of entering a linker command file, a platform file describing the memory map is supplied (see Figure 3.7). This platform can be viewed and modified by selecting (while in the CCS Debug mode) **Tools → RTSC Tools → Platform → New**. Figure 3.7 to Figure 3.12 are self-explanatory on how to generate a new target platform from a seed platform. Once a platform is created, the seed platform has to be replaced by the new platform, as shown in Figure 3.13.

Add the path shown in Figure 3.12 (**C:\Users\eend\myRepositor\packages**) using the Add button shown in Figure 3.13, then select the platform **myBoard**.

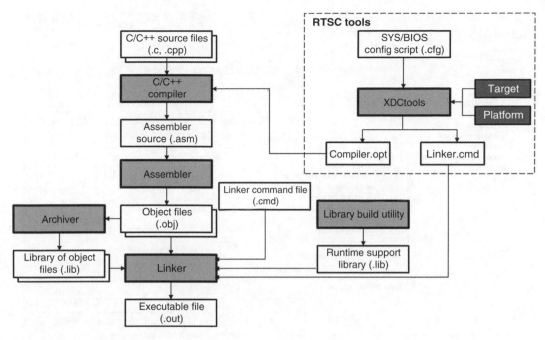

Figure 3.5 RTSC tools [6].

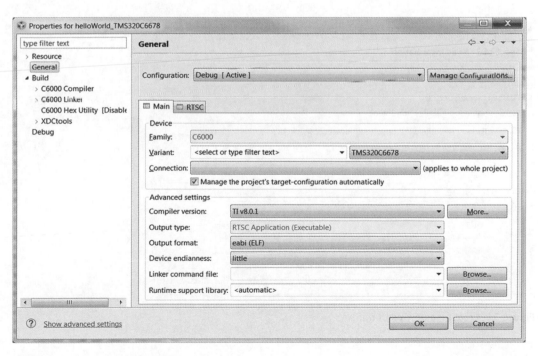

Figure 3.6 Entering a linker command file.

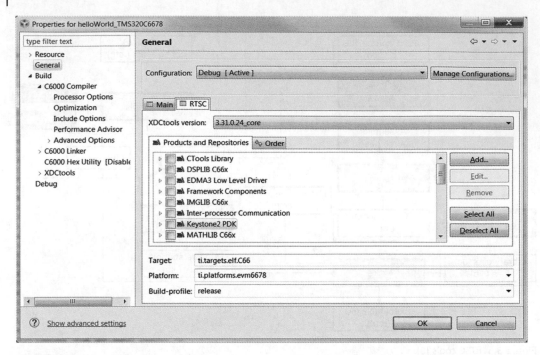

Figure 3.7 Platform selection.

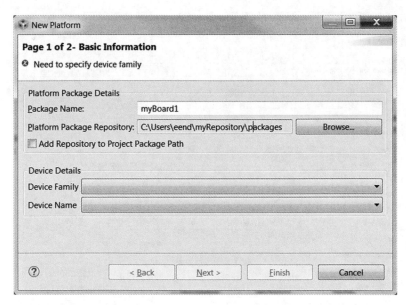

Figure 3.8 Creating a new platform.

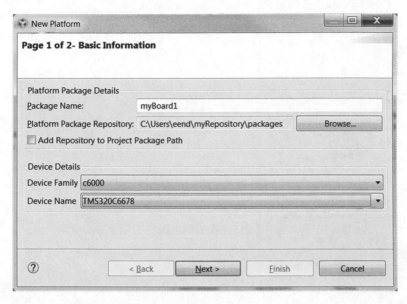

Figure 3.9 Selecting the device family and device name for the new platform.

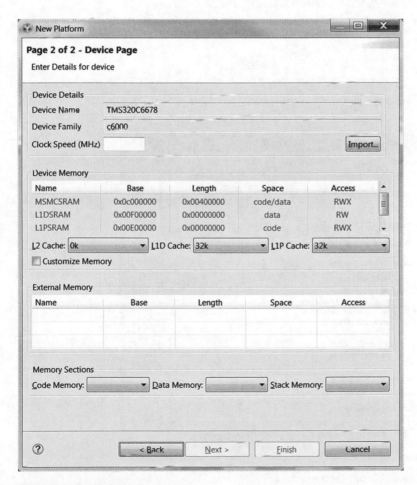

Figure 3.10 Device page.

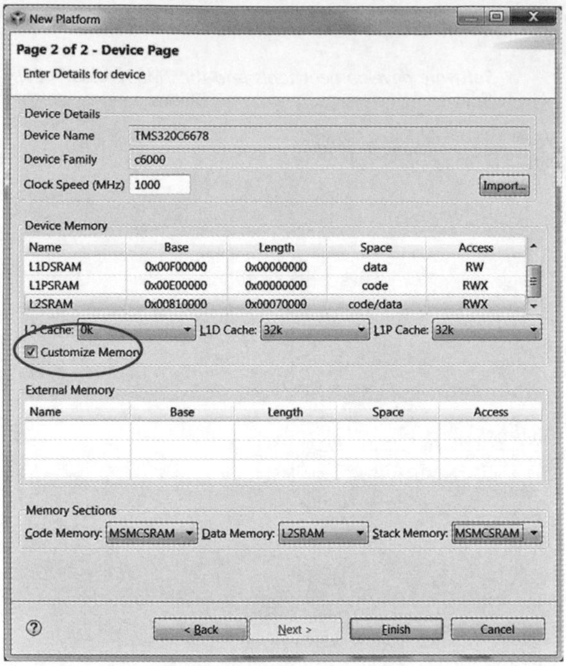

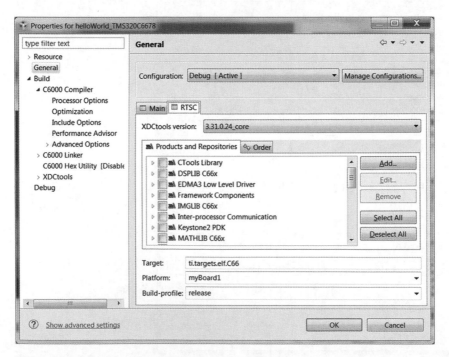

Figure 3.11 How to modify the new platform.

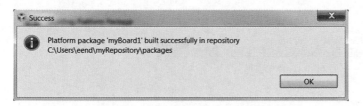

Figure 3.12 Output when a successful platform is generated.

Figure 3.13 Selecting the new platform for the project.

3.2.6 KeyStone Multicore Software Development Kit

The KeyStone System-on-Chip (SoC) is a very powerful and complex processor. To ease its use and reduce time-to-market, TI has developed foundation software called the Multicore Software Development Kit (MCSDK); see Figure 3.14 [7]. The MCSDK, with the supported EVMs, provide out-of-box demos with source code that can be modified, libraries and device drivers. TI also provides a platform development kit with low-level software drivers, libraries and chip support software for peripherals supported by the KeyStone [8].

3.3 Hardware development tools

EVMs are relatively low-cost demonstration boards. They allow one to evaluate the performance of the processors. The EVMs are platforms that contain the processor, some peripherals, expansion connectors and emulators. In the case of the TMDXEVM6678LE EVM, there are two emulators, one on-board (XDS100) and one on a mezzanine (XDS560), that use the JTAG (Joint Test Action Group) emulator header, as shown in Figure 3.15a. The KeyStone II has a Mezzanine XDS200 emulator connected to the JTAG header, as shown in Figure 3.15b.

Note: In this book, two EVMs are used: the TMDXEVM6678LE and the K2EVM-HK. For a complete description, please refer to Refs. [7] and [11–13].

The basic layouts of both EVMs are shown in Figure 3.16.

3.3.1 EVM features

The key features of the TMDXEVM6678L or TMDXEVM6678LE EVM are [16]:

- TI's multicore DSP – TMS320C6678
- 512 MB of double data rate type 3 (DDR3)-1333 memory
- 64 MB of NAND ('not AND') flash
- 16 MB of serial peripheral interface (SPI) NOR ('not OR') flash
- Two Gigabit Ethernet ports supporting 10/100/1000 Mbps data rates – one Advanced Mezzanine Card (AMC) connector and one RJ-45 connector
- 170-pin B + -style AMC interface containing serial RapidIO (SRIO), PCI Express (PCIe), Gigabit Ethernet and time-division multiplexing (TDM)
- High-performance connector for HyperLink
- 128 KB inter-integrated circuit (I2C) electrically erasable programmable read-only memory (EEPROM) for booting
- Two user light-emitting diodes (LEDs), five banks of dual in-line package (DIP) switches and four software-controlled LEDs
- RS232 serial interface on a 3-pin header or Universal Asynchronous Receiver/Transmitter (UART) over a mini-USB connector
- External memory interface (EMIF), timer, SPI and UART on 80-pin expansion header
- On-board XDS100 type emulation using high-speed USB 2.0 interface
- TI 60-pin JTAG header to support all external emulator types
- Module Management Controller (MMC) for the Intelligent Platform Management Interface (IPMI)
- Optional XDS560v2 System Trace Emulator mezzanine card
- Powered by DC power-brick adaptor (12 V/3.0 A) or AMC Carrier backplane
- PICMG® AMC.0 R2.0 single-width, full-height AMC module.

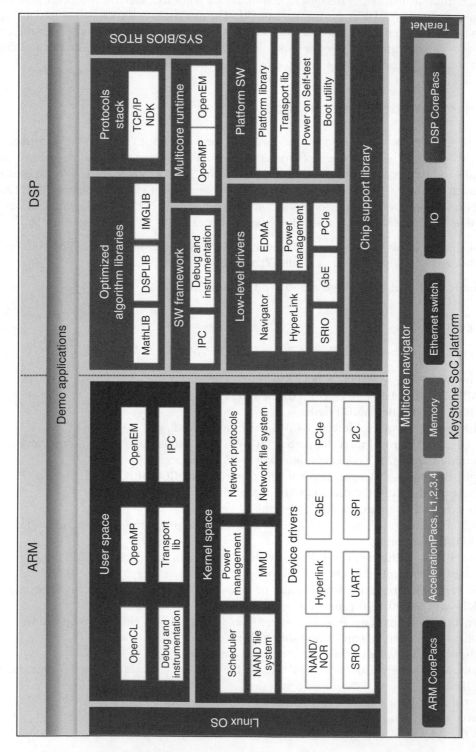

Figure 3.14 Multicore Software Development Kit (MCSDK) [9, 10].

(a)

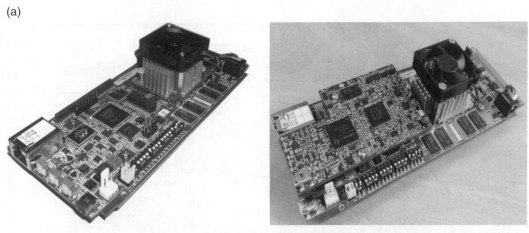

TMS320C6678 EVM without and with an emulator

(b)

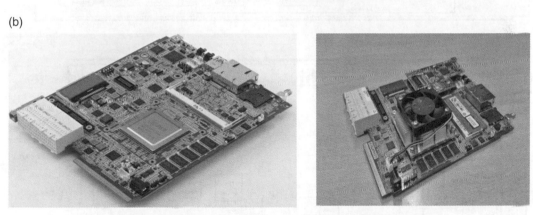

KeyStone II EVM without and with an emulator

Figure 3.15 The TMS320C6678 and the KeyStone II EVMs. (a) TMS320C6678 EVM without and with an emulator; (b) KeyStone II EVM without and with an emulator.

The key features of the KeyStone II EVM are [15]:

- TI's 8-core DSP and 4-core ARM SoC
- 1024/2048 MB of DDR3-1600 memory on board
- 2048 MB of DDR3-1333 error-correcting code (ECC) small-outline dual-inline memory module (SO-DIMM)
- 512 MB of NAND flash
- 16 MB SPI NOR flash
- Four Gigabit Ethernet ports supporting 10/100/1000 Mbps data rate
- AMC connector and two RJ-45 connectors
- 170-pin B + -style AMC interface containing SRIO, PCIe, Gigabit Ethernet, Architecture Antenna Interface 2 (AIF2) and TDM
- Two 160-pin ZD + -style universal reversible Turing machine (uRTM) interfaces containing HyperLink, AIF2 and XGMII (not supported for all EVMs)
- 128 KB I2C EEPROM for booting

(a)

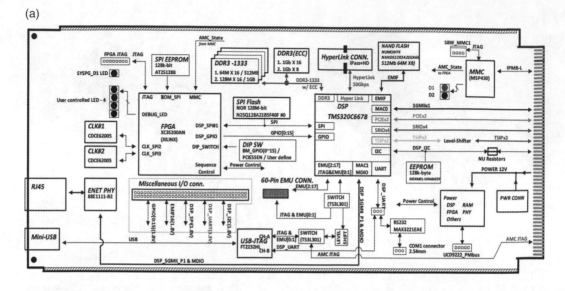

(b)

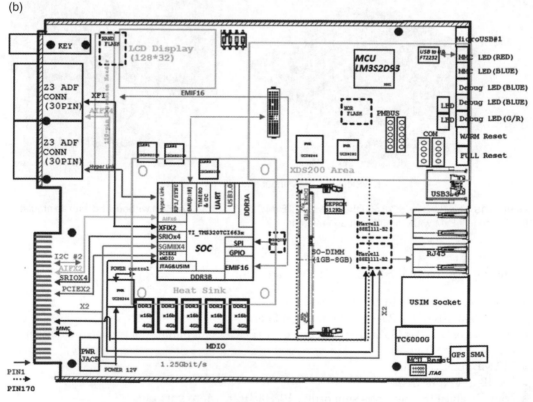

Figure 3.16 EVM layout. (a) TMS320C6678L [14]; (b) KeyStone II [15].

- Four user LEDs, one bank of DIP switches and three software-controlled LEDs
- Two RS232 serial interfaces on a 4-pin header or a UART over a mini-USB connector
- EMIF, timer, I2C, SPI and UART on a 120-pin expansion header
- One USB 3.0 port supporting a 5 Gbps data rate
- MIPI 60-pin JTAG header to support all external emulator types

- LCD display for debugging state
- Microcontroller unit (MCU) for the IPMI
- Optional XDS200 System Trace Emulator mezzanine card
- Powered by DC power-brick adaptor (12 V/7.0 A) or AMC Carrier backplane
- PICMG® AMC.0 R2.0 and uTCA.4 R1.0 double-width, full-height AMC module.

3.4 Laboratory experiments based on the C6678 EVM: introduction to Code Composer Studio (CCS)

All laboratories experiments have been tested, and solutions are provided.
   File location

Chapter_3_Code:\

3.4.1 Software and hardware requirements

1) CCS version 6.0 or higher: CCS6.0 (see Figure 3.17)
2) PC with the following hardware and software:

	Minimum	Recommended
Memory	1 GB	4 GB
Disk space	300 MB	2 GB
Processor	1.5 GHz single core	Dual core

3) Operating system requirements for the PC:
    a) Windows. Windows XP, 7, 8 or 10.
    b) Linux. Details on the Linux distributions supported are available in Ref. [17].
4) A TMX320C6678 EVM. The EVM used in this laboratory experiment is based on the TMS320C6678 EVM module shown in Figure 3.18.

Figure 3.17 Code Composer Studio (CCS).

TMDXEVM6678L
TMDX320C6678 Evaluation Module

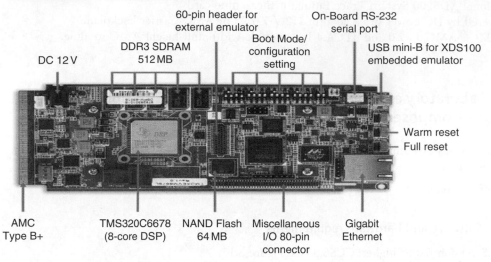

Figure 3.18 The TMS320C6678 EVM.

3.4.1.1   Key features

Hardware features	Software features	Kit contents
• Single wide AMC-like form factor • Single TMS320C6670 multicore processor • 512 MB DDR3 • 128 MB NAND flash • 1 MB I2C EEPROM for local boot (remote boot possible) • Two 10/100/1000 Ethernet ports on board • RS232 UART • 2 user-programmable LEDs and DIP switches • 14-pin JTAG emulator header • Embedded JTAG emulation with USB host interface • Board-specific Code Composer Studio Integrated Development Environment • Simple setup • Design files such as Orcad and Gerber	• Power-on self-test in EEPROM at 0x50 address (POST) • Intermediate boot loader in EEPROM at 0x51 address (IBL) • High-performance DSP utility applications in NOR (HUA)	• TMX320C6670 evaluation module • Power adapter and power cord • USB cable for on-board JTAG emulation (XDS100v1) • Ethernet cable • RS-232 serial cable • Software (DVD) and documentation

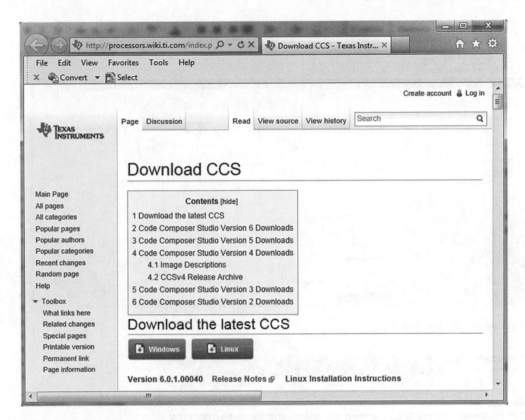

Figure 3.19 CCS download page.

3.4.1.2 Download sites

Follow these links to access the sites required.

- Wiki: Code Composer Studio. Information on how to more effectively use the CCS.
  http://processors.wiki.ti.com/index.php/Category:Code_Composer_Studio_v6
- Download site. All current and archived product images.
  http://processors.wiki.ti.com/index.php/Download_CCS
- HELP: having trouble? Where to go for downloads, upgrades, licencing and subscription help.
  http://www.ti.com/lsds/ti/software-help.page
- System requirements. Details on the minimum and recommended system requirements.
  http://processors.wiki.ti.com/index.php/System_Requirements
- Subscription information. Details on the CCS subscription service (no subscription required).
  http://www.ti.com/tool/ccssub

For this teaching material, download the Windows version shown in Figure 3.19. You will be prompted to register with TI.com as shown in Figure 3.20. Once registered, you will be given access to download the CCS as shown in Figure 3.21.

3.4.2 Laboratory experiments with the CCS6

These laboratory experiments mainly provide an introduction to the CCS, implementation of a dot product (**dotp**), how to use the CCS clock to benchmark code and how to download code to separate cores and run them on the TMS320C6678 EVM.

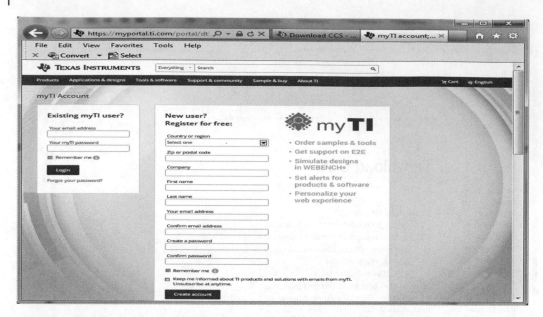

Figure 3.20 Registration with myTI.

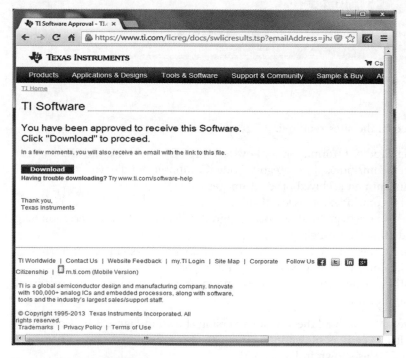

Figure 3.21 CCS download.

File locations for this chapter:

Chapter_3_Code\dotp
Chapter_3_Code\Print_Functions

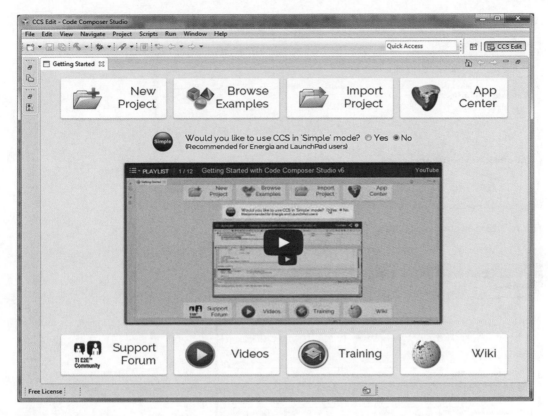

Figure 3.22 CCS starting window.

3.4.2.1 Introduction to CCS

The aim of this laboratory exercise is to become familiar with the DSP tools and to be introduced to programming the TMS320C66xx SoC DSP.

This section shows how to use development tools, create a project and modify the compiler switches in order to achieve the best performance.

Starting the experiment

1) Launch CCS. Launch CCS by double-clicking on the desktop icon of your PC or using: **Start > All Programs > Code Composer Studio 6.0.0** (or a later version). You should see a CCS window, as shown in Figure 3.22, if you run the CCS for the first time; or the screen will look like Figure 3.23 if the CCS has been used before.

   Note: You have to be logged in as an administrator if you require updates.

2) Create a new project. A project stores all the information needed to build an individual program or library, including:
   - File names of source code and object libraries
   - Build-tool options
   - File dependencies
   - A build-tool version used to build the project.

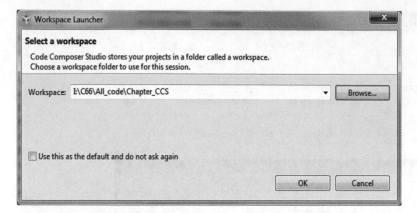

Figure 3.23 Selecting a workspace location.

Figure 3.24 Lab1 basic project settings.

A. Select **File > New > CCS Project** (see Figure 3.24).

Note: **DO NOT press 'Finish' until you have configured your project.**

- Set up the project settings according to the screenshot given in Figure 3.24.
- Insert the target.
- Select the device family and variant.
- Select the connection (the XDS560v2-USB Mezzanine Emulator is used).
- Select the compiler version.
- Choose the **Hello Word** template.

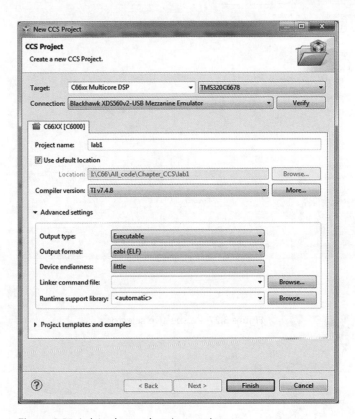

Figure 3.25 Lab1 advanced project settings.

B. Check **Advanced Settings** (see Figure 3.25). Even though this dialogue is HIDDEN and called **Advanced**, it is CRITICAL to check this to make sure you are creating a project with the right tools, endianness and output format.

Note: If you import a project and you do not have the right compiler or XDCtools version, you can download them separately using these links:

https://www-a.ti.com/downloads/sds_support/TICodegenerationTools/download.htm
http://downloads.ti.com/dsps/dsps_public_sw/sdo_sb/targetcontent/rtsc/

However, you may have to register with TI first. Once the tools are downloaded and installed, restart the CCS so that the tools will be automatically updated.

Press 'Finish' to confirm the settings. The window in Figure 3.26 will then appear.

3) Edit perspective (see Figure 3.27). Once you create a project, you will have two default perspectives, one for editing (**CCS Edit**) and one for debugging (**CCS Debug**). Make sure you select the appropriate perspective for what you want to do.

4) Adding a DSP target configuration (see Figure 3.28). The target configuration is used by the debugger. You can have one project with many target configurations and therefore many platforms (e.g. a simulator or an EVM). This is very convenient as no modification to your project is required when you change emulators.

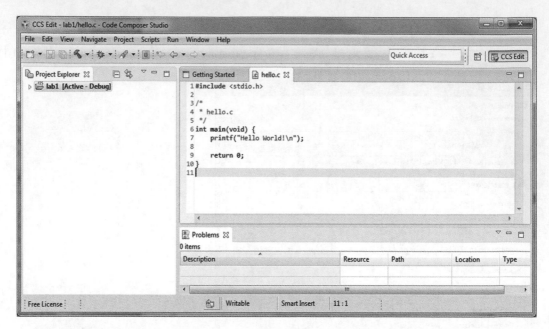

Figure 3.26 Default view.

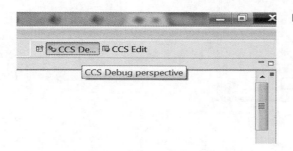

Figure 3.27 Perspective selector.

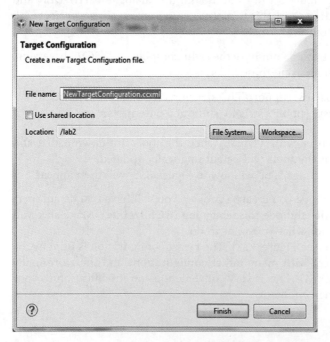

Figure 3.28 Naming a target configuration.

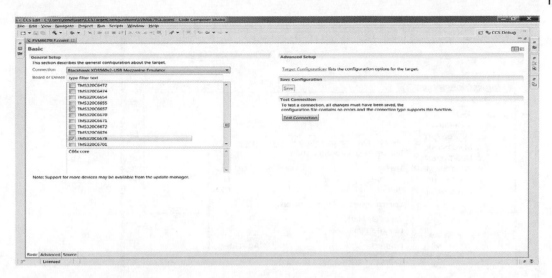

Figure 3.29 Selecting the appropriate emulator for the target configuration.

Once you have defined a project, you can add a configuration to this project using these commands:

- Select the project in Project Explorer, and then select the **File > New > Target Configuration File** command from the main menu.

  Type a file name, and click **Finish**.

  Select **File > Save As** (💾) to record your target configuration selections.

  Open the target configuration and set it for the **Blackhawk XDS560v2-USB Mezzanine Emulator** and the **TMS320C6678**, as shown in Figure 3.29.

  Select the **Advanced** tab and explore the **CPU Properties**. Then, save and close the file.

Note: You can add multiple target configurations to your projects (this is very useful if you are using different platforms). However, only one target configuration can be active at any time. You can also choose a default target so that you don't have to add a target configuration to each project.

To add a new target configuration to the list of target configurations or to select a new configuration for a project, do the following:

a) Select **View > Target Configurations**.
b) Right-click on **User Defined**.
c) Select the appropriate target.
d) Select the appropriate function and close the window (see Figure 3.30).

By selecting one of the commands from the list, a display of the **New Target Configuration** dialogue will appear as shown in Figure 3.30.

You can now see all the configurations created and their locations. Now you can select the appropriate configuration. In this laboratory, you will be using the configuration that you just created.

Once the target configuration is completed and closed, it will be automatically added to your project and set to **Active**.

Go back to the Project Explorer. If your Project Explorer is not visible, then select **Project Explorer** (see Figure 3.31).

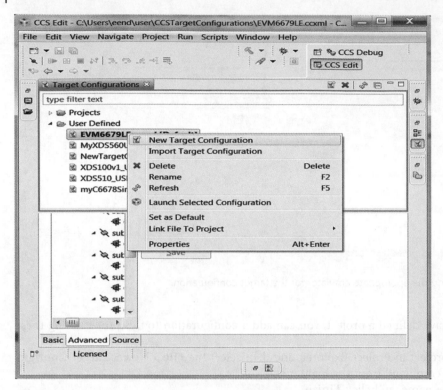

Figure 3.30 Selecting the appropriate target configuration.

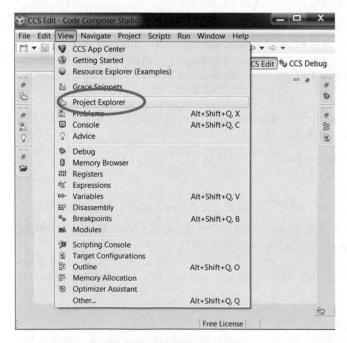

Figure 3.31 Selecting the Project Explorer.

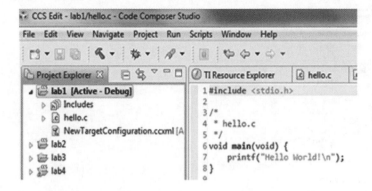

Figure 3.32 Building a project.

5) Building and loading the code (see Figure 3.32)

Notice that the project is highlighted and there is [**Active – Debug**] next to the project name. This means:

Highlighted: The project is active (only one project can be active at a time).

Active: The project is active.

Debug: The project is in the debug mode.

Right-click on the project and select **Build Configurations** → **Set Active**.

Click on **Release**, and see **Debug** change to **Release** in the project. This is how you change build configurations (the set of build options) when you build your code.

Note: Build configurations such as **Debug, Release** and others that you create will not contain the same build options, like levels of optimisation and debug symbols; specify **file search paths** for libraries (**-l**) and **include search paths** (**-i**) for included directories.

Change the build configuration back to **Debug**. Near the top left-hand corner of CCS, you will see the build **Hammer** and the **Bug**:

The **Hammer** allows you to change the build configuration (Figure 3.33 and Figure 3.34) and build your project.

The **Bug** allows you to debug the code.

Select the **Debug** mode, and start debugging your code.

If you simply click the bug, it will build whichever configuration is set as the default (either **Debug** or **Release**). It will always do an *incremental* build (i.e. build only those files that have changed since the last build; it is much faster this way).

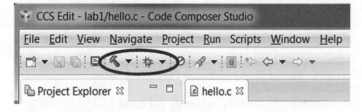

Figure 3.33 Building and debugging.

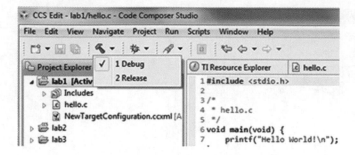

Figure 3.34 Changing the configuration option.

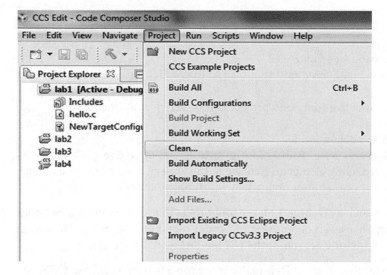

Figure 3.35 Build types.

You can also specify a build type as shown in Figure 3.35.

There are three kinds of builds:

- Build
- Clean build
- Rebuild.

  Incremental and clean builds can be done over a specific set of projects or the workspace as a whole. Specific files and folders cannot be built. There are two ways that builds can be performed:

- Automatic builds are performed as resources are saved. Automatic builds are always incremental and always operate over the entire workspace. You can configure your preferences (**Window > Preferences > General > Workspace**) to perform builds automatically on resource modification.
- Manual builds are initiated when you explicitly select a menu item or press the equivalent shortcut key. Manual builds can be either clean or incremental and can operate over collections of projects or the entire workspace.

6) Running the project. To run the project, press the bug as shown in Figure 3.33. If the EVM is connected properly and the project successfully built, then the window shown in Figure 3.36 will appear.

Figure 3.36 Launching the **Debug** session.

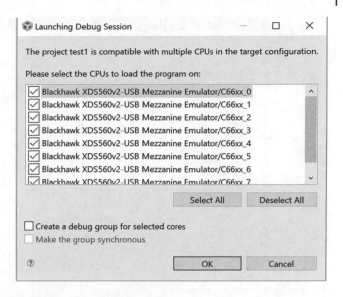

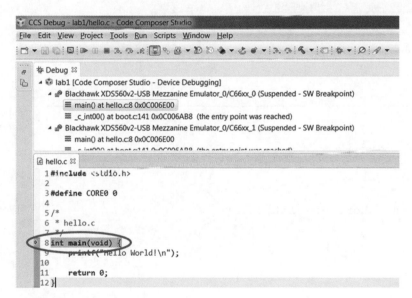

Figure 3.37 Running the project.

Once you have successfully built your project, the **Debug** window will appear as shown in Figure 3.37.

You can observe the CDT Build Console to see the compilation feedback. If the window is not visible, choose **View > Console**. In fact, **View** lets you select a list of windows that you would like to display; see Figure 3.38.

3.4.2.2 Implementation of a DOTP algorithm

Task 1: Implement the dotp function in C language

Use the starting code given in **dotp.c** (Figure 3.39) to implement a **dotp** function ($y = \sum a_i x_i$).

1) In this example, the operating system (**SYS/BIOS**) will be used, and therefore it needs to be installed. To do so, download the latest version from Ref. [18].

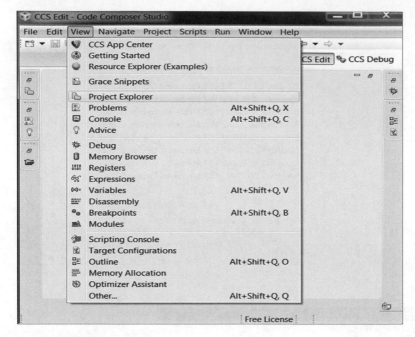

Figure 3.38 View functions.

```
void main()
{
      dotp(a, x, COUNT);   // matrix multiplication
      System_printf("y   = %d \n",  y);
}

int dotp(short *m, short *n, int count)
{
      int acc,i;
      //to be completed
      return acc;
}
```

Figure 3.39 dotp.c: Source code to be completed.

2) Copy the project **Chapter_3_Code\dotp** to a directory called **...\\dotp**.
3) Build and load your project (the project should build without errors). If the console is not visible, you can open it by selecting: **View > Console.**
4) Change the directive to **CCS Edit**, open the **dotp.c** file and complete it.
5) Build and run the project.

Check the value of **y** and write it here:

   y = _____?

The answer should be: **y** = 2829056 decimal.

Add another function just below the **System_printf("y = %d \n", y)** in order to print **y** in hexadecimal format.

> **y** = _____?

The answer should be: **y** = 2b2b00 hexadecimal.

Solution. See file **dotp_solution.txt**.

Task 2: Using System_sprintf()

This function is identical to **printf** except that the output is copied to the specified character buffer but followed by a terminating '**\0**' character.

1) Copy the project **Chapter_3_Code\Print_Functions\print** to a directory called **...\\PRINT**.
2) Build and load your project (the project should build without errors). If the console is not visible, you can open it by using: **View > Console**.
3) Change the directive to **CCS Edit** and open the **dotp.c** file.
4) Create two buffers of 30 characters each (**buf1** and **buf2**).
5) Create two character buffers, **s1** and **s2**, and initialise them with **Hello** and **Print**, respectively.
6) Use the two following instructions to add the content of buffers **s1** and **s2** to **buf1** and **buf2**, respectively:

```
System_sprintf(buf1, "First output : %s\n", s1);
System_sprintf(buf2, "Second output: %s\n", s2);
```

7) Use the following instructions to print the contents of both **buf1** and **buf2**:

```
System_printf(buf1);
System_printf(buf2);
```

8) Build and run the project. The console should open, and the output should be:

```
[C66xx_0] First output: Hello
Second output: Print
y = 2829056
```

Solution. See file**/print/dotp_solution.txt**.

3.4.3 Profiling using the clock

This section describes how to set up and use the profile clock in CCS to count instruction cycles between two locations in the code. Since the CCS profiler has some limitations when profiling on hardware, the profile clock is one of the suggested alternate options.

1) Load the project. Select and load the project that needs to be profiled. In this example, select: **Chapter_3_Code\Profiling\myprofiling**
2) Enable the profile clock. In the **Debug** perspective, go to the menu and select **Run → Clock → Enable**. This will add a clock icon and cycle counter to the status bar (shown in Figure 3.40).

Figure 3.40 Clock icon and cycle count.

Profile Clock <CPU Execute Cycles>

Figure 3.41 Clock setup.

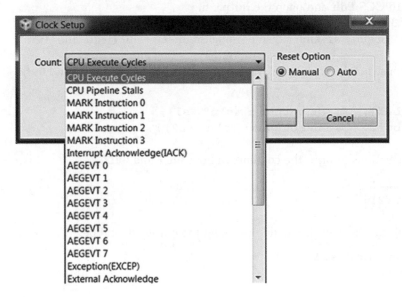

Figure 3.42 Clock setup.

3) Set up the profile clock. Once the clock is enabled, in the **Debug** perspective, go to the menu and select **Run** → **Clock** → **Setup**. This will bring up the clock setup dialog; see Figure 3.41. In the clock setup dialog box, you can specify the event you want to count in the drop-down list of the count field. Depending on your device, cycles may be the only option listed. However, some device drivers make use of the on-chip analysis capabilities and may allow profiling other events. With the KeyStone, we have the following options: Figure 3.42 and Figure 3.43:

4) Reset the profile clock and measure the number of cycles. Set three breakpoints at the following locations:

```
→ y = dotp(a, x, COUNT);
→ System_printf("y = %d \n", y);
→
```

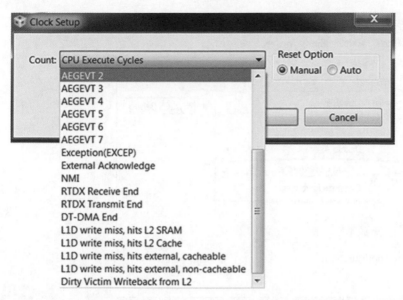

Figure 3.43 Clock setup.

Run the code to the first breakpoint, then double-click on the clock value in the status bar to reset it to zero. Next, run to the second point (by pressing the green arrow or typing F8), and record how long it takes to run the **dotp()** function (7807 cycles). Now reset the clock, run the code again and note how long it takes to run the **Sytem_printf()** function (4133 cycles).

3.4.4 Considerations when measuring time

Some cores have a 1:1 relationship between the clock and the CPU cycles; therefore, a simple instruction like **NOP** located in the internal memory should just jump one unit in the counter. However, if the code is located in the external memory, the CPU will have to wait several cycles until the instruction is fetched to its internal pipeline (caused by wait states and stalls). This translates to additional clock cycles measured by the profile clock.

Similarly, certain instructions require additional CPU cycles to complete their execution if they access memory (store and load: e.g. **STW** and **LDW**), branch to other parts of the code (branch: e.g. **B**) or do not execute at all (conditional instructions in the TMS320C66x ISA: e.g. **[A0] MPY A1,A2,A4**).

For the instructions that access memory, keep in mind that other peripherals (DMA, HPI) or cores (in the case of SoC devices) may be accessing the same region at the same time, which can cause a bus contention and make the CPU wait until it is allowed to fetch the data/instruction.

Lastly, if the software under evaluation contains interrupt requests, keep in mind that the cycle count may increase significantly if an interrupt is serviced in the middle of the region under evaluation.

3.5 Loading different applications to different cores

So far, we have seen that by default an application is automatically loaded to all cores. However, it is sometimes desirable to load different applications to specific core or cores.

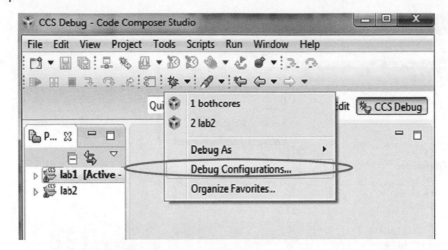

Figure 3.44 Selecting **Debug** configurations.

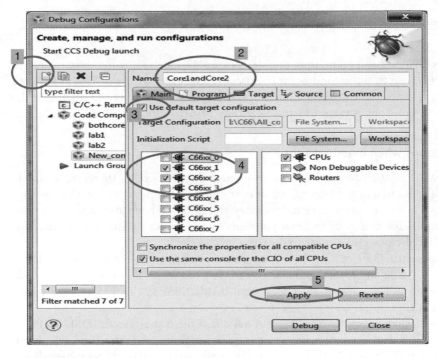

Figure 3.45 Setting the **Debug** configuration.

For instance, if we have two projects **lab1** and **lab2** and we want to load **lab1** to **Core 1** and **lab 2** to **Core 2**, the following procedure can be followed:

1) Build **lab1** and **lab2** separately.
2) Select **Debug Configurations** as shown in Figure 3.44.
3) Create a new configuration as illustrated in Figure 3.45 to Figure 3.49.
4) Group the two cores as shown in Figure 3.50.
5) Select **Group core(s)** as shown in Figure 3.51, and run the projects. The output is shown in Figure 3.52.

Figure 3.46 Setting the device.

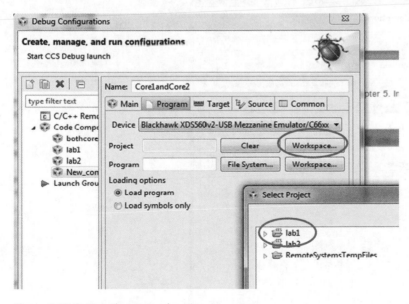

Figure 3.47 Setting the project location.

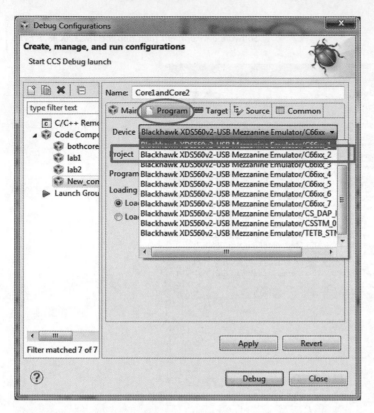

Figure 3.48 Setting **Core 2**.

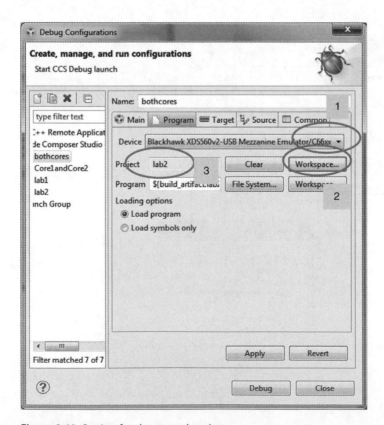

Figure 3.49 Setting for the second project.

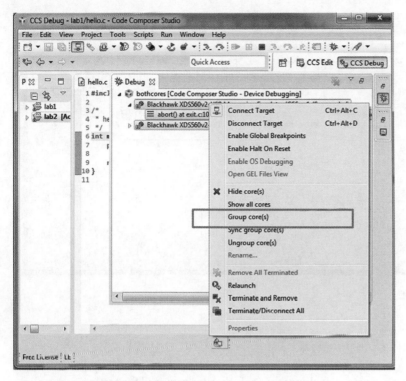

Figure 3.50 Grouping the cores.

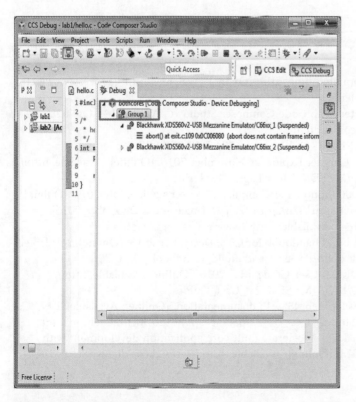

Figure 3.51 Grouping the cores.

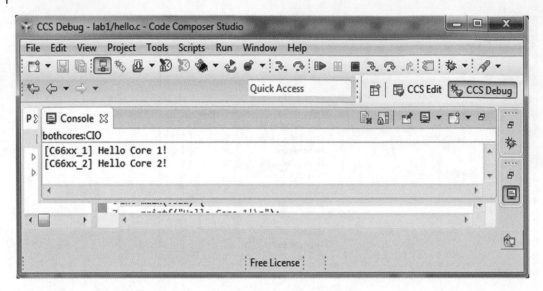

Figure 3.52 Console output.

3.6 Conclusion

This chapter describes the software development tools that are required for testing the applications used in this book. It provides a step-by-step description of the installation and use of the Code Composer Studio (CCS).

References

1 Texas Instruments, Category:Compiler, June 2014. [Online]. Available: http://processors.wiki.ti.com/index.php/Category:Compiler.
2 Texas Instruments, Getting Started with Code Composer Studio v6, April 2014. [Online]. Available: https://www.youtube.com/watch?v=uAb5MScflEo&index=1&list=PL3NIKJ0FKtw4w_bK7FASz6RrTZb8PD3j5.
3 Texas Instruments, MCSDK UG Chapter Exploring, November 2016. [Online]. Available: http://processors.wiki.ti.com/index.php/MCSDK_UG_Chapter_Exploring.
4 Texas Instruments, TMS320C6000 Optimizing Compiler v7.4 user's guide, July 2012. [Online]. Available: http://www.ti.com/lit/ug/spru187u/spru187u.pdf. [Accessed 2 December 2016].
5 Eclipse, RTSC home page, [Online]. Available: http://www.eclipse.org/rtsc/.
6 Texas Instruments, Projects and build handbook for CCS, December 2016. [Online]. Available: http://processors.wiki.ti.com/index.php/Projects_and_Build_Handbook_for_CCS.
7 Texas Instruments, BIOS MCSDK 2.0 User Guide, May 2016. [Online]. Available: http://processors.wiki.ti.com/index.php/BIOS_MCSDK_2.0_User_Guide.
8 Texas Instruments, Platform development kit API documentation, [Online]. Available: file:///C:/TI/MCSDK_3_0_0_12/pdk_KeyStone2_3_00_01_12/packages/API%20Documentation.html.
9 T. Flanagan, Z. Lin and S. Narnakaje, Accelerate multicore application development with KeyStone software, February 2013. [Online]. Available: http://www.ti.com/lit/wp/spry231/spry231.pdf.

10 Texas Instruments, SYS/BIOS and Linux Multicore Software Development Kits (MCSDK) for C66x, C647x, C645x processors - BIOSLINUXMCSDK, [Online]. Available: http://www.ti.com/tool/bioslinuxmcsdk.

11 Advantech Co. Ltd., EVM documentation, [Online]. Available: http://www2.advantech.com/Support/TI-EVM/EVMK2HX_sd4.aspx.

12 Texas Instruments, BIOS MCSDK 2.0 getting started guide, May 2016. [Online]. Available: http://processors.wiki.ti.com/index.php/BIOS_MCSDK_2.0_Getting_Started_Guide.

13 Texas Instruments, XDS200 Texas Instruments wiki, December 2016. [Online]. Available: http://processors.wiki.ti.com/index.php/XDS200#Updating_the_XDS200_firmware.

14 Advantech Co. Ltd., EVM documentation (TMDXEVM6678L/LE Rev 0.5), [Online]. Available: http://www2.advantech.com/Support/TI-EVM/6678le_sd.aspx.

15 Texas Instruments, Keystone 2 EVM technical reference manual version 1.0, March 2013. [Online]. Available: http://wfcache.advantech.com/www/support/TI-EVM/download/XTCIEVMK2X_Technical_Reference_Manual_Rev1_0.pdf.

16 Advantech, TMDXEVM6678L EVM technical reference manual version 1.0, April 2011. [Online]. Available: http://wfcache.advantech.com/www/support/TI-EVM/download/TMDXEVM6678L_Technical_Reference_Manual_1V00.pdf.

17 Texas Instruments, Linux host support CCSv6, [Online]. Available: http://processors.wiki.ti.com/index.php/Linux_Host_Support_CCSv6. [Accessed January 2016].

18 Texas Instruments, SYS/BIOS product releases, [Online]. Available: http://downloads.ti.com/dsps/dsps_public_sw/sdo_sb/targetcontent/bios/sysbios/.

4

Numerical issues

4.1 Introduction

The majority of digital signal processing (DSP) algorithms are based on the dot product (dotp; $dotp = \sum_{k=0}^{N} a_k x_k$). The *dotp* is an approximation of $\int_{a}^{b} a_i f(x_i)$, which means that by moving to the digital domain we have already started introducing errors. These errors come from the sampling rate, and that is not a problem if the application's bandwidth can be handled by the processor and the other errors come from the fact that numbers cannot be represented with an infinite number of bits. Depending on the application, the dotp function can be implemented in fixed- or floating-point arithmetic. For applications that require high dynamic precision and accuracy, floating-point arithmetic is the best choice. However, it is important to note that even with floating-point arithmetic, calculation errors are unavoidable since processors have a limited number of bits. When dealing with multicores, the effect of parallelising a function can also introduce further calculation errors. Using numerical analysis to obtain an approximate solution while maintaining reasonable bounds on errors is out of the scope of this book, and the reader is referred to Refs. [1] and [2]. Digital signal processors (also DSPs) which process digitalised data use fixed-point and/or floating-point arithmetic. These processors can be divided into two categories: fixed-point and floating-point processors. As their names suggest, fixed-point processors

Multicore DSP: From Algorithms to Real-time Implementation on the TMS320C66x SoC, First Edition. Naim Dahnoun.
© 2018 John Wiley & Sons Ltd. Published 2018 by John Wiley & Sons Ltd.
Companion website: www.wiley.com/go/dahnoun/multicoredsp

use fixed-point arithmetic, have a smaller form factor, consume less power and are less expensive; however, in general they require a longer development time than the floating-point processors. Floating-point processors use floating-point arithmetic. In order to take full advantage of both fixed- and floating-point, high-end processors, like the KeyStone family, support both fixed- and floating-point arithmetic on an instruction-by-instruction basis while still maintaining a high clock speed, which means that one instruction could be written in one format and the next one could be written in a different format, with the two different instructions running at high speed.

Different applications use different numerical formats.

Table 4.1 shows typical fixed- and floating-point applications. It is important to note at this stage that the high data rate and the high precision required for the new communication standards like Long-Term Evolution (LTE) are now easily supported by the KeyStone floating-point formats. Section 4.2 shows the various formats supported by the KeyStone processors.

The fixed-point and floating-point capabilities of the KeyStone offer the following:

1) Floating-point instructions
    A) Single-precision complex multiplication
    B) Vector multiplication
    C) Single-precision vector addition and subtraction
    D) Vector conversion of single-precision floating point to or from an integer
    E) Double-precision floating-point arithmetic for addition, subtraction, multiplication, division and conversion to or from an integer.
2) Fixed-point instructions
    A) Complex vector and matrix multiplications, such as DCMPY for vector, and CMATM-PYR1 for matrix multiplications
    B) Real vector multiplications
    C) Enhanced *dot product* (dotp) calculation
    D) Vector addition and subtraction
    E) Vector shift
    F) Vector comparison
    G) Vector packing and unpacking.

4.2 Fixed- and floating-point representations

This section explains the fixed- and floating-point representations that are crucial for implementation, especially for fixed points.

Table 4.1 Examples of fixed- and floating-point applications

Applications suitable for fixed point	Applications suitable for floating point
• Portable devices • Image and video • Automotive • Mobile base station	• High-performance computing (HPC) • Radars • Professional audio • Medical • Robotics • Scientific instrumentation • Wireless communication standard, such as Long-Term Evolution (LTE)

Table 4.2 4-bit unsigned integer numbers

	Binary number a_3 a_2 a_1 a_0	Decimal equivalent
	0 0 0 0	0
	0 0 0 1	1
	0 0 1 0	2
	0 0 1 1	3
	0 1 0 0	4
	0 1 0 1	5
	0 1 1 0	6
	0 1 1 1	7
	1 0 0 0	8
	1 0 0 1	9
	1 0 1 0	10
	1 0 1 1	11
	1 1 0 0	12
	1 1 0 0	13
	1 1 1 0	14
	1 1 1 1	15

Unsigned integer numbers

4.2.1 Fixed-point arithmetic

The fixed-point format can represent three types of data: unsigned integers, signed integers or fractional numbers (signed).

4.2.1.1 Unsigned integer

An unsigned x integer number that can be represented with N-bit is shown as follows:

$$x = a_{N-1}2^{N-1} + \ldots + a_2 2^2 + a_1 2^1 + a_0 2^0$$

where a_{N-1}, a_{N-2}, $\ldots$ a_1 *and* a_0 are represented by 0 or 1.

The dynamic range for x is $2^N - 1$. With 16-bit representation ($N = 16$), the dynamic range will be $(2^{16} - 1)$ *or* 65,535. As an example, 4-bit unsigned numbers are shown in Table 4.2.

Table 4.3 4-bit signed integer numbers

	a_3 a_2 a_1 a_0	Decimal equivalent
	0 0 0 0	0
	0 0 0 1	1
	0 0 1 0	2
	0 0 1 1	3
	0 1 0 0	4
	0 1 0 1	5
	0 1 1 0	6
Signed integer numbers	0 1 1 1	7
	1 0 0 0	−8
	1 0 0 1	−7
	1 0 1 0	−6
	1 0 1 1	−5
	1 1 0 0	−4
	1 1 0 0	−3
	1 1 1 0	−2
	1 1 1 1	−1

4.2.1.2 Signed integer

Signed integer numbers are similar to unsigned ones, except that the last bit is a negative number, as shown here:

$$x = -a_{N-1}2^{N-1} + \ldots + a_2 2^2 + a_1 2^1 + a_0 2^0$$

where a_{N-1}, a_{N-2}, ... a_1 *and* a_0 are represented by 0 or 1.

The dynamic range for x is $2^N - 1$. With 16-bit representation ($N = 16$), the dynamic range will be from -2^{15} *to* $+ \left(2^{15} - 1\right)$. As an example, 4-bit signed numbers are shown in Table 4.3.

4.2.1.3 Fractional numbers

As stated earlier, the *dotp* equation is the basis of many DSP algorithms. However, if we use signed or unsigned integer numbers, the dotp will overflow after a few multiplications or additions. For instance, if we multiply only two numbers (like 256 ∗ 250) and use 16-bit, overflow will occur. The same will apply to additions; if you add 32,768 + 32,768, an overflow will occur. To reduce this overflow, the following solutions can be used:

- Saturate the results.
- Use double precision for the results.

- Use fractional arithmetic.
- Use floating-point arithmetic.

Although saturation can be very useful in some cases, it is not used very often in practical applications. The DSP cores on KeyStone processors have no means of automatically detecting overflow, and therefore it is up to the programmer to test for overflow which is time-consuming. Some processors support a hardware overflow flag that can generate an interrupt or exception when an overflow occurs. The other option is to use double precision for storing the results, but this is not very useful for a recursive algorithm (e.g. an infinite impulse response (IIR) filter) or when data need to be stored in a peripheral with lower precision, for instance when trying to send a 32-bit result to a 16-bit digital-to-analogue converter. The third option is to use fractional numbers. This format is very interesting since a fractional number multiplied by another fractional number will result in a smaller fractional number, and therefore no overflow will occur. Precision loss or overflow can still occur when truncating or rounding is used.

The fractional representation of an N-bit number is shown in Figure 4.1.

As an example, let's take a 4-bit fractional number as shown in Table 4.4.

The largest number for an N-bit number is X (01111 ... 111) which can be represented by:

$$X = 1 - 2^{-(N-1)}$$

and the smallest number is Y (100000...00) which can be represented by: $Y = -1$. Consider the multiplication of two 4-bit fractional numbers **a** and **b** shown in Figure 4.2. The result should be 1.110, as highlighted in Figure 4.2. Also, notice the sign extension bit.

The processor will produce the result as it is (11110100), but it is up to the programmer to decide what to do with it. For instance, the programmer can decide to keep the whole result (1.110 1000) or just 4-bit (1.110). In any case, the programmer has to perform the shift left by one bit. The shift left by one bit will apply to any number of bits used, as shown in Figure 4.3. The format is often expressed by Qx, where x is the number of fractional bits. For instance, in Figure 4.2, **a** and **b** are represented by Q3 and the result is Q6. And, in Figure 4.3, we have a multiplication of Q15 by a Q15, resulting in a Q30 and then scaled to Q15.

4.2.2 Floating-point arithmetic

The floating-point data types provide representation of very small and very large numbers, and that is due to the exponent part of the number. The 32-bit and the 64-bit floating-point IEEE 754 standards are defined as shown in Figure 4.4 and Figure 4.5, respectively. The sign bit s can be 0 or 1. The exponent e is represented by an unsigned integer, and the mantissa m is represented by a fractional number. A 64-bit floating-point IEEE 754 format number (double precision) is similar to the 32-bit format, except that the exponent is 11-bit and the mantissa is 52-bit.

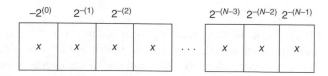

$x = 1$ or 0.

Figure 4.1 *N*-bit fractional number representation.

Table 4.4 4-bit fractional numbers

Binary number	a_3 (−1)	a_2 (0.5)	a_1 (0.25)	a_0 (0.125)	Decimal equivalent
Fractional numbers	0	0	0	0	0
	0	0	0	1	0.125
	0	0	1	0	0.25
	0	0	1	1	0.375
	0	1	0	0	0.5
	0	1	0	1	0.625
	0	1	1	0	0.75
	0	1	1	1	0.875
	1	0	0	0	−1
	1	0	0	1	−0.875
	1	0	1	0	−0.75
	1	0	1	1	−0.625
	1	1	0	0	−0.5
	1	1	1	0	−0.25
	1	1	1	1	−0.125
	0	0	0	0	0

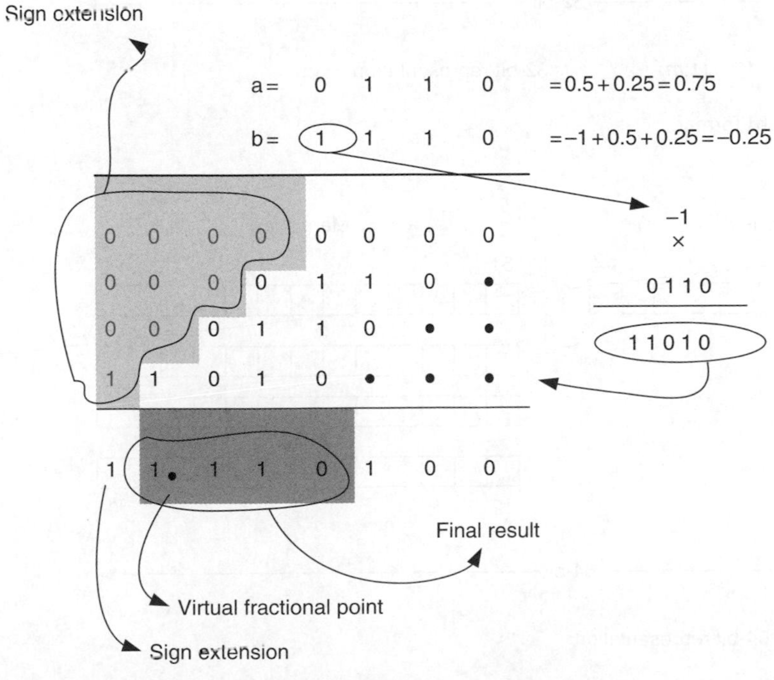

Figure 4.2 Binary multiplication of two fractional numbers.

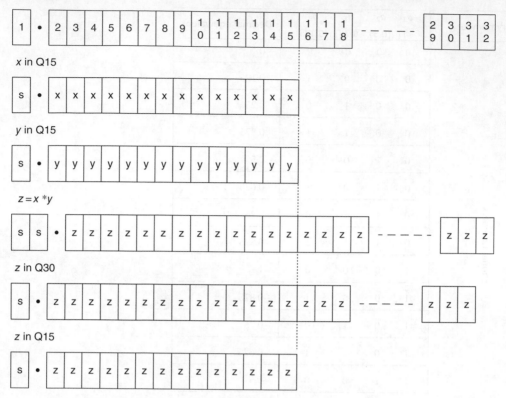

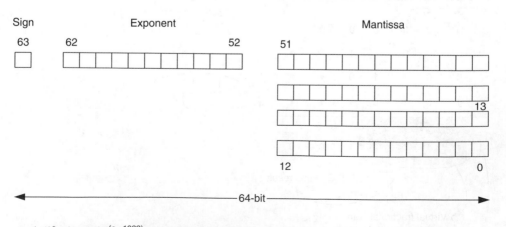

Figure 4.3 15-bit * 15-bit resulting in *Q*30 and *Q*15 formats.

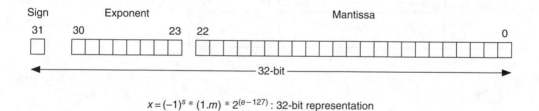

$$x = (-1)^s * (1.m) * 2^{(e-127)} : \text{32-bit representation}$$

Figure 4.4 32-bit IEEE standard format.

$$x = (-1)^s * (1.m) * 2^{(e-1023)} : \text{64-bit representation}$$

Figure 4.5 64-bit IEEE standard format.

Examples: 32-bit floating-point numbers

Example 1
Sign = 1
Exponent = 127
Mantissa = 0

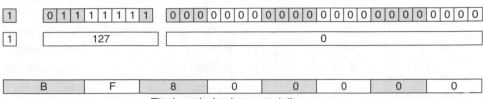

(−1)¹ * (1.0) * 2^(127−127) = −1: The decimal equivalent

$(-1)^1 * (1.0) * 2^{(127-127)} = -1$: The decimal equivalent

Example 2
Sign = 0
Exponent = 2^7 −1
Mantissa = 2^22

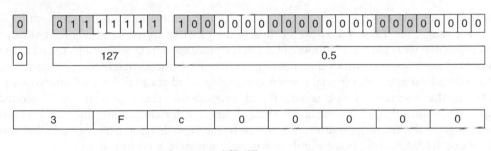

$(-1)^0 * (1.5) * 2^{(127-127)} = 1.5$

4.2.2.1 Special numbers for the 32-bit and 64-bit floating-point formats

In IEEE formats, the exponent fields of all of the 0 s and all of the 1 s represent special values (zero, plus infinity, minus infinity, not a number (NaN) and denormalised); see Table 4.5.

Zero cannot be represented by the equations shown in Figure 4.4 and Figure 4.5. Zero is a special number and is represented by a number with all bit fields of the exponent and the mantissa of the 0 s. NaN represents a value that is undefined and unpresentable, for instance the division 0/0, the multiplication $0 \times \pm\infty$ and so on. This is represented by all exponent fields of the 1 s and the mantissa non-zero. This example represents a NaN:

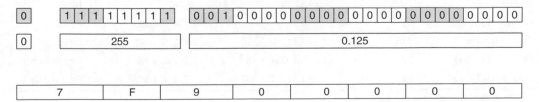

Table 4.5 Special numbers

Sign	Exponent (hexadecimal)		Mantissa (hexadecimal)	Description
	Single (8-bit)	Double precision (11-bit)		
0	00	000	000000	Positive zero
1	00	000	000000	Negative zero
0	FF	3FF	000000	Positive infinity
1	FF	3FF	000000	Negative infinity
0	FF	3FF	Non-zero	Not a number (NaN)

4.3 Dynamic range and accuracy

Floating-point processors are ideally suited for handling operations on numbers with a large dynamic range; see Table 4.6. However, due to the fact that numbers in any format can only be represented with a limited number of bits, the number's accuracy will be affected. Fixed-point numbers have a lower accuracy than floating-point ones. During implementation, accuracy has to be taken into account, especially when comparing numbers in different floating-point formats, as the accuracy is not constant and depends on the exponent (as illustrated in Figure 4.6). Note that as the exponent changes linearly, for the same value of the mantissa, the change in the floating-point number is not linear. It is also important to note that mixing 32-bit and 64-bit floating-point numbers will not increase the accuracy.

Table 4.6 Numerical format used for the KeyStone

Data definition	Number of bits	Minimum number	Maximum number	Data types
Unsigned char	8	0	255	Integer
Signed char	8	−128	127	Integer
Unsigned short	16	0	65,535	Integer
Signed short	16	−32,768	32,767	Integer
Unsigned integer	32	0	4,294,967,295	Integer
Signed integer	32	−2,147,483,648	2,147,483,647	Integer
Float (IEEE 754)	32	−3.4028E + 38	3.4028E + 38	Real number
Double (IEEE 754)	64	−1.7977E + 308	1.7977E + 308	Real number

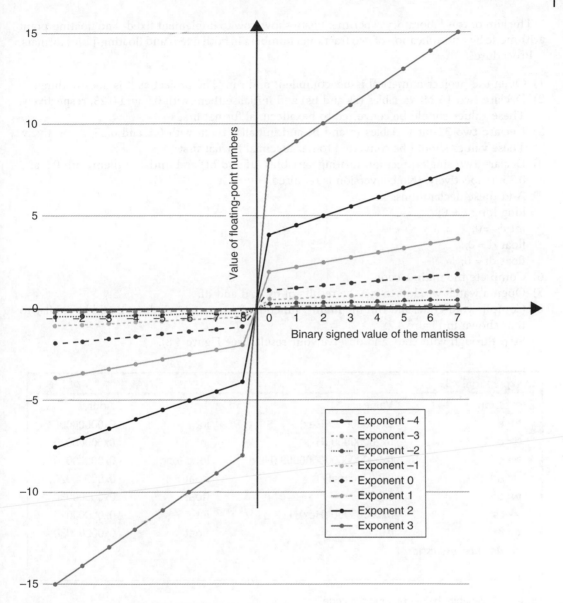

Figure 4.6 Accuracy of the 32-bit floating-point number.

4.4   Laboratory exercise

Project location:

\Chapter_4_Code\Numerical_Issues

  Solution:

\Chapter_4_Code\Solution\Numerical_Issues

The aim of this laboratory experiment is to show how to implement fixed- and floating-point arithmetic by multiplication of two fractional numbers in both fixed- and floating-point formats.
  Procedure:

1) Open the project numerical issue, compile it and run. The project as it is does nothing.
2) Declare two 16-bit variables (**as** and **bs**) and initialise them with 0.5 and 0.25, respectively. These values should be converted to hexadecimal format first.
3) Declare two 32-bit variables (**a** and **b**) and initialise them with 0.5 and 0.25, respectively. These values should be converted to hexadecimal format first.
4) Declare two single-precision floating variables (**af** and **bf**) and initialise them with 0.5 and 0.25, respectively. (No conversion is required.)
5) Add these declarations:
   long long c = 0;
   int cs =0;
   float d = 0;
   float df = 0;
6) Complete the **main.c** code.
7) Open a watch window to see the results of **cs**, **c**, **d** and **df**.
8) Compile your code, and run the **main()** function. Your watch window should be similar to that shown in Figure 4.7.
9) Step through your code and observe your results (see Figure 4.8).

∞ Expressions ⊠			
Expression	Value	Type	Address
(×)= a	0x40000000 (Hex)	int	0x0000000C
(×)= b	0x20000000 (Hex)	int	0x00000010
(×)= c	0x0000000000000000 (Hex)	long long	0x00000020
(×)= d	0.0 (Decimal)	float	0x0000002C
(×)= df	0.0	float	0x00000030
(×)= cs	0x00000000 (Hex)	int	0x00000028
(×)= df	0.0	float	0x00000030
✚ Add new expression			

Figure 4.7 Variables before running the code.

∞ Expressions ⊠			
Expression	Value	Type	Address
(×)= a	0x40000000 (Hex)	int	0x0000000C
(×)= b	0x20000000 (Hex)	int	0x00000010
(×)= c	0x0800000000000000 (Hex)	long long	0x00000020
(×)= d	5.764608e+17 (Decimal)	float	0x0000002C
(×)= df	0.125	float	0x00000030
(×)= cs	0x08000000 (Hex)	int	0x00000028
✚ Add new expression			

Figure 4.8 Final results.

4.5   Conclusion

This chapter explained how fixed and floating points are represented and how to handle binary arithmetic. It provided examples showing how to display various data formats using the Code Composer Studio.

References

1 R. L. Burden and J. D. Faires, *Numerical Analysis*, 10th ed., Cengage Learning, 2016.
2 S. C. Chapra and R. P. Canale, *Numerical Methods for Engineers*, 7th ed., McGraw-Hill Education, 2016.

5

Software optimisation

CHAPTER MENU

5.1   Introduction

Software optimisation is the process of manipulating software code to achieve one or a combination of the following goals, depending on the application: faster code execution, smaller code size and low power consumption.

To implement efficient software on a multicore processor, the programmer must be familiar with the processor architecture (that includes the CPU, the memory and any peripheral or

Multicore DSP: From Algorithms to Real-time Implementation on the TMS320C66x SoC, First Edition. Naim Dahnoun.
© 2018 John Wiley & Sons Ltd. Published 2018 by John Wiley & Sons Ltd.
Companion website: www.wiley.com/go/dahnoun/multicoredsp

coprocessor to be used); the language(s) (i.e. C and assembly) used; the programming model to efficiently distribute software (tasks) across all cores, for example OpenMP (see Chapter 12) or OpenCL (see Chapter 13); the operating system used (see Chapter 7); the compiler, assembler and linear assembler features and the code that they generate. In addition to these, the development tools (see Chapter 3) and the debugging tools (see Chapter 10) can also have a significant impact on the development time and the quality of the implemented application. The preferred and most supported high-level language used for programming DSP processors is the ANSI C language. In fact, most DSP manufacturers support the ANSI C language.

Code written in C is in general portable and not processor-specific. However, code written in assembly runs faster and consumes less memory than code written in C, but it is processor-specific, not portable, time consuming, prone to errors and very difficult to maintain.

Code optimisation for multicore processors can be performed at different levels, and each level may have a different impact on the optimisation. That said, higher levels in general have higher impacts but may require rewriting the lower levels if any change is required at the higher level. Different levels starting from the higher are as follows:

1) Efficient algorithm. Before trying to implement and optimise an application, one should try to optimise the algorithm itself and make it more efficient. In the worst case, an algorithm can be completely rewritten in order to achieve the performance required.
2) Data size. Try to minimise the data to be processed if possible. For instance, in an automotive application where lanes on the road need to be detected, one can select only the region of interest (ROI) for processing and discard the rest.
3) Data structure. Choosing the right data structure for the right application will provide an efficient way of accessing data and therefore improve the performance. For instance, if data are to be inserted or removed from an array, using linked lists will be more efficient than using arrays.
4) Software scalability. Efficiently distributing the software functionality of the application/ algorithm across the available cores. This is the main task for making an application scalable on a multicore processor.

There are two methods for writing software for multicore:
a) Each core has a different code to run.
b) Each core has the same code to run.

The preferred method is the second since it is easier to implement, as can be seen in various examples in this book. However, this is normally dictated by the algorithm [1].

Software scalability can be hindered by the many factors that are discussed in Section 5.2.

It was shown in Chapter 1 that the performance improvement by parallelising a code will depend on the code that cannot be parallelised (Amdahl's law); therefore, one should analyse its code before attempting to scale it.
5) Single-core code optimisation. If we assume that scaling offers some performance improvement, then any improvement on a single core will be noticeable. For instance, if we consider that an application can be completely parallelised, doubling the processing speed of an algorithm on a core will result in doubling the performance of the application. Of course, this is an ideal condition, and in a practical situation one should take into account the performance of the serial code too. Code optimisation for a single core is described in Section 5.3.

5.2   Hindrance to software scalability for a multicore processor

Software scalability can be hindered by many factors:

1) Algorithm parallelisation. Not all algorithms can be partially or fully parallelised. This again depends on the application and the algorithm written for that application. For instance, many applications were not written for parallel computing, and therefore running them on multicore processors will not increase the performance.
2) Data sharing. Sharing data can degrade the performance of a multicore processor since data can be protected by locks, a mutex or any other method of synchronisation.
3) Memory access. Cores share some memory, and access to this shared memory can create contention. To prevent this, one should use local memory when possible and reduce the frequency to access the shared memory.
4) Peripheral access. A peripheral that can be accessed by many cores at the same time can result in a bottleneck. In addition to this, peripherals share the same buses as the memory and will also result in performance degradation.
5) Cache coherency. DSP cores on the KeyStone have both local cache and shared cache, and false sharing (different cores writing to different variables located in the same cache line) can be an issue if data alignment is not performed properly.
6) Synchronisation overhead. Tasks can be dependent, and one task may be waiting for intermediate results before proceeding. This usually is handled by some synchronisation mechanisms like locks which may degrade the overall performance.
7) Workload imbalance. Generally, cores may use the same code but operate on different data. This may create a load imbalance, and therefore some cores will be idling while waiting for other cores to complete. Workload imbalance can also happen in heterogeneous processors like the KeyStone II where, for instance, one of the ARM cores can wait for the DSP cores to finish; the solution to overcome this is shown in Ref. [2]. Degradation due to load imbalance can be reduced by either reducing the load of each core or further parallelising the serial code. However, creating more tasks than cores can also hinder the performance.

For the rest of this chapter, only the optimisation of code running on a single processor is considered.

5.3   Single-core code optimisation procedure

Figure 5.1 illustrates the procedure for optimising the software code. In the first step, the developer must make sure that the algorithm to be implemented is fully functional and 'optimised' at the algorithm level. In the second step, the algorithm can be implemented in ANSI C language without any optimisation. If the code is operational and the execution speed is adequate, then there is no need to develop the code further. However, if the code is functional but the execution time is not satisfactory, then the code will need to be further optimised. If all optimisations supported by the compiler still do not produce a satisfactory result, the developer needs to progress to the next step, which involves data alignment, memory management, use of the cache, the EDMA and coprocessor(s) if necessary. If this is still not sufficient, proceed to the next step by replacing pertinent instructions with intrinsic. If this is still insufficient, then identify the slow code and rewrite it in linear assembly. In general, only

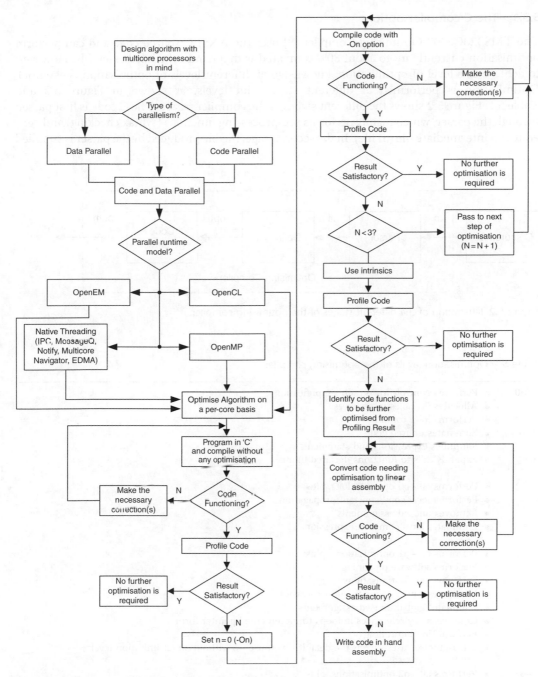

Figure 5.1 Optimisation flow procedure.

some functions need to be implemented in linear assembly; these functions can be determined by the profiler. If the results obtained are still not satisfactory, the developer can move
to the final stage and code the critical part of the algorithm in hand-scheduled assembly
language.

5.3.1 The C compiler options

The TMS320C6000 Optimising Compiler [3] uses the ANSI C source code and can perform optimisation currently up to about 80% compared with a hand-scheduled assembly. However, to achieve this level of optimisation, knowledge of different levels of optimisation is essential. Optimisation is performed at different stages and levels, as shown in Figure 5.2 and Table 5.1. Figure 5.2 shows the different stages of the compiler passes. The C code is first passed through the parser, which mainly performs pre-processing functions (syntax checking) and generates an intermediate file (**.if** file). In the second stage, the file produced by the parser is supplied

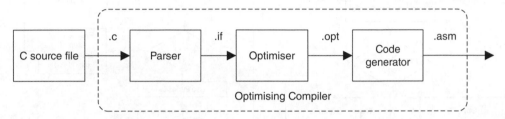

Figure 5.2 Illustration of the different stages of the optimising compiler.

Table 5.1 Optimisation levels of the optimising compiler

−o0	• Performs control-flow-graph simplification • Allocates variables to registers • Performs loop rotation • Eliminates unused code • Simplifies expressions and statements • Expands calls to functions declared inline
−o1	• Performs all −o0 optimisations, plus: • Performs local copy/constant propagation • Removes unused assignments • Eliminates local common expressions
−o2	• Performs all −o1 optimisations, plus: • Performs software pipelining • Performs loop optimisations • Eliminates global common sub-expressions • Eliminates global unused assignments • Converts array references in loops to incremented pointer form • Performs loop unrolling • The optimiser uses −o2 as the default if you use −o without an optimisation level.
−o3	• Performs all −o2 optimisations, plus: • Removes all functions that are never called • Simplifies functions with return values that are never used • Inlines calls to small functions • Reorders function declarations so that the attributes of called functions are known when the caller is optimised • Propagates arguments into function bodies when all calls pass the same value in the same argument position • Identifies file-level variable characteristics

Table 5.2 Parser and optimiser options summary

A. Most common options that control the parser	
Option	**Description**
−pf	Generates function prototype listing file
−pk	Allows K&R compatibility (does not apply to C++ code)
−pl	Generates pre-processed listing
−pm	Combines source files to perform program-level optimisation

B. Most common options that control the optimiser	
Option	**Description**
−o0	Optimises register usage
−o1	Uses **−o0** optimisations and optimises locally
−o2 (or **−o**)	Uses **−o1** optimisations and optimises globally
−o3	Uses **−o2** optimisations and optimises file

to the optimiser, which performs most of the optimisation, and produces the **.opt** file. In the third and last stage, the **.opt** file is supplied to the code generator that produces the assembly code.

In order to have further control over the parser or optimiser, different options are available as shown in Table 5.2.

5.4 Interfacing C with intrinsics, linear assembly and assembly

In general, the application can be written with a combination of C language, intrinsics, assembly or linear assembly when optimisation is required. Therefore, understanding how to interface these languages is very important.

5.4.1 Intrinsics

Due to the fact that single instructions can operate on multiple data (SIMD operations), expressing some operations in C language will not produce efficient code. Consider, for example, the **dotp** algorithm shown here:

```
//dotp function
for (i = 0; i < count; i++) {
      sum + = m[i] * n[i];
   }
```

It is not possible to express multiple multiplication and addition instructions in C language that take advantage of the available SIMD instructions.

The TMS320C6000 compiler allows the use of some functions called *intrinsics* that are identical to the assembly instructions. When using these intrinsics, it is possible to directly call an assembly language statement from the C code. The intrinsics are automatically inlined by the compiler but can be disabled. Refer to the 'Optimising C Compiler' in Ref. [3] for more details.

The list of instruction functions supported is described in the 'Optimising C Compiler' [3].

As an example, the **dotp** function can be rewritten as:

```
int dotpC_intrinsic(short *m, short *n, int count)
{       int sum = 0, sum1 = 0, sum2 = 0, sum3 = 0, sum4 = 0;
        int i;
for (i = 0; i < count; i + =4)

        {
sum1 + = _mpy (_lo(_memd8_const(&n[i])), _lo(_memd8_const(&m[i])));
sum2 + = _mpyh(_lo(_memd8_const(&n[i])), _lo(_memd8_const(&m[i])));
sum3 + = _mpy (_hi(_memd8_const(&n[i])), _hi(_memd8_const(&m[i])));
sum4 + = _mpyh(_hi(_memd8_const(&n[i])), _hi(_memd8_const(&m[i])));

        }
                sum = sum1 + sum2 + sum3 + sum4;
                return sum;
                }
```

where:

_memd8_const(): Equivalent to LDNDW: loads 8 bytes (64-bit) of unaligned data.

_lo: Returns the low 32-bit register of a 64-bit register pair.

_hi: Returns the high 32-bit register of a 64-bit register pair.

_mpy: Equivalent to MPY: multiplies two 16-bit least significant bits (LSBs) of two 32-bit registers and return the results in a 32-bit register.

_mpyh: Equivalent to MPYH: multiplies two 16-bit most significant bits (MSBs) of two 32-bit registers and returns the results in a 32-bit register.

The other possibility is to use the inline assembly language embedded in the C code as shown here:

```
asm ("MV       0x440,B2");
asm ("MVLKH    0x7,B2");
asm ("MVC      B2,AMR");
asm ("; insert a comment here");
```

The **asm** statement can be useful for debugging, as one can insert comments in the compiler output.

5.4.2 Interfacing C and assembly

The C and assembly functions may exchange data. Therefore, code interfacing requires a means of handing off data and control info, and some rules of handling shared registers.

Table 5.3 Registers use

A registers	Number	B registers
	0	
	1	
	2	
	3	**ret addr**
arg1/r_val	4	**arg2**
	5	
arg3	6	**arg4**
	7	
arg5	8	**arg6**
	9	
arg7	10	**arg8**
	11	
arg9	12	**arg10**
	13	
	14	
–	–	–
–	–	–
–	–	–
	30	
	31	

Figure 5.5 shows an example of a C function calling an assembly function. The C function passes the argument variables **a**, **b** and **c** via registers. The protocol between C code and assembly is that the arguments are passed in a specific order: for instance, Argument 1 should be passed in A4, Argument 2 should be passed in B4 and so on (see Table 5.3).

Passing more than one 32-bit argument, the compiler will concatenate two or four consecutive registers from the same side, as shown in the examples here:

Example 1:

```
int func(int a, double b, float c, long double d);
```

- Argument a will be passed in A4.
- Argument b will be passed using B5:B4.
- Argument c will be passed using A6.
- Argument d will be passed using B7:B6.

Example 2: Passing a 128-bit argument

```
__x128_t  myquad,y3;
int a1 = 0x00000000;
int b1 = 0x11111111;
int c1 = 0x22222222;
int d1 = 0x33333333;
Void main()
{
    __x128_t myquad = _ito128(a1, b1, c1, d1);//Pack values into
    a __x128_t
    //at this point myquad = 0x00000000111111112222222233333333;
y3 = dotpsa2(a, x, COUNT,myquad);

}

        .global dotpsa2
dotpsa2     .cproc ap, xp, cnt,z4:z3:z2:z1
            .reg a, x, prod, y,z
            zero y
loop
-
-
.return z4:z3:z2:z1
    .endproc
```

1) Set a breakpoint in **dotpsa2** as shown in Figure 5.3, and run the code. At this stage, you can see that z4:z3:z2:z1 are passed with A11:A10:A9:A8.
2) To observe the return value, in the debug mode, select **View → Registers → RegisterPairs**, run the code up to the breakpoint and verify that data are passes A11:A10:A9:A8, as shown in Figure 5.4.

C code and assembly code share the same resources (registers etc.). The C code will use some or all of the registers. The assembly code may also require the use of some or all registers. If nothing is done, then on return to the C code, some of the variables may have been changed by the assembly code. The solution is for both the C code and assembly code to be responsible for saving some registers if they need to use them.

Which register to be saved by the C code and which to be saved by the assembly code are specified in Figure 5.6. The registers are split in this way in order to keep compatibility with previous devices. If one needs to pass a large number of arguments, then organise data in an array and pass only the pointer of the first argument.

If the return value is 32-bit, it will be returned in A4. If it is 128-bit, it will be returned in A7:A6: A5:A4.

Before calling the assembly code, the compiler will record the return address in B3. Therefore, if this register is to be used, it must be saved first and restored before returning.

```
 1
 2              .global dotpsa2
 3
 4 dotpsa2 .cproc   ap, xp, cnt,z4:z3:z2:z1
 5          .reg    a, x, prod, y,z
 6
 7          zero    y
 8
 9
10 loop
11
12          ldh     *ap++, a
13          ldh     *xp++, x
14          mpy     a, x, prod
15          add     prod, y, y
16          sub     cnt, 1, cnt
17 [cnt]    b       loop
18
19 ;        .return z4:z3:z2:z1
20
21      .return z4:z3:z2:z1
22          .endproc
23
```

Figure 5.3 **dotpsa2** file.

Name	Value	Descrip
▷ Core Registers		
▷ ControlRegisters		
◢ RegisterPairs		
A1_A0	0x0080B7E000000001	Core
A3_A2	0x0000000000000000	Core
A5_A4	0xFFFFFFFF00811DC8	Core
A7_A6	0x0000000000000100	Core
A9_A8	0x2222222233333333	Core
A11_A10	0x0000000011111111	Core
A13_A12	0x0000000000000000	Core
A15_A14	0x0000000000000000	Core
A17_A16	0x0000000000811D54	Core
A19_A18	0x0000002000811D24	Core
A21_A20	0x0000000000000000	Core
A23_A22	0x0000000000000000	Core
A25_A24	0x0000000000000000	Core
A27_A26	0x0000000000000000	Core
A29_A28	0x0000000000000000	Core
A31_A30	0x0000000000812648	Core

Figure 5.4 Viewing register pairs.

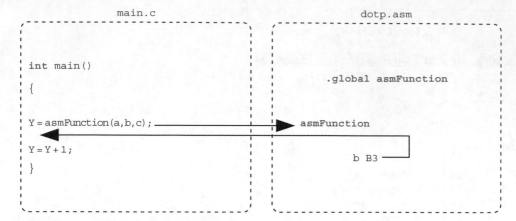

Figure 5.5 Interfacing C and assembly.

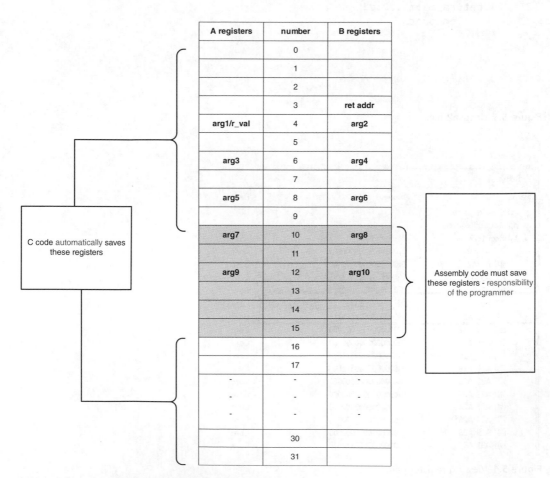

Figure 5.6 Automatic and manual saving of registers.

5.5 Assembly optimisation

To develop an appreciation of how to optimise code, let us optimise an FIR filter algorithm, which is represented by Equation (5.1):

$$y[n] = \sum_{k=1}^{N} h[k]x[n-k] \tag{5.1}$$

For simplicity, we can rewrite Equation (5.1) by assuming that we can reorder samples at each sampling instant. This will lead to Equation (5.2).

$$y[n] = \sum_{i=1}^{N} h[i]x[i] \tag{5.2}$$

To implement Equation (5.2), we need to perform the following steps:

1) Load the samples **x[i]**.
2) Load the coefficients **h[i]**.
3) Multiply **x[i]** and **h[i]**.
4) Add (**x[i]**. **h[i]**) to the current content of the accumulator.
5) Repeat steps (1) to (4) $N-1$ times.
6) Store the value in the accumulator to y.

These can be interpreted in TMS320C66x code as shown in Program 5.1.

Program 5.1:   Assembly code for implementing an FIR filter

```
        MVK    .S1   0,B0        ; Initialise the loop counter
        MVK    .S1   0,A5        ; Initialise the accumulator
loop    LDH    .D1   *A8++,A2    ; Load the samples x[i]
        LDH    .D1   *A9++,A3    ; Load the coefficients h[i]
        NOP          4           ; Add 'nop 4' because the LDH has a
                                    latency of 5
        MPY    .M1   A2,A3,A4    ; Multiply x[i] and h[i]
        NOP                      ; Multiply has a latency of 2 cycles
        ADD    .L1   A4,A5,A5    ; Add 'x[i].h[i]' to the accumulator
[B0]    SUB    .L2   B0,1,B0     ; ⎫
[B0]    B      .S1   loop        ; ⎬ loop overhead
        NOP          5           ; ⎭ The branch has a latency of 6 cycles
```

If we represent the flow of instructions on a cycle-by-cycle basis as shown in Table 5.4, we can see that for each cycle, at most one of the units is active and therefore the code as it is written is not optimised. It is clear from Table 5.4 that in order to optimise the code, we need to:

1) Use instructions in parallel, which means that multiple units will be operating on the same cycle.
2) Remove the NOPs (put code in place of NOPs).
3) Unroll the loop (See Section 5.5.3).
4) Use word or double-word access instead of half-word access (see Section 5.5.4).

Table 5.4 Iteration interval table for an FIR filter

Source/cycle	.D1	.D2	.L1	.L2	.M1	.M2	.S1	.S2	NOP
1	LDH								
2	LDH								
3									NOP
4									NOP
5									NOP
6									NOP
7					MPY				
8									NOP
9			ADD						
10				SUB					
11							B		
12									NOP
13									NOP
14									NOP
15									NOP
16									NOP

Let us now take each case separately and try to apply it to the code shown above.

5.5.1 Parallel instructions

Looking at Table 5.4, we see that the **.D2** unit is unused, and therefore the LDH instruction in Cycle 2 can be moved to be executed in Cycle 1 in the **.D2** unit. This can be written as:

```
   LDH .D1 *A8++,A2
|| LDH .D2 *B9++,B3 ; Notice that the registers come from the register
                    ; file B since.D2 is now used.
```

The **SUB** instruction in Cycle 10 could also be moved to Cycle 9, and this can be written as:

```
   ADD  .L1  A1,A2,A1
|| SUB  .L2  B10,1,B10
```

The other instructions cannot be put in parallel since the result of one unit is used as an input to the following unit. In general, up to eight instructions can be put in parallel, and therefore to achieve the current maximum performance all eight units should be used in parallel.

Note: For maximum performance, the Execute Packet (instructions to be executed in the same cycle) should contain eight instructions.

5.5.2 Removing the NOPs

Ten cycles have been 'wasted' using NOP instructions in the code in Table 5.4. To optimise the code further, the NOP instructions can be replaced by useful code. Since the **SUB** and the **B** (branch) instructions are independent of the rest of the code, then by rearranging some of the code, some NOPs can be eliminated as shown in Program 5.2.

Program 5.2:   Assembly code for an FIR filter

```
loop    LDH    .D1    *A8++,A2    ; Load the samples x(i)

        LDH    .D1    *A9++,A3

[B0]    SUB    .L2    B0,1,B0

[B0]    B      .S1    loop

        NOP           2           ; the 5 NOPs required for the branch
                                    instruction are replaced by (NOP 2, MPY
                                    and NOP)

        MPY    .M1    A2,B3,A4

        NOP

        ADD    .L1    A4,A5,A5

                                  ←The branch occurs here
```

Notice that the **ADD .L1** and **SUB .L2** are not used in parallel, since the **SUB** instruction has moved up with the branch instruction, and only three NOPs instead of ten are being used.

5.5.3 Loop unrolling

The **SUB** and **B** instructions consume at least two extra cycles per iteration (this is known as the *branch overhead*). If instead of looping using the **SUB** and **B** instructions, we simply replicate the code unlooped, the branch overhead can be removed completely and the code can be reduced by at least two instructions per iteration. It is clear that with loop unrolling, the code size has increased (see Program 5.3).

Program 5.3:   Unlooped code

```
        LDH       .D1      *A8++,A2        ; Start of iteration 1

        LDH       .D1      *B9++,B3

        NOP                4

        MPY       .M1X     A2,B3,A4        ; Use of cross-path

        NOP
```

```
        ADD       .L1        A4,A5,A5
        LDH       .D1        *A8++,A2          ; Start of iteration 2
        LDH       .D1        *A9++,A3
        NOP                  4
        MPY       .M1        A2,B3,A4
        NOP
        ADD       .L1        A4,A5,A5
;                 :
;                 :
;                 :
        LDH       .D1        *A8++,A2          ; Start of iteration n
        LDH       .D1        *A9++,A3
        NOP                  4
        MPY       .M1        A2,B3,A4
        NOP
        ADD       .L1        A4,A5,A5
```

5.5.4   Double-Word Access

The TMS320C66x devices have two 64-bit data buses for data memory access, and therefore two 64-bit data can be loaded into the registers at any one time. In addition, the TMS320CC66x devices have variants of the multiplication instruction to support different operations (see Chapter 2). Using these two features, the previous code can be rewritten as shown in Program 5.4.

Program 5.4:   Double-word access

```
loop:
        lddw    *ap++, a1h:a11
        lddw    *xp++, x1h:x11

        dotp4h a1h:a11, x1h:x11, dSum

        add     dSum, rsum, rsum
[cnt] sub       cnt, 4, cnt
[cnt] b         loop
```

By loading double words and using the **DOTP4H** instruction, the execution time has been reduced, since in each iteration four 16-by-16-bit multiplications are performed.

5.5.5   Optimisation summary

This section has shown that there are four complementary methods for code optimisation. Using instructions in parallel, filling the delay slots or replacing NOPs with useful code and using the load word (LDDW) instruction increases the performance and reduces the code size. However, by using the loop-unrolling method, the performance improves at the cost of a larger code size. Filling NOPs by reshuffling instructions can be a very tedious task. However, this chapter will show that by using software-pipelining procedures, it can be simplified and optimised.

5.6 Software pipelining

The main objective of software pipelining is to optimise code associated with loops. The loop code is optimised by scheduling instructions in parallel and eliminating or replacing the NOPs with useful code. Due to the facts that multiple units are available on the C6x devices and also that instructions have different latency, code optimisation can be a complex task. However, by using the compiler options −o2 or −o3 as shown in Section 5.3, or by using the assembler optimiser as shown in Section 5.7, the burden of software pipelining can be left to the tools. To define the problem, let us return to the FIR code from Program 5.5.

Program 5.5:   Un-optimised assembly code

```
        LDH     .D1
||      LDH     .D2
        NOP             4
        MPY     .M1
        NOP
        ADD     .L1
;               :
;               :
;               :
        LDH     .D1
||      LDH     .D2
        NOP             4
        MPY     .M1
        NOP
        ADD     .L1
```

If we consider a table representing all units for *all* cycle numbers and fill the appropriate boxes with the appropriate instructions, we can form a clear view of the resources used for each cycle (see Table 5.5). It is clear that each loop iteration takes eight cycles and at most one or two units are used. However, if we advance each loop by seven cycles, as shown in Table 5.6, the code still executes properly. From Cycles 8 to 10, four units are used by the code, and they execute in parallel. In this case, we can say that we have a single-cycle loop. As can be seen from Table 5.6, the code can be split into three sections (prologue, kernel and epilogue).

As the name suggests, software pipelining is the process of putting code in a pipeline, as shown in Table 5.6 and described in Table 5.7. Software pipelining is only concerned with loops since the repeatability of the code is exploited. It is evident from Table 5.6 that the loop kernel iterates the same code for each cycle.

Table 5.5 Iteration interval table for an FIR filter

Unit/cycle	.L1	.L2	.M1	.M2	.S1	.S2	.D1	.D2	NOP
(1)	LDH	LDH							
(2)									NOP
(3)									NOP
(4)									NOP
(5)									NOP
(6)			MPY						
(7)									NOP
(8)					ADD				
(9)	LDH	LDH							
(10)									NOP
(11)									NOP
(12)									NOP
(13)									NOP
(14)			MPY						
(15)									NOP
(16)					ADD				
(17)	LDH	LDH							
(18)									NOP
(19)									NOP
(20)									NOP
(21)									NOP
(22)			MPY						
(23)									NOP
(24)					ADD				
(25)	LDH	LDH							

Table 5.5 (Continued)

Unit/cycle	.L1	.L2	.M1	.M2	.S1	.S2	.D1	.D2	NOP
(26)									NOP
(27)									NOP
(28)									NOP
(29)									NOP
(30)			MPY						
(31)									NOP
(32)					ADD				
(33)	LDH	LDH							
(34)									NOP
(35)									NOP
(36)									NOP
(37)									NOP
(38)			MPY						
(39)									NOP
(40)					ADD				
(41)	LDH	LDH							
(42)									NOP
(43)									NOP
(44)									NOP
(45)									NOP
(46)			MPY						
(47)									NOP
(48)					ADD				

Table 5.6 Iteration interval table for an FIR filter

	Unit/cycle	.D1	.D2	.M1	.M2	.S1	.S2	.L1	.L2
Prologue	(1)	LDH	LDH						
	(2)	LDH	LDH						
	(3)	LDH	LDH						
	(4)	LDH	LDH						
	(5)	LDH	LDH						
	(6)	LDH	LDH	MPY					
	(7)	LDH	LDH	MPY					
Loop kernel	(8)	LDH	LDH	MPY				ADD	
	(9)	LDH	LDH	MPY				ADD	
	(10)	LDH	LDH	MPY				ADD	
Epilogue	(11)			MPY				ADD	
	(12)			MPY				ADD	
	(13)			MPY				ADD	
	(14)			MPY				ADD	
	(15)			MPY				ADD	
	(16)							ADD	
	(17)							ADD	

Table 5.7 Different sections of the code

1. Prologue	In this section, the code is building up, and its length is the length of the unroll loop minus one. In this case, it is 7 (=8 − 1).
2. Loop or kernel	Each execute packet in this section contains all instructions required for executing one loop.
3. Epilogue	Contains the rest of the code necessary for completing the algorithm

5.6.1 Software-pipelining procedure

To optimise code as shown in this chapter can be a very tedious task, especially when the loop code does not fit in a single cycle. To simplify code optimisation, it is suggested that:

1) The code is written in linear assembly fashion. This provides a clear view of the algorithm. There is no need to specify the units, registers or delay slots (NOPs), as these will be taken care of in the last two steps.
2) The algorithm is drawn on a dependency graph to illustrate the flow of data of the algorithm.
3) List the resources (functional units, registers and cross-paths) required to determine the minimum number of cycles required of each loop.
4) Create a scheduling table that shows instructions executing on the appropriate units, on a cycle-by-cycle basis. This table is drawn with the help of the dependency graph.
5) Generate the final assembly code.

To gain experience of hand optimisation using software pipelining, an FIR code is taken as an example. The five steps are shown in the remainder of this section.

5.6.1.1 Writing linear assembly code

The code shown here does not specify any unit or delay slots. Furthermore, all the registers are represented by symbolic names which make the code more readable.

```
Loop        LDH    *p_to_a,a
            LDH    *p_to_b,bfs
            MPY    a,b,prod
[count]     ADD    sum,prod,sum
            SUB    count,1,count
[count]     B      Loop
```

5.6.1.2 Creating a dependency graph

Before creating the dependency graph, the algorithm first needs to be written in linear assembly language as shown above. Creating the dependency graph consists of four steps:

Step 1: Draw the nodes and paths. In this step, each instruction is represented by a node and the node is represented by a circle. Outside the circle the instruction is written, and inside the circle the register holding the result is written. The nodes are then connected by paths showing the data flow (condition paths are represented by dashed lines). This is shown in Figure 5.7.
Step 2: Write the number of cycles it takes for each instruction to complete executing. The LDH takes five cycles, the **MPY** takes two cycles, the **ADD** and **SUB** take one cycle each and the **B** instruction takes six cycles to complete executing. The number of cycles should be written along the associated data path. This is shown in Figure 5.8.
Step 3: Assign functional units to each node. Since each node represents an instruction, it is advantageous to start allocating units to instructions which require a specific unit. In other words, start by allocating units to nodes associated with load, store and branch. We do not need to be concerned with the multiply instruction, since multiplication can only be performed in .**M** units. This is shown in Figure 5.9. At this stage, the units have been specified but not the side.
Step 4: Data-path partitioning. To optimise code, we need to make sure that a maximum number of units are used with a minimum of cross-paths (see Chapter 2). To make this visible from the dependency graph, a line is drawn on the graph to separate the two sides (see Figure 5.10).

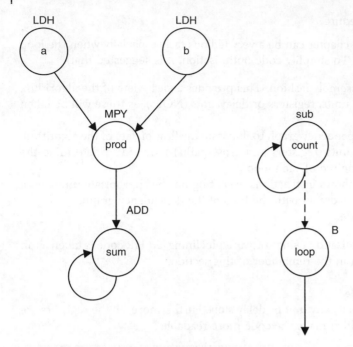

Figure 5.7 Dependency graph of an FIR filter.

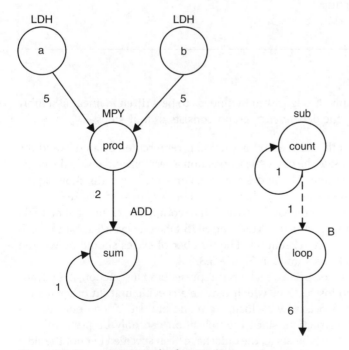

Figure 5.8 Dependency graph of an FIR filter.

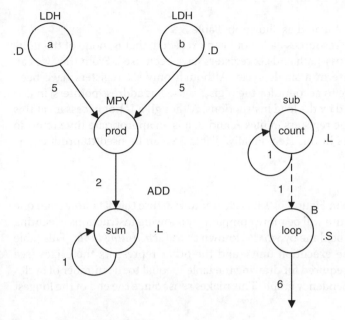

Figure 5.9 Dependency graph of an FIR filter.

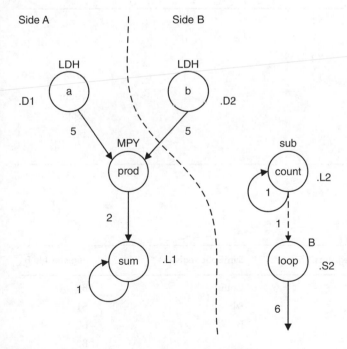

Figure 5.10 Final dependency graph of an FIR filter.

5.6.1.3 Resource allocation

In this step, all the resources are tabulated as shown in Table 5.8.

It is clear from Table 5.8 that the resources have not been exceeded, that is, none of the units have been re-used, and only one cross-path and six registers have been used. From this, we can conclude that the code can execute in a single cycle. Although only six registers have been named in this example, we still have to account for the registers used as address pointers; in this case, we have the two addresses used by the load instructions. Although it is not necessary at this stage to link the registers used to the registers in files A and B, it is an appropriate time to do so since we are dealing with resources in this step. Finally, Table 5.9 can be used to produce the register allocation.

5.6.1.4 Scheduling table

From the dependency graph shown in Figure 5.10, it is clearer to visualise the data flow from one unit to the other. However, the picture will be more complete by showing instructions executing on a cycle-by-cycle basis. This can be done by what is known as the *scheduling table*. This table has two entries: one represents the execution units, and the other represents the cycles (see Table 5.10). The number of cycles required for drawing the table is equal to the number of cycles found in the longest path of the dependency graph. This makes sense since the end of the longest

Table 5.8 Resource allocation

Units available	Number used	Cross-paths available	Number used	Register used
.L1	1	X1	1	sum
.S1	1	T1	0	
.D1	1	X2	0	a
.M1	1	T2	1	prod
.L2	1			count
.S2	1			loop
.D2	1			b
.M2	0			

Table 5.9 Register allocation

Register file A	Symbolic registers	Symbolic registers	Register file B
A0		Count	B0
A1	&a	&b	B1
A2	A	B	B2
A3	Prod		B3
A4	Sum		B4
A5			B5

Table 5.10 Scheduling table

Cycle/units	Cycle 1	Cycle2	Cycle3	Cycle4	Cycle5	Cycle6	Cycle7	Cycle8
.L1								
.L2								
.S1								
.S2								
.M1								
.M2								
.D1								
.D2								

path represents the end of the algorithm. In this example, the maximum number of cycles is eight (=5 + 2 + 1).

From Figure 5.10, one can complete Table 5.10 to generate the final scheduling table as shown in Table 5.11. One notices that on the first cycle, the two loads are executed (**fill cycle1/.D1** and **cycle 1/.D2**). In order to supply the multiplication unit with the destination contents of the load instructions, the multiplication operation has to be delayed by five cycles (**fill cycle 6/M.1**). Two cycles after the multiplication, the addition can be processed (**fill cycle 8/.L1**).

Now that we have finished with the main part of the dependency graph, let's move on to the program control part. In this case, we would like to branch at the beginning of the program as soon as the addition is performed. To do so, we need to schedule the branch instruction so that it executes just after the **ADD** instruction (Cycle 9). Since the branch instruction has a latency of six cycles, it is scheduled in Cycle 3 (**fill cycle 3/.S2**).

The **SUB** instruction should occur one cycle before the branch instruction, and therefore should be scheduled in Cycle 2. So far we have determined the cycles in which each instruction starts to be active. From that cycle, the same instruction is repeated for all the other cycles. In Cycle 8, a single-cycle loop is achieved, and hence the next cycles are identical. In practical situations, the loop count is a finite number, and in order to include it in the scheduling table, we need to create the epilogue.

The epilogue can be created by removing the loop overhead (**B** and **SUB** instructions) and instructions from the prologue of the main loop code on a cycle-by-cycle basis. For example, to create the epilogue for Cycle 9, we need to perform the subtraction shown in Table 5.11.

5.6.1.5 Generating assembly code

From Table 5.11, we can generate the assembly code as shown in Program 5.6. Notice that the single-cycle loop can be repeated n times ($n = N - 7$), and the total number of iterations will be equal to N. This shows that the loop count is not always equal to the number of algorithm iterations.

Table 5.11 Scheduling table

Single-cycle loop	Subtract	Loop overhead	Subtract	Prologue	Result: Epilogue
LDH || LDH || MPY || ADD || SUB || B	(minus) –	SUB || B	(minus) –	LDH || LDH	MPY || ADD

	Prologue							Loop	Epilogue						
Cycle / Unit	1	2	3	4	5	6	7	8	9	10	11	12	13	14	15
.D1	LDH	LDH	LDH	LDH	LDH	LDH	LDH	LDH							
.D2	LDH	LDH	LDH	LDH	LDH	LDH	LDH	LDH							
.L1								ADD	ADD	ADD	ADD	ADD	ADD	ADD	ADD
.L2		SUB	SUB	SUB	SUB	SUB	SUB	SUB							
.S1															
.S2			B	B	B	B	B	B							
.M1						MPY	MPY	MPY	MPY	MPY	MPY	MPY	MPY		
.M2															

Program 5.6: Code obtained from the scheduling table

```
;Cycle 1
        LDH   .D1   *A1++,A2
  ||    LDH   .D2   *B1++,B2

;Cycle 2
        LDH   .D1   *A1++,A2
  ||          LDH   .D2   *B1++,B2
  ||          SUB   .L2   B0,1,B0

;Cycle 3- 4 and 5
          LDH   .D1   *A1++,A2
  ||            LDH   .D2   *B1++,B2
  ||    [B0]    SUB   .L2   B0,1,B0
                B     .S2   Loop

;Cycle 6 and 7
      LDH   .D1   *A1++,A2
  ||          LDH   .D2   *B1++,B2
  ||    [B0]   SUB   .L2   B0,1,B0
  ||    [B0]   B     .S2   Loop
  ||           MPY   .M1x  A2,B2,A3

;Cycle 8 to N
              LDH   .D1   *A1++,A2
  ||          LDH   .D2   *B1++,B2
  ||    [B0]  SUB   .L2   B0,1,B0
  ||    [B0]  B     .S2   Loop
  ||          MPY   .M1x  A2,D2,A3
  ||          ADD   .L1   A4,A3,A4

;Cycle N + 1 to N + 4
              MPY   .M1x  A2,B2,A3
  ||          ADD   .L1   A4,A3,A4
;Cycle N+ 6 TO N + 7
              ADD   .L1   A4,A3,A4
```

5.7 Linear assembly

In this chapter, it has been shown that code optimisation for loops can be achieved by the software-pipelining technique. This has been done the hard way using pen and paper. However, with the assembly optimiser, optimisation for loops can be made very simple. The tools accept code that is written in a linear fashion without considering the delay slots or even specifying the functional units, and by using symbolic variable names instead of registers as shown in Program 5.7.

Program 5.7:   Linear assembly code representing an FIR filter

```
        ZERO   sum
loop    LDH    *p_to_a,a
        LDH    *p_to_b,b
        MPY    a,b,prod
        ADD    sum,prod,sum
        SUB    B0,1,B0
        B      loop
```

For the tools to understand which part of the code is written in linear assembly, two directives are required: the first indicating the start of the code (**.cproc**) and the second indicating the end of the code (**.endproc**). The tools also require that all symbolic registers (except registers declared in **.cproc**) must be declared using the **.reg** directive as shown in Program 5.8.

Program 5.8:   Linear assembly code

```
        .proc p_to_a, p_to_b
        .reg a, b, prod, sum
        ZERO   sum
loop
        LDH    *p_to_a,a
        LDH    *p_to_b,b
        MPY    a,b,prod
        ADD    sum,prod,sum
        B      loop
        .endproc
```

5.7.1   Hand optimisation of the dotp function using linear assembly

It has been shown in Chapter 2 that in order to make maximum use of the units and therefore improve performance, one should exploit the SIMD operations available with the TMS320C66x. Before hand writing code in assembly or linear assembly, one needs to know which SIMD instructions are available. These instructions have been highlighted in Chapter 2. It is clear from the previous examples that none of the SIMD instructions was used.

The following examples show how to improve the performance by using some SIMD instructions for performing the **dot** product.

1) 2-way 16-bit multiplications using **dotp2** instruction. Exploiting the SIMD **DOTP2** instruction illustrated in Figure 5.11, the dependency graph can be written for the **dotp** function as shown in Figure 5.12. The handwritten code can be shown as in Figure 5.13.

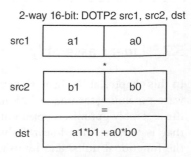

Figure 5.11 **DOTP2** instruction.

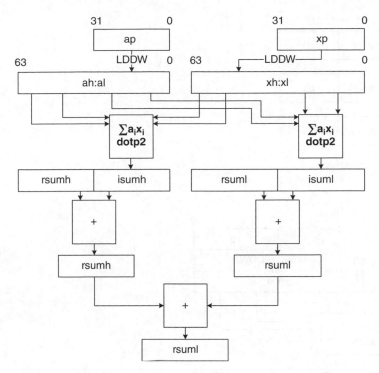

Figure 5.12 Dependency graph of the **dotp** function using **dotp2** instructions.

```
        ;LDDW_DOTP2sa.sa

        .global         LDDW_DOTP2sa


dotpLDDW_DOTP2sa: .cproc ap, xp, cnt
        .reg    ah:al, xh:xl, isuml, isumh, rsuml, rsumh

        zero    rsuml
        zero    rsumh

loop: .trip 10, 10, 4
        lddw    *ap++, ah:al
        lddw    *xp++, xh:xl
        dotp2 al, xl, isuml
        dotp2 ah, xh, isumh
        add             isuml, rsuml, rsuml
        add             isumh, rsumh, rsumh
        sub             cnt, 4, cnt
  [cnt]b          loop

        add     rsuml, rsumh, rsumh
        .return rsumh
        .endproc
```

Figure 5.13 dotp function implemented with **dotp2** instructions.

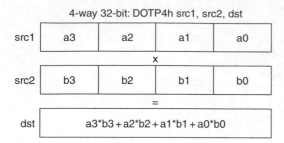

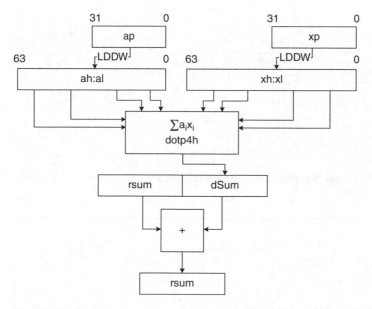

Figure 5.14 dotp4h instruction functionality.

Figure 5.15 Dependency diagram for the **dotp** function.

2) 4-way 16-bit multiplications using **dotp4h** instruction. Exploiting the SIMD **DOTP4H** instruction illustrated in Figure 5.14, the dependency graph can be written for the **dotp4h** function as shown in Figure 5.15. The handwritten code can be written as in Figure 5.16.
3) 8-way 16-bit multiplications using **ddotp4h** instruction. Exploiting the SIMD **DDOTP4H** instruction illustrated in Figure 5.17, the dependency graph can be written for the **ddotp4h** function as shown in Figure 5.18. The handwritten code can be written as in Figure 5.19.
4) Two 8-way 16-bit multiplications using **DDOTP4H** instruction. Exploiting the SIMD **DDOTP4H** instruction illustrated in Figure 5.20, the dependency graph can be written for the **ddotp4h** function as shown in Figure 5.21. The handwritten code can be written as in Figure 5.22.

With one **ddotp4h** on each side (**.M1** and **.M2**), one would expect to double the performance. However, due to the limitation of the load instructions (that can load a maximum of a double word each), the performance will be less than that with a **DDOTP4H** used on only one side.

```
        .global dotp4h

dotp4h:    .cproc ap, xp, cnt
           .reg   a1h:a1l, x1h:x1l, dSum, rsum

           zero   rsum
loop:

           lddw   *ap++, a1h:a1l
           lddw   *xp++, x1h:x1l

           dotp4h a1h:a1l, x1h:x1l, dSum

           add            dSum, rsum, rsum

[cnt] sub              cnt, 4, cnt
[cnt] b                loop

           .return rsum
           .endproc
```

Figure 5.16 dotp implemented with **dotp4h** instructions.

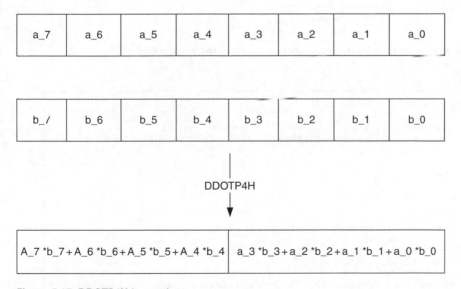

Figure 5.17 DDOTP4H instruction.

The code has been hand optimised, but further optimisation is required. So far, nothing has been done about the memory access. If this optimised code is located in the external memory, the optimisation will have no effect if the cache is not used. Furthermore, if data are not aligned, the CPU may have to stall as the **.D1** and **.D2** unit try to access the same memory bank.

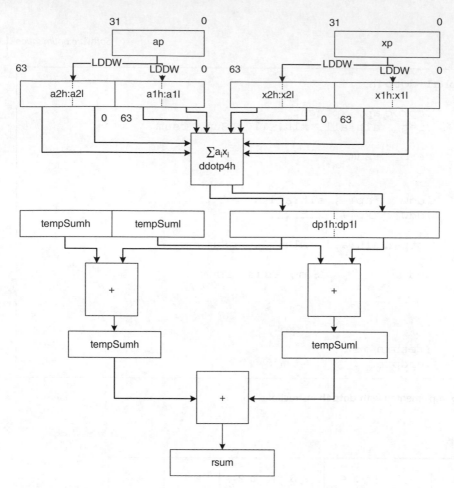

Figure 5.18 Dependency diagram for the **dotp** function using ddotp4h instruction.

```
        .global ddotp4h

ddotp4h:        .cproc ap, xp, cnt
                .reg   a1h:a1l, x1h:x1l, a2h:a2l, x2h:x2l, dp1h:dp1l,  tempSuml,
                tempSumh, rsum

                zero   rsum
                zero   tempSuml
                zero   tempSumh

loop:
                lddw   *ap++, a1h:a1l
                lddw   *xp++, x1h:x1l
                lddw   *ap++, a2h:a2l
                lddw   *xp++, x2h:x2l

                ddotp4h        a1h:a1l:a2h:a2l, x1h:x1l:x2h:x2l, dp1h:dp1l

                add            dp1h, tempSumh, tempSumh
                add            dp1l, tempSuml, tempSuml

[cnt] sub            cnt, 8, cnt
[cnt] b              loop

                add    tempSuml, tempSumh, rsum

                .return rsum
                .endproc
```

Figure 5.19 **dotp** implemented with **ddotp4h** instructions.

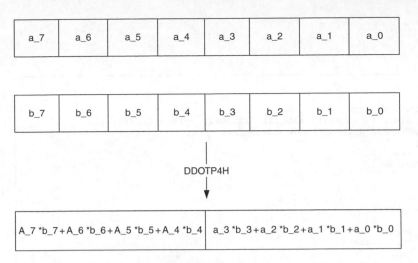

Figure 5.20 Illustration of the **DDOTP4H** instruction.

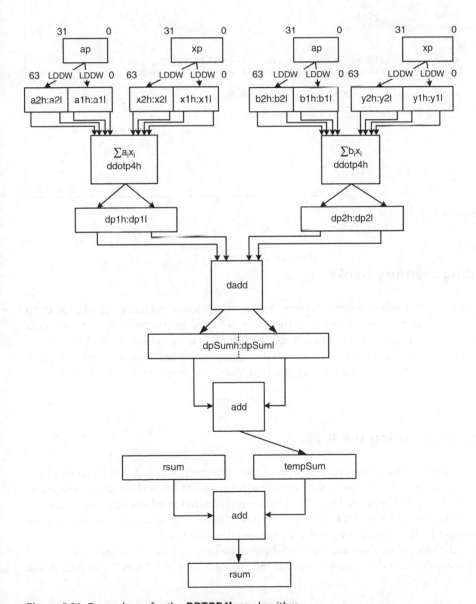

Figure 5.21 Dependency for the **DDTOP4h.sa** algorithm.

```
        .global ddotp4h2

ddotp4h2:   .cproc ap, xp, cnt
            .reg  a1h:a1l, x1h:x1l, a2h:a2l, x2h:x2l, dp1h:dp1l,b1h:b1l,
y1h:y1l, b2h:b2l, y2h:y2l, dp2h:dp2l, dpSumh:dpSuml, tempSum, rsum

            zero  rsum
            zero tempSum

;If you change this loop to only load 64 bits at a time and use a single dotp4h,
;then it takes about 130 cycles, but if you load 128 bits, and have two ddotp4h,
;then takes about 167 cycles
loop:
            lddw    *ap++, a1h:a1l
            lddw    *xp++, x1h:x1l
            lddw    *ap++, a2h:a2l
            lddw    *xp++, x2h:x2l

            lddw    *ap++, b1h:b1l
            lddw    *xp++, y1h:y1l
            lddw    *ap++, b2h:b2l
            lddw    *xp++, y2h:y2l

            ddotp4h     a1h:a1l:a2h:a2l, x1h:x1l:x2h:x2l, dp1h:dp1l
            ddotp4h     b1h:b1l:b2h:b2l, y1h:y1l:y2h:y2l, dp2h:dp2l

            dadd  dp1h:dp1l, dp2h:dp2l, dpSumh:dpSuml
            add         dpSumh, dpSuml, tempSum
            add         tempSum, rsum, rsum

[cnt] sub         cnt, 16, cnt
[cnt] b           loop

            .return rsum
            .endproc
```

Figure 5.22 **ddotp4h2.sa**.

5.8   Avoiding memory banks

TMS320C66x has eight (bank 0 to bank 7) memory banks as shown in Figure 5.23. In the **dotp** functions shown here, it has been shown that there are two load instructions that try to access data at the same time through **.D1** and **.D2**. These data can be as wide as two words (LDDW). To avoid memory conflict, the data can be made to start at different banks. In the example shown in Figure 5.23, the data are 16-byte data aligned. To align data, the pragmas shown in Figure 5.23 can be used.

5.9   Optimisation using the tools

This chapter has shown how to optimise code 'by hand' (the hard way!). It will be shown here that, by passing the right information to the compiler, the optimisation will be carried out automatically and efficiently by the tools. Once the programmer knows what information to pass to the compiler and knows the TMS320C66x CPU architecture, it will be a matter of passing the correct information to the compiler in order to achieve the best results.

Let's examine the algorithm shown in Figure 5.24 and see how to pass this information to the compiler. If at compile time the compiler has no information on the COUNT variable, it will

```
#pragma DATA_ALIGN(a1, 16);
#pragma DATA_ALIGN(x1, 16);
```

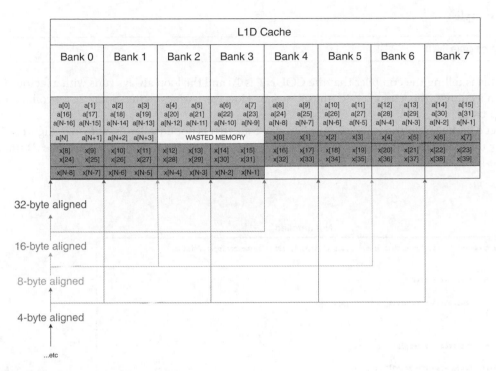

Figure 5.23 TMS320C66x memory banks.

```
for (i=0; i < COUNT; i++)
  {
    c[i] = a[i] +  b[i]);
  }
```

Figure 5.24 dotp code to optimise.

generate a zero-trip loop test (will slightly increase the code size) and generate two programs, one pipelined and one not to pipeline, because the compiler does not know how many times the loop is going to iterate. The loop could be iterating once only, for instance.

To tell the compiler how many times the loop must iterate, the **#pragma** function shown here should be used:

```
#pragma MUST_ITERATE(lower_bound, upper_bound, factor)
```

By passing this information to the compiler, it will be able to decide to pipeline the loop or not. There are three parameters to supply when using the **MUST_ITERATE**:

lower_bound: This defines the minimum possible total iterations of the loop. Cannot pipeline without this.

upper_bound: (Optional) This defines the maximum possible total iterations of the loop.
factor: This tells the compiler that the total iteration is always an integer multiple of this number (good for unrolling the loop).

Example:

```
#pragma MUST_ITERATE(10,, 2)
```

This is telling the compiler that the COUNT is 10 and the loop always runs with a factor of 2. This allows the compiler to unroll twice. However, what is the point of unrolling the loop twice?

Let's first examine code that is unrolled and code that is not unrolled, as shown in Figure 5.25.

By unrolling the loop, better load balancing and usage of the units (in this case, the **.D** units) have been achieved (see Figure 5.25).

Not unrolled

```
void Loop(int * restrict output, int * restrict input1, int * restrict input2, int n)
{
    int i;
    for (i=0; i<n; i++)
    {
        output[i] = input1[i] + input2[i];
    }
}

//An excerpt from its compiler feedback

;* Partitioned Resource Bound(*) : 2
;* Resource Partition:
;*              A-side    B-side
;* .L units     0         0
;* .S units     0         1
;* .D units     2*        1
;* .M units     0         0
```

Unrolled

```
void Loop(int * restrict output, int * restrict input1, int * restrict input2, int n)
{
    int i;
    for (i=0; i<n; i+=2)
    {
        output[i] = input1[i] + input2[i];
        output[i+1] = input1[i+1] + input2[i+1];
    }
}

//An excerpt from its compiler feedback

;* Partitioned Resource Bound(*) : 3
;* Resource Partition:
;*              A-side    B-side
;* .L units     0         0
;* .S units     1         0
;* .D units     3*        3*
;* .M units     0         0
```

Figure 5.25 Load balancing by unrolling a loop.

Figure 5.26 Loop-carried dependency graph.

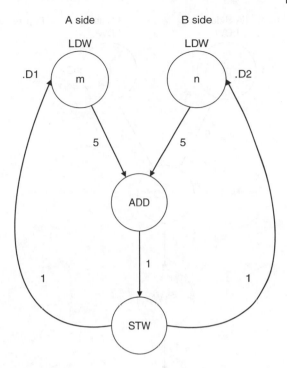

The compiler has another reason not to pipeline. In the example shown in Figure 5.26, the next load cannot be issued until 7 (5 + 1 + 1) cycles later. However, on the diagram shown in Figure 5.27, there is no data dependency, and therefore data load can be issued every cycle. The information that tells the compiler that there is no dependency can be passed to the compiler by using the **restrict** keyword as shown in this example:

```
void func1(int* restrict a, int* restrict b, int*restrict c)
```

The **restrict** keyword can also be used for data arrays as shown in this example:

```
void myfnt(int c[restrict], int a[restrict]), int b[restrict])
```

If the compiler does not unroll because the information supplied was not enough, one can tell the compiler to unroll by using the **UNROLL** keyword as shown in this example:

```
#pragma MUST_ITERATE(10,, 2)
#pragma UNROLL(2)//this tells the compiler to unroll the loop twice
```

The **#pragma MUST_ITERATE()** must be used when using the **UNROLL()**, and if unrolling is not desired, then use a '1' as a parameter as shown in this code:

```
#pragma UNROLL(1)
```

It has been shown here that SIMD can improve the performance. These SIMD instructions may require data to be packets. Before loading these data, data have to be aligned properly and

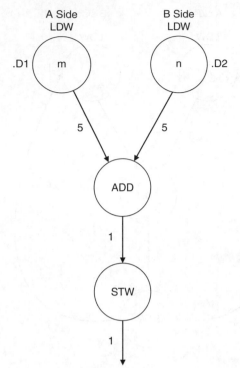

A Side
LDW

B Side
LDW

Figure 5.27 Loop-carried dependency graph.

the compiler needs to be told that these data are aligned. To tell the compiler that data are aligned, use the **_nassert** keyword; and to align data to a specific boundary, use the **DATA_ALIGN** keyword. See the examples here:

Example 1: Aligning data (**#pragma DATA_ALIGN (symbol, constant)**). The constant must be a power of 2. The maximum alignment is 32,768 [3].

Example:

```
#pragma DATA_ALIGN(a, 8)//tell the compiler to align the data a.
                        //8 means 8 bytes (double words).
#pragma DATA_ALIGN(x1, 16);//16 mains 16 bytes (quad words).
```

Example: Telling the compiler that data are actually aligned.

```
_nassert(((int)a& 0x1) == 0);//a is a half word aligned
_nassert(((int)b& 0x3) == 0);//b is a word aligned
_nassert(((int)c& 0x7) == 0);//c is a double word aligned
_nassert(((int)d& 0xF) == 0);//d is a quad word aligned
```

The compiler does not know if the data are aligned or not unless *program-level optimisation* **O3** is selected. By selecting program-level optimisation, the compiler will combine all source files to have better visibility of all codes before performing maximum optimisation.

More information on the compiler switches can be found in Ref. [3].

5.10   Laboratory experiments

Experiment 1:

dotp implementation

  Project location:

\Chapter_5_Code\dotp_ALL_SA

  a) Open the project **dotp_ALL_SA** and explore the code.
  b) Run the project and verify the results as shown in Figure 5.28.

Experiment 2:

$$y = \sum_{i=0}^{N} [a_{2i} B + a_{2i+1} C] \text{ implementation.}$$

1) Draw the dependency diagram for implementing the function **y** shown in Equation [5.1] using LDDW to load a_i:

$$y = \sum_{i=0}^{N} [a_{2i} B + a_{2i+1} C] \qquad [5.1]$$

where a_i are stored in the memory, and B and C are constants stored in registers. The implementation should be on the TMS320C66x processor, considering a_i, B, C, N and y are 16-bit integers declared as follows:

```
int a[] = {A_ARRAY};
int B = 6;
int C = 7;
int N = 128;
int y = 0;
```

```
                                                          —    □    ✕

▣ Console ✕                                    ▣ ▣ | ⧉ ▣ ▾ ⧉ ▾ ▭ ▭
dotp_ALL_SA:CIO
[C66xx_0] Compile Flags: -O2 --disable:sploop, offset x
COUNT is equal to  = 256
y = 2829056        [function: dotp0]       [310 cycles] (software pipelining disabl
y = 2829056        [function: dotp1]       [117 cycles] (DATA_ALIGN + offset)
y = 2829056        [function: dotp1]       [117 cycles] (DATA_MEM_BANK)
y = 2829056        [function: dotp2]       [117 cycles]
y = 2829056        [function: dotp4h]      [129 cycles]
y = 2829056        [function: ddotp4h]     [129 cycles]
y = 2829056        [function: ddotp4h2]    [145 cycles]
y = 2829056        [dotpC_intrinsic]       [194 cycles]
y = 2829056.0000       [function: mpyspdot]    [569 cycles]
I finished!
```

Figure 5.28 Console output showing all results.

The dependency diagram should show that the algorithm can be implemented in a single-cycle loop.

2) Write the scheduling diagram for the dependency diagram obtained in 1.
3) Write the optimised pipeline code for **y** using the scheduling diagram obtained in 2.

Solutions:

1) See Figure 5.29.
2) See Table 5.12.
3) See Program 5.9. The solution can be found in:
   Chapter_5_Code\addition

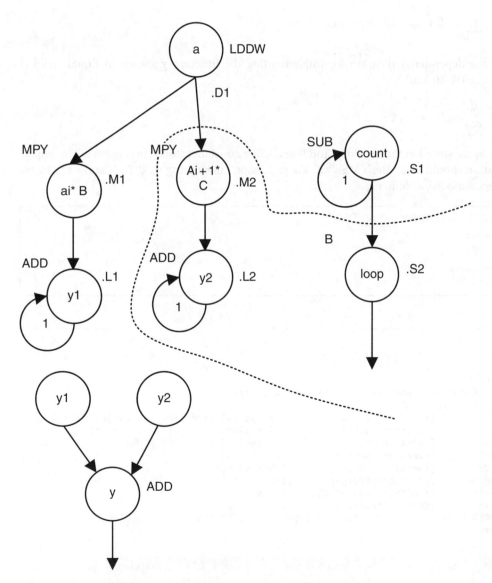

Figure 5.29 Dependency graph.

Table 5.12 Scheduling table

	1	2	3	4	5	6	7	8	
D1	LDDW	X	X	X	X	X	X	X	
M1						MPY	X	X	
L1								ADD	
M2						MPY	X	X	
L2								ADD	
S1		SUB	X	X	X	X	X	X	
S2			B	X	X	X	X	X	
D2									

Program 5.9: Assembly code for the implementation of **y**

```
.global Addition_In_ASM

Addition_In_ASM

            MV      A6,B5
            MV B4, A6

            MV      B6,A1
            Zero A8
            zero b6
;1
            LDDW .D1      *A4++, A11:A10
;2
            LDDW .D1      *A4++, A11:A10
||          [A1]  SUB     .S1        A1,1,A1
;3
            LDDW .D1      *A4++, A11:A10
||          [A1]  SUB     .S1        A1,1,A1
||          [A1]  B       .S2        LOOP
;4
            LDDW .D1      *A4++, A11:A10
||          [A1]  SUB     .S1        A1,1,A1
||          [A1]  B       .S2        LOOP
;5
            LDDW .D1      *A4++, A11:A10
||          [A1]  B       .S2        LOOP
;6
            LDDW .D1      *A4++, A11:A10
```

```
||              MPY    .M1       A10,A6,A7
||              MPY    .M2       A11,B5,B8
||    [A1]      SUB    .S1       A1,1,A1
||    [A1]      B      .S2       LOOP
;7
                LDDW   .D1       *A4++, A11:A10
||              MPY    .M1       A10,A6,A7
||              MPY    .M2       A11,B5,B8
||    [A1]      SUB    .S1       A1,1,A1
||    [A1]      B      .S2       LOOP
;8

LOOP            LDDW   .D1       *A4++, A11:A10
||              MPY    .M1       A10,A6,A7
||              ADD    .L1       A7,A8,A8
||              MPY    .M2       A11,B5,B8
||              ADD    .L2       B8,B6,B6
||    [A1]      SUB    .S1       A1,1,A1
||    [A1]      B      .S2       LOOP

                ADD a8,b6,a4
                B B3
                nop 5
```

5.11 Conclusion

Optimising an application on a multicore system-on-chip (SoC) can be performed at four levels. Level 1 is the algorithm-level optimisation. Level 2 is task-level optimisation, where tasks are distributed to the available cores so they can run in parallel. The third level is at the instruction level (using the maximum number of units), and the fourth level is at the data level (using SIMD instruction).

A single-core benchmark alone is meaningless if the memory structure is not taken into consideration. For instance, if the L3 cache is used, other cores may be competing for the same resources (L3 and bus bandwidth), and therefore the performance will be degraded. It also has been shown that having a larger-way SIMD will not necessarily improve the performance if data cannot be loaded fast enough.

References

1 S.-K. Chen, T.-J. Lin and C.-W. Liu, Parallel object detection on multicore platforms, in *Workshop on Signal Processing Systems (SiPS)*, Tampere, Finland, 2009.
2 V. Kumar, A. Sbîrlea, A. Jayaraj, Z. Budimlić, D. Majeti and V. Sarkar, Heterogeneous work-stealing across CPU and DSP cores, in *High Performance Extreme Computing Conference (HPEC), 2015 IEEE*, Massachusetts, USA, 2015.
3 Texas Instruments, TMS320C6000 Optimizing Compiler v8.1.x user's guide, January 2016. [Online]. Available: http://www.ti.com/lit/ug/sprui04a/sprui04a.pdf.

6

The TMS320C66x interrupts

6.1   Introduction

As with most microprocessors, the TMS320C66x allows normal program flow to be interrupted. In response to the interruption, the CPU finishes executing the current instruction(s) and branches to a procedure which services the interrupt. To service an interrupt, the user or the system must save the contents of the registers and the context of the current process, then service the interrupt task, restore the registers and the context of the process, and finally resume the original process (see Figure 6.1). The interrupt can come from an external device, an internal peripheral or simply a special instruction in the program.

There are four types of interrupts on the TMS320CC66x CPUs. These are the two non-maskable interrupts (**Reset** and **NMI**) and maskable interrupts (**EXCEP** and **INT4–INT15**) (see Figure 6.2 and Table 6.1). The interrupt controllers described in this chapter allow events to be mapped to any of the input interrupts from **INT4** to **INT15**.

Due to the sheer number of events available (hundreds) and the low number of interrupts that the CPU, Enhanced Direct Memory Access (EDMA) or hyperlink can handle, some events are aggregated first by the chip-level interrupt controllers (CICs or CpIntcs) to generate the secondary events. The other events are unchanged and are called *primary events*. Secondary events are infrequent events and are routed to the CIC first to offload the interrupt controller (INTC) as shown in Figure 6.3.

Each processor has a fixed number of CICs. For instance, the TMS320C6678 has four CICs and the 66AK2H14/12 has only three CICs; see Figure 6.4 or 6.5. As can be seen from these

Multicore DSP: From Algorithms to Real-time Implementation on the TMS320C66x SoC, First Edition. Naim Dahnoun.
© 2018 John Wiley & Sons Ltd. Published 2018 by John Wiley & Sons Ltd.
Companion website: www.wiley.com/go/dahnoun/multicoredsp

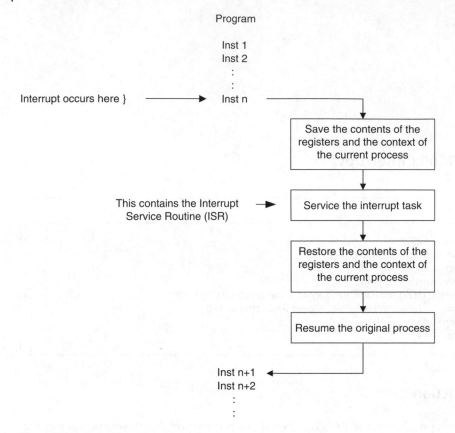

Figure 6.1 Interrupt response procedure.

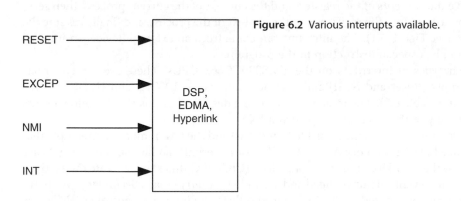

Figure 6.2 Various interrupts available.

figures, each event (primary or secondary) is mapped to a specific core or peripheral. However, some events are broadcast to many cores.

The primary task of the user is to identify which event or events are to be programmed to generate an interrupt. Each event is identified by a number and described in the user guide. A sample of events available is shown in Table 6.2. Let's now examine the CIC.

Table 6.1 Interrupt sources and priority

Type	Interrupt name	Priority
Non-maskable	$\overline{\text{RESET}}$	Highest
	NMI	
Maskable	EXCEP	
	INT4	
	INT5	
	INT6	
	INT7	
	INT8	
	INT9	
	INT10	
	INT11	
	INT12	
	INT13	
	INT14	
	INT15	Lowest

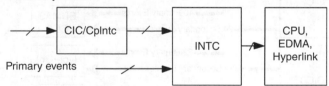

Secondary events

CIC/CpIntc

Primary events

INTC

CPU, EDMA, Hyperlink

CIC = CpIntc : Chip Level Interrupt Controller
INTC : Interrupt Controller

Figure 6.3 Interrupts: the big picture.

6.1.1 Chip-level interrupt controller

A CIC (see Figures 6.3, 6.4 and 6.5) accepts system-level events (see datasheet for a particular device) and combines them to generate secondary events to the interrupt controller, as shown in Figure 6.3. Figure 6.4 shows that the TMS320C6678 has four CICs (CIC0, CIC1, CIC2 and CIC3) responding to various events (some events can be found in different CICs).

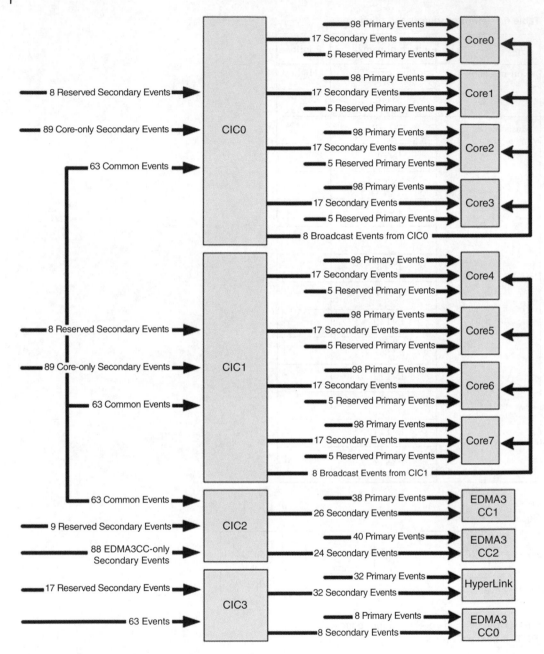

Figure 6.4 CIC controllers for the TMS32C6678 [1].

The CICs are composed of:

1) An enabler. The enabler is used to enable or disable an event. This event will be logged in the interrupt status register. If the event was enabled and the event happened, the Enabled Status will be set. If the event happens and the event was disabled, the event will be logged in the Raw Status and not in the Enable Status; see Figure 6.6. The Enabled Status can also be set by software, which is very convenient for debugging.

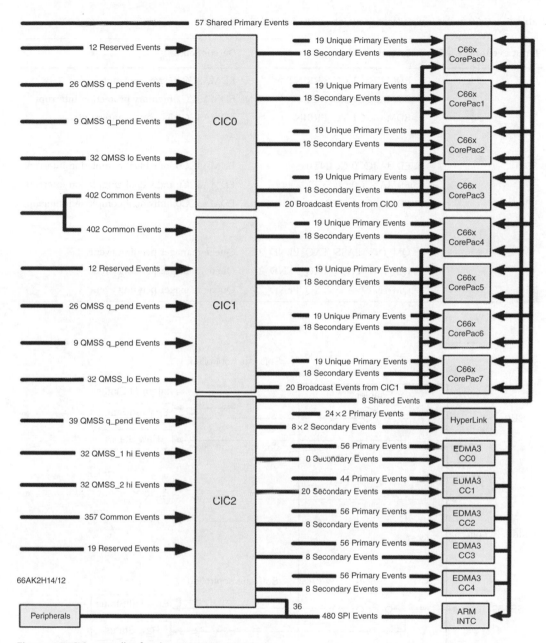

Figure 6.5 CIC controller for the 66AK2H14/12 [2].

CIC0 registers for the TMS320C6678 are located in address 0x02600000 as shown in Table 6.3. The interrupt status register is offset by 0x20 as shown in Table 6.4.

As an example, if System Event 4 needs to be enabled, there are two options:

A) Use the configuration file:

```
CpIntc.sysInts[4].enable = true;
```

Table 6.2 CIC0 event inputs (secondary interrupts for TMS320C66x CorePacs) [1]

Input event no. on CIC	System interrupt	Description
0	EDMA3CC1 CC_ERRINT	EDMA3CC1 error interrupt
1	EDMA3CC1 CC_MPINT	EDMA3CC1 memory protection interrupt
2	EDMA3CC1 TC_ERRINT0	EDMA3CC1 TC0 error interrupt
–	–	–
–	–	–
38	EDMA3CC0 CCINT0	EDMA3CC0 individual completion interrupt
39	EDMA3CC0 CCINT1	EDMA3CC0 individual completion interrupt
40	EDMA3CC0 CCINT2	EDMA3CC0 individual completion interrupt
–	–	–
–	–	–
157	QM_INT_PASS_TXQ_PEND_23	Queue manager pending event
158	QM_INT_PASS_TXQ_PEND_24	Queue manager pending event
159	QM_INT_PASS_TXQ_PEND_25	Queue manager pending event

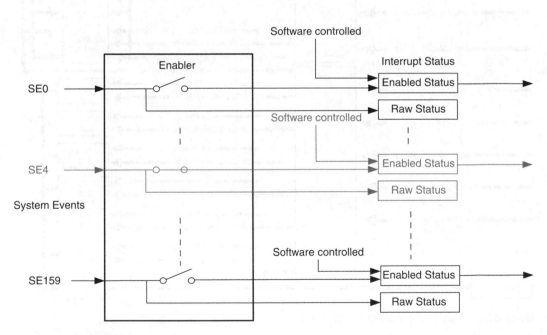

Figure 6.6 The enabler functionality.

Table 6.3 Memory location of the CIC0 and CIC1 for the TMS320C6678 [1]

02600000	02601FFF	0 02600000	0 02601FFF	8K	Chip Interrupt Controller (CIC) 0
02602000	02603FFF	0 02602000	0 02603FFF	8K	Reserved
02604000	02605FFF	0 02604000	0 02605FFF	8K	Chip Interrupt Controller (CIC) 1
02606000	02607FFF	0 02606000	0 02607FFF	8K	Reserved

Table 6.4 CIC register offsets

Address Offset	Register
0×000	Revision Register
0×004	Control Register
0×008 – 0×00C	Reserved
0×010	Global Enable Register
0×014 – 0×01C	Reserved
0×020	System Interrupt Status Indexed Set Register
0×024	System Interrupt Status Indexed Clear Register
0×028	System Interrupt Enable Indexed Set Register
0×02C	System Interrupt Enable Indexed Clear Register

31	10	9	0
Reserved		INDEX	
R-0		R/W-0	

Legend: R = Read only; R/W = Read/Write; -n = value after reset

Figure 6.7 System Interrupt Status Indexed Set Register (**STATUS_SET_INDEX_REG**).

B) Use software:

```
*(unsigned int *) (0x02600000 + 0x20) = 0x4;
```

To set Event 4, the index in the System Interrupt Status Indexed Set Register (**STATUS_SET_INDEX_REG**) (see Figure 6.7) needs to be set to 4. The index represents the event number. The index is a 10-bit number, and therefore 1024 events can be represented.

2) A channel mapper. The aim of the CICs is to combine events. This is achieved by the combiner of each CIC that groups events. Each enabled event can select a channel as follows:

```
CpIntc.sysInts[event number].hostInt = channel number;
```

Example:

```
CpIntc.sysInts[4].hostInt = 0; //setting event 4 to channel 0
```

This is illustrated in Figure 6.8. The event to channel mapping is made through the Channel Interrupt Map Register (**CH_MAP_REGx**) illustrated in Figure 6.9 and viewed using the CCS (see Figure 6.10).

3) Host interrupt mapper. Each channel must be mapped to an interrupt. However, the host mapping is fixed, that is:
Channel 0 is mapped to Interrupt 0.
Channel 1 is mapped to Interrupt 1.
Channel 2 is mapped to Interrupt 2 and so on.

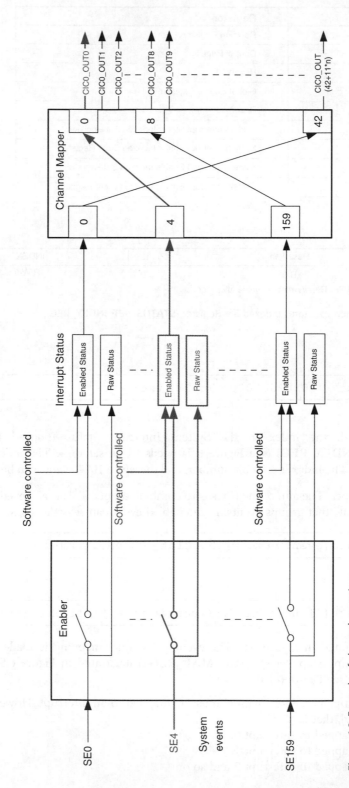

Figure 6.8 Example of channel mapping.

Interrupt Channel Map Registers (CH_MAP_REGx)

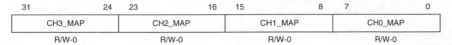

31	24	23	16	15	8	7	0
CH3_MAP		CH2_MAP		CH1_MAP		CH0_MAP	
R/W-0		R/W-0		R/W-0		R/W-0	

Legend: R = Read only; −*n* = value after reset

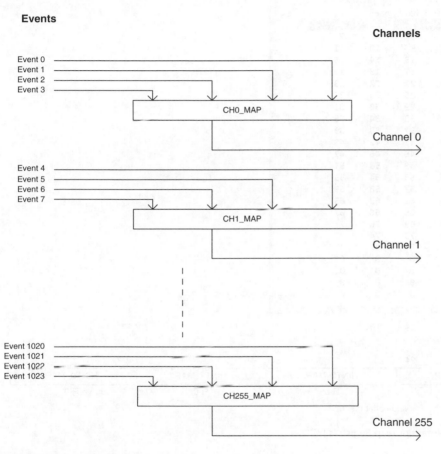

Figure 6.9 Default mapping.

The channel number will depend on the device used. Each channel has a register. The mapping of the channels to host interrupts is fixed (one-to-one mapping). Each of the four channels has a register to define their host interrupts, and the register is read-only; see Figure 6.11.

6.2 The interrupt controller

The INTC, shown in Figure 6.3 and detailed in Figure 6.16, is composed of:

1) Event combiner. As its name suggests, the combiner combines many events to produce one event. There are four groups of events that can be combined (see Figure 6.17). Notice that events cannot be grouped randomly. As an example, enabled events within one group can be

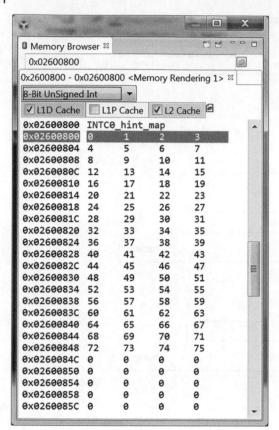

Figure 6.10 Host interrupt mapping for the CIC0 viewed with the CCS.

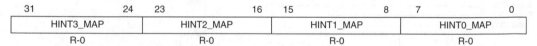

31	24	23	16	15	8	7	0
HINT3_MAP		HINT2_MAP		HINT1_MAP		HINT0_MAP	
R-0		R-0		R-0		R-0	

Legend: R = Read only; −*n* = value after reset

Figure 6.11 Host Interrupt Map Registers [3].

```
// combine Group 2 to event 2 and map this event to interrupt 6
EventCombiner.eventGroupHwiNum[2] = 6;

// event 90 and 89 will call hwicombine_GPIO14_GPIO15
EventCombiner.events[90].fxn = "&hwicombine_GPIO14_GPIO15";
EventCombiner.events[89].fxn = "&hwicombine_GPIO14_GPIO15";
// Unmask event 89 and 90 (GPIO14 and GPIO15)
EventCombiner.events[90].unmask = true;
EventCombiner.events[89].unmask = true;
```

Figure 6.12 Configuration script.

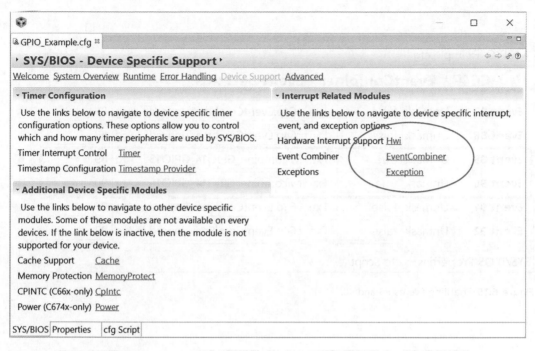

Figure 6.13 Accessing the event combiner.

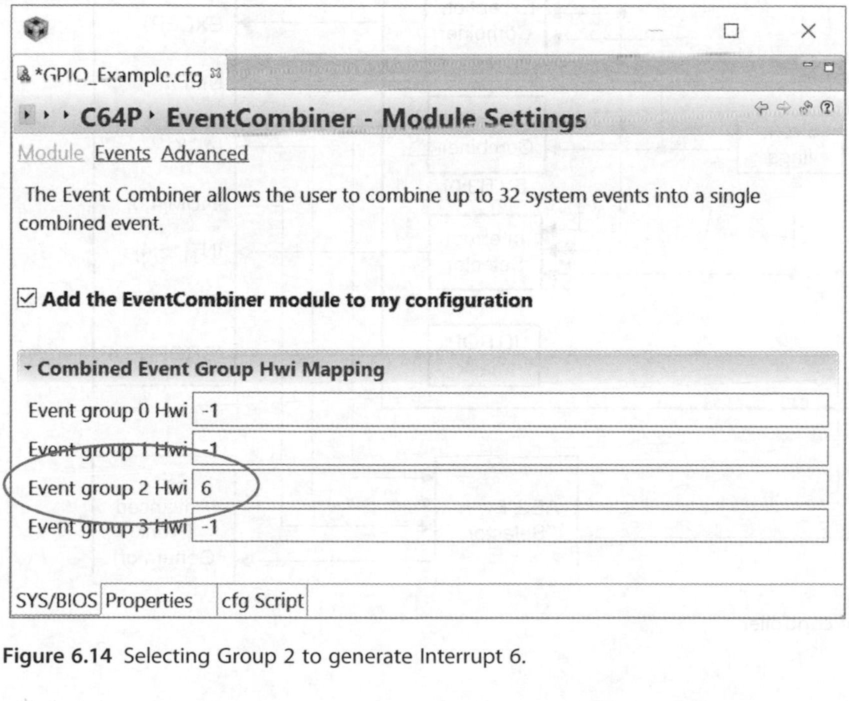

Figure 6.14 Selecting Group 2 to generate Interrupt 6.

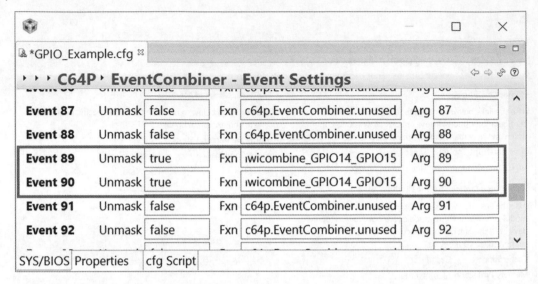

Figure 6.15 Enabling Events 89 and 90.

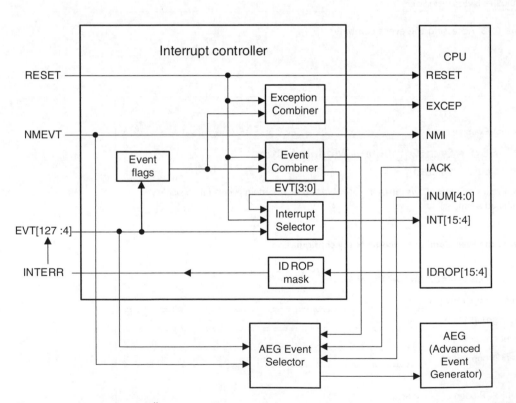

Figure 6.16 Interrupt controller.

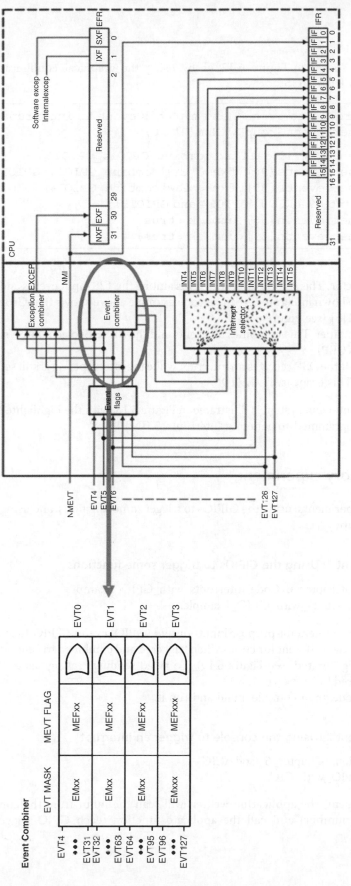

Figure 6.17 Event combiner.

combined as shown in Figure 6.12 or by using the graphical configuration shown in Figure 6.13 to Figure 6.15.

```
//combine Group 2 to event 2 and map this event to interrupt 6
EventCombiner.eventGroupHwiNum[2] = 6;

//event 90 and 89 will call hwicombine_GPIO14_GPIO15
EventCombiner.events[90].fxn = "&hwicombine_GPIO14_GPIO15";
EventCombiner.events[89].fxn = "&hwicombine_GPIO14_GPIO15";
//Unmask event 89 and 90 (GPIO14 and GPIO15)
EventCombiner.events[90].unmask = true;
EventCombiner.events[89].unmask = true;
```

2) Interrupt selector. This selects up to 12 events among the 128 input events and maps them to **INT[15:4]** as shown in Figure 6.16. The code example above shows that Group 2 is selected to link to **INT[6]** (see Figure 6.14).
3) Exception combiner. This combines all events and the **RESET** to generate a single hardware exception (**EXCEP**).
4) IDROP mask. If the DSP receives an interrupt while the interrupt flag is still set, then an event is generated. This event is the **EVT96**.

The complete interrupt system is illustrated in Figure 6.18, with the highlighted System Event 4 (SE4) being programmed to generate Interrupt 15 (**INT15**).

6.3   Laboratory experiment

There are two experiments: using the GIPIOs to trigger some functions, and using the console to trigger an interrupt.

6.3.1   Experiment 1: Using the GIPIOs to trigger some functions

Project location: \Chapter_6_Code\Interrupts_with_GPIO_Example
Project name: Interrupts_with_GPIO_Example.

In this example, two general-purpose input–output (GPIO) events (GPIO 14 and GPIO15) are made to generate two different functions when triggered and combined to generate one function if either event is generated (see Figure 6.19). To avoid sending real signals to the GPIOs, the events are triggered by software.
Examine the code, then compile, build and run it.

6.3.2   Experiment 2: Using the console to trigger an interrupt

Project file location: \Chapter_6_Code\CIC
Project name: GPIO_with_CIC.

In this experiment, the application written will scan the input from the console (the user inputs a **GPIO** number) and call the applicate function (each GPIO triggers a different

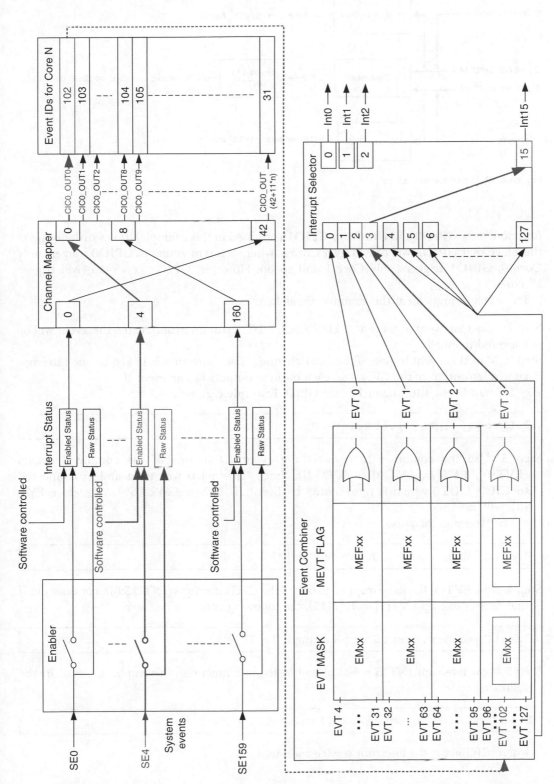

Figure 6.18 Overall functionality of the interrupt mechanism.

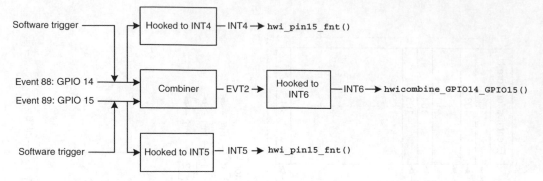

Figure 6.19 Experimental setup.

function; only **GPIO0** and **GPIO8** to **GPIO15** are used in this example). It is worth noting at this stage that **GPIOn** will only trigger Core *n*, where *n* is 0 to 7; that is, **GPIO0** triggers only Core 0, **GPIO1** triggers only Core 1 and so on. However, **GPIOx** ($x = 8$–15) will trigger all cores.

The steps to implement this example are as follows:

Step 1. Select an event (e.g. Event 4: EDMA3CC1 TC2 error interrupt). Later, this event will be triggered manually.

Step 2. Map this event to one of the 1024 channels. Each interrupt is linked to one interrupt controller output of the CICx, and each of these outputs has an event ID.

To map Event 4 to Channel 0 (see **GPIO_Example.cfg**), use:

```
CpIntc.sysInts[4].hostInt = 0;
```

Step 3. Use the combiner to connect the event ID to one of the combined events (**EVT[0]**, **EVT**[1], **EVT[2]** or **EVT[3]**). Since Channel 0 was used and is connected to **CIC0_OUT0** which is represented by Event ID 102 (see SPUR691), therefore **EVT [3]** will be used.

To program this, use:

```
EventCombiner.eventGroupHwiNum[3]
```

Step 4. Now **EVT[3]** has to be mapped to one of the CPU interrupts (**INT[15:0]**). For example, if one desires to map **EVT[3]** to **INT[15]**, the following code can be used:

```
EventCombiner.eventGroupHwiNum[3] = 15;
```

Step 5. Now interrupt **INT15** needs to be hooked to a function. This can be achieved by the following code:

```
EventCombiner.events[102].fxn = "&CIC_evt";
```

where **CIC_evt** is the interrupt service routine.

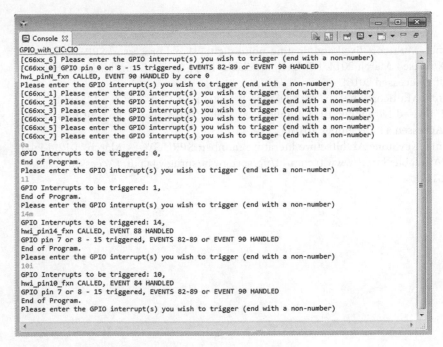

Figure 6.20 Console output showing the functions called.

Step 6. After completing the setup, the event has to be enabled. This is accomplished by the following code:

```
EventCombiner.events[102].unmask = true;
```

Step 7. In order to run this example, an event can be set by software. To do so, one needs to identify the interrupt status index register address to set the corresponding status of the system interrupt as shown in Section 6.1.1.

Step 8. Open, build and load the project.

Step 9. Group all cores so that they will all run at the same time.

Step 10. Run all cores.

Step 11. Enter 0 (followed by any letter) to simulate a **GPIO0** signal. Enter 1 (followed by any letter) to simulate a **GPIO1** signal (this has not been programmed, and therefore no output function call is performed). Enter 14 (followed by any letter) to simulate **GPIO14** (followed by any letter) to simulate **GPIO14** and so on. See output in Figure 6.20.

6.4 Conclusion

This chapter shows how the interrupt controller events and the CIC work, and how to program them to respond to events. The examples given use the GPIO pins to provide the interrupts. To avoid sending real signals to the GPIOs, the interrupts have been set by software.

References

1 Texas Instruments, Multicore fixed and floating-point digital signal processor: SPRS691E, November 2010, revised March 2014. [Online]. Available: http://www.ti.com/lit/ds/symlink/tms320c6678.pdf. [Accessed 2016].

2 Texas Instruments, Multicore DSP+ARM KeyStone II System-on-Chip (SoC): SPRS866E, November 2012, revised November 2013. [Online]. Available: http://www.ti.com/lit/ds/symlink/66ak2h12.pdf. [Accessed 11 December 2016].

3 Texas Instruments, KeyStone Architecture literature number: SPRUGW4A March 2012, March 2012. [Online]. Available: http://www.ti.com/lit/ug/sprugw4a/sprugw4a.pdf. [Accessed 11 December 2016].

7

Real-time operating system: TI-RTOS

Multicore DSP: From Algorithms to Real-time Implementation on the TMS320C66x SoC, First Edition. Naim Dahnoun.
© 2018 John Wiley & Sons Ltd. Published 2018 by John Wiley & Sons Ltd.
Companion website: www.wiley.com/go/dahnoun/multicoredsp

7.1 Introduction

A computer system or an embedded system is composed of hardware and software elements. The software part can play an important role and may include an operating system (OS) that provides low-level services for applications to run efficiently. These low-level services can be used for:

1) Performing multitasking operations (and therefore reducing application complexities)
2) Initialising I/Os
3) Management of memory
4) Handling a file system
5) Abstracting hardware.

There are many OSs available. These include MS_DOS, Android, BSD, iOS, Linux, OS X, QNX, Microsoft Windows, WP (Windows Phone) and IBM z/OS.

OSs can be grouped as:

1) Multi-user operating systems. Multi-user OSs allow two or more users to run programs at the same time.
2) Multi-tasking operating systems. Multitasking OSs allow more than one program to run concurrently.
3) Distributed operating systems. In a distributed system, the OS is distributed over physically separated hardware that is networked. The main advantage of distributed OSs is the concept of transparency. However, due to the network-introduced delay, timing can be an issue.
4) Embedded operating systems. Embedded OSs are designed to be used in embedded computer systems where memory size and other resources are limited.
5) Real-time operating system (RTOS). An RTOS is a multitasking OS which executes applications in real time, is deterministic (precise timing) and has minimal interrupt latency and minimal thread switching.

In addition to the functionality of the generic OS, for the RTOS, timing and memory size are critical. Therefore:

1) Interrupt latency and task switching are minimal.
2) Better memory management is achieved.
3) An RTOS may also include real-time debugging tools.

7.2 TI-RTOS

TI-RTOS is a real-time operating system that provides a low footprint and includes various modules that provide deterministic pre-emptive multithreading and synchronisation services, interrupt handling, memory management, instrumentation, communication and a File System; see Figure 7.1 and Figure 7.2. The TI-RTOS is scalable in the sense that only the components required for a specific application are selected. In addition, each component can be further minimised, in order to provide further scalability.

In the remainder of this chapter, real-time scheduling and dynamic memory management will be studied.

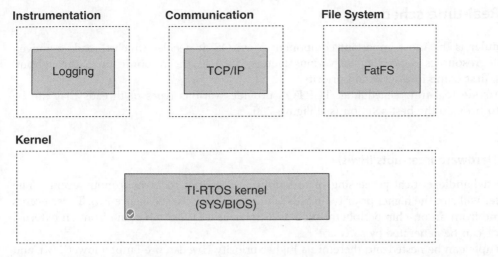

Figure 7.1 TI-RTOS components.

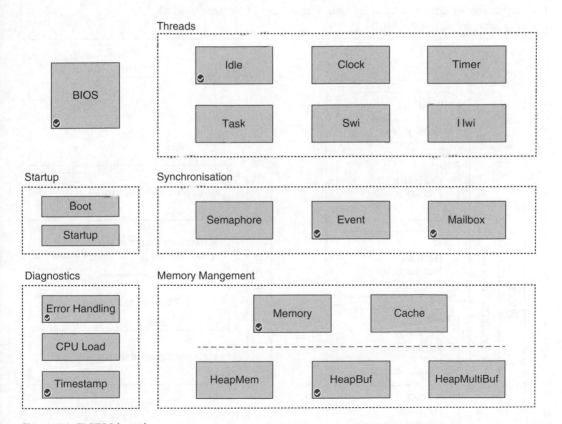

Figure 7.2 TI-RTOS kernel.

7.3 Real-time scheduling

A scheduler is the most important component of an OS. It decides which threads are given access to resources depending on various factors, such as the response time, throughput, fairness, first come first serve and priority.

To provide real-time scheduling, TI-RTOS provides various types of threads that can be mixed to provide the best solution (see Figure 7.3).

7.3.1 Hardware interrupts (Hwis)

Hwis can handle critical processing of functions in response to asynchronous events. The scheduler will run the higher priority threads which are the Hwis (see Figure 7.3). These events can come from an on-chip peripheral such as a timer or a DMA, can come from an external device or can be generated by software.

Interrupts can be nested and therefore a higher priority Hwi can pre-empt a lower Hwi, and hence the stack should be kept large enough to handle the appropriate number of nested interrupts. This can be achieved by setting the stack size in the configuration script as shown below; the programmer should not rely on the default setting.

```
/* System stack size in bytes (used by ISRs and Swis) */
Program.stack = 0x1000;
```

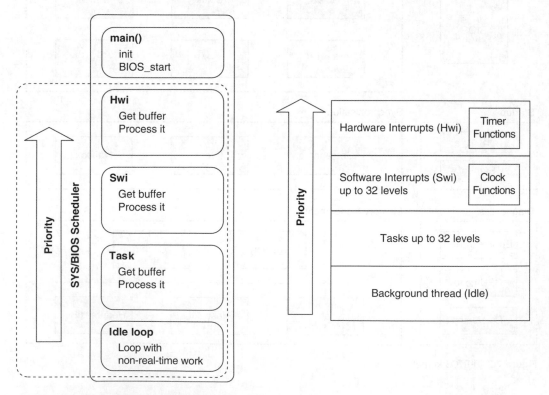

Figure 7.3 Various threads available for the TI-RTOS.

Hwi service routines should be used when the response time is critical. For instance, data may be overwritten if the deadline is not met or a critical function should be completed before any other thread could resume. The use of Hwis is suggested for a deadline range of 5 microseconds. This corresponds to a frequency of 200 kHz.

7.3.1.1 Setting an Hwi

Before setting an interrupt, the user can start writing the Interrupt Service Routine (ISR). For instance, the user can create an ISR called **myHWI()** as shown here:

```
void myHWI (UArg arg0, UArg arg1)
{
System_printf ("I am processing myHWI \n");
Semaphore_post (semaphore0);
}
```

The next step is to associate the ISR function with a particular interrupt, so that, when that interrupt occurs, the ISR is called. This can be configured either statically or dynamically. Static configuration has the advantage of minimising users' errors and is much faster to set up and modify.

The example below shows how to associate Timer 0 to an ISR and also set some other functionality of the dispatcher and the stack management, using a static configuration. For dynamic configuration, please refer to Ref. [1].

To set a hardware interrupt, the following steps are required:

1) Open the configuration file *.cfg, and select **Hwi** as highlighted in Figure 7.4.
2) Select ☑ **Add the portable Hwi management module to my configuration**, and fill the appropriate functions for the dispatcher and the stack management, then select instance Module Instance Advanced, as shown in Figure 7.5.
3) For each ISR, use a handle, type the ISR function name and select the interrupt number (see Chapter 6, 'The TMS320C66x interrupts'). For the TMS320C6xx Hwis, priorities are fixed and therefore any number typed will be ignored. The **Event Id** is the actual interrupt source. 64 corresponds to internal Timer 0; see Figure 7.6, Figure 7.7 and Figure 7.8. The user can also select which interrupt to disable when an interrupt is taken, as described further here and shown in Figure 7.6, where:
   MaskingOption_NONE: No interrupts are disabled, and therefore any interrupt can interrupt the current interrupt.
   MaskingOption_ALL: All interrupts are disabled, and therefore no interrupt can interrupt the current interrupt.
   MaskingOption_SELF: Only the current interrupt is disabled.
   MaskingOption_BITMASK: The user can supply interrupt enable masks.
   MaskingOption_LOWER: All current and lower priority interrupts are disabled.

Once the configuration is performed, a configuration script is generated; see Figure 7.7. It is worth noting that the configuration script can be modified manually. However, the changes will not be reflected in the graphical interface.

7.3.1.2 Hwi hook functions

Each thread has a life cycle which occurs during booting, creation, entering, ending or deleting of a thread. The user can optionally insert a function called the *hook function* that can be run during

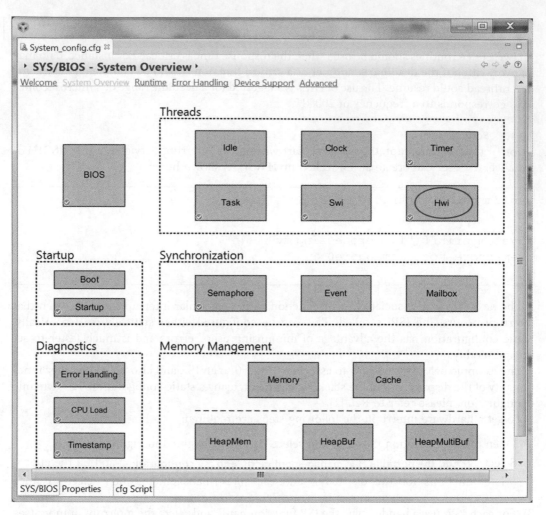

Figure 7.4 SYS/BIOS configuration file.

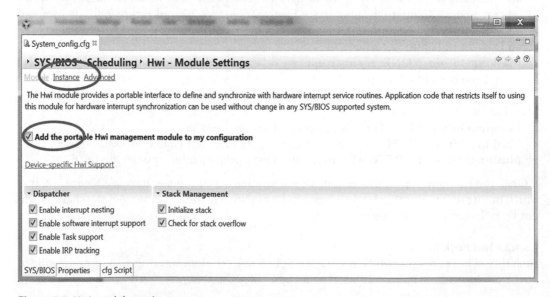

Figure 7.5 Hwi module settings.

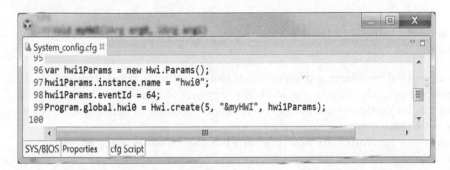

Figure 7.6 Hwi instance settings.

```
95
96 var hwi1Params = new Hwi.Params();
97 hwi1Params.instance.name = "hwi0";
98 hwi1Params.eventId = 64;
99 Program.global.hwi0 = Hwi.create(5, "&myHWI", hwi1Params);
100
```

SYS/BIOS Properties cfg Script

Figure 7.7 Configuration script generated.

64	TINTLn [9]	Local timer interrupt low
65	TINTHn [9]	Local timer interrupt high
66	TINT8L	Timer interrupt low
67	TINT8H	Timer interrupt high
68	TINT9L	Timer interrupt low
69	TINT9H	Timer interrupt high
70	TINT10L	Timer interrupt low
71	TINT10H	Timer interrupt high
72	TINT11L	Timer interrupt low
73	TINT11H	Timer interrupt high
74	TINT12L	Timer interrupt low
75	TINT12H	Timer interrupt high
76	TINT13L	Timer interrupt low
77	TINT13H	Timer interrupt high
78	TINT14L	Timer interrupt low
79	TINT14H	Timer interrupt high
80	TINT15L	Timer interrupt low
81	TINT15H	Timer interrupt high

Figure 7.8 Timers events IDs [1].

```
Hwi.addHookSet({
     registerFxn: '&myRegister1',
     createFxn: '&myCreate1',
     beginFxn: '&myBegin1',
     endFxn: '&myEnd1',
});
```

Figure 7.9 Definition of a hook set.

```
Hwi.addHookSet({
     /*registerFxn: '&myRegister1',*/
     /*createFxn: '&myCreate1', */
     beginFxn: '&myBegin1',
     endFxn: '&myEnd1',
});
```

Figure 7.10 Definition of a hook set with only two elements.

a specific life cycle of a thread; see Figure 7.9. This property of hooking enhances the functionality of the RTOS and the application.

1) A function can be run at boot time and before the **main()** function is called. This is referred to as the **Register** mode. The **Register** mode is the very first to run.
2) A function can run during the creation of an Hwi function (statically or dynamically). This is referred to as the **Create** mode.
3) A function can run just prior to entering an Hwi ISR. This is referred to as the **Begin** mode.
4) A function can run at the end of an ISR. This is referred to as the **END** mode.
5) A function can run after runtime deletion of an Hwi by using the function **Hwi_delete()**. This is referred to as the **Delete** mode.

These hook functions are supported for Hwis, Swis and task objects. The user can create many hook sets as shown in Figure 7.10. However, not all elements of a hook set have to be used. For instance, a user may decide to only use one hook function just before beginning an Hwi and one hook function after finishing an Hwi. In this case, the hook set can be defined as shown in Figure 7.10.

Figure 7.11 (configuration file) and Figure 7.12 (C code) show examples with two hook sets where not all functions are used. For instance, in Hook set 1, **createFxn** is not used. Figure 7.13 shows that the program counter is pointing to the **main()** function, and Figure 7.14 shows that the two register functions **myRegister1_HWI()** and **myRegister2_HWI()** have run before the application reached the function **main()**.

By setting a breaking point in the Hwi (see Figure 7.15), setting the breakpoint count to 3 (as shown in Figure 7.16) and running the code, the output shown in Figure 7.17 reveals that the hook function **myBegin1()** runs before the Hwi function **myHWI()**, and the hook function **myEnd1()** runs after the Hwi function **myHWI()** has completed. The black arrows shown in Figure 7.17 indicate the location of the **myBegin1()** and **myEnd1()** functions.

```
/* Define and add two Hwi HookSets
 * Notice, no deleteFxn is provided.
 */
var Hwi = xdc.useModule('ti.sysbios.hal.Hwi');
/* Hook Set 1 */
Hwi.addHookSet({
      registerFxn: '&myRegister1_HWI',
      /*createFxn: '&myCreate1',*/
      beginFxn: '&myBegin1',
      endFxn: '&myEnd1',
});

/* Hook Set 2 */
Hwi.addHookSet({
      registerFxn: '&myRegister2_HWI',
      /*createFxn: '&myCreate1',
beginFxn: '&myBegin1',
endFxn: '&myEnd1',*/
});
```

Figure 7.11 Configuration code setting two Hwi hook sets.

```
/* Hwi HOOK functions setup*/

/* ======== myRegister1 ========
* invoked during Hwi module startup before main()
* for each HookSet */
Void myRegister1_HWI(Int hookSetId)
{
System_printf("This is the Hwi myRegister1_HWI  before reaching main: assigned
hookSet Id = %d\n",
hookSetId);
//myHookSetId1 = hookSetId;
}

/* ======== myRegister2 ========
* invoked during Hwi module startup before main()
* for each HookSet */
Void myRegister2_HWI(Int hookSetId)
{
System_printf("This is the Hwi myRegister2_HWI  before reaching main: assigned
hookSet Id = %d\n",
hookSetId);
}

/* ======== myBegin1 ========
* invoked before Timer Hwi func */
Void myBegin1(Hwi_Handle myHWI)
{
System_printf("myBegin1:\n");
}

/* ======== myEnd1 ========
* invoked after Timer Hwi func */
Void myEnd1(Hwi_Handle myHWI)
{
System_printf("myEnd1\n");
}
```

Figure 7.12 C code defining the hook functions.

```
38 /*
39 *    ======== main ========
40 */
41 Void main()
42 {
43
44      Clock_start(clock1);
45
46      BIOS_start();
47 }
```

Figure 7.13 Program counter in **main()**.

Module	OutputBuffer	Raw
entry		
This is the HWI myRegister1_HWI before reaching main: assigned hookSet Id = 0		
This is the HWI myRegister2_HWI before reaching main: assigned hookSet Id = 1		

Figure 7.14 Output showing **myRegister1_HWI()** and **myRegister2_HWI()** run before the application reaches **main()**.

```
198 Void myHWI(UArg arg0, UArg arg1) {
199     int EventId;
200
201     System_printf("I am processing myHWI \n");
202     Semaphore_post(semaphore0);
203
204     System_printf("the semaphore0 is    = %d \n",  Semaphore_getCount(semaphore0));
205
206     EventId= Hwi_getEventId(5);  // You must add this: #include <ti/sysbios/family/c64p/Hwi.h>
207     // as it is not supported by the C66
208     System_printf("the EventId is    = %d \n",  EventId);
209
210 }
211
```

Figure 7.15 Breakpoint set in **myHWI**.

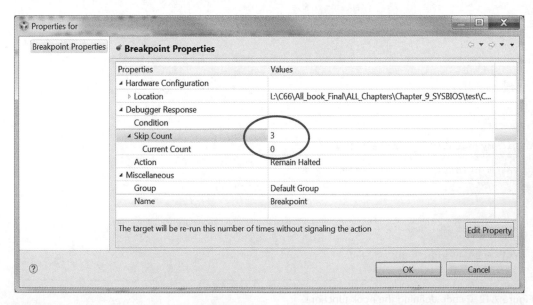

Figure 7.16 Setting the breakpoint counter to 3.

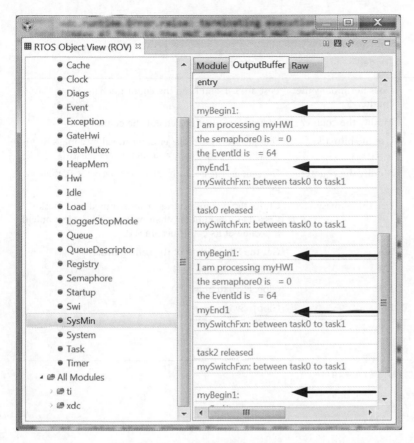

Figure 7.17 Output showing the hook functions running before and after the Hwi.

The complete project can be found in:

\Chapter_7_Code\Events\CLK_SWI_TASK_HWI_with_Hook

7.3.2 Software interrupts (Swis), including clock, periodic or single-shot functions

Swis are triggered by software application programming interfaces (APIs). This allows Hwis to defer less critical functions to a lower priority thread so that any Hwis will not be delayed.

Use Swis when data dependency is relaxed. Data should be ready before posting an Swi. The use of an Swi is suggested for a deadline range of 100 microseconds. This corresponds to a frequency of 10 kHz.

When memory size is a constraint, use Swis as they all share the same stack. Swis have priorities between those of Hwis and task priorities. The various APIs for manipulating a Swi are described in Table 7.1.

7.3.3 Tasks

Use tasks when you have functions with complex interdependency and data sharing. Task objects are designed to wait/pend for a signal (semaphores) before starting or resuming execution, and each task has its own stack and therefore consumes more memory than Swis. While

Table 7.1 Swi APIs

Swi APIs (posting condition)	Trigger	Swi description
Swi_post() (always post)	Does not modify the counter	Post an Swi, and keep the count unchanged.
Swi_inc() (always post)	Modify the counter.	Post an Swi, then increment the count.
Swi_or() (always post)	Use the bitmask.	Sets the bits in the trigger determined by a mask that is passed as a parameter, and then posts the Swi.
Swi_dec() (if count becomes zero)	Modify the counter.	Decrement the count if the count becomes 0, then post an Swi.
Swi_andn() (if count becomes zero)	Use the bitmask.	Clears the bits in the trigger determined by a mask passed as a parameter, then posts an Swi object only if the value of its count becomes 0.
Swi_getPri()	NA	Get the Swi priority of the calling Swi.
Swi_enable()	NA	Global Swi enable
Swi_disable()	NA	Global Swi disable
Swi_restore()	NA	Global Swi restore

```
void task0Fxn(UArg arg0, UArg arg1)
{
    System_printf("entering task0 epilog \n",

            while (1){

                // wait here for the semaphore
                Semaphore_pend(semaphore0,  BIOS_WAIT_FOREVER );

                /* start process from here when the semaphore is available*/
                System_printf("task0 unblocked\n");

            }

    System_printf("entering task0 epilog \n");

}
```

Figure 7.18 A task structure.

Hwis and Swis eventually run to completion and may or may not be called again, a task is designed to run in a loop, as shown in Figure 7.18. There are 32 levels of priorities for the tasks. These can be set by the user, and the default number of tasks is 16. The number 16 has been chosen in order to be compatible with other processors. Figure 7.19 shows how to set the number of priorities.

A task can be in one of four possible states or modes of execution, as shown in Figure 7.20.

Task_Mode_RUNNING. The CPU is executing the task.
Task_Mode_READY. The task is waiting for its turn to run.
Task_Mode_BLOCKED. The task is waiting for an event before running.
Task_Mode_TERMINATED. The task is 'terminated' and will not execute again.

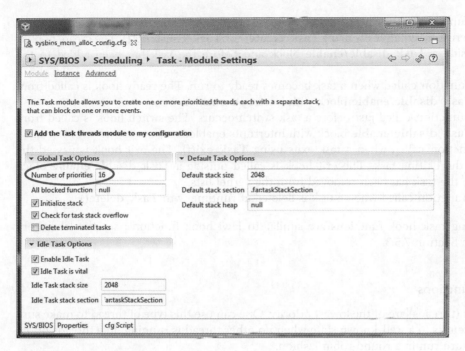

Figure 7.19 Setting the number of priorities for the tasks.

```
typedef enum Task_Mode {

Task_Mode_RUNNING// Task is currently executing
Task_Mode_READY,// ready to run
Task_Mode_BLOCKED,// waiting for signal
Task_Mode_TERMINATED,// terminated
Task_Mode_INACTIVE // set to be inactive
} Task_Mode;
```

Figure 7.20 Task modes.

Task_Mode_INACTIVE. The task is inactive. The user can set a task to be inactive so that it does not run after being created. The task is inactive when its priority is equal to −1 and is in a pre-ready state. By changing its priority, this task can be put in **Task_Mode_Ready**.

The priority of a task can be changed at runtime by using the **Task_setPri()** API.

7.3.3.1 Task hook functions
Similar to Hwi modules, task modules also have hook functions [1]:

1) Register. A function called before any statically created tasks are initialised at runtime. The register hook is called at boot time before the **main()** function and before interrupts are enabled.

2) Create. A function called when a task is created. This includes tasks that are created statically and those created dynamically using **Task_create()** or **Task_construct()**. The create hook is called outside a **Task_disable/enable** block and before the task has been added to the ready list.
3) Ready. A function called when a task becomes ready to run. The ready hook is called from within a **Task_disable/enable** block with interrupts enabled.
4) Switch. A function called just before a task switch occurs. The switch hook is called from within a **Task_disable/enable** block with interrupts enabled.
5) Exit. A function called when a task exits using **Task_exit()**. The exit hook is passed the handle of the exiting task. The exit hook is called outside a **Task_disable/enable** block and before the task has been removed from the kernel lists.
6) Delete. A function called when a task is deleted at runtime with **Task_delete()**.

Programming task hook functions are similar to Hwi hook functions; see the laboratory experiment in Section 7.5.3.

7.3.4   Idle functions

Idle functions (threads) are of the lowest priority. One can use this type of thread to make sure that your system is in a well-known state when no other thread is running. Idle threads of the same priority are run in a round-robin fashion.

An idle thread runs until:

1) It relinquishes control.
2) It is pre-empted by a higher priority thread.
3) Or it has consumed its time slice when running in a round-robin fashion.

An idle task should never be made to block.

7.3.5   Clock functions

Use Clock functions when you want a function to run at a rate based on a multiple of the interrupt rate of the peripheral that is driving the clock tick. Clock functions can be configured to execute either periodically or just once (single shot). These functions run as Swi functions.

7.3.6   Timer functions

Timer functions are run within the context of Hwi threads and have the priority of the timer interrupts. These threads run as Hwi functions.

7.3.7   Synchronisation

For an OS, there are two requirements for synchronisation: one is synchronising threads and the other is synchronising access to resources. In this chapter, semaphores and events are described. Gates, mailbox and queues can also be used.

7.3.7.1 Semaphores

Semaphores are simply variables that the OS uses for synchronising tasks. The semaphores (variables) can be either binary or positive integers, and the APIs used are the same for both types. However, binary semaphores are more time efficient than counting semaphores.

As seen in Section 7.3.3 and Figure 7.18, the tasks are designed to be synchronised by semaphores. The synchronisation is required for sharing resources.

The semaphores are easy to use, and there are two main semaphore APIs that can be used for synchronising tasks: **Semaphore_pend** and **Semaphore_post**.

7.3.7.2 Semaphore_pend

If the semaphore count is greater than zero (which means a resource is available), the **Semaphore_pend()** decrements the count and returns TRUE. If the semaphore count is zero (unavailable), this function suspends execution of the current task until **post()** is called or the timeout expires.

A timeout value of **BIOS_WAIT_FOREVER** causes the task to wait indefinitely for its semaphore to be posted. A timeout value of **BIOS_NO_WAIT** causes **Semaphore_pend()** to return immediately.

The **Semaphore_pend()** API can be used as follows:

```
        //wait here for the semaphore
Semaphore_pend(semaphore0, BIOS_WAIT_FOREVER);
```

7.3.7.3 Semaphore_post

Semaphore_post() is used to increment the count and therefore is used to signal the availability of a resource. When the semaphore is used, it readies the first task waiting for the semaphore. If no task is waiting, this function simply increments the semaphore count and returns.

The **Semaphore_post()** API can be used as follows:

```
//post/increment the semaphore counter
Semaphore_post(semaphore0);
```

7.3.7.4 How to configure the semaphores

Figure 7.21, Figure 7.22 and Figure 7.23 are self-explanatory for setting a semaphore.

7.3.8 Events

Events are similar to semaphores. However, they have the added advantage of allowing multiple conditions to happen before a waiting thread can be released.

To set an event, follow the instructions as shown from Figure 7.24, Figure 7.25 and Figure 7.26. The generated configuration is shown in Figure 7.27.

Figure 7.28 shows how to make a task pend for two events. These events can be posted as shown in Figure 7.29. See the laboratory experiment in Section 7.5.4.

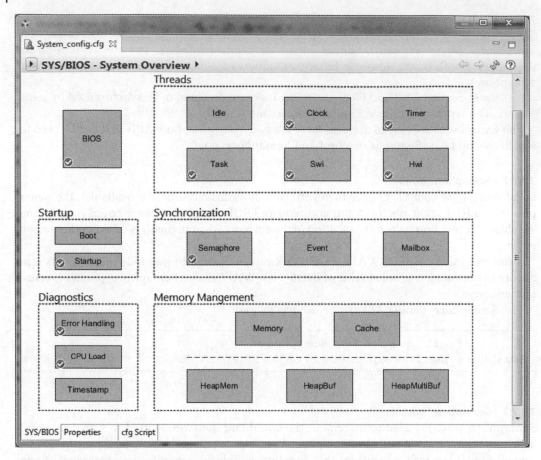

Figure 7.21 Selecting semaphores for setups.

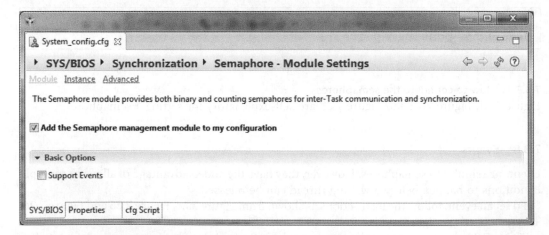

Figure 7.22 Selecting a semaphore module.

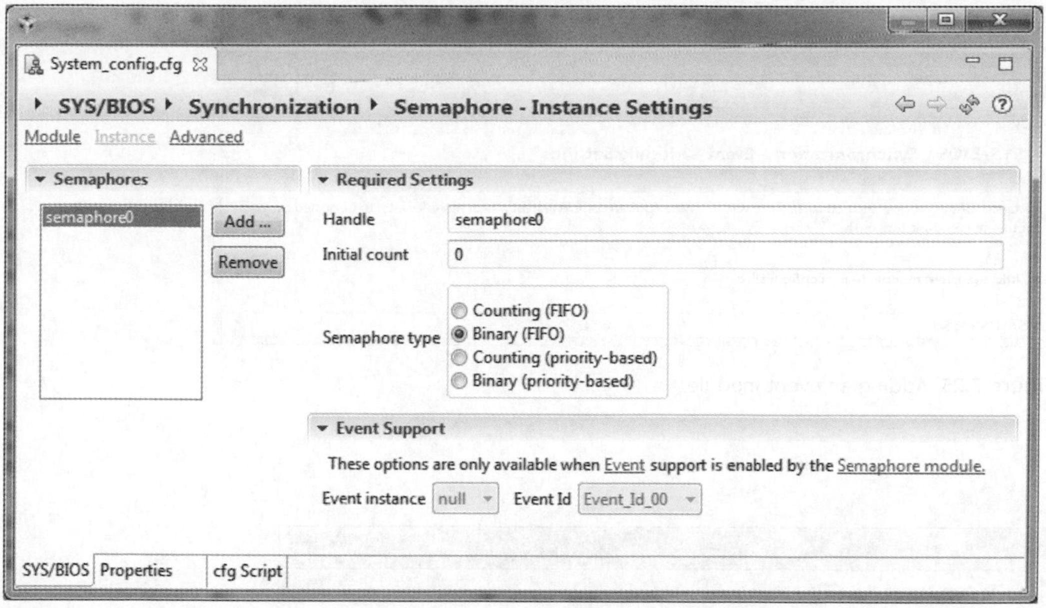

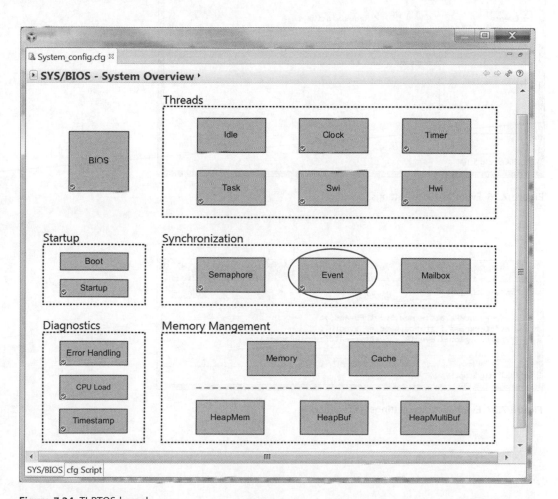

Figure 7.23 Instance settings.

Figure 7.24 TI-RTOS kernel.

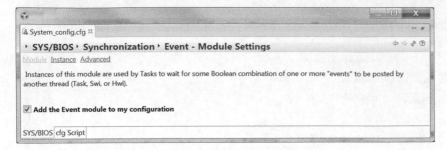

Figure 7.25 Adding an event module.

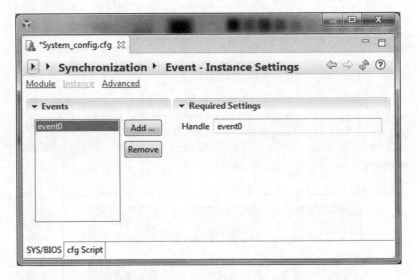

Figure 7.26 Event instance settings.

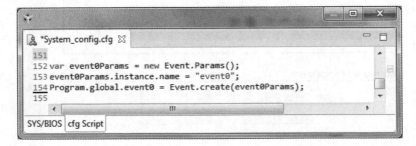

Figure 7.27 Event instance settings generated.

```
/*
 *   ======== task0Fxn ========
 */

// UInt Event_pend(Event_Handle handle, UInt andMask, UInt orMask, UInt timeout);

void task0Fxn(UArg arg0, UArg arg1)
{

    UInt all_events;

    while (1){

        /* wait for (Event_Id_00 & Event_Id_01) */
        all_events = Event_pend(event0,
                Event_Id_00 + Event_Id_01,   /* andMask */
                NULL,                        /* orMask, not used*/
                BIOS_WAIT_FOREVER);

        System_printf("task0 unblocked and all events occured\n");
    }

}
```

Figure 7.28 A task synchronised by events.

```
void myHWI(UArg arg0, UArg arg1) {
    int EventId;



    Event_post(event0, Event_Id_00);

    Event_post(event0, Event_Id_01);

}
```

Figure 7.29 Hwi posting two events.

7.3.9 Summary

A comparison of thread characteristics for the KeyStone devices is shown in Table 7.2.

7.4 Dynamic memory management

Dynamic memory allocation must be avoided as it consumes a large number of cycles and it can be non-deterministic. However, for an embedded system when memory is scarce, dynamic memory allocation is inevitable.

There are two ways to use dynamic memory allocation; one is by using the stack allocation, and the other is by using the heap allocation.

Table 7.2 Comparison of thread characteristics for the KeyStone devices [1]

Characteristic	Hardware interrupt	Software interrupt	Task	Idle
Priority	Highest	2nd highest	2nd lowest	Lowest
Number of priority levels	Family/device specific	Up to 32	Up to 32	1
Can yield and pend	No; runs to completion except for pre-emption	No; runs to completion except for pre-emption	Yes	Should not pend. Pending would disable all registered idle threads.
Execution states	Inactive, ready, running	Inactive, ready, running	Ready, running, blocked, terminated	Ready, running
Thread scheduler disabled by	**Hwi_disable()**	**Swi_disable()**	**Task_disable()**	Program exit
Posted or made ready to run by	Interrupt occurs	**Swi_post()**, **Swi_andn()**, **Swi_dec()**, **Swi_inc()**	**Task_create()** and various **Swi_or()** task synchronisation mechanisms (events, semaphores, mailboxes)	**main()** exits and no other thread is currently running.
Stack used	System stack (1 per program)	System stack (1 per program)	Task stack (1 per task)	Task stack used by default; see Note (1) at bottom of table.
Context saved when pre-empts other thread	Entire context minus saved-by callee registers (as defined by the TI C compiler) is saved to system.	Certain registers saved to system	Entire context saved to task stack	NA
Context saved when blocked	NA	NA	Saves the saved-by callee registers (see the optimising compiler user's guide for your platform)	NA
Share data with thread via	Streams, lists, pipes, global variables	Streams, lists, pipes, global variables	Streams, lists, pipes, gates, mailboxes, message queues, global variables	Streams, lists, pipes, global variables
Synchronise with thread via	NA	Swi trigger	Semaphores, events, mailboxes	NA
Function hooks	Yes: register, create, begin, end, delete	Yes: register, create, ready, begin, end, delete	Yes: register, create, ready, switch, exit, delete	No
Static creation	Yes	Yes	Yes	Yes
Dynamic creation	Yes	Yes	Yes	No
Dynamically change priority	See Note (2) at bottom of table.	Yes	Yes	No
Implicit logging	Interrupt event	Post, begin, end	Switch, yield, ready, exit	None
Implicit statistics	None	None	None	None

Note: (1) If you disable the task manager, idle threads use the system stack. (2) Some devices allow hardware interrupt priorities to be modified.
NA = Not applicable.

7.4.1   Stack allocation

The stack is simple to use as it is managed by the OS and no intervention is required by the programmer. However, the stack allocation is not practical when the total amount of memory used by an application does not fit in the memory available. It is useful at this stage to know that the stack memory known as the *system stack* is located in the **.stack** section.

The C/C++ C6000 compiler uses the stack to:

- Save function return addresses.
- Allocate local variables.
- Pass arguments to functions.
- Save temporary results.

7.4.2   Heap allocation

The heap is a memory region that is mainly controlled by the programmer. To allocate memory on the heap, functions like **malloc**, **calloc** and **realloc** can be used. These functions rely on a proper setup on the heap. Sometimes different heaps are required in order to improve the memory usage. To achieve this, the SYS/BIOS provides the heap modules. The heap modules are dynamic memory managers that manage specific memories. Memory can be allocated from a global pool or heap that is defined in the **.sysmem** section.

7.4.3   Heap implementation

Selecting the right heap location is very important, as discussed in Section 7.4.2. However, how the heap is implemented is another issue that must be considered when speed is important. The SYS/BIOS offers four different implementations.

7.4.3.1   *HeapMin* implementation

HeapMin implementation provides a very small footprint. However, the memory allocated cannot be freed, and therefore it is not applicable for applications that require memory swapping.

7.4.3.2   *HeapMem* implementation

HeapMem implementation can provide both memory allocation and deallocation. This implementation is flexible as the memory allocated can be of any size and also provides memory protection by using **Gates** (see Chapter 9, 'Inter-Processor Communication'). However, this implementation is not deterministic as the search for free memory has to go through the entire linked list where the free memory is stored. This implementation also may cause memory leaks as the freed memories appear in different locations and therefore contiguous memory allocation may not be available. Figure 7.30 shows an example of memory leak. In Figure 7.30a, memory has been allocated and 4 k of memory is left. However, in Figure 7.30b, 8 k was freed and if any memory allocation requests more than 4 k of memory, it will not be allocated since the memory is fragmented. This type of fragmentation is called external.

To set a **HeapMem** object graphically, two steps are required:

1) Open the configuration file as shown in Figure 7.31 and select **HeapMem**.
2) Complete the instance setting as shown in Figure 7.32 or Figure 7.33. Notice that the heap used, **myHeap**, is defined somewhere; see Figure 7.35.

(a)

(b)

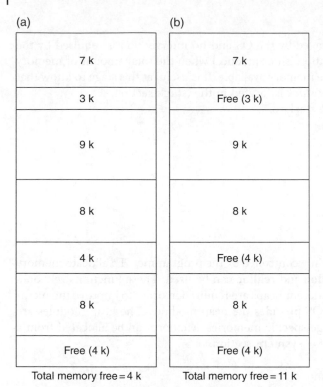

Total memory free = 4 k

Total memory free = 11 k

Figure 7.30 Example of an external memory fragmentation.

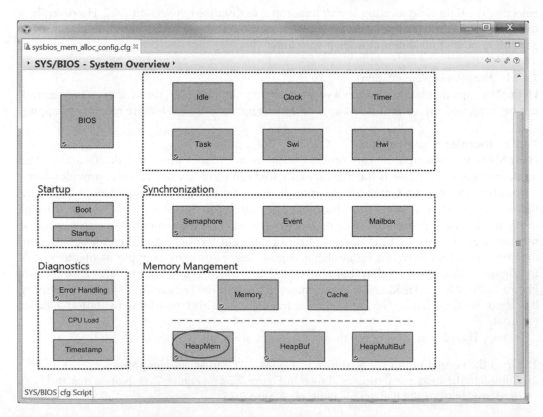

Figure 7.31 Setting the **HeapMem**.

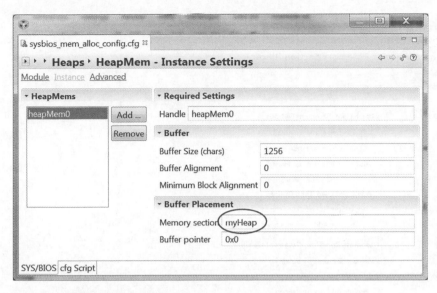

Figure 7.32 HeapMem instance settings.

```
73 var heapMem0Params = new HeapMem.Params();
74 heapMem0Params.instance.name = "heapMem0";
75 heapMem0Params.size = 1256;
76 heapMem0Params.sectionName = "myHeap";
77 Program.global.heapMem0 = HeapMem.create(heapMem0Params);
78
```

Figure 7.33 Code generated from Figure 7.32.

To use **HeapMem**, one can use the following code:

```
/*Alloc and free using another heap: heapMem) */
buf2 = Memory_alloc(heapMem0, 128, 0, &eb);
```

heapMem0 specifies the use of **HeapMem**. The heap itself is specified in **myHeap**.

If a **NULL** is specified as shown below, the default heap will be used. The default heap is defined as shown in Figure 7.35. Figure 7.34 shows how to link a section to a physical memory.

```
/*Alloc and free using the default Heap */
buf1 = Memory_alloc(NULL, 128, 0, &eb);
```

7.4.3.3 *HeapBuf* implementation

HeapBuf is designed to allocate memory from fixed blocks. **HeapBuf** implementation is fast and deterministic since the search for a new block is easier. If all memory allocations are of the same size, the fragmentation will not occur as illustrated in Figure 7.36a and 7.36b. Since all blocks are of the same size, allocation of a new block can fit in any place.

If the application requires similar block sizes, use of **HeapBuf** can be a good choice. However, in a practical situation, not all memory allocations require blocks of the same size; this may cause memory leaks as each block may contain unused memory, as shown in Figure 7.36c.

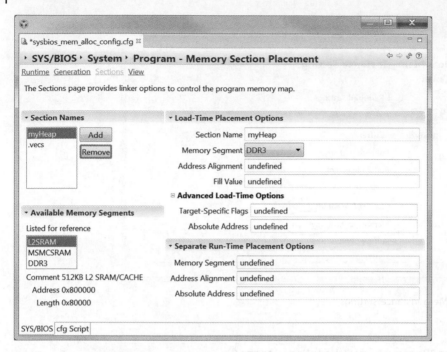

Figure 7.34 Setting **myHeap** section to be in the DDR.

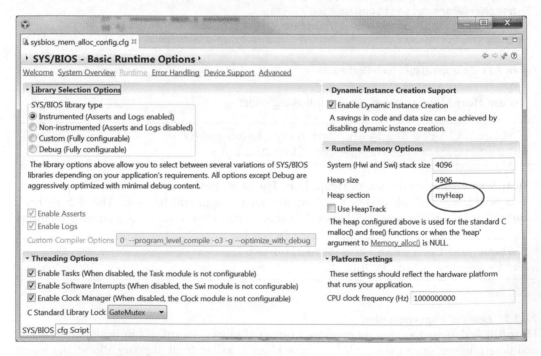

Figure 7.35 Setting the default heap.

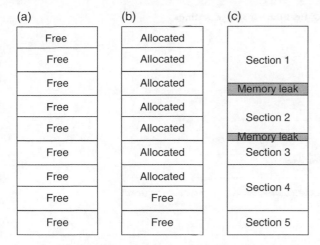

Figure 7.36 HeapBuf with fixed blocks.

Allocation from and freeing to a **HeapBuf** instance are non-blocking and always take the same time (deterministic). The drawback of the **HeapBuf** is that all the buffers have to be of the same size. To remedy this problem, make use of the **heapMultiBuf** described below. The configuration of **HeapBuf** is shown in Figure 7.37 through Figure 7.42.

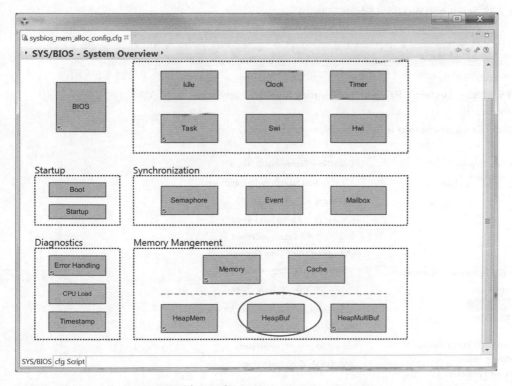

Figure 7.37 Selecting the **HeapBuf** for configuration.

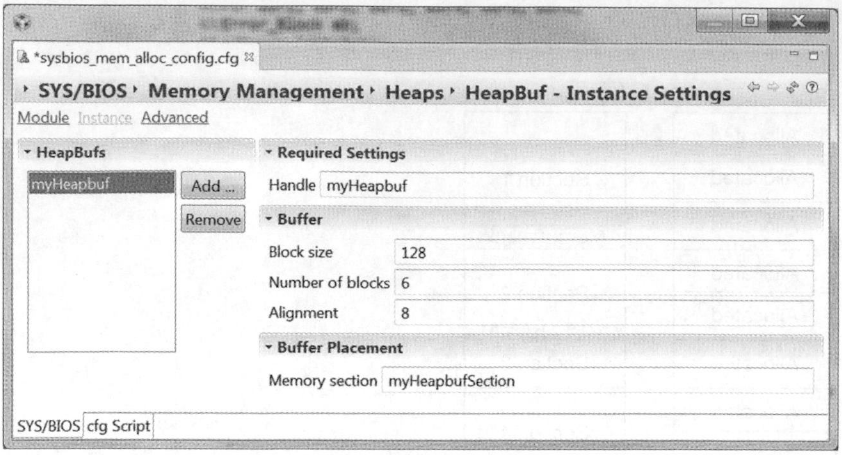

Figure 7.38 Configuration of the **HeapBuf**.

```
/* Create a heap using HeapBuf */
var heapBufParams = new HeapBuf.Params;
heapBufParams.blockSize = 128;
heapBufParams.numBlocks = 6;
heapBufParams.align = 8;
heapBufParams.sectionName = "myHeapbufSection";
heapBufParams.instance.name = "myHeapbuf";
Program.global.myHeapbuf = HeapBuf.create(heapBufParams);
```

Figure 7.39 Script obtained from Figure 7.38.

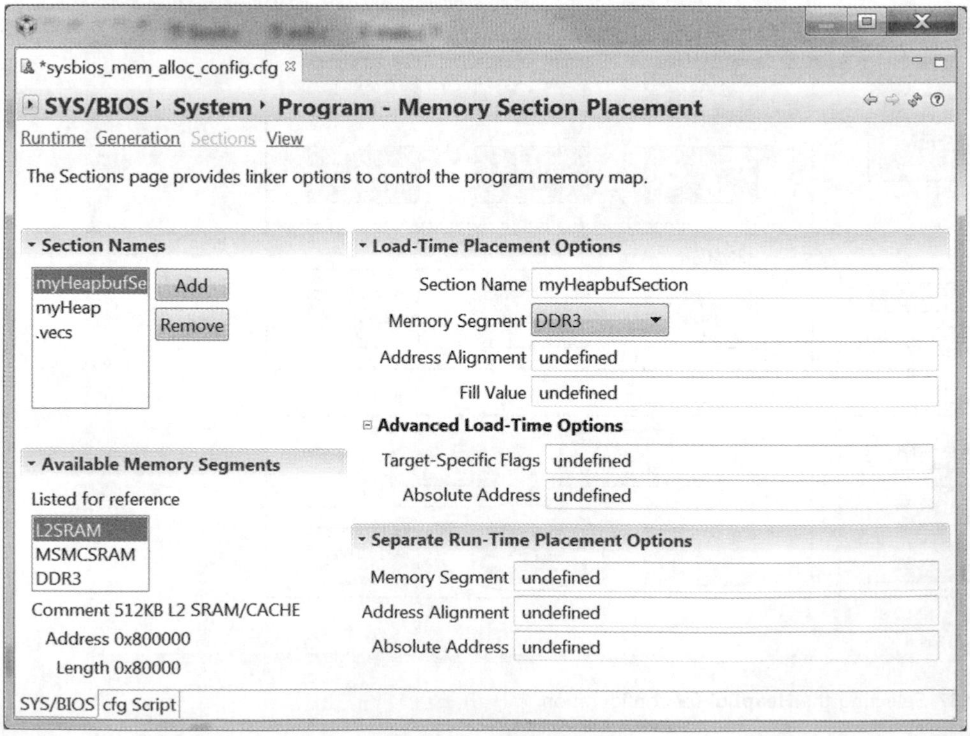

Figure 7.40 **myHeapSection** allocation in DDR3.

```
Program.sectMap["myHeapbufSection"] = new Program.SectionSpec();

Program.sectMap["myHeapbufSection"].loadSegment = "DDR3";
```

Figure 7.41 Code generated from Figure 7.40.

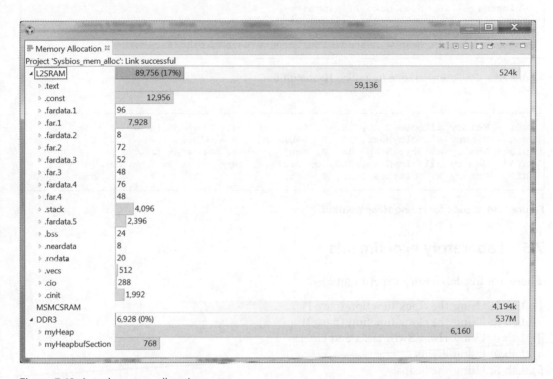

Figure 7.42 Actual memory allocation.

myHeapBufSection has been allocated in the DDR3 as shown in Figure 7.40 and Figure 7.41.

To verify the locations of sections, one can use the **View > Memory** allocation as shown in Figure 7.42.

7.4.3.4 *HeapMultiBuf* implementation

HeapMultiBuf extends the capability of the **HeapBuf** by providing multiple **HeapBuf** with different block sizes, alignments and number of blocks. In this case, the application asks for any block size (from the blocks selected) and the **HeapMultiBuf** will take care of allocating a block of memory from the appropriate block.

For instance, the user can create three buffers (**buf_1**, **buf_2** and **buf_3**) with **buf_1** containing two blocks of 16 bytes, **buf_2** containing one block of 32 bytes and **buf_3** containing one block of 128 bytes. The user can then ask for any buffer size (16, 32 or 128 bytes). If the user then asks for three buffers of 16 bytes (and we have only two), then there is the possibility to borrow from the higher block size (that is, blocks from **buf_2** if available and, if not, it will borrow a block from **buf_3** if available). If no block remains, the allocation will fail. The code illustrating this example is shown in Figure 7.43 and Figure 7.44, and the source code can be found in:

\Chapter_7_Code\Sysbios_mem_alloc

```
var HeapMultiBuf = xdc.useModule('ti.sysbios.heaps.HeapMultiBuf');
/* HeapMultiBuf without blockBorrowing. */
/* Create as a global variable to access it from C Code. */
var heapMultiBufParams = new HeapMultiBuf.Params();
heapMultiBufParams.numBufs = 3;
heapMultiBufParams.blockBorrow = true; this to allow or not allow borrowing
heapMultiBufParams.bufParams =
[{blockSize: 16, numBlocks:2, align: 0},
{blockSize: 32, numBlocks:1, align: 0},
{blockSize: 128, numBlocks:1, align: 0}];
Program.global.myHeap = HeapMultiBuf.create(heapMultiBufParams);
```

Figure 7.43 Configuration code for testing **HeapMultiBuf**.

```
Buf1 = Memory_alloc(myHeap, 16, 0, &eb); // take a buffer from buf_1
Buf2 = Memory_alloc(myHeap, 16, 0, &eb); /// take a buffer from buf_1
Buf3 = Memory_alloc(myHeap, 16, 0, &eb); // take a buffer from buf_2
Buf4 = Memory_alloc(myHeap, 128, 0, &eb); // take a buffer from buf_3
buf5 = Memory_alloc(myHeap, 32, 0, &eb); // this will fail as buffers are used
```

Figure 7.44 C code for testing **HeapMultiBuf**.

7.5 Laboratory experiments

There are five laboratory experiments:

1) Lab 1: Using the clock function (part 1)
2) Lab 2: Using the clock function (part 2)
3) Lab 3: Using Hwis, Swis, tasks and clocks
4) Lab 4: Using events
5) Lab 5: Using the heaps.

7.5.1 Lab 1: Manual setup of the clock (part 1)

Build, run and analyse the project **clock1** located in:

\Chapter_7_Code\clock1

In **clock.c**, the **clk0Fxn** is called after a certain time **mytimeout**. In this case, the **clk0Fxn** will be called for the first time after two ticks. The clock period is five ticks; see Figure 7.45.

The output should be as shown in Figure 7.46. Notice that the **clk0Fxn** starts after two cycles and runs after every five cycles.

7.5.2 Lab 2: Manual setup of the clock (part 2)

Export the **clock1** project and rename the exported project as **clock2**. Build and run the **clock2** project to make sure the project is working properly.

Modify **clock.c**, so that the **clk0Fxn** will be called for the first time after five ticks.

Add another clock (**clk1**) as a one-shot clock instance with **mytimeout** = 21 ticks, which forces the SYS/BIOS to exit.

The output should be as shown in Figure 7.47.

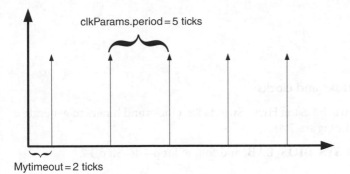

Figure 7.45 Timing of the **clk0Fxn** function.

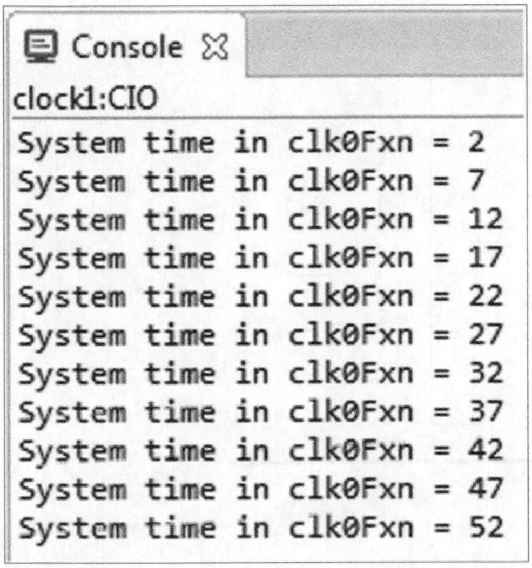

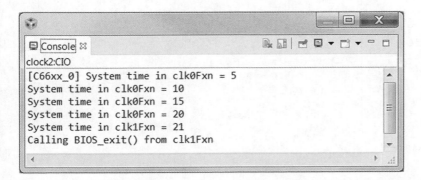

Figure 7.46 Console output.

Figure 7.47 Output of project.

The solutions can be found in:

\Chapter_7_Code\Solutions\clock2

7.5.3 Lab 3: Using Hwis, Swis, tasks and clocks

In this laboratory experiment, you will be using Hwis, Swis, tasks, clocks and timers to generate a sequence of events that is shown in Figure 7.48.

Procedure to follow. Open project **SYS_BIOS_LAB**, and follow Step 1 to Step 14.

1) In **SWI_HWI_TASK_CLK.c**, uncomment lines 27 to 29 as shown in Figure 7.49.
2) Create a software interrupt 0 (**swi0**) and a software interrupt 1 (**swi1**) as shown in Figure 7.50, Figure 7.51 and Figure 7.52.

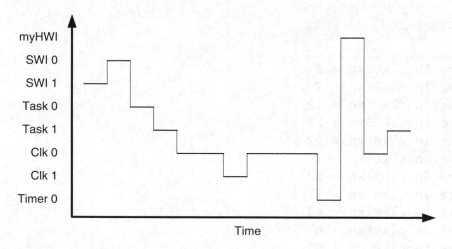

Figure 7.48 Sequence of events required.

```
23 *    ======== main ========
24 */
25 Void main()
26 {
27 //    Swi_post(swi0);
28 //    Swi_post(swi1);
29 //    Clock_start(clk1);
30
31      BIOS_start();
32 }
33
```

Figure 7.49 Code to uncomment.

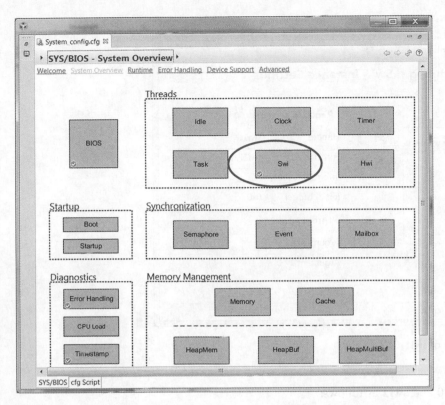

Figure 7.50 Select the Swi for configuration.

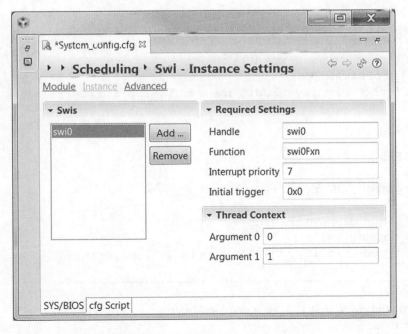

Figure 7.51 Setting of **swi0**.

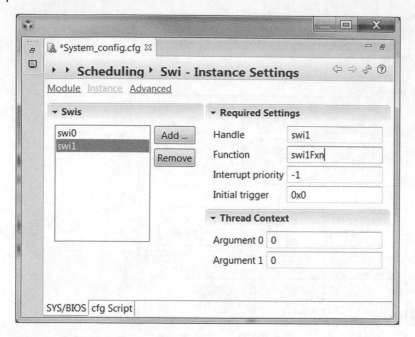

Figure 7.52 Setting of **swi1**.

3) Initialise **swi0** and **swi1** as follows:

```
/*
 * ======== swi0Fxn ========
 */
Void swi0Fxn(UArg arg0, UArg arg1)
{  UInt priority;
   System_printf("Running swi0Fxn\n");
   priority= Swi_getPri(swi0) ;
   System_printf("the Priority of SWI0 is  = %d \n", priority);
}

/*
 * ======== swi1Fxn ========
 */
Void swi1Fxn(UArg arg0, UArg arg1)
{  UInt priority;
   System_printf("Running swi1Fxn\n");
   priority= Swi_getPri(swi1) ;
   System_printf("the Priority of SWI1 is  = %d \n", priority);
}
```

4) Create clock 0 (**clk0**) and clock 1(**clk1**) using a graphical interface as shown in Figure 7.53, Figure 7.54 and Figure 7.55.
5) Make sure the clock setting is as shown in Figure 7.56.

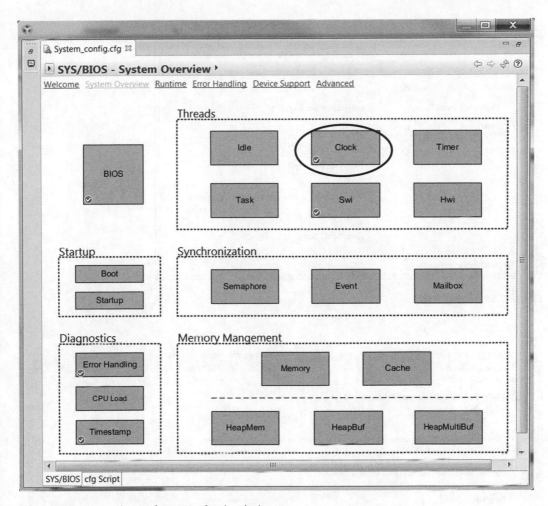

Figure 7.53 Setting the configuration for the clock.

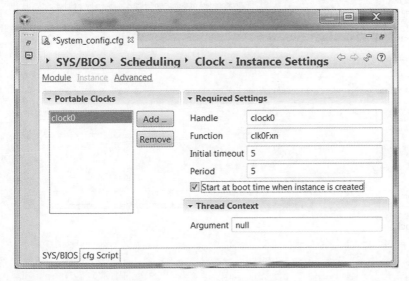

Figure 7.54 Setting **clock0**.

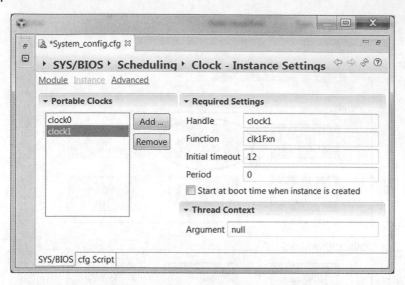

Figure 7.55 Setting **clock1**.

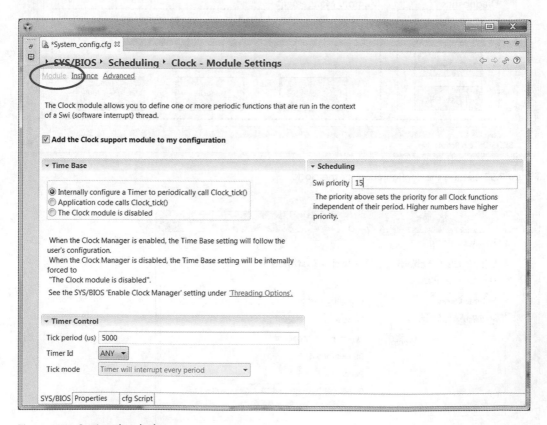

Figure 7.56 Setting the clock.

6) Display **clk0** and **clk1** as follows:

```
/*
 *  ======== clk0Fxn ========
 */
void clk0Fxn(UArg arg0)
{
      System_printf("Running clk0Fxn\n");
}

/*
 *  ======== clk1Fxn ========
 */
void clk1Fxn(UArg arg0)
{
      System_printf("Running clk1Fxn to finish\n");
}
```

7) Create a semaphore (**sem0**) using a graphical interface as shown in Figure 7.57, Figure 7.58 and Figure 7.59.
8) Create task 0 (**task0**) and task 1 (**task1**) as shown in Figure 7.60 to Figure 7.65.

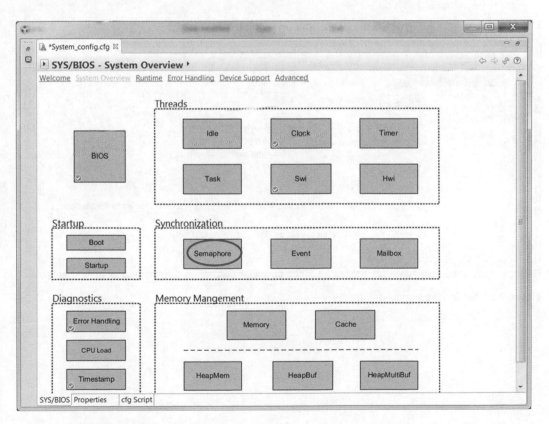

Figure 7.57 Setting the configuration for the semaphore.

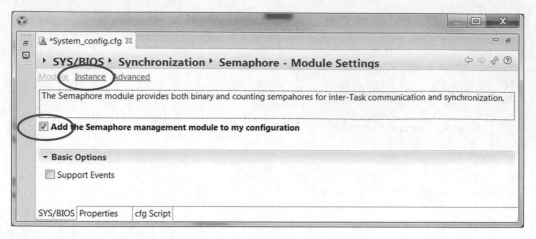

Figure 7.58 Adding a semaphore.

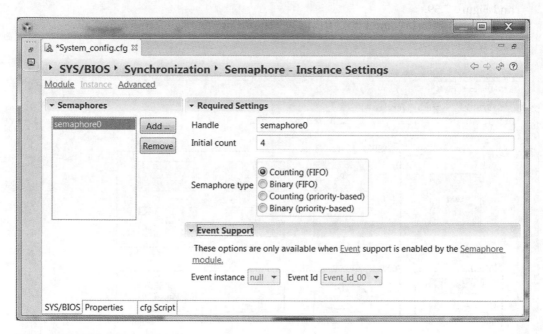

Figure 7.59 Setting the semaphore.

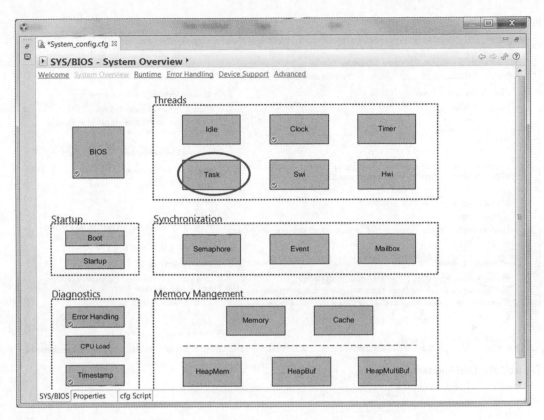

Figure 7.60 Setting the configuration for the tasks.

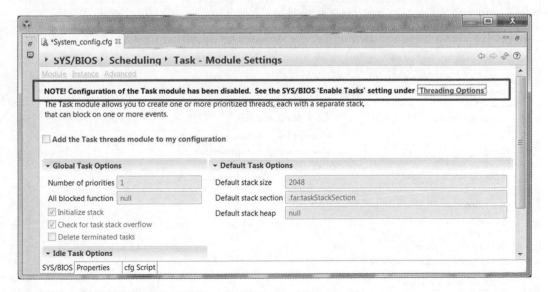

Figure 7.61 Warning that the tasks are not enabled.

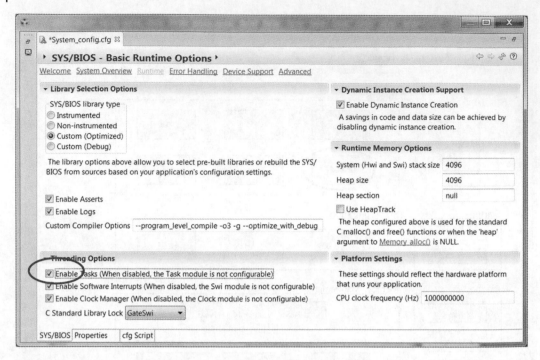

Figure 7.62 Enabling tasks.

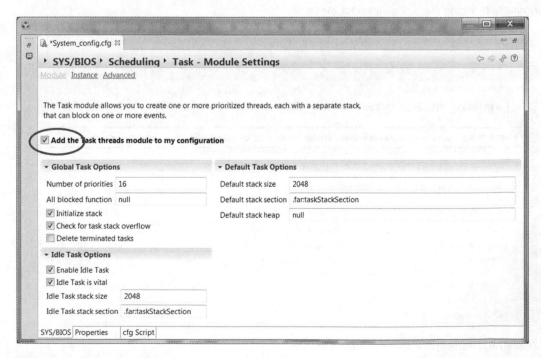

Figure 7.63 Adding threads modules.

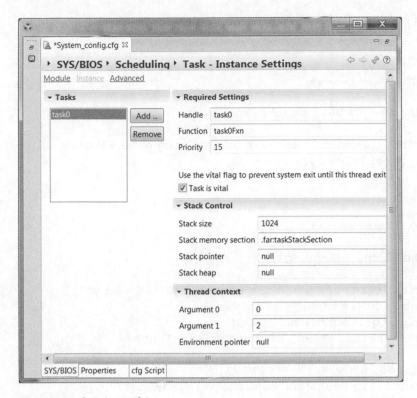

Figure 7.64 Creating **task0**.

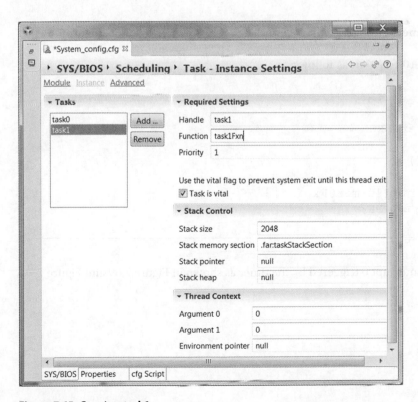

Figure 7.65 Creating **task1**.

9) Initialise the tasks as follows (in **SWI_HWI_TASK_CLK.c**):

```
/*
 *  ======== task0Fxn ========
 */
void task0Fxn(UArg arg0, UArg arg1)
{
      System_printf("Running task0Fxn and setting sem0 =0\n");
      Semaphore_reset(sem0, 0);
}


/*
 *  ======== task1Fxn ========
 */
void task1Fxn(UArg arg0, UArg arg1)
{
      System_printf("in task 1 and the semaphore0 is  = %d \n",
   Semaphore_getCount(sem0));
      if (Semaphore_getCount(sem0) == 0) {
             System_printf("Sem0 blocked in task1\n");
           }
Semaphore_pend(sem0, BIOS_WAIT_FOREVER);
      System_printf("in task1 and semaphore released\n");
}
```

10) Create a timer (**timer0**) using a graphical interface as shown in Figure 7.66, Figure 7.67 and Figure 7.68.
11) Initialise the **timer0** function as follows:

```
/*
 *  ======== timer0Fxn ========
 */
void timer0Fxn()
{
     System_printf("timer0Fxn\n");
     finishFlag = TRUE;
}
```

12) Create hardware interrupt 0 triggered by the timer as shown in Figure 7.69 and Figure 7.70.

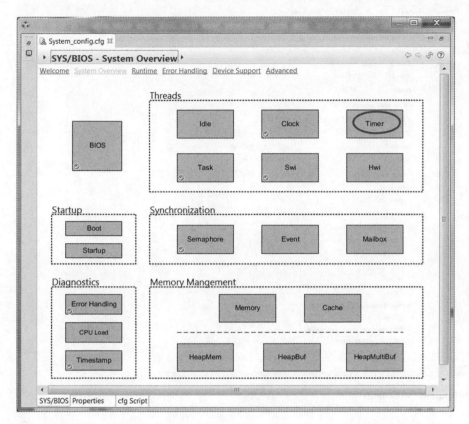

Figure 7.66 Setting the configuration for the timer.

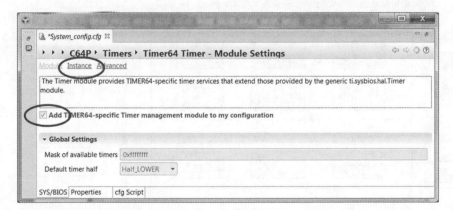

Figure 7.67 Adding a timer instance.

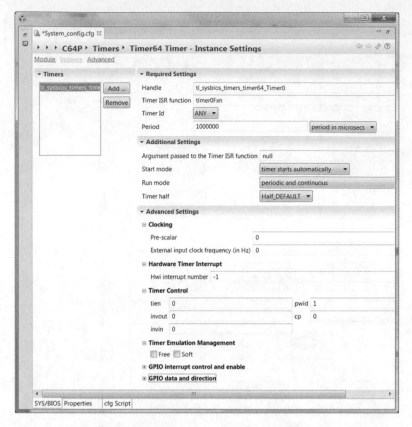

Figure 7.68 Configuring the timer.

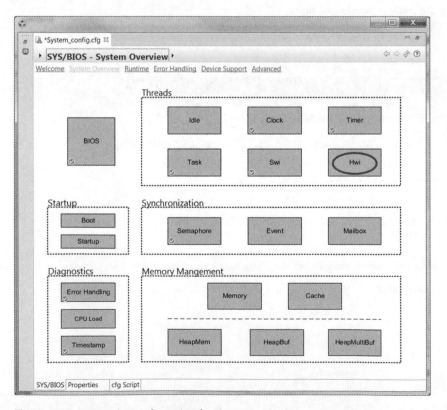

Figure 7.69 Setting the configuration for the Hwi.

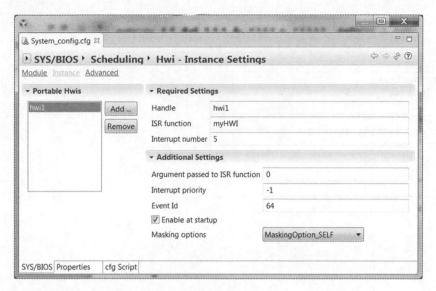

Figure 7.70 Setting **hwi1**.

13) Initialise **hwi0** as follows:

```
void myHWI (UArg arg0, UArg arg1)
{int EventId;
System_printf ("Running myHWI and posting sem0 \n");
Semaphore_post (sem0);
EventId = Hwi_getEventId (5); //You must add this: #include
//<ti/sysbios/family/c64p/Hwi.h>
//as it is not supported by the C66
System_printf ("the EventId is = %d \n", EventId);
}
```

14) Final console printout should be as shown in Figure 7.71.

The solution can be found in:

\Chapter_7_Code\Solutions\CLK_SWI_TASK_HWI

7.5.4   Lab 4: Using events

1) Load and run the project located in:
   \Chapter_7_Code\Event2\CLK_SWI_TASK_HWI_with_Events
2) Explore the **task0Fxn()** in **main.c** file and the **System_config.cfg**.
3) Run the code and notice that **task0Fxn()** was unblocked as shown in Figure 7.72.
4) Uncomment one **Event_post()** as shown in Figure 7.73, and then rebuild, run and explore
   the console output. In this case, the **task0Fxn()** will not be unblocked.

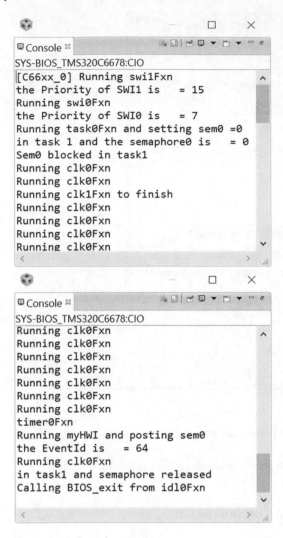

Figure 7.71 Console output.

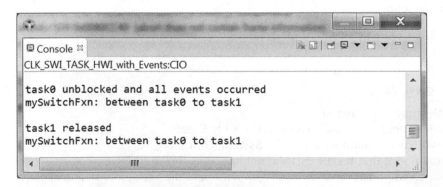

Figure 7.72 Console output when both events occur.

```
*main.c ⊠
197
198 Void myHWI(UArg arg0, UArg arg1) {
199     int EventId;
200
201     System_printf("I am processing myHWI \n");
202     Semaphore_post(semaphore0);
203
204     Event_post(event0, Event_Id_00);
205
206 //  Event_post(event0, Event_Id_01);
207
```

Figure 7.73 How to post an event.

7.5.5 Lab 5: Using the heaps

In this example, **HeapMem**, **HeapBuf** and **HeapMultiBuf** have been implemented.

The file is located in:

\Chapter_7_Code\Sysbios_mem_alloc

Project: **Sysbios_mem_alloc.pjt**.

Rebuild it and explore the **main.c** file and the **sysbios_mem_alloc_config.cfg** file. Explore the buffer locations as shown in Figure 7.74.

Step through the code and verify that the memory buffers are located as expected.

If you uncomment the code between lines 129 and 133 (as shown in Figure 7.75) in order to try to allocate a memory section, an error will be generated since all the memory available has been used; see Figure 7.76.

Expression	Type	Value	Address
buf1	void *	0x800004E8	0x00815720
buf2	void *	0x80000000	0x00815724
buf3	void *	0x80001810	0x00815728
buf4	void *	0x80001890	0x0081572C
buf5	void *	0x008138B0	0x00815730
buf6	void *	0x008138C0	0x00815734
buf7	void *	0x00813800	0x00815738
buf8	void *	0x008138D0	0x0081573C
Add new expression			

Figure 7.74 Buffer locations.

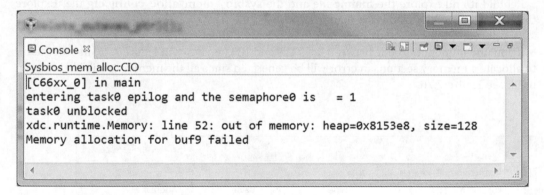

```
*main.c ⊠
127      }
128
129    /*  buf9 = Memory_alloc(myHeap, 128, 0, &eb);
130
131       if (buf9 == NULL) {
132           System_abort("Memory allocation for buf9 failed");
133       }*/
134
135       Memory_free(NULL, buf1, 128);
136       Memory_free(heapMem0, buf2, 128);
137       Memory_free(myHeapbuf, buf3, 128);
138       Memory_free(myHeapbuf, buf4, 128);
139       Memory_free(myHeap, buf5, 16);
140       Memory_free(myHeap, buf6, 16);
141       Memory_free(myHeap, buf7, 128);
142       Memory_free(myHeap, buf8, 32);
143
144       System_printf("All allocations succesful");
145    }
146
```

Figure 7.75 Setting a breakpoint and checking the memory allocation.

```
Console ⊠
Sysbios_mem_alloc:CIO
[C66xx_0] in main
entering task0 epilog and the semaphore0 is    = 1
task0 unblocked
xdc.runtime.Memory: line 52: out of memory: heap=0x8153e8, size=128
Memory allocation for buf9 failed
```

Figure 7.76 Error generated since all memory is in use.

7.6   Conclusion

TI-RTOS is a real-time scalable, low-footprint operating system that enables faster applications by providing various functionalities such as scheduling and data management. Using the TI-RTOS also has the benefit of making applications modifiable and/or upgradable with a minimum of effort. This is an advantage when time-to-market is imperative. In this chapter, key features have been studied and examples have been provided.

References

1 Texas Instruments, SYS/BIOS (TI-RTOS Kernel) v6.46 user's guide, June 2016. [Online]. Available: http://www.ti.com/lit/ug/spruex3q/spruex3q.pdf.

References (further reading)

2 Texas Instruments, Multicore fixed and floating-point digital signal processor, March 2014. [Online]. Available: http://www.ti.com/lit/ds/symlink/tms320c6678.pdf.

8

Enhanced Direct Memory Access (EDMA3) controller

8.1   Introduction

There are two methods for transferring data from one part of the memory or peripheral to another. These methods are:

1) CPU transfer
2) Direct Memory Access (DMA) transfer.

Using the CPU to transfer data is very simple (using load and store instructions), but it is time-consuming as the CPU will not be free to perform other tasks while transferring data.

Multicore DSP: From Algorithms to Real-time Implementation on the TMS320C66x SoC, First Edition. Naim Dahnoun.
© 2018 John Wiley & Sons Ltd. Published 2018 by John Wiley & Sons Ltd.
Companion website: www.wiley.com/go/dahnoun/multicoredsp

If a DMA is used, then the CPU only needs to configure the DMA. Whilst the transfer is taking place, the CPU is then free to perform other operations. The KeyStone devices have an Enhanced Direct Memory Access version 3 (EDMA3) controller that is very flexible, as demonstrated later in this chapter. The EDMA has the following features that will be explored in this chapter:

1) Transferring data from one memory location to another location in two or three cycles (e.g. transfer from a DRAM memory or serial port to an L2 SRAM)
2) In addition to only transferring data, the EDMA3 can also perform data sorting.
3) The EDMA3 is capable of performing one-dimensional (1D) (A-synchronised transfer) or 2D transfers (AB-synchronised transfer).
4) The transfers can be initiated by event(s), the CPU or the completion of another transfer.
5) The source and destination addresses can be indexed independently.
6) Various transfers can be linked.
7) Chaining of multiple transfers using one event is also supported.
8) The EDMA3 can generate an interrupt when a transfer is completed or when an error occurs.
9) Support memory protection.
10) Support many transfer requests per DMA.

To check the performance of the EDMA3 on the KeyStone I and II, consult Ref. [1].

Before embarking on using the EDMA, one needs to find out how many EDMAs are available on a specific device and how they operate. From Figure 8.1 and Figure 8.2, one can see that the TMS320C66AK2H12 has five EDMA controllers (engines) and the TMS320C6678 has three EDMA controllers.

8.2 Type of DMAs available

There are three types of DMAs available in the KeyStone architectures: EDMA, internal DMA (IDMA) and peripheral, or packet, DMA (PKDMA).

1) EDMA3. Enhanced DMA handles **m** DMA channels and **n** QDMA channels, depending on the device (see Figure 8.3):
   - DMA. **m** channels that can be triggered manually, by events or chained.
   - QDMA. **n** channels of Quick DMA (QDMA) triggered by writing to a trigger word. The TMS320C6678 is composed of:
   - 1x EDMA3 (16 independent channels) (DSP/2 clock rate)
   - 2x EDMA3 (64 independent channels) (DSP/3 clock rate)
   - 1x QDMA per EDMA3 (total 3 QDMAs) (3x 8 channels in total) (see Figure 8.4).
2) Internal DMA. Two IDMAs, channel 0 (**IDMA0**) and channel 1 (**IDMA1**), are used to move data within a CorePac, which consists of moving data between Level 1 program (**L1P**), Level 1 data (**L1D**) and Level 2 (**L2**) memories, or in the external peripheral configuration (CFG) memory. Channel 0, which is higher priority than Channel 1, is used for fast programming of peripheral configuration registers through the CFG bus; and Channel 1 is used for data transfer between **L1P**, **L1D** and **L2**. The IDMA also can provide interrupt to the DSP on transfer completion [2] (see Figure 8.5).

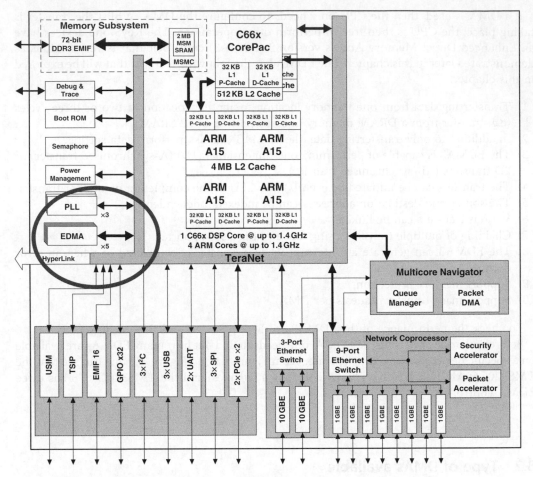

Figure 8.1 TMS320C66AK2H12 functional block diagram.

3) Peripheral, or packet, DMAs. Each PKDMA is composed of a transmit DMA (TxDMA) and receive DMA (RxDMA) that are located in some peripherals (e.g. SRIO, EMAc and FFTC) that are used to move data around; see Chapter 14 (Multicore Navigator).

8.3 EDMA controllers architecture

Each EDMA controller is composed of two parts: the EDMA Channel Controller (EDMA3CC) and the DMA Transfer Controller (EDMA3TC) (see Figure 8.6).

8.3.1 The EDMA3 Channel Controller (EDMA3CC)

The EDMA3CC services events (external, manual, chained and QDMA) and is responsible for submitting transfer requests to the transfer controllers.

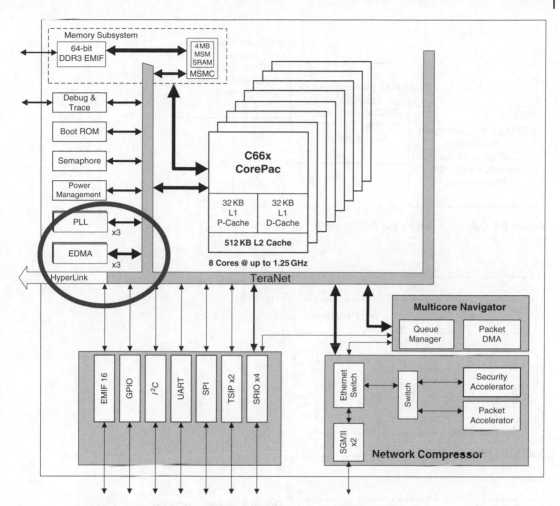

Figure 8.2 TMS32C6678 functional block diagram.

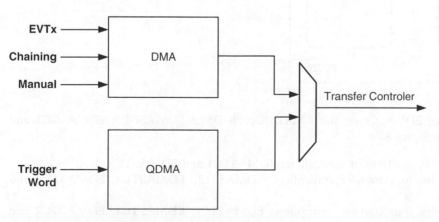

Figure 8.3 DMA and QDMA within an EDMA.

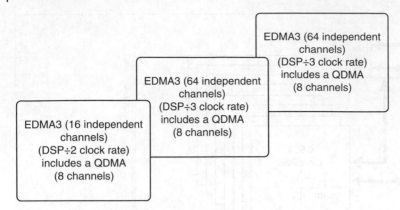

Figure 8.4 DMA channels for the TMS320C6678.

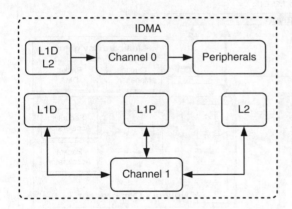

Figure 8.5 IDMA Channel 0 and Channel 1 functions.

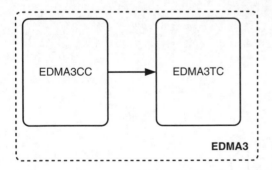

Figure 8.6 EDMA controller.

There are three EDM3CCs on the TMS320C6678 DSPs: EDMA3CC0, EDMA3CC1 and EDMA3CC2 (see Figure 8.7):

1) EDMA3CC0 has two transfer controllers: EDMA3TC1 and EDMA3TC2.
2) EDMA3CC1 has four transfer controllers: EDMA3TC0, EDMA3TC1, EDMA3TC2 and EDMA3TC3.
3) EDMA3CC2 has four transfer controllers: EDMA3TC0, EDMA3TC1, EDMA3TC2 and EDMA3TC3.

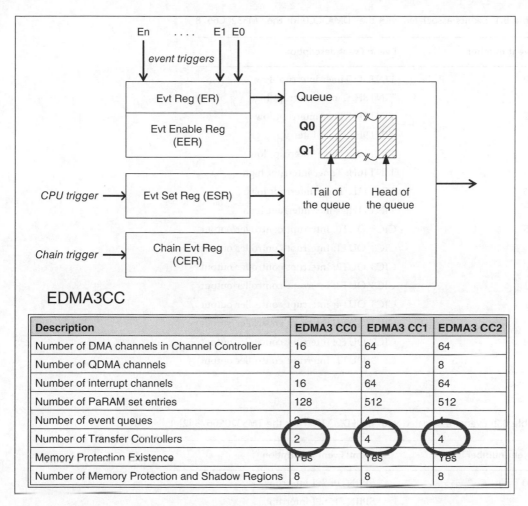

Figure 8.7 EDMA3 channel controller (EDMA3CC).

Description	EDMA3 CC0	EDMA3 CC1	EDMA3 CC2
Number of DMA channels in Channel Controller	16	64	64
Number of QDMA channels	8	8	8
Number of interrupt channels	16	64	64
Number of PaRAM set entries	128	512	512
Number of event queues	2	4	4
Number of Transfer Controllers	2	4	4
Memory Protection Existence	Yes	Yes	Yes
Number of Memory Protection and Shadow Regions	8	8	8

Each EDMA3CC has a number of channels (see Figure 8.7). For instance, EDMA3CC0 has 16 channels, and each channel is associated with a specific hardware event (see Table 8.1, Table 8.2 and Table 8.3, which are associated with EDMA3CC0, EDMA3CC1 and EDMA3CC2, respectively).

Not all EDMAs receive the same events. In fact, each EDMA responds to some particular events (see Figure 8.8). Therefore, a user must choose the right EDMA controller for the event to be used. For instance, if the event to trigger the EDMA is GPINT0, then the EDMA controller with the EDMA3CC1 or EDMA3CC2 could be used (see Table 8.2 and Table 8.3). Table 8.1, Table 8.2 and Table 8.3 show all events associated with the EDMA3CC0, EDMA3CC1 and EDMA3CC2, respectively. For a complete description, refer to Ref. [3].

Once an event is received and recognised, it is recorded in the Event Register (ER); and in order for an event to take effect, it should be enabled by setting the appropriate bit in the Event Enable Register (EER). Each logged event is queued in one of the queues, and each queue is 16 levels deep. There are only two queues for the EDMA3CC0 and four queues each for the EDMA3CC1 and EDMA3CC2 for the TMS320C6678 (see Figure 8.7). The EDMA3CC

Table 8.1 Events associated with the EDMA3CC0 for the TMS320C6678

Event number	Event: Event description
0	TINT8L: Timer interrupt low
1	TINT8H: Timer interrupt high
2	TINT9L: Timer interrupt low
3	TINT9H: Timer interrupt high
4	TINT10L: Timer interrupt low
5	TINT10H: Timer interrupt high
6	TINT11L: Timer interrupt low
7	TINT11H: Timer interrupt high
8	CIC3_OUT0: Interrupt controller output
9	CIC3_OUT1: Interrupt controller output
10	CIC3_OUT2: Interrupt controller output
11	CIC3_OUT3: Interrupt controller output
12	CIC3_OUT4: Interrupt controller output
13	CIC3_OUT5: Interrupt controller output
14	CIC3_OUT6: Interrupt controller output
15	CIC3_OUT7: Interrupt controller output

Table 8.2 Events associated with the EDMA3CC1 for the TMS320C6678 [2]

Event number	Event: Event description
0	SPIINT0: SPI interrupt
1	SPIINT1: SPI interrupt
2	SPIXEVT: Transmit event
3	SPIREVT: Receive event
4	I2CREVT: I2C receive event
5	I2CXEVT: I2C transmit event
6	GPINT0: GPIO interrupt
7	GPINT1: GPIO interrupt
8	GPINT2: GPIO Interrupt
9	GPINT3: GPIO interrupt
10	GPINT4: GPIO interrupt
11	GPINT5: GPIO interrupt
12	GPINT6: GPIO interrupt
13	GPINT7: GPIO interrupt
14	SEMINT0: Semaphore interrupt
15	SEMINT1: Semaphore interrupt
16	SEMINT2: Semaphore interrupt

Table 8.2 (Continued)

Event number	Event: Event description
17	SEMINT3: Semaphore interrupt
18	SEMINT4: Semaphore interrupt
19	SEMINT5: Semaphore interrupt
21	SEMINT7: Semaphore interrupt
22	TINT8L: Timer interrupt low
23	TINT8H: Timer interrupt high
24	TINT9L: Timer interrupt low
25	TINT9H: Timer interrupt high
26	TINT10L: Timer interrupt low
27	TINT10H: Timer interrupt high
	–
	–
36	TINT15L: Timer interrupt low
37	TINT15H: Timer interrupt high
38	CIC2_OUT48: Interrupt controller output
39	CIC2_OUT49: Interrupt controller output
40	URXEVT: UART receive event
41	UTXEVT: UART transmit event
42	CIC2_OUT22: Interrupt controller output
44	CIC2_OUT24: Interrupt controller output
	–
	–
61	CIC2_OUT41: Interrupt controller output
62	CIC2_OUT42: Interrupt controller output
63	CIC2_OUT43: Interrupt controller output

Table 8.3 Events associated with the EDMA3CC2 for the TMS320C6678 [3]

Event number	Event: Event description
0	SPIINT0: SPI interrupt
1	SPIINT1: SPI interrupt
2	SPIXEVT: Transmit event
3	SPIREVT: Receive event
4	I2CREVT: I2C receive event
5	I2CXEVT: I2C transmit event

(Continued)

Table 8.3 (Continued)

Event number	Event: Event description
6	GPINT0: GPIO interrupt
7	GPINT1: GPIO interrupt
8	GPINT2: GPIO Interrupt
9	GPINT3: GPIO interrupt
10	GPINT4: GPIO interrupt
11	GPINT5: GPIO interrupt
12	GPINT6: GPIO interrupt
13	GPINT7: GPIO interrupt
14	SEMINT0: Semaphore interrupt
15	SEMINT1: Semaphore interrupt
	–
	–
20	SEMINT6: Semaphore interrupt
21	SEMINT7: Semaphore interrupt
22	TINT8L: Timer interrupt low
23	TINT8H: Timer interrupt high
24	TINT9L: Timer interrupt low
25	TINT9H: Timer interrupt high
26	TINT10L: Timer interrupt low
27	TINT10H: Timer interrupt high
28	TINT11L: Timer interrupt low
29	TINT11H: Timer interrupt high
	–
	–
36	TINT15L: Timer interrupt low
37	TINT15H: Timer interrupt high
38	CIC2_OUT48: Interrupt controller output
39	CIC2_OUT49: Interrupt controller output
40	URXEVT: UART receive event
41	UTXEVT: UART transmit event
42	CIC2_OUT22: Interrupt controller output
43	CIC2_OUT23: Interrupt controller output
	–
	–
51	CIC2_OUT31: Interrupt controller output
52	CIC2_OUT32: Interrupt controller output
	–
	–
62	CIC2_OUT42: Interrupt controller output
63	CIC2_OUT43: Interrupt controller output

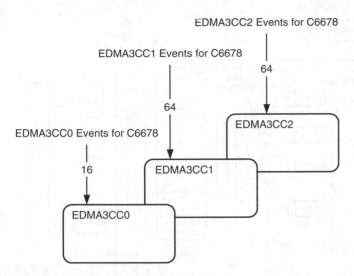

Figure 8.8 TMS320C6678 EDMA3 events.

prioritises and queues the event in the appropriate event queue. When the event reaches the head of the queue, it is evaluated for submission as a transfer request (TR) to the transfer controller (see Figure 8.10).

8.3.2 The EDMA3 transfer controller (EDMA3TC)

The EDMA3CC submits TRs to the EDMA3TC. The EDMA3TC is the engine which actually performs the data movement. Each EDMA3CC has a number of EDMA3TCs, as shown in Figure 8.9.

8.3.3 EDMA prioritisation

If many events are used at the same time, concurrent events may happen; therefore, prioritisation should be understood for proper operation of the EDMA. The prioritisation is handled at different levels as shown in Figure 8.10.

At the EDMA controller level, there are three levels of priority.

Figure 8.9 Transfer controllers.

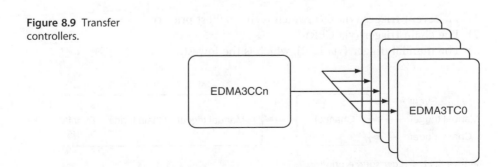

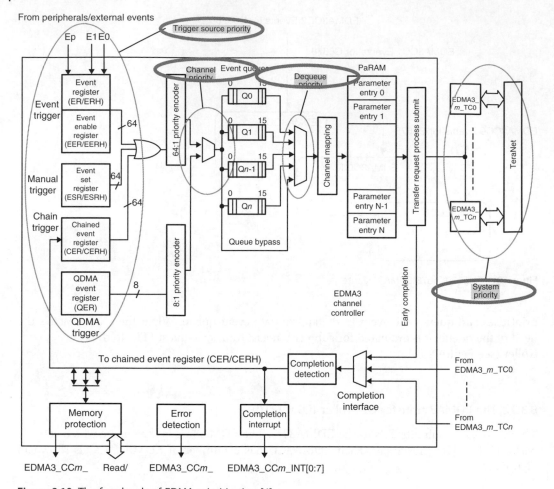

Figure 8.10 The four levels of EDMA prioritisation [4].

8.3.3.1 Trigger source priority

A DMA channel can be associated with more than one trigger; for instance, it can be triggered by its own source, by a manual trigger or by a chain trigger.

If a DMA channel is associated with more than one trigger source and if multiple events are set simultaneously for the same channel (**ER.En** = 1, **ESR.En** = 1 and **CER.En** = 1), then the EDMA3CC always services these events in the following priority order (see Figure 8.11):

1) The event trigger (via ER), which is the highest priority
2) The chain trigger (via CER)
3) The manual trigger (via ESR), which is the lowest.

Figure 8.11 Trigger source priority.

8.3.3.2 Channel priority

DMA and QDMA events can occur simultaneously. For events arriving simultaneously, the event associated with the lowest channel number is prioritised for submission to the event queues.

For DMA events, Channel 0 has the highest priority and channel n has the lowest priority, where n is the number of DMA channels supported in the EDMA3CC.

For QDMA events, channel 0 has the highest priority and channel m has the lowest priority, where m is the number of QDMA channels supported in the EDMA3CC.

If a DMA and QDMA event occur simultaneously, the DMA event always has prioritisation against the QDMA event for submission to the event queues.

In conclusion, the user cannot change the events' priority.

8.3.3.3 Dequeue priority

For submission of a TR to the transfer controller, events need to be dequeued from the event queues. A lower numbered queue has a higher dequeuing priority than a higher numbered queue. For example, if the TMS320C6678 has only two queues Q0 and Q1, then the TRs associated with events in Q0 will get submitted to TC0 prior to any TRs associated with events in Q1 getting submitted to TC1.

8.3.3.4 System (transfer controller) priority

At the system level, the priority takes place at the Teranet where other masters' peripherals (e.g. DSP cores and SRIO) submit their requests.

8.4 Parameter RAM (PaRAM)

All information needed by an EDMA or QDMA controller for a transfer (e.g. source/destination addresses, count and indexes) is programmed in a parameter RAM table (PaRAM) that is located within the EDMA3CC, referred to as the PaRAM (see Figure 8.10).

The PaRAM table is segmented into multiple PaRAM sets and must be initialised to desired values before it is used. Each PaRAM set includes eight 4-byte PaRAM set entries (32 bytes in total per PaRAM set) (see Figure 8.12). The PaRAM structure supports flexible ping-pong, circular buffering, channel chaining and auto-reloading (linking).

8.4.1 Channel options parameter (OPT)

The configuration options can be found in the OPT register that is described in Ref. [4] and summarised in Figure 8.13. Examples using the OPT register fields are used in the laboratories (see Section 8.9).

8.5 Transfer synchronisation dimensions

The EDMA3 has two transfer synchronisation dimensions.

- A-synchronised. Each event triggers the transfer of one element (or of a single array of ACNT bytes).
- AB-synchronised. Each event triggers the transfer of one frame (or of BCNT arrays of ACNT bytes) (see Figure 8.14).

PaRAM Set

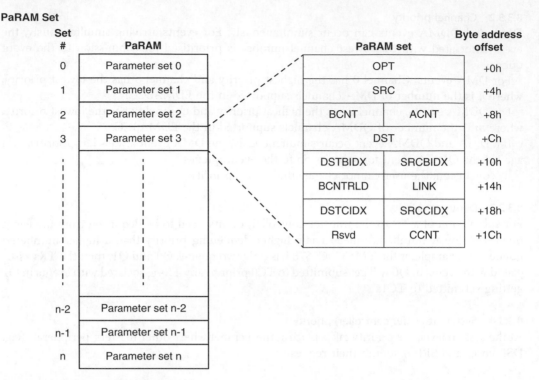

Figure 8.12 Parameter Ram (PaRam).

8.5.1 A – Synchronisation

With an A-sync mode, an event triggers the transfer of one element of size ACNT bytes. In this case, the arrays are of size ACNT (see Figure 8.15).

8.5.2 AB – Synchronisation

With an AB-sync mode, an event triggers the transfer of one frame of size ACNT * BCNT bytes (see Figure 8.16).

8.6 Simple EDMA transfer

A simple example of EDMA transfer is shown in Figure 8.17. In the example, an EDMA channel is programmed to transfer data from source 1 (**Src1**) to a destination (**Dst1**), and when the transfer is finished, the EDMA starts a callback function. Before programming the EDMA, one needs to decide on the transfer type to be used, as described in Section 8.5. In this example, only a single trigger and one array will be sufficient for transferring data. This is illustrated in Figure 8.18, which shows that only one frame and one array are being used (Frame 1 and Array 1). Laboratory 1 in Section 8.9.1 deals with a complete implementation.

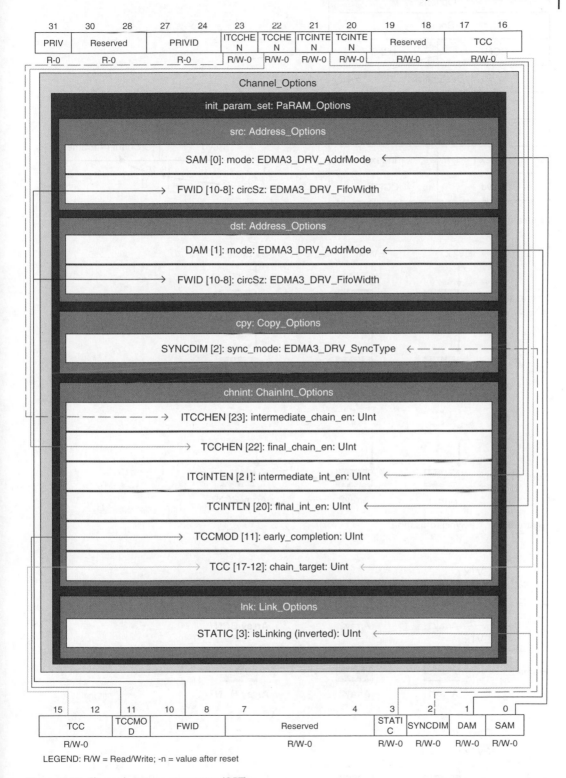

Figure 8.13 Channel options parameter (OPT).

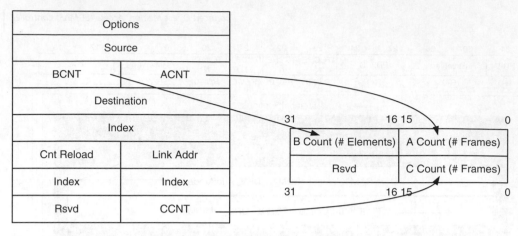

Figure 8.14 Transfer configuration.

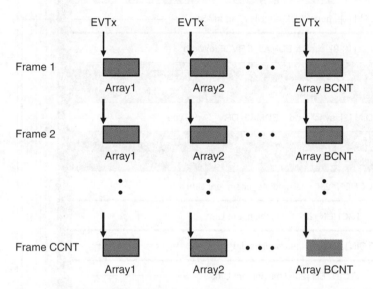

Figure 8.15 A – Synchronisation.

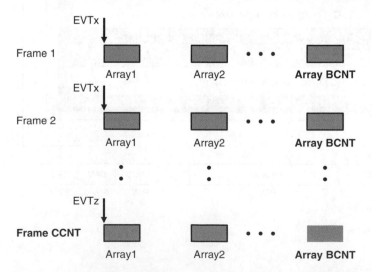

Figure 8.16 AB – Synchronisation.

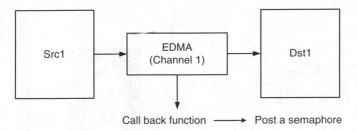

Figure 8.17 Simple EDMA transfer.

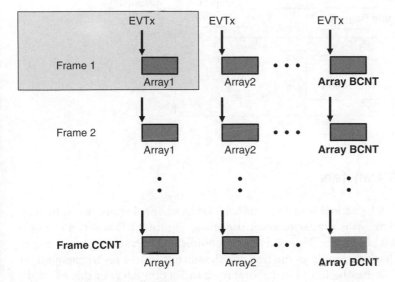

Figure 8.18 One block transfer.

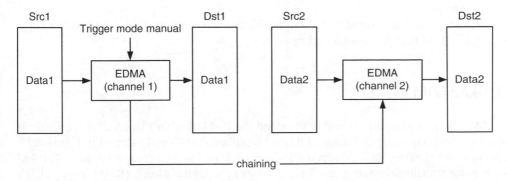

Figure 8.19 Chaining two EDMAs: example.

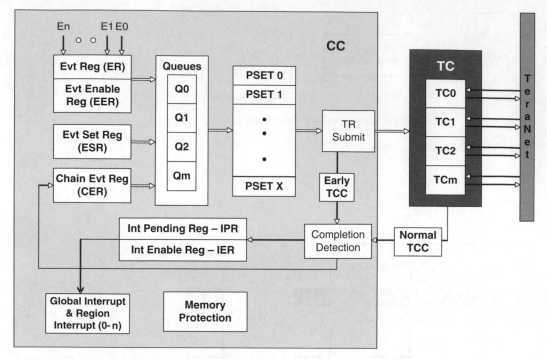

Figure 8.20 Early and normal transfer triggers.

8.7 Chaining EDMA transfers

After completion of an EDMA channel transfer, another EDMA channel transfer can be triggered. This triggering mechanism is similar to event triggering. Figure 8.19 shows an example of two EDMAs chained; that is, when EDMA Channel 1 completes the transfer, it will trigger Channel 2 and when this channel completes the transfer, the next channel may trigger another transfer. A complete example can be found in Laboratory 2 in Section 8.9.2; in this example; Channel 1 triggers Channel 2 when it completes its transfer, and Channel 2 is programmed to not trigger any other channel.

It can be seen from Figure 8.20 that chaining or triggering the second channel can be used early (Early TCC) which means the next transfer is started before the first transfer took place (submitted), or normally (Normal TCC) which means the new transfer is started after the transfer has been completed.

The EDMA is very flexible and offers other possibilities like chaining on the last transfer completion, intermediate transfer completion or both.

8.8 Linked EDMAs

The EDMA3 also provides the possibility to reload the PaRAM of an EDMA after a transfer is completed without requiring CPU intervention. The linked EDMA is illustrated in Figure 8.21. Figure 8.21a shows that two PaRAM sets (Parameter set 2 and Parameter set x) are used. PaRAM set 2 is used for transferring data from Source 1 (**Src1**) to Destination 1 (**Dst1**). Figure 8.21b shows that when this transfer completes, Parameter set x is loaded automatically into the

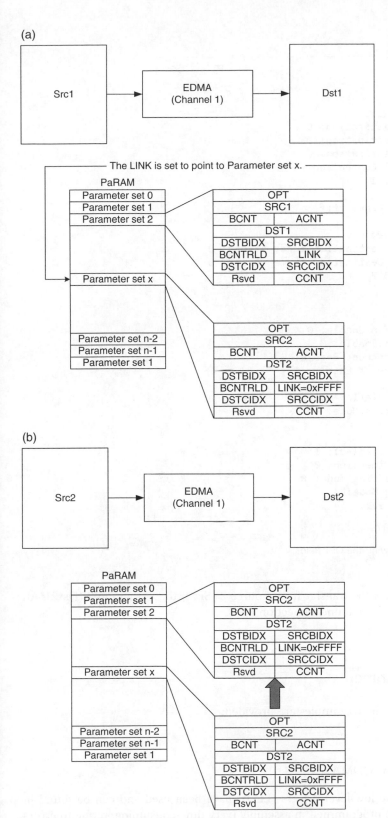

Figure 8.21 Linked EDMA (a) before Channel 1 completes the transfer and (b) after Channel 1 completes the transfer.

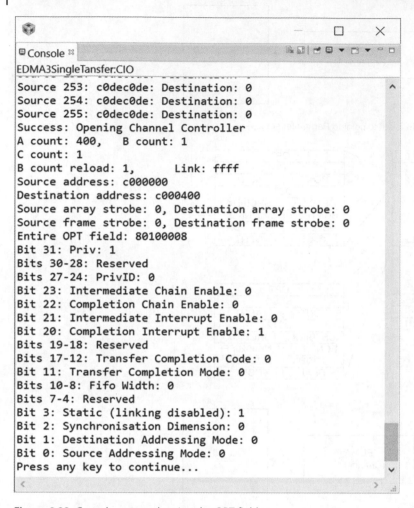

Figure 8.22 Console output showing the OPT fields.

Channel 1 PaRAM to perform the transfer from Source 2 (**Src2**) to Destination 2 (**Dst2**). An example of linked EDMAs can be found in Section 8.9.3.

8.9 Laboratory experiments

In these laboratory sessions, three examples are provided:

1) A simple transfer
2) A chaining transfer
3) A linked transfer using the QDMA3.

In these laboratories, some low-level drivers (LLDs) have been used and can be found in Ref. [5]. The EDMA3 can be programmed in assembly (very time-consuming, prone to errors and not portable), using the Chip Support Library (CSL) or the LLD.

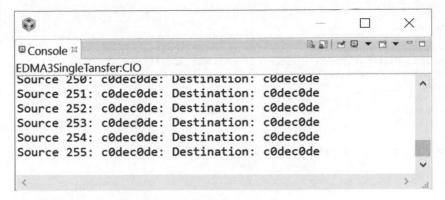

Figure 8.23 Console showing the data have been transferred.

8.9.1 Laboratory 1: Simple EDMA transfer

File location:

Chapter_8_Code\EDMA3SingleTansfer

Step 1. Build and load the project, explore the code and run the project.
Step 2. The output console displays the content of the source buffer and PaRAM as shown in Figure 8.22. Analyse the results and press continue.
Step 3. The console shows data that have been transferred (see Figure 8.23).

Referring to 'Section B.1 Setting Up a Transfer' in Ref. [4], identify each step in the project.

8.9.2 Laboratory 2: EDMA chaining transfer

Chapter_8_Code\EDMA3ChainedTransfer

Step 1. Build and load the project, and explore the code.
Step 2. Identify for each channel where the 'trigger mode' is set, where to specify the chaining and where to specify the channel to chain to. The answers are shown in Figure 8.24.
Step 3. Run the project and observe the initialisation of the source arrays as shown in Figure 8.25.

```
          channel 1 (configuration)
1.Trigger mode:
chopt-> trig_mode = EDMA3_DRV_TRIG_MODE_MANUAL;
//Trigger the EDMA Manually(CPU)
2.propt-> chnint .final_chain_en = TRUE;
//Chain on block completion
3.Chain target:
propt-> chnint . chain_target = channel2;
//Chain to second channel if enabled
```

```
          channel 2 (configuration)
1.Trigger mode:
//Not specify as the chaining is taking place.
2.propt-> chnint. final_chain_en = TRUE;
//Chain on block completion
3.Chain target:
propt2-> chnint. chain_target = 0;
//Chain to channel 0 if enabled
```

Figure 8.24 Channel 1 and Channel 2 configurations.

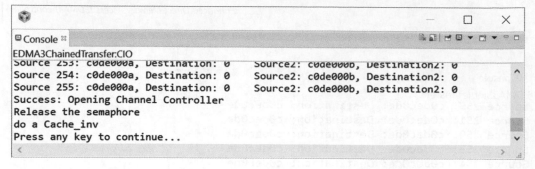

```
Console 🖾                                                        ▣ ▣ | ▨ ▣ ▾ ▣ ▾ ▫ ▫
EDMA3ChainedTransfer:CIO
Source 253: c0de000a, Destination: 0     Source2: c0de000b, Destination2: 0
Source 254: c0de000a, Destination: 0     Source2: c0de000b, Destination2: 0
Source 255: c0de000a, Destination: 0     Source2: c0de000b, Destination2: 0
Success: Opening Channel Controller
Release the semaphore
do a Cache_inv
Press any key to continue...
```

Figure 8.25 Console showing initialisation of the source arrays.

```
Console 🖾                                                        ▣ ▣ | ▨ ▣ ▾ ▣ ▾ ▫ ▫
EDMA3ChainedTransfer:CIO
Source 251: c0de000a, Destination: c0de000a     Source2: c0de000b, Destination2: c0de000b
Source 252: c0de000a, Destination: c0de000a     Source2: c0de000b, Destination2: c0de000b
Source 253: c0de000a, Destination: c0de000a     Source2: c0de000b, Destination2: c0de000b
Source 254: c0de000a, Destination: c0de000a     Source2: c0de000b, Destination2: c0de000b
Source 255: c0de000a, Destination: c0de000a     Source2: c0de000b, Destination2: c0de000b
```

Figure 8.26 Console showing the data have been transferred.

```
Console 🖾                                                        ▣ ▣ | ▨ ▣ ▾ ▣ ▾ ▫ ▫
EDMA3LinkedQDMA:CIO
Source 252: c0de000a, Destination: 0     Source2: c0de000b, Destination2: 0
Source 253: c0de000a, Destination: 0     Source2: c0de000b, Destination2: 0
Source 254: c0de000a, Destination: 0     Source2: c0de000b, Destination2: 0
Source 255: c0de000a, Destination: 0     Source2: c0de000b, Destination2: 0
Success: Opening Channel Controller
A count: 400,    B count: 1
C count: 1
B count reload: 1,        Link: ffff
Source address: c000800
Destination address: c000c00
Source array strobe: 0, Destination array strobe: 0
Source frame strobe: 0, Destination frame strobe: 0
Entire OPT field: 8000000c
Bit 31: Priv: 1
Bits 30-28: Reserved
Bits 27-24: PrivID: 0
Bit 23: Intermediate Chain Enable: 0
Bit 22: Completion Chain Enable: 0
Bit 21: Intermediate Interrupt Enable: 0
Bit 20: Completion Interrupt Enable: 0
Bits 19-18: Reserved
Bits 17-12: Transfer Completion Code: 0
Bit 11: Transfer Completion Mode: 0
Bits 10-8: Fifo Width: 0
Bits 7-4: Reserved
Bit 3: Static (linking disabled): 1
Bit 2: Synchronisation Dimension: 1
Bit 1: Destination Addressing Mode: 0
Bit 0: Source Addressing Mode: 0
Press any key to continue...
```

Figure 8.27 Console output showing the OPT fields.

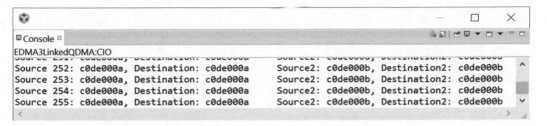

Figure 8.28 Console showing the data have been transferred.

Step 4. Press any key on the console and observe that both destinations have been updated (see Figure 8.26).

8.9.3   Laboratory 3: EDMA link transfer

File location:

Chapter_8_Code\EDMA3LinkedQDMA

Step 1. Build and load the project, and explore the code.
Step 2. Explain the OPT fields shown in Figure 8.27.
Step 3. Run the project, and observe that the data are transferred from source to destination and from Source2 to Destination2 (see Figure 8.28).

8.10   Conclusion

The EDMA3 is very flexible but can be very intimidating to use. However, using this chapter in conjunction with the datasheet and the examples provided, it can be easy to use and modify to fit many applications. The most important transfer scenarios (single, linked and chained transfers) have been implemented. The ping-pong transfer is also implemented in Chapter 18 (FFT implementation).

References

1 Texas Instruments, EDMA FAQ for KeystoneI/II devices, [Online]. Available: http://wiki.tiprocessors.com/index.php/EDMA_FAQ_for_KeystoneI/II_devices.
2 Texas Instruments, TMS320C66x DSP CorePac user's guide, [Online]. Available: http://www.ti.com/lit/ug/sprugw0c/sprugw0c.pdf.
3 Texas Instruments, Multicore fixed and floating-point digital signal processor, SPRS691E, [Online]. Available: http://www.ti.com/lit/ds/symlink/tms320c6678.pdf.
4 Texas Instruments, KeyStone Architecture Enhanced Direct Memory Access (EDMA3) controller user's guide, [Online]. Available: http://www.ti.com/lit/ug/sprugs5b/sprugs5b.pdf.
5 Texas Instruments, Programming the EDMA3 using the low-level driver (LLD), [Online]. Available: http://processors.wiki.ti.com/index.php/Programming_the_EDMA3_using_the_Low-Level_Driver_(LLD). [Accessed 2017].

9

Inter-Processor Communication (IPC)

Multicore DSP: From Algorithms to Real-time Implementation on the TMS320C66x SoC, First Edition. Naim Dahnoun.
© 2018 John Wiley & Sons Ltd. Published 2018 by John Wiley & Sons Ltd.
Companion website: www.wiley.com/go/dahnoun/multicoredsp

9.1   Introduction

Having a processor with multicores has many advantages, as is seen in previous chapters. However, not having the cores communicating very efficiently at high speed will make the processor very inefficient.

Algorithms are normally divided into threads that run on different cores. Depending on the application, these threads may exchange data at specific times, and therefore problems of data exchange and synchronisation between threads will emerge. High bandwidth and low latency are necessary for real-time application. To achieve these, communication mechanisms have to be put in place, and this is known as Inter-Processor Communication (IPC). IPC can be used mainly for data sharing between processes running on a single core or different cores to speed the execution of an application. However, this may require synchronisation.

The Texas Instruments (TI) IPC can be used to communicate between threads on the same processor, be it single-core or multicore, and to communicate between threads on different processors that support SYS/BIOS, Linux, Android and QNX operating systems (OSs) [1]. The performance of the IPC will depend on the OS used. The link in Ref. [2] shows how to build a benchmark for each OS.

There are mainly two models of IPC, the **Shared Memory** model and the **Message Passing** (or message queue) model. Since any core can have full access to the device memory, senders and receivers can also communicate data by exchanging pointers only without actually exchanging data. In this case, the sender writes data to a specific memory, notifies the receiver and sends the pointer. The receiver then accesses the memory and when it finishes with it, the receiver notifies the sender. Finally, in order to increase the performance, TI also introduced the multicore Navigator for data movement. The Navigator uses hardware queues and DMAs to move data. The Navigator is introduced in Chapter 14.

In the shared memory model, the processor processes, communicates or shares data through a shared memory as shown in Figure 9.1. This model provides a fast and simple communication mechanism since only load and store instructions can be used and the OS is not involved. However, to keep data secure (memory coherency) so that one processor cannot overwrite data

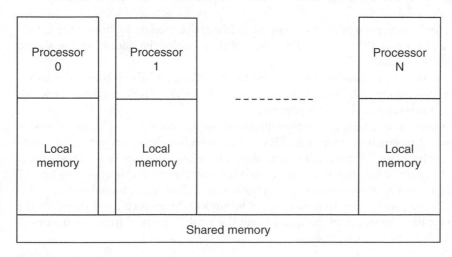

Figure 9.1 Shared memory model.

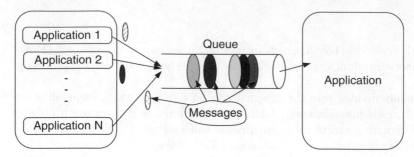

Figure 9.2 Example of a message queue mechanism.

Shared memory	Message Queue
• Simple to programme.	• Easy to manage complex applications.
• High performance.	• Multiple queues can be usedwith different data types.
• Synchronisation is required.	• Can be synchronous or asynchronous.
• Data can be corrupted.	• Fixed size header and variable length messages supported.

Figure 9.3 Features for the shared memory and message queue IPCs.

(accidentally or asynchronously) in the shared memory, a synchronisation mechanism has to be implemented by the user.

In this scheme, for instance, a sender sends a message (data) to a receiver by storing the data in shared memory and notifies the receiver. The receiver then copies the data from the shared buffer and notifies the sender.

The shared memory implementation is the easiest to implement and also is one of the fastest as it minimises the processing and the storage overhead. Both the SYS/BIOS and the IPC use this method for passing messages and synchronisation.

The message queue allows different unsynchronised applications to exchange messages (data). It is important to emphasise that applications may be unsynchronised but the communication can be synchronised. In this model, an application writes a message to a named queue, and the same application or another application reads the message from this queue. Figure 9.2 and Figure 9.3 illustrate the message-passing mechanism and show some basic features.

The message queues provide the asynchronous communication protocol, and therefore the sender can keep sending messages to the queues and the reader can read from the queues at the same time.

In general, there are many mechanisms to implement an IPC. The most common are:

- Shared memory
- TCP
- Named pipe
- File mapping
- Mail slots
- MSMQ (Microsoft Queue Solution).

9.2 Texas Instruments IPC

For TI processors, and particularly KeyStone I and II, the IPC allows communication between processors and peripherals. These work transparently in both uni-processor and multiprocessor configurations.

For the KeyStone processors, the IPC is supported by the components or modules shown in Figure 9.4 and described in Table 9.1. After installing the IPC package, the source code for these modules can be found in:

C:\TI\ipc_3_35_01_07\packages\ti\sdo\ipc (after installing the ipc package)

These components serve different purposes that are explained in this chapter; some are for data exchange, some for synchronisation and some for shared memory configuration.

Which component(s) to use will depend on the application, the type of data to be communicated, the method of synchronisation, the bandwidth, latency in communicating between two threads and ease of use. These components can be used independently or be used by another

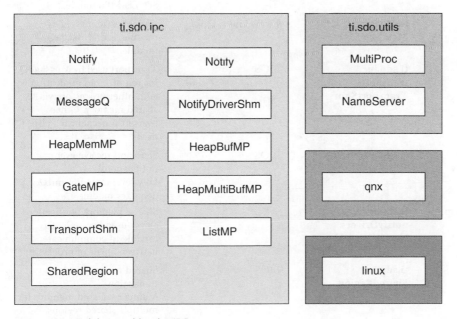

Figure 9.4 Modules used by the IPC.

Table 9.1 Modules used by the IPC [1]

Module	Module path	Operating system supported	Description
GateMP Module	ti.sdo.ipc.gates	BIOS, Linux, QNX	Protects a critical section: Manages gates for mutual exclusion of shared resources by multiple processors and threads.
HeapBufMP Module	ti.sdo.ipc.heaps. HeapBufMP	BIOS	Multi-processor memory allocator: Fixed-sized shared memory heaps. Similar to SYS/BIOS's **ti.sysbios.heaps.HeapBuf** module, but with some configuration differences.
HeapMemMP Module	ti.sdo.ipc.heaps. HeapMemMP	BIOS	Multi-processor memory allocator: Variable-sized shared memory heaps.
HeapMultiBufMP Module	ti.sdo.ipc.heaps. HeapMultiBufMP	BIOS	Multi-processor memory allocator: Multiple fixed-sized shared memory heaps.
Ipc Module	ti.sdo.ipc.Ipc	BIOS, Linux, QNX	IPC manager: Provides **Ipc_start()** function and allows **startup** sequence configuration.
ListMP Module	ti.sdo.ipc.ListMP	BIOS	Doubly linked list for shared memory, multi-processor applications. Very similar to the **ti.sdo.utils.List** module. See ListMP module.
MessageQ Module	ti.sdo.ipc. MessageQ	BIOS, Linux, QNX	Variable-size messaging module. See **MessageQ** module.
TransportShm	ti.sdo.ipc. transports. TransportShm	BIOS	Transport used by **MessageQ** for remote communication with other processors via shared memory. Other transport mechanisms also exist.
Notify Module	ti.sdo.ipc.Notify	BIOS	Send and receive event notifications: Low-level interrupt mux/demuxer module.
NotifyDriverShm	ti.sdo.ipc. notifyDrivers. NotifyDriverShm		Shared memory notification driver used by the **Notify** module to communicate between a pair of processors
SharedRegion Module	ti.sdo.ipc. SharedRegion	BIOS	Shared memory address translation. Maintains shared memory for multiple shared regions.
MultiProc	ti.sdo.utils	BIOS, Linux, QNX	Processor identification
NameServer	ti.sdo.utils	BIOS, Linux, QNX	Distributed name/value database

Source: Courtesy of Texas Instruments.

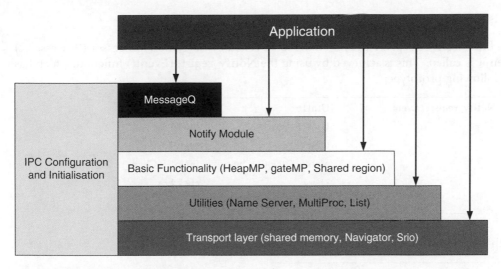

Figure 9.5 Components dependency.

component as shown in Figure 9.5. For instance, an application can use the **Notify** module which itself can use one of the basic functions like **HeapMP**, **GateMP** and so on. In this chapter, we will review all these components and highlight the advantages and disadvantages.

9.3 Notify module

The **Notify** module is mainly used for synchronisation as it abstracts physical hardware interrupts into multiple logical events. It can be used to send a 32-bit message (payload) between processors without the user dealing with the interrupts for notification or synchronisation. In this sense, it is the simplest IPC mechanism for communicating data.

How does it work? Assume we have two processors (Processor 1 and Processor 2) that would like to communicate with each other. Processor 1 sends a message to Processor 2, such as asking it to perform a certain task, and when Processor 2 finishes it will notify Processor 1 (see Figure 9.6).

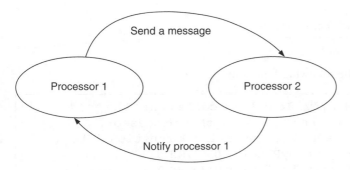

Figure 9.6 Notify module functionality.

This is achieved with two steps:

1) The first step is to register the functions to be called by the remote processor (Processor 2) when it is called. This is achieved by using the **Notify_registerEvent()** function, which has the following prototype:

Int Notify_registerEvent	(	UInt16	**procId**
		UInt16	**lineId**
		UInt32	**eventId**
		Notify_FnNotifyCbck	**fnNotifyCbck**
		UArg	**cbckArg**
	)		

With the following parameters:

	Parameters
procId	Remote processor ID
lineId	Line ID (0 for most systems)
eventId	Event ID
fnNotifyCbck	Pointer to callback function
cbckArg	Callback function argument

Returns

Notify status:
- **Notify_S_SUCCESS**. Event successfully registered.
- **Notify_E_MEMORY**. Failed to register due to memory error.

As an example, we can register an event as follows:

procId	**myProcId**
lineId	**0**
eventId	**EVENTID**
fnNotifyCbck	myFxn1
cbckArg	0x1010 //Callback function argument

This can be implemented by the following function:

```
/* Register myFxn1 with Notify. It will be called when the sender
 * sends event number EVENTID to line #0 on this processor.
 * The argument 0x1010 is passed to the callback function. */
status = Notify_registerEvent(myProcId, 0, EVENTID,
            (Notify_FnNotifyCbck)myFxn1, 0x1010);
```

2) The second step is to send an event. This is achieved by the **Notify_sendEvent()** function. The prototype of the **Notify_sentEvent()** function is as follows:

```
Int Notify_sendEvent
(          UInt16 procId,
           UInt16 lineId,
           UInt32 eventId,
           UInt32
           payload,
           Bool
           waitClear
)
```

With the following parameters:

Parameters

[in]	**procId**	Remote processor ID
[in]	**lineId**	Line ID
[in]	**eventId**	Event ID (the user can specify any event ID)
[in]	**payload**	Payload to be sent along with the event
[in]	**waitClear**	Indicates whether to spin waiting for the remote core to process previous events

Returns

Notify status:

- **Notify_E_EVTNOTREGISTERED**. Event has no registered **callback** functions.
- **Notify_E_NOTINITIALIZED**. Remote driver has not yet been initialised.
- **Notify_E_EVTDISABLED**. Remote event is disabled.
- **Notify_E_TIMEOUT**. Timeout occurred (when **waitClear** is TRUE).
- **Notify_S_SUCCESS**. Event successfully sent.

Note: [in] = Input.

For example, if we want to tell the remote processor to run the registered function, we can do so by sending an event as shown here:

```
#define EVENT 5
Notify_sendEvent(myProcId, 0, EVENT, 0xbbbb, TRUE);
```

Notice that when we send an event, we do not specify the function, but instead we specify the **eventID** that has been set by the **Notify_sentEvent()** function as shown above.

The function **Notify_sendEvent ()** sends an event (**EVENT**) to a processor (**myProcId**) and a line ID (0). A payload (0xbbbb) is the fourth argument.

Once the event is sent, at the destination processor, the **callback** functions that were registered with the **Notify_register_Event** with the associated **eventId** and source processor ID are called. In the example shown above, we have registered the function (**myFxn1**).

What would happen if you send an event to the remote processor and would like to send another event to the same processor? Would you wait for the remote processor to notify you, or would the system take care of it?

The answer depends on the setup of the **Notify** drivers. The IPC provides several **Notify** driver implementations. All processors involved in the notification process must use the same driver.

There are two types of **Notify** driver, and both can only use a shared memory for transport:

1) **Shared memory notify driver (NotifyDriverShm)**
This is the default one, and no setup is required. In this mode, each event has one pending notification in the shared memory.

When using **NotifyDriverShm**, a **waitClear** value of TRUE indicates that, if an event was previously sent to the same **eventId**, **sendEvent** should spin until the previous event has been acknowledged by the remote processor. If **waitClear** is FALSE, a pending event with the same **eventId** will be overwritten by the event currently being sent.

In this mode, each event can be disabled/enabled as shown here:

```
Notify_disableEvent(myProcId, 0, EVENT);
Notify_enableEvent(myProcId, 0, EVENT);
```

2) **Circular buffer notify driver (NotifyDriverCirc)**
This **Notify** driver uses a circular buffer (**NotifyDriverCirc**) in the shared memory to store all the notifications.

With this mode, single events cannot be disabled or enabled, and events can be dropped if global notifications are disabled by the receiver. However, the latency is lower than that of **NotifyDriverShm**.

9.3.1 Laboratory experiment

Laboratory objectives: in this laboratory session, you will learn how to use the **Notify**'s APIs in a single core and explore the following functions:

- **Notify_register**
- **Notify_unregister**
- **Notify_disableEvent**
- **Notify_enableEvent**
- **Notify_sendEvent**.

File location:

\Chapter_9_Code\Notify\NotifySingleProcessor

Tasks:

1) Open project **NotifySingleProcessor**.
2) Build and run the project.
3) Verify the output console (see Figure 9.8).
4) Explore the file **notify_loopback.c** shown in Figure 9.7.

9.4 MessageQ

The **MessageQ** component may be used for data transfer and messaging. Messages of variable lengths (that make programming easier) can be exchanged between processors and are sent

```
/*
 *  =======- notify_loopback.c ========
 *  This program demonstrates the functionality of the Notify module on a
 *  single processor.
 *
 *  The purpose of this example is to show the usage of Notify APIs. All
 *  events are registered and sent locally.
 *
 *  Initially two functions are registered for an event. This is to
 *  show that multiple functions can be registered. Each function
 *  will be passed its specified "arg".
 *
 *  Functions demonstrated:
 *   - Notify_register
 *   - Notify_unregister
 *   - Notify_disableEvent
 *   - Notify_enableEvent
 *   - Notify_sendEvent
 *
 *  See notify_loopback.k file for expected output.
 */

#include <xdc/std.h>
#include <xdc/runtime/System.h>
#include <ti/ipc/MultiProc.h>
#include <ti/ipc/Notify.h>

#include <xdc/cfg/global.h>

/* Event number to be used in the example */
#define EVENT   5

/*
 *  ======== myFxn1 ========
 */
Void myFxn1(UInt16 procId, UInt16 lineId, UInt32 eventNo, UArg arg,
            UInt32 payload)
{
    UInt32 *theArg = (UInt32 *)arg;
    System_printf("I am running myFxn1: eventNo: #%d, arg: %d, payload: %x\n",
                  eventNo, *theArg, payload);
}

/*
 *  ======== myFxn2 ========
 */
Void myFxn2(UInt16 procId, UInt16 lineId, UInt32 eventNo, UArg arg,
            UInt32 payload)
{
    UInt32 *theArg = (UInt32 *)arg;
    System_printf("I am running myFxn2: eventNo: #%d, arg: %d, payload: %x\n",
                  eventNo, *theArg, payload);
}

/*
 *  ======== main ========
 */
Int main(Int argc, Char* argv[])
{
    UInt32 myArg1 = 12345;
    UInt32 myArg2 = 67890;
    UInt16 myProcId = MultiProc_self();
    Int status;
```

Figure 9.7 notify_loopback.c file.

```
    /* Register the functions to be called */
    System_printf("Registering myFxn1 & myArg1 to event #%d..\n", EVENT);
    Notify_registerEvent(myProcId, 0, EVENT,
                         (Notify_FnNotifyCbck)myFxn1, (UArg)&myArg1);

    System_printf("Registering myFxn2 & myArg2 to event #%d..\n", EVENT);
    Notify_registerEvent(myProcId, 0, EVENT,
                         (Notify_FnNotifyCbck)myFxn2, (UArg)&myArg2);

    /* Send an event */
    System_printf("Sending event #%d (myFxn1 and myFxn2 should run)\n", EVENT);
    Notify_sendEvent(myProcId, 0, EVENT, 0xaaaaaa, TRUE);

    /* Unregister one of the functions */
    System_printf("Unregistering myFxn1 + myArg1\n");
    status = Notify_unregisterEvent(myProcId, 0, EVENT,
                                    (Notify_FnNotifyCbck)myFxn1,
                                    (UArg)&myArg1);
    if (status < 0) {
        System_abort("Listener not found! (THIS IS UNEXPECTED)\n");
    }

    /* Send an event */
    System_printf("Sending event #%d (myFxn2 should run)\n", EVENT);
    Notify_sendEvent(myProcId, 0, EVENT, 0xbbbbb, TRUE);

    /* Disable event */
    System_printf("Disabling event #%d:\n", EVENT);
    Notify_disableEvent(myProcId, 0, EVENT);

    /* Send an event (nothing should happen) */
    System_printf("Sending event #%d (nothing should happen)\n", EVENT);

    /* Enable event */
    System_printf("Enabling event #%d:\n", EVENT);
    Notify_enableEvent(myProcId, 0, EVENT);

    /* Send an event */
    System_printf("Sending event #%d (myFxn2 should run)\n", EVENT);
    Notify_sendEvent(myProcId, 0, EVENT, 0xbbbbb, TRUE);

    System_printf("Test completed\n");
    return (0);
}
```

Figure 9.7 (Continued)

through queues. Each queue is identified by a unique name. The message component can only be used when we have one or multiple writers and only one reader. However, the reader can read/write from/to several queues as shown in Figure 9.9. **MessageQ** objects enable zero-copy, variable-length message passing.

9.4.1 MessageQ protocol

The protocol for the **MessageQ** is as follows: the reader creates both a **MessageQ** and a message; the writer opens the queue, allocates a memory space for the message to be sent and then sends the message; and, finally, the reader reads the message. Table 9.2 shows the **MessageQ** functions.

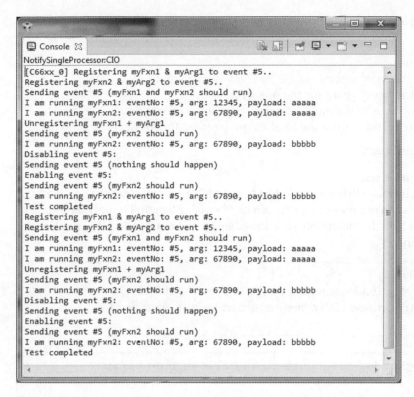

```
Console ⊠                                    
NotifySingleProcessor:CIO
[C66xx_0] Registering myFxn1 & myArg1 to event #5..
Registering myFxn2 & myArg2 to event #5..
Sending event #5 (myFxn1 and myFxn2 should run)
I am running myFxn1: eventNo: #5, arg: 12345, payload: aaaaa
I am running myFxn2: eventNo: #5, arg: 67890, payload: aaaaa
Unregistering myFxn1 + myArg1
Sending event #5 (myFxn2 should run)
I am running myFxn2: eventNo: #5, arg: 67890, payload: bbbbb
Disabling event #5:
Sending event #5 (nothing should happen)
Enabling event #5:
Sending event #5 (myFxn2 should run)
I am running myFxn2: eventNo: #5, arg: 67890, payload: bbbbb
Test completed
Registering myFxn1 & myArg1 to event #5..
Registering myFxn2 & myArg2 to event #5..
Sending event #5 (myFxn1 and myFxn2 should run)
I am running myFxn1: eventNo: #5, arg: 12345, payload: aaaaa
I am running myFxn2: eventNo: #5, arg: 67890, payload: aaaaa
Unregistering myFxn1 + myArg1
Sending event #5 (myFxn2 should run)
I am running myFxn2: eventNo: #5, arg: 67890, payload: bbbbb
Disabling event #5:
Sending event #5 (nothing should happen)
Enabling event #5:
Sending event #5 (myFxn2 should run)
I am running myFxn2: eventNo: #5, arg: 67890, payload: bbbbb
Test completed
```

Figure 9.8 Output console.

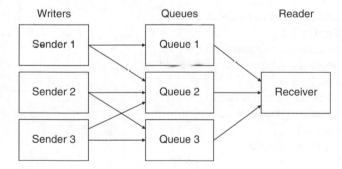

Figure 9.9 Example of a **MessageQ** sender/receiver topology.

Table 9.2 Main **MessageQ** functions

Writer threads call	The reader thread calls
• **MessageQ_open()**	• **MessageQ_create()**
• **MessageQ_alloc()**	• **MessageQ_get()**
• **MessageQ_put()**	• **MessageQ_free()**
• **MessageQ_close()**	• **MessageQ_delete()**

The following steps show how to use the **MessageQ**.

1) Create a message to be sent.
2) Create a message queue.
3) The writer opens a message queue.
4) The writer allocates memory space for the message.
5) The writer puts a message in the message queue (sending a message).
6) The receiver gets the message from the local queue (reading the message).

These steps are detailed here:

1) **Create a message to be sent**
   Let's first create a message that we would like to send from one processor to another via a message queue. Messages in a message queue can be of variable length. The only requirement is that the first field in the definition of a message must be a **MessageQ_MsgHeader** structure as shown here:

```
typedef struct MyMsg {
      MessageQ_MsgHeader header; // always required.
      //Application specific message can be entered here.
} MyMsg;
```

```
The MessageQ_MsgHeader structure is as follows (see ti/ipc/
MessageQ.h):
typedef struct {
      Bits32 reserved0; // reserved for List.elem->next
      Bits32 reserved1; // reserved for List.elem->prev
      Bits32 msgSize; // message size
      Bits16 flags; // bitmask of different flags
      Bits16 msgId; // message id
      Bits16 dstId; // destination queue id
      Bits16 dstProc; // destination processor id
      Bits16 replyId; // reply id
      Bits16 replyProc; // reply processor
      Bits16 srcProc; // source processor
      Bits16 heapId; // heap id
      Bits16 seqNum; // sequence number
      Bits16 reserved; // reserved
} MessageQ_MsgHeader;
typedef MessageQ_MsgHeader *MessageQ_Msg;
```

As an example, let's have the following message:

```
typedef struct myMsg {
      MessageQ_MsgHeader header; // always required.
      msgType messageType;
      Int tokenCount;
}myMsg;
```

2) **Create a message queue**

To create a **MessageQ** instance, two parameters and the name of the queue are required, as well as how the queue is synchronised, as shown in this structure:

```
MessageQ_Handle MessageQ_create
(        String          name,
         const MessageQ_Params *    params
)
```

Parameters

[in]	**name**	Name of the queue//make sure the name is unique.
[in]	**params**	Initialised **MessageQ** parameters

Returns

MessageQ handle

Each message queue has its own synchroniser object; see 'Thread Synchronisation' in Section 9.4.3. For a NULL parameter, by default the synchroniser is a **SyncSem**.

The following code creates a **MessageQ** named **localQueueName**, and the synchronisation used is **SyncSem**.

```
messageQ = MessageQ_create(localQueueName, NULL);
```

3) **The writer opens a message queue**

The writer thread opens the created message queue in order to have access to it. This can be achieved by using the following instruction:

```
int MessageQ_open (String name, MessageQ_QueueId *queueId);
```

where:

name [input parameter]: is the name of the created queue.

The name can be in any processor. If the name is found, the **queueId** (output parameter) is filled and the function **MessageQ_open()** will return a **MessageQ_S_SUCCESS**.

If the name is not found, the function **MessageQ_open()** will return a **Message Q_E_NOTFOUND**.

```
status = MessageQ_open(remoteQueueName, &msgQueueIds[coreCount]);
```

4) **The writer allocates memory space for the message**

Once the message has been created, the **MessageQ_alloc()** function can be used to allocate a memory space for the message from the heap that is associated to **HEAP_ID.** The location of the memory can be specified as shown here, for example:

```
msg = MessageQ_alloc(HEAP_ID, sizeof(myMsg));
```

The **HEAP_ID** will be used to get the actual heap as shown here:

```
MessageQ_registerHeap((IHeap_Handle)SharedRegion_getHeap(0),
HEAP_ID);
```

5) **The writer puts a message in the message queue (sending a message)**
Now that the message queue is created and the message is created and allocated, the message can be sent by the using the **MessageQ_put** function shown here. The **MessageQ_put** requires two parameters, the **queueId** and **msg**, also shown here:

```
int MessageQ_put
  (MessageQ_QueueId queueId,
   MessageQ_Msg msg
   )
```

Parameters

Two parameters are required:

[in]	**queueId**	The destination **MessageQ**
[in]	**Msg**	The message to be sent

Returns

Status of the call:

- **MessageQ_S_SUCCESS** denotes success.
- **MessageQ_E_FAIL** denotes failure. The **MessageQ_put** was not successful. The caller still owns the message.
  The **queueId** could have been obtained with one of the following functions:

- **MessageQ_open()**
- **MessageQ_getReplyQueue()**
- **MessageQ_getDstQueue()**.

The **MessageQ_put** can be used as follows:

```
status = MessageQ_put(msgQueueIds[1], msg);
```

6) **The receiver gets the message from the local queue (reading the message)**
To read the message, the reader thread can use the following instruction:

```
int MessageQ_get(MessageQ_Handle handle, MessageQ_Msg *msg, UInt
timeout)
```

Parameters

[in]	**handle**	**MessageQ** handle that was created in (2)
[out]	**msg**	Pointer to the arriving message location
[in]	**timeout**	Maximum duration to wait for a message in microseconds

Returns

MessageQ status:

- **MessageQ_S_SUCCESS**: Message successfully returned.
- **MessageQ_E_TIMEOUT**: **MessageQ_get()** timed out.
- **MessageQ_E_UNBLOCKED**: **MessageQ_get()** was unblocked.
- **MessageQ_E_FAIL**: A general failure has occurred.

The message will be returned in the **msg**. If the message is not present, the thread will block on the synchronisation object used until the synchroniser sends a signal or a timeout occurs.

If a timeout of zero is specified, the function returns immediately; and if no message is available, the **msg** is set to NULL and the status is **MessageQ_E_TIMEOUT**.

The **MessageQ_get()** function can be used as follows:

```
status = MessageQ_get(messageQ, &msg, MessageQ_FOREVER);
```

When sending a message, the message queue ID can be incorporated into the message; this will allow the remote processor to extract the message queue and be used to reply. This is illustrated here:

```
//On one core:
dm = MessageQ_alloc(0, (sizeof(structDataMsg) + N * sizeof(char)));
// embed the reply queue into the message
MessageQ_setReplyQueue(localQ, (MessageQ_Msg) dm);
// send the message
status = MessageQ_put(remoteQ[0], (MessageQ_Msg) dm);

//On the second core:
// retrieve the message queue Id (remoteQ)
remoteQ - MessageQ_getReplyQueue(dm);
// send message back

MessageQ_put(remoteQ, (MessageQ_Msg) dm);
```

A complete example showing the ARM–DSP communication with the KeyStone II is shown in Section 9.10.

9.4.2 Message priority

There are three priority levels: normal, high and urgent. Tasks of normal priority will be sent through a normal queue channel, and the high-priority tasks will be sent through a high-priority queue channel. The urgent-priority tasks will be put at the head of the high-priority queue as shown in Figure 9.11. To change the message queue priority, the message queue header should be initialised as shown in Figure 9.10.

9.4.3 Thread synchronisation

Thread synchronisation is implemented by two different mechanisms: (1) a signal synch is sent to the writer, and (2) the wait function is used at the reader (see Figure 9.12).

```
#define MessageQ_setMsgPri
(
msg,
priority
)
(((MessageQ_Msg) (msg))->flags = ((priority) & MessageQ_PRIORITYMASK))
//Sets the message priority of a message.

Parameters:

        [in]            msg      Message of type MessageQ_Msg

        [in]            priority Priority of message to be set.



typedef enum {
MessageQ_NORMALPRI   = 0,
MessageQ_HIGHPRI     = 1,
MessageQ_RESERVEDPRI = 2,
MessageQ_URGENTPRI   = 3
             } MessageQ_Priority;



#define MessageQ_PRIORITYMASK 0x3: Mask to extract priority setting.
```

Figure 9.10 Message priority settings.

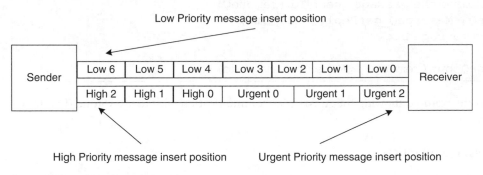

Figure 9.11 Priority illustration.

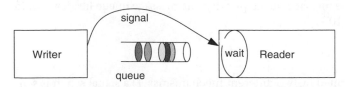

Figure 9.12 Synchronisation between the writer and the reader.

After the message is placed into the queue by using **MessageQ_put()**, the queue's **MessageQ_Params::synchronizer** signal function is called. The synchroniser waits in the **MessageQ_get** if there are no messages present.

The **MessageQ** supports different thread models such as **Hwi**, **Swi** or **task**, as described in the SYS/BIOS chapter.

Each **MessageQ** instance would get its own synchroniser object that is selected during the initialisation of the **MessageQ** parameter initialisation, as shown here:

MessageQ_Params_init(&messageQParams);
messageQParams.synchronizer = SyncSem_Handle_upCast(syncSemHandle);

When using a **MessageQ_create("Name", NULL)** with the second parameter as NULL, by default the synchroniser is a **SyncSem**.

The following are **ISync** implementations provided by XDCtools and SYS/BIOS [3]:

- **xdc.runtime.knl.SyncNull**. The **signal()** and **wait()** functions do nothing. Basically, this implementation allows for polling.
- **xdc.runtime.knl.SyncSemThread**. An implementation built using the **xdc.runtime.knl. Semaphore** module, which is a binary semaphore.
- **xdc.runtime.knl.SyncGeneric.xdc**. This implementation allows you to use custom **signal()** and **wait()** functions as needed.
- **ti.sysbios.syncs.SyncSem**. An implementation built using the **ti.sysbios.ipc.Semaphore** module. The **signal()** function runs a **Semaphore_post()**. The **wait()** function runs a **Semaphore_pend()**. See Example 9.1.
- **ti.sysbios.syncs.SyncSwi**. An implementation built using the **ti.sysbios.knl.Swi** module. The **signal()** function runs a **Swi_post()**. The **wait()** function is implemented as a **Swi** function and returns FALSE if the timeout elapses. See Figure 9.13 and Example 9.2.

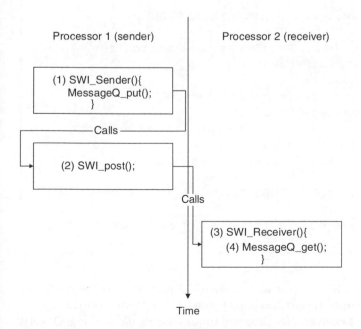

Figure 9.13 Illustration of synchronisation when using **Swi**s.

- **ti.sysbios.syncs.SyncEvent**. An implementation built using the **ti.sysbios.ipc.Event** module. The **signal()** function runs an **Event_post()**. The **wait()** function does nothing and returns FALSE if the timeout elapses. This implementation allows waiting on multiple events.

Example 9.1: Explicit use of **SyncSem** as a synchroniser [4]

```
/* Create a message queue. Using SyncSem as the synchronizer */
#include <ti/sysbios/syncs/SyncSem.h>
...
MessageQ_Params  messageQParams;
SyncSem_Handle    syncSemHandle;

/* Create a message queue using SyncSem as synchronizer */
syncSemHandle = SyncSem_create(NULL, NULL);
MessageQ_Params_init(&messageQParams);
messageQParams.synchronizer = SyncSem_Handle_upCast
(syncSemHandle);
messageQ = MessageQ_create(CORE1_MESSAGEQNAME,
&messageQParams, NULL);
```

Example 9.2: An example of explicitly using **SyncSwi** which is non-blocking

```
Swi_Params_init(&swiParams);
swiParams.priority = 1;
swiHandle = Swi_create(swi1_func, &swiParams, NULL);

/* Create a message queue. Using SyncSwi as the synchronizer */
SyncSwi_Params_init(&syncSwiParams);
syncSwiParams.swi = swiHandle;
syncSwiHandle = SyncSwi_create(&syncSwiParams, NULL);
MessageQ_Params_init(&messageQParams);
// unconditionally move one level up the inheritance hierarchy
//ISync_Handle SyncSwi_Handle_upCast( SyncSwi_Handle handle );
messageQParams.synchronizer = SyncSwi_Handle_upCast
(syncSwiHandle);
messageQ = MessageQ_create(CORE0_MESSAGEQNAME, &messageQParams);
if (messageQ == NULL) {
      System_abort("MessageQ_create failed\n" );
}
```

Figure 9.13 shows an example with two software interrupt functions, **Swi_Sender()** and **Swi_Receiver()**. When the **Swi_Sender()** calls **MessageQ_put()**, a **Swi_Post()** API is automatically called. Also, when the **Swi_Receiver()** is scheduled to run, it calls the **MessageQ_get()** function. **MessageQ_get()** will block until a timeout occurs.

9.5   ListMP module

Since we are dealing with messages and message queues, linked lists are very handy since they provide dynamic data structures for allocating memory while the program is running. Insertion and deletion node operations are easily implemented with linked lists. Linked lists are very convenient for linear data structures such as stacks and queues and also are easy to implement. However, they do not support synchronisation, and therefore, if needed for example, they can be built using the **Notify** module. The data to exchange must first be put in a buffer that is pushed to the shared memory. This shared memory can be configured statically or dynamically. Shared data are not protected, and therefore gates (as shown in Section 9.6) can be used with **ListMP** for data protection.

The linked lists also come with some drawbacks: they waste memory (need to store structure and data), do not allow random access and are not deterministic as each memory access takes a different amount of time, depending on the data location in the linked list.

The functions given in Table 9.3 are available for the **ListMP** module.

There are three main steps required for using the **ListMP**: initialise, create and open.

1) Initialise the **ListMP** parameters using **ListMP_Params_init** (**ListMP_Params *params**). The data fields for the parameter structure **params** are:

Type	Attribute	Comments
GateMP_Handle	gate	Using the default value of NULL will result in the use of the **GateMP** system gate for context protection.
String	name	**name** is the name of the instance, if not NULL. It must be unique among all **ListMP** instances in the entire system. When creating a new heap, it is necessary to supply an instance name.
UInt16	regionId	**SharedRegion** ID. The shared memory is divided into shared regions, and each shared region is represented by a **regionId**. This **regionId** is where this shared instance is to be placed.

Table 9.3 Functions for the **ListMP** module

ListMP_Params_init()	Initialises **ListMP** parameters
ListMP_create()	Creates and initialises **ListMP** module
ListMP_close()	Closes an opened **ListMP** instance
ListMP_delete()	Deletes a **ListMP** instance
ListMP_open()	Opens a created **ListMP** instance
ListMP_empty()	Tests for an empty **ListMP**
ListMP_getHead()	Gets the element from the front of the **ListMP**
ListMP_getTail()	Gets the element from the end of the **ListMP**
ListMP_insert()	Inserts element into a **ListMP** at the current location
ListMP_next()	Returns the next element in the **ListMP** (non-atomic)
ListMP_prev()	Returns the previous element in the **ListMP** (non-atomic)
ListMP_putHead()	Puts an element at the head of the **ListMP**
ListMP_putTail()	Puts an element at the end of the **ListMP**
ListMP_remove()	Removes the current element from the middle of the **ListMP**

As an example, we can use the following code to declare a structure of type **ListMP_Parms** and initialise it.

ListMP_Params_init(&params); this function will initialise the **params** structure with the default values

```
params.gate = gateHandle; // initialise the gate attribute with
gateHandle
params.name = "myListMP"; // initialise the name attribute with
myListMP
params.regionId = 0;    // initialise the region id, default 0
```

2) Once the **ListMP** parameters are initialised, a **ListMP** instance can be created and initialised by using:

```
ListMP_Handle ListMP_create(const ListMP_Params * params)
//This function will return a ListMP handle if found as shown below:
handle1 = ListMP_create(&params);
```

3) Once a **ListMP** instance is created and initialised, it can be opened by the processor or thread that created it or a remote processor or thread. For example:

```
(ListMP_open("myListMP", &handle1, NULL);
```

9.6   GateMP module

GateMPs are used to protect reads/writes to a shared resource, such as shared memory. This applies to both local and remote processors. For example, when a task enters a **GateMP**, this task will be protected and therefore will not be pre-empted by other tasks (running on local or remote processors).

To use **GateMP**, the gate parameters have to be initialised, a gate has to be created, then the gate is entered (for protection) and finally the gate is left as shown in Figure 9.14. The main APIs used for **GateMP** are:

- **GateMP_create**: Create a new instance.
- **GateMP_open**: Open an existing instance.
- **GateMP_enter**: Acquire the gate.
- **GateMP_leave**: Release the gate.

9.6.1   Initialising a GateMP parameter structure

The **GateMP** structure is as follows:

```
void GateMP_Params_init (GateMP_Params *params)
```

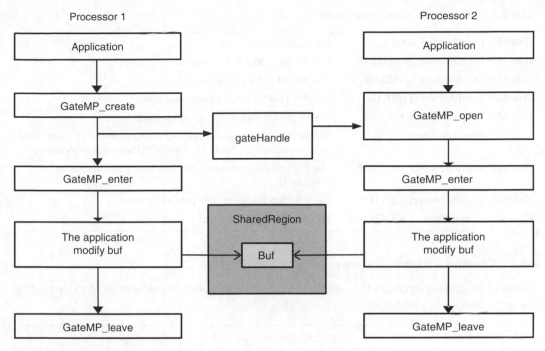

Figure 9.14 How **GateMP** is used.

The data fields for the **GateMP** parameter structure, **GateMP_Params**, are given here.

Data fields for the **GateMP_Params**:

```
String name // name of the instance
UInt16 regionId // shared region
GateMP_LocalProtect localProtect // Local protection level
GateMP_RemoteProtect     remoteProtect // Remote protection level
```

Example:

```
GateMP_Params_init(&gparams); // initialise the gparams with the
default value
gparams.remoteProtect = GateMP_RemoteProtect_SYSTEM;
//see below the type of remote gate.
gparams.name = "myGate";
```

9.6.1.1 Types of gate protection

There are two types of gate protection. One type is local (**GateMP_LocalProtect**), and the other type is remote (**GateMP_RemoteProtect**). Each type supports different levels of protection; see Table 9.4.

Table 9.4 Local and remote protection levels

GateMP_LocalProtect_NONE	Use no local protection.
GateMP_LocalProtect_INTERRUPT	Use the **INTERRUPT** local protection level.
GateMP_LocalProtect_TASKLET	Use the **TASKLET** local protection level.
GateMP_LocalProtect_THREAD	Use the **THREAD** local protection level.
GateMP_LocalProtect_PROCESS	Use the **PROCESS** local protection level.
GateMP_RemoteProtect_NONE	No remote protection; the **GateMP** instance will exclusively offer local protection configured in **GateMP_Params.localProtect**.
GateMP_RemoteProtect_SYSTEM	Use the **SYSTEM** remote protection level (default remote protection).
GateMP_RemoteProtect_CUSTOM1	Use the **CUSTOM1** remote protection level.
GateMP_RemoteProtect_CUSTOM2	Use the **CUSTOM2** remote protection level.

9.6.2 Creating a GateMP instance

Once the parameter structure is available, the creation of a **GateMP** instance can be achieved by using the following instruction:

```
GateMP_Handle  GateMP_create (const GateMP_Params * params)
```

Example:

```
gateHandle = GateMP_create(&gparams);
```

9.6.3 Entering a GateMP

To enter a **GateMP**, use the following instruction:

```
IArg GateMP_enter(GateMP_Handle handle)
```

Example:

```
IArg key;
key = GateMP_enter(gateHandle);
```

When you enter the gate, you keep the **key**, and when you leave you return the **key**. The **key** is needed for **GateMP_leave** as shown in Section 9.6.4.

9.6.4 Leaving a gate

For someone else to use the **key**, it must be released. This is achieved by using **GateMP_leave** as shown here:

```
Void GateMP_leave (GateMP_Handle handle,
     IArg key
       )
```

Parameters

[in]	**handle**	**GateMP handle**
[in]	**key**	**key** returned from **GateMP_enter**

9.6.5   The list of functions that can be used by GateMP

Other functions that can be used by **GateMP**:

Int	**GateMP_close** (**GateMP_Handle** *handlePtr)	
	Close an opened gate.	
GateMP_Handle	**GateMP_create** (const **GateMP_Params** *params)	
	Create a **GateMP** instance.	
Int	**GateMP_delete** (**GateMP_Handle** *handlePtr)	
	Delete a created **GateMP** instance.	
GateMP_Handle	**GateMP_getDefaultRemote** (**Void**)	
	Get the default remote gate.	
GateMP_LocalProtect	**GateMP_getLocalProtect** (**GateMP_Handle** handle)	
	Get the local protection level configured in a **GateMP** instance.	
GateMP_RemoteProtect	**GateMP_getRemoteProtect** (**GateMP_Handle** handle)	
	Get the remote protection level configured in a **GateMP** instance.	
Int	**GateMP_open** (**String** name, **GateMP_Handle** *handlePtr)	
	Open a created **GateMP** by name.	
Void	**GateMP_Params_init** (**GateMP_Params** *params)	
	Initialise **GateMP** parameters **struct**.	
IArg	**GateMP_enter** (**GateMP_Handle** handle)	
	Enter the **GateMP**.	
Void	**GateMP_leave** (**GateMP_Handle** handle, **IArg** key)	
	Leave the **GateMP**.	

The example in Table 9.5 shows how **GateMP** can be used between two cores. See the laboratory experiment in Section 9.9.2.

9.7   Multi-processor Memory Allocation: HeapBufMP, HeapMemMP and HeapMultiBufMP

Chapter 7 showed that there are three modules – The heap package include **HeapBuf** (fixed-size buffers), **HeapMem** (variable-sized buffers), or **HeapMultiBuf** (multiple fixed-size buffers) – for memory management that have the advantages of being fast, having a small footprint and being more deterministic than **malloc()**, which is not suitable for an embedded system as it causes memory fragmentation, is slow and may make debugging more complicated. The only difference between HeapBuf, HeapMem or HeapMultiBuf and HeapBufMP, HeapMemMP or HeapMultiBufMP is that the latter are extended for shared memory in a multi-processor environment.

Table 9.5 Example showing how to use **GateMP**

Core0	Core1

#include < xdc/std.h>
#include < ti/ipc/GateMP.h>
GateMP_Params gparams;
GateMP_Handle gateHandle;
GateMP_Params_init(&gparams);
gparams.name = "myGate";
gparams.localProtect = GateMP_LocalProtect_NONE;
gparams.remoteProtect = GateMP_RemoteProtect_SYSTEM;
gateHandle = **GateMP_create**(gparams);
GateMP_enter(gateHandle);
/∗ function to modify the share memory ∗/
GateMP_leave(gateHandle);
	#include < xdc/std.h>
	#include < ti/ipc/GateMP.h>
	GateMP_Handle gateHandle;
	GateMP_open("myGate", &gateHandle);
	GateMP_enter(gateHandle);
	/* function to modify the share memory */
	GateMP_leave(gateHandle);

When using the IPC, a **ShareRegion** always has a heap. The cores acquire the handle for this heap and then allocate a memory from this heap.

- **HeapBufMP**. All buffers allocated from the **HeapBufMP** are of fixed sizes.
- **HeapMemMP**. All buffers allocated from the **HeapBuff** are of variable sizes.
- **HeapMultiBufMP**. All buffers allocated from the **HeapMultiBufMP** are of variable sizes but internally allocated from fixed-size blocks. If the block that is required is not exactly a multiple number of the fixed buffer, then it will add an extra buffer if the option **BlockBorrow** is set to 1.

The initialisation parameter required for each instance is shown in the remainder of this section.

9.7.1 HeapBuf_Params

```
typedef struct HeapBuf_Params {
// Instance config-params structure
   IInstance_Params *instance;
   // Common per-instance configs
   SizeT align;   // Alignment (in MAUs) of each block
   SizeT blockSize;   // Size (in MAUs) of each block
   Ptr buf;   // User supplied buffer; for dynamic creates only
   Memory_Size bufSize;   // Size (in MAUs) of the entire buffer; for
   dynamic creates only
   UInt numBlocks;   // Number of fixed-size blocks
} HeapBuf_Params;

Void HeapBuf_Params_init(HeapBuf_Params *params);
// Initialise this config-params structure with supplier-specified
   defaults before instance creation
```

9.7.2 HeapMem_Params

```
typedef struct HeapBuf_Params {
// Instance config-params structure
    IInstance_Params *instance;
    // Common per-instance configs
    SizeT align;    // Alignment (in MAUs) of each block
    SizeT blockSize;    // Size (in MAUs) of each block
    Ptr buf;    // User supplied buffer; for dynamic creates only
    Memory_Size bufSize;    // Size (in MAUs) of the entire buffer; for
    dynamic creates only
    UInt numBlocks;    // Number of fixed-size blocks
} HeapBuf_Params;

Void HeapBuf_Params_init(HeapBuf_Params *params);
// Initialise this config-params structure with supplier-specified
    defaults before instance creation
```

9.7.3 HeapMultiBuf_Params

```
typedef struct HeapMultiBuf_Params {
// Instance config-params structure
IInstance_Params *instance;  // Common per-instance configs
Bool blockBorrow;                  // Turn block borrowing on (true) or
                                   off (false)
HeapBuf_Params bufParams[];  // Config parameters for each buffer
Int numBufs;                       // Number of memory buffers
} HeapMultiBuf_Params;
Void HeapMultiBuf_Params_init(HeapMultiBuf_Params *params);
// This initialises this config-params structure with supplier-
    specified
//defaults before instance creation
```

9.7.4 Configuration example for HeapMultiBuf

The following configuration code creates a **HeapMultiBuf** instance which manages three pools of ten blocks each, with block sizes of 64, 128 and 256 [5].

```
var HeapMultiBuf = xdc.useModule('ti.sysbios.heaps.HeapMultiBuf');
var HeapBuf = xdc.useModule('ti.sysbios.heaps.HeapBuf');

// Create parameter structure for HeapMultiBuf instance.
var hmbParams = new HeapMultiBuf.Params();
hmbParams.numBufs = 3;

// Create the parameter structures for each of the three
```

```
// HeapBufs to be managed by the HeapMultiBuf instance.
hmbParams.bufParams.$add(new HeapBuf.Params());
hmbParams.bufParams[0].blockSize = 64;
hmbParams.bufParams[0].numBlocks = 10;
hmbParams.bufParams.$add(new HeapBuf.Params());
hmbParams.bufParams[1].blockSize = 128;
hmbParams.bufParams[1].numBlocks = 10;
hmbParams.bufParams.$add(new HeapBuf.Params());
hmbParams.bufParams[2].blockSize = 256;
hmbParams.bufParams[2].numBlocks = 10;

// Create the HeapMultiBuf instance, and assign the global handle
// 'multiBufHeap' to it. Add '#include <xdc/cfg/global.h>' to your
// .c file to reference the instance by this handle.
Program.global.multiBufHeap = HeapMultiBuf.create(hmbParams);
```

The **HeapMemMP_create** call initialises the shared memory as needed and creates an instance. Once an instance is created, a **HeapMemMP_open** can be performed. The **HeapMemMP_open** is used to gain access to the same **HeapMemMP** instance. Generally, an instance is created on one processor and opened on the other processor(s) [6]; see Figure 9.15.

Example:

```
IHeap_Handle heap;
Ptr ptr;
heap = (IHeap_Handle)SharedRegion_getHeap(0); // get the heap handle
ptr = Memory_alloc(heap, 100, 0, NULL); // allocate memory from heap
Ptr    Memory_alloc
(          IHeap_Handle heap,
           SizeT        size,
           SizeT        align,
           Ptr          eb
)
```

The above code allocates the specified number of bytes [7].

Parameters

heap	Handle to the heap from which the memory is to be allocated. Specify NULL to allocate from local memory.
size	Amount of memory to be allocated
align	Alignment constraints (power of 2)
eb	Not used. Pass as NULL.

Return values

Pointer	Success: Pointer to allocated buffer
NULL	Failed to allocate memory

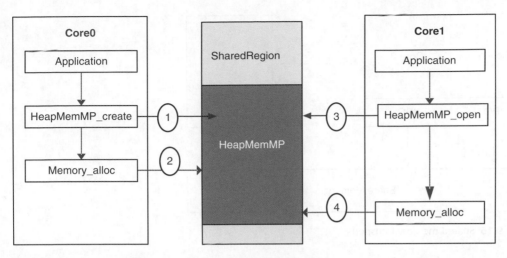

Figure 9.15 Allocation and using a shared memory.

What happens if the heap is not NULL and you would like to allocate memory using **HeapBufMP**?

By using **#include < xdc/cfg/global.h >** in the C file, you will get all global object definitions from the **config file**.

If a heap named **myHeap** is created statically, for instance by using **HeapMem.Create()**, by using **Program.global** (as **Program.global.myHeap = HeapMem.create();**), you will get **myHeap** defined in < **xdc/cfg/global.h** > just as it has been defined statically.

9.8 Transport mechanisms for the IPC

Shared memory is the default transport mechanism for passing data from one core to another on a single device; see Figure 9.16. However, the Navigator and the SRIO (see Figure 9.17) are also available. Other mechanisms of transport such as over the Hyperlink are possible but were not supported when this book was written. Further details on the transport can be found in Ref. [8].

9.9 Laboratory experiments with KeyStone I

9.9.1 Laboratory 1: Using MessageQ with multiple cores

File location:

\Chapter_9_Code\IpcSharedMem
Project: **IpcSharedMem**.

This laboratory experiment will demonstrate the use of the IPC.

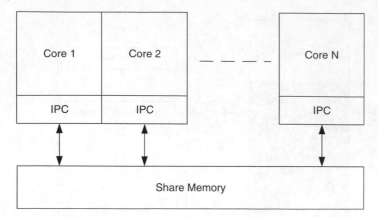

Figure 9.16 Shared memory transport.

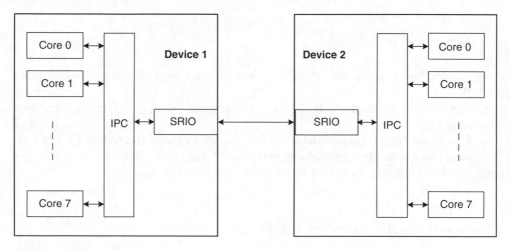

Figure 9.17 SRIO transport.

9.9.1.1 Overview

In this example, we will use **MessageQ** to pass a token message from one core to other randomly chosen cores. A total of **NUM_MESSAGES** will be passed. There are three types of messages that we will use in this example, **MSG_TOKEN**, **MSG_ACK** and **MSG_DONE**:

MSG_TOKEN. The token message is passed between the randomly chosen cores.

MSG_ACK. When a core receives the token, it will send an acknowledge message back to the core that sent it the token.

MSG_DONE. Once **NUM_MESSAGE** passes have occurred, whichever core has the token will send a **DONE** message to all other cores, letting them know to exit.

Starting the laboratory experiment:

1) Connect your EVM to your PC, and power your EVM.

Open the CCS, import the project **IpcSharedMem** and verify the following: the files are available, your compiler version is **TI v7.4.2** or higher and the correct emulator is as shown in Figure 9.18.

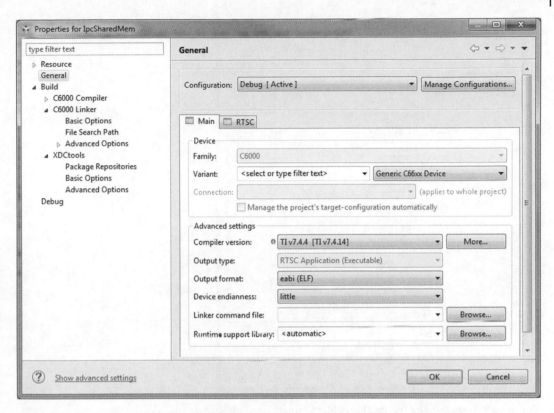

Figure 9.18 Properties used for the **IpcSharedMem** project.

2) Build the project (**Project > Build All**, or **Project > Clean**, then **Project > Build**). If no error is shown, that means your tools are updated for this project.
3) Now click on the RTSC tab and verify that there are no errors; see Figure 9.19.
4) Click on the bug ![bug icon] ▼ to load the code, then in the debug window select all the cores and group them as shown in Figure 9.20.
5) Click on Group 1 and press run; see Figure 9.21.
6) Observe the output; see Figure 9.22.
7) Examine the code. Use this link **C:\ti\ipc_3_21_00_07\docs\doxygen\html\index.html** to search for the IPC function definitions; see Figure 9.23.

9.9.2 Laboratory 2: Using ListMP, ShareRegion and GateMP

File location:

\Chapter_9_Code\ListMP_2Cores
Project Name: **ListMP_2Cores**.

In this laboratory experiment, the **ListMP**, **ShareRegion** and **GateMP** modules are used. In this laboratory, two cores are used. **Core0** creates a **LinkList**; starts filling it from the tail with 0, 2, 4, 6 and 8; and prints the list. **Core1** continues to fill the **LinkList** with odd numbers 1, 3, 5, 7 and 9 and prints the list again.

Figure 9.19 Tools used for the **IpcSharedMem** project.

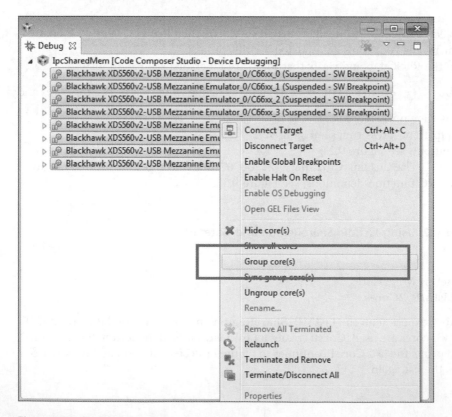

Figure 9.20 Grouping the cores.

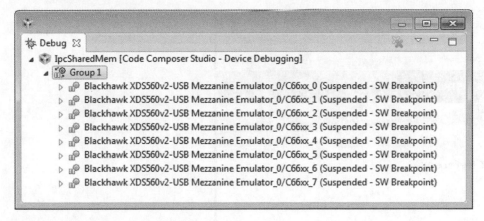

Figure 9.21 All cores grouped.

Figure 9.22 Output console.

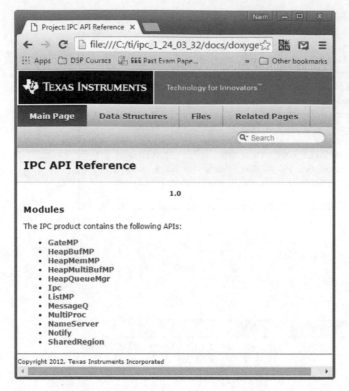

Figure 9.23 IPC API references.

To achieve these, the following tasks have to be performed:

1) Create two tasks. Each should run on one core (**cpu0Task** and **cpu1Task**); see the function **main().**
2) Create a data structure that needs to be inserted into the **LinkList**, as shown here:

```
typedef struct MyStructure {
    ListMP_Elem elem;
    Int scratch[30]; // make sure that the structure fits in a cache line
    Int flag;
            } MyStructure;
```

3) Create a parameter list as shown here:

```
ListMP_Params_init(&params);
params.gate = gateHandle;
params.name = "myListMP";
params.regionId = 1;
```

4) Create a memory region where the list is going to be put. In this example, two regions (**region 0** and **region 1**) have been created for demonstration:

```
Program.global.shmBase0 = 0xC000000;
Program.global.shmSize0 = 0x200000;

mem = [];

mem[0] =
    {
    base: Program.global.shmBase0,
        len: Program.global.shmSize0,
        ownerProcId: 0,
        name: "shared_mem",
        isValid: true
    }

SharedRegion.setEntryMeta(0, mem[0]);

Program.global.shmBase1 = 0xc200000;
Program.global.shmSize1= 0x200000;

mem[1] =
    {
    base: Program.global.shmBase1,
        len: Program.global.shmSize1,
        ownerProcId: 0,
        name: "shared_mem1",
        isValid: true
    }
SharedRegion.setEntryMeta(1, mem[1]);
```

5) Allocate a buffer where the **LinkList** will be allocated:

```
buf = Memory_alloc(SharedRegion_getHeap(1), sizeof(MyStructure) *
COUNT, 128,NULL);
```

    In this case, **regionId** 1 is used.

6) Create a parameter list as shown here:

```
GateMP_Params_init(&gparams);
gparams.remoteProtect = GateMP_RemoteProtect_SYSTEM;
gparams.name = "myGate";
gateHandle = GateMP_create(&gparams);
```

7) Use **GateMP_enter()** and **GateMP_leave()** before and after putting the structure in the list since **ListMP_put_Tail()** functions are not atomic. See code here:

```
key = GateMP_enter(gateHandle);
/* Add 0, 2, 4, 6, 8 */
for (i = 0; i < COUNT; i = i + 2)    {
      ListMP_putTail(handle1, (ListMP_Elem *) &(buf[i]));
}
GateMP_leave(gateHandle, key);
```

8) Create and initialise the **ListMP** as shown here:

```
handle1 = ListMP_create(&params);
```

9) Use the functions for the **ListMP**:

```
key = GateMP_enter(gateHandle);
/* Add 0, 2, 4, 6, 8 */
for (i = 0; i < COUNT; i = i + 2)    {
      ListMP_putTail(handle1, (ListMP_Elem *) &(buf[i]));
}
GateMP_leave(gateHandle, key);
```

10) Explore the files **listmp_2Core.c** and **listmp_2Core_static.cfg**.
11) Build the project and group the two cores as shown in Figure 9.24, then run the project. The output should be as shown in Figure 9.25.

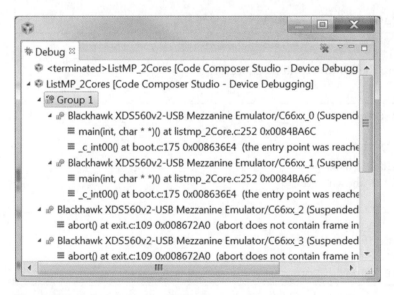

Figure 9.24 Grouping **Core0** and **Core1**.

```
Console ⊠
ListMP_2Cores:CIO
[C66xx_0] Started IPC

I am running in Core:   = 0

The RegionID used is:   = 1
ListMP instance is empty
ListMP instance has been filled
Print the List: 0 2 4 6 8
Print the list in reverse: 8 6 4 2 0
ListMP instance FALSE
ListMP instance is NOT empty
[C66xx_1] Started IPC

I am running in Core:   = 1
ListMP instance is NOT empty
Print the List: 0 2 4 6 8 1 3 5 7 9
Print the list in reverse: 9 7 5 3 1 8 6 4 2 0

Remove the first two elements by using ListMP_getHead: 0 2
Print the List: 4 6 8 1 3 5 7 9
Print the list in reverse: 9 7 5 3 1 8 6 4
```

Figure 9.25 Console output.

9.10 Laboratory experiments with KeyStone II

Copy all files from \Chapter_9_Code\KS2\Chapter_MessageQ to your Virtual Machine (/home/Chapter_MessageQ).

9.10.1 Laboratory experiment 1: Transferring a block of data

Project location:

/home/Chapter_MessageQ/lab1_data

In this example, the setup and communication between the ARM and DSP using the **MessageQ** method are performed; see Figure 9.26.

9.10.1.1 Set the connection between the host (PC) and the KeyStone
The hardware setup is shown in Figure 9.27. The port addresses on the PC will change every time a serial device is connected. Open the device manager and find out the port addresses. In Figure 9.28, the port addresses found for the serial ports are **COM7** and **COM8**.

In Figure 9.29, **PuTTY** as a terminal emulator has been used. Open two terminals with **COM7** and **COM8** (that will depend on your settings, as it is very likely that you will have different ports), and set them as shown in Figure 9.29.

Start the VMware as shown in Figure 9.30. If the code is stored on a USB flash drive, disconnect it and connect it again, and press OK as shown in Figure 9.31. Make sure that the connections between the PC and the EVM are established by pressing the icon highlighted in Figure 9.32. If the connection cannot be established, one can create a new connection by following the steps shown in Figure 9.33 through Figure 9.37.

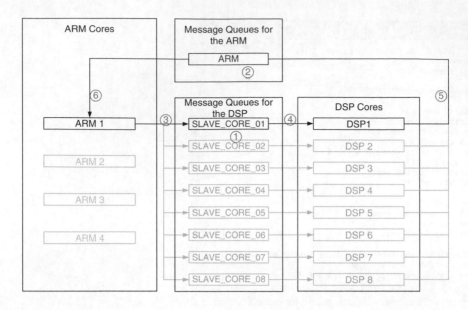

Figure 9.26 Illustration of the IPC communication using the **MessageQ**.

ARM side:

2. MessageQ_create();
   create ARM queues (named "arm")

3. MessageQ_open();
   open DSP queues "SLAVE_CORE_0#" as remote queue
   MessageQ_setReplyQueue();
   set local queue "arm" as reply queue
   MessageQ_put();
   send message (image data) to "SLAVE_CORE_0#"

6. MessageQ.get();
   get reply message

DSP side:

1. MessageQ_create();
   create DSP queues (named "SLAVE_CORE_0#")

4. MessageQ_get();
   get message (image data), can now process the image

5. MessageQ.getReplyQueue();
   get the writers' queue as remote queue, which is "arm"
   MessageQ.put();
   send reply message to "arm"

Open a terminal (**ctrl + alt + t**), and type **ifconfig** to check the IP address as shown in Figure 9.38. Go back to **PuTTY** and wait for the booting process to complete as shown in Figure 9.39. Check the IP address of the KeyStone II by typing **ifconfig** as shown in Figure 9.40.

Use FileZilla to transfer the file between EVM and VMware. Fill the host with **sftp://10.42.0.25** and username with **root** as shown in Figure 9.41.

9.10.1.2 Explore the ARM code

Explore the ARM code shown here and located in:

/home/Chapter_MessageQ/lab1_data/lab1_ARM/main.cpp

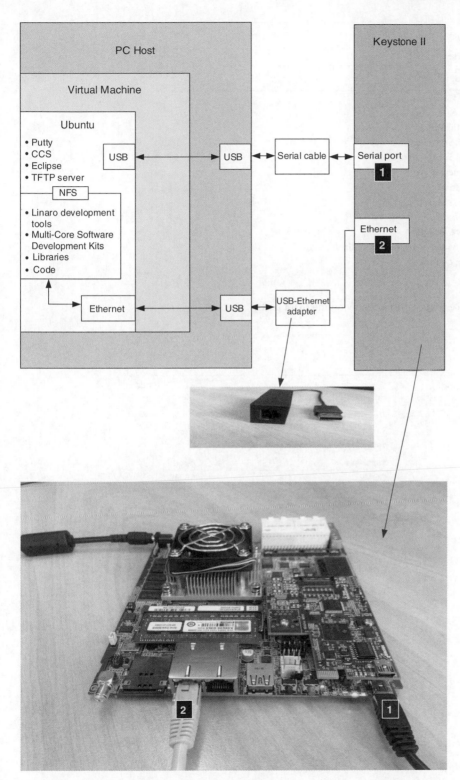

Figure 9.27 KeyStone II EVM setup.

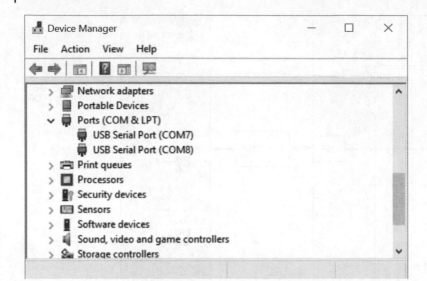

Figure 9.28 Device manager for identifying the **COM** ports.

Figure 9.29 Setting up the **COM** ports.

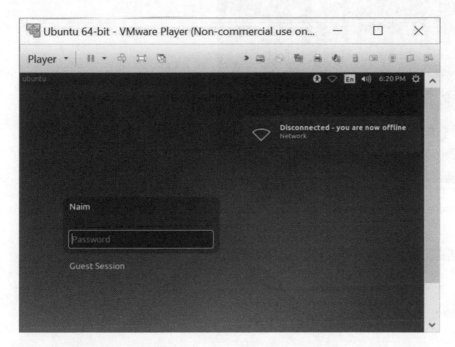

Figure 9.30 VMware.

Figure 9.31 Connecting the flash drive.

Figure 9.32 Establishing the connection between the PC and the EVM.

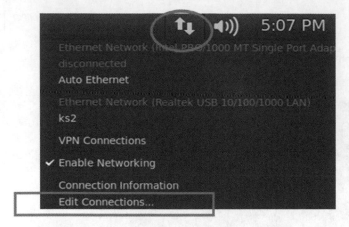

Figure 9.33 Edit connections.

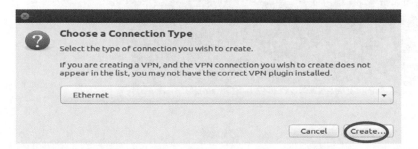

Figure 9.34 Choose a connection type.

Editing K2H

Connection name: K2H

General | Ethernet | 802.1x Security | IPv4 Settings | IPv6 Settings

Device MAC address: 00:E0:4C:68:0E:41 (eth7)

Cloned MAC address:
00:E0:4C:68:0E:41 (eth7)

MTU:
00:0C:29:C0:62:D4 (eth0)

Cancel   Save...

Figure 9.35 Select a MAC address and a connection name.

Figure 9.36 Select **Shared to other computers**.

Figure 9.37 Output when the connection is made.

Figure 9.38 Use **ifconfig** to check the IP addresses.

Figure 9.39 PuTTY terminal after booting.

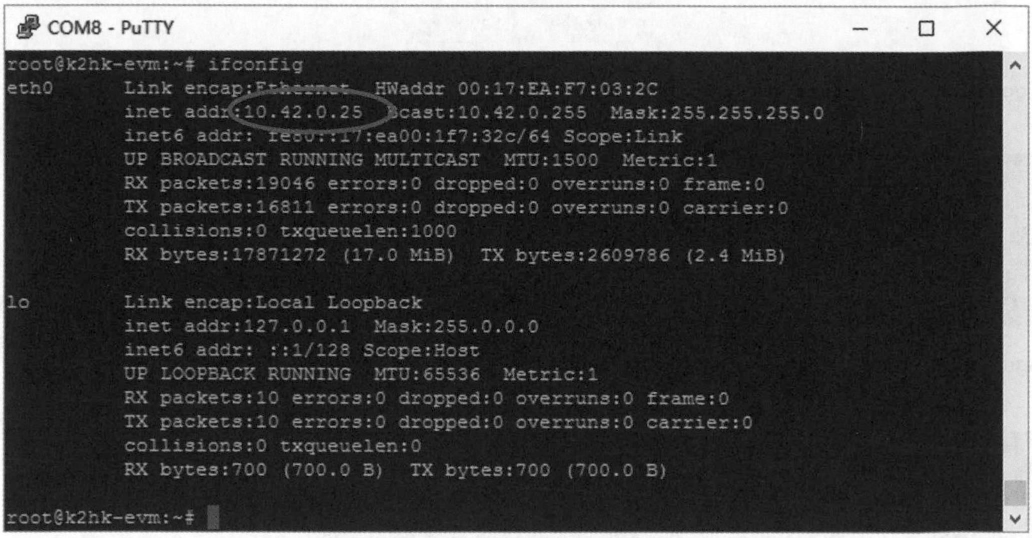

Figure 9.40 Finding the IP address for EVM.

Figure 9.41 Setting up FileZilla.

```
/*
 * main.c
 */
#include <stdio.h>
#include <stdlib.h>
#include <string.h>
#include <sys/types.h>
#include <sys/time.h>
#include <unistd.h>
#include <time.h>
#include <sys/time.h>
#include <ti/ipc/Std.h>
#include <ti/ipc/Ipc.h>
#include <ti/ipc/MessageQ.h>
#include "Left1.h"
#define Height 223
#define Width 280
typedef unsigned char byte;
typedef struct {
  MessageQ_MsgHeader h;
  int totalSize;
  short size;
  byte data[];
} structDataMsg;
typedef structDataMsg *DataMsg;
#define DATA_LEN 256
#define DSP_NUM_CORES 8
#define DSP_OFFSET 1
char *localQueueName = "arm";
MessageQ_QueueId remoteQ[DSP_NUM_CORES];
struct timeval tv1, tv2;

int main(int argc, char **argv) {
  gettimeofday(&tv1, NULL);
  int i;
  int status;
  printf("Starting IPC\n");
  Ipc_start();
  MessageQ_Params mqp;
  printf("init\n");
  MessageQ_Params_init(&mqp);
  MessageQ_Handle localQ = MessageQ_create(localQueueName, &mqp);
  char remoteQName[14];
  i = 0;
  DataMsg dm;
  byte buffer[] = { Left };
  byte buffer_out[Height*Width] = {0};
  int filelen = sizeof(buffer) / sizeof(byte);

  int offset = 0;
  int sizeLeft = filelen;
```

```
    snprintf(remoteQName, 14, "SLAVE_CORE_01");
    do {
      printf("Opening Queue %i\n", i + DSP_OFFSET);
    status = MessageQ_open(remoteQName, &(remoteQ[i]));
    if (status < 0) {
     printf("Error opening queue %s: %i\n", remoteQName, status);
    }
   } while (status < 0);
  int sending = 1;
  while(sizeLeft > 0) {
   short pcktLen = sizeLeft > DATA_LEN ? DATA_LEN : sizeLeft;
   dm = (DataMsg) MessageQ_alloc(0, (sizeof(structDataMsg) + sizeof
(int) + sizeof(short) + DATA_LEN * sizeof(char)));
   if(sending)
   {
    dm->totalSize = filelen;
    dm->size = pcktLen;
    memcpy(dm->data, buffer + offset, dm->size);
    offset += dm->size;
    sizeLeft -= dm->size;
    if(sizeLeft==0)
    {
     sizeLeft=filelen;
     sending=0;
     offset=0;
    }
   }
   else
   {
    dm->totalSize = -1;
    dm->size = pcktLen;
    memcpy(dm->data, buffer + offset, dm->size);
    offset += dm->size;
    sizeLeft -= dm->size;
   }
   MessageQ_setReplyQueue(localQ, (MessageQ_Msg) dm);
   status = MessageQ_put(remoteQ[i], (MessageQ_Msg) dm);
   if (status == MessageQ_S_SUCCESS) {
    int revStatus = -1;
    do {
     revStatus = MessageQ_get(localQ, (MessageQ_Msg *) &dm, 1000);
    } while (revStatus != MessageQ_S_SUCCESS);

    if(!sending) memcpy(buffer_out + offset - pcktLen, dm->data,
    dm->size);

   } else {
    printf("Message not sent for some reason (err %i)\n", status);
   }
  }
```

```
gettimeofday(&tv2, NULL);
printf ("Total time = %f seconds\n", (double) (tv2.tv_usec - tv1.
tv_usec) / 1000000 + (double) (tv2.tv_sec - tv1.tv_sec));
FILE *f = fopen("output.pgm", "wb");
fprintf(f, "P2\n%i %i %i\n", Width, Height,255);
for (int y=0; y<Height; y++)
{
  for (int x=0; x<Width; x++)
  {
    fprintf(f,"%d ",buffer_out[y*Width+x]);
  }
}
fclose(f);


Ipc_stop();
return 0;
}
```

9.10.1.3 Explore the DSP code
Explore the DSP code shown here and located in:

/home/Chapter_MessageQ/lab1_data/lab1_DSP/main.cpp

```
#include <xdc/std.h>
#include <xdc/cfg/global.h>

#include <xdc/runtime/Types.h>
#include <xdc/runtime/System.h>
#include <xdc/runtime/Memory.h>
#include <xdc/runtime/IHeap.h>
#include <xdc/runtime/Timestamp.h>
#include <xdc/runtime/Startup.h>

#include <ti/sysbios/BIOS.h>
#include <ti/sysbios/knl/Semaphore.h>
#include <ti/sysbios/knl/Task.h>

#include <ti/ipc/MultiProc.h>
#include <ti/ipc/Ipc.h>
#include <ti/ipc/Notify.h>
#include <ti/ipc/MessageQ.h>
#include <ti/ipc/SharedRegion.h>
#include <ti/ipc/transports/TransportRpmsg.h>

#include <c6x.h>
```

```c
#include <stdio.h>
#include <stdlib.h>
#include <string.h>

#define Height 223
#define Width 280

#define DATA_LEN 256

void tsk_fxn(UArg a0, UArg a1);

typedef unsigned char byte;

typedef struct {
 MessageQ_MsgHeader h;
 int totalSize;
 short size;
 byte data[];
} structDataMsg;

typedef structDataMsg *DataMsg;
int selfId;
int numCores;
char localQueueName[14];
char *remoteQueueName = "arm";

extern "C" {
Void myStartup() {
 int dsp_number = DNUM;
 MultiProc_setLocalId(dsp_number + 1);
}
}
void main() {
 selfId = MultiProc_self();
 numCores = MultiProc_getNumProcessors();
 snprintf(localQueueName, 14, "SLAVE_CORE_%02i", selfId);

 Task_Params tp;
 Task_Params_init(&tp);
 tp.vitalTaskFlag = true;
 tp.priority = 1;
 tp.affinity = selfId;
 Task_create(&tsk_fxn, &tp, NULL);
 BIOS_start();
 volatile int b = 1;
 while (b)
  ;
 return;
}
void edge(unsigned char *in, unsigned char *out)
{
```

```c
    int valX, valY = 0;
    int x, y = 0;
    int res = 0;
    for(y = 0; y < Height; y++)
    {
      for(x = 0; x < Width; x++)
      {
        valX = 0;
        valY = 0;
        if((x != 0)&&(x != Width-1)&&(y != 0)&&(y != Height-1)){
            // j = -1
            valX += in[(y-1)*Width+x-1];
            valX += in[(y-1)*Width+x]*2;
            valX += in[(y-1)*Width+x+1];

            //j = 0;
            valX += in[(y)*Width+x-1]*0;
            valX += in[(y)*Width+x]*0;
            valX += in[(y)*Width+x+1]*0;
            // j = 1
            valX += in[(y+1)*Width+x-1]*(-1);
            valX += in[(y+1)*Width+x]*(-2);
            valX += in[(y+1)*Width+x+1]*(-1);
            // j = -1
            valY += in[(y-1)*Width+x-1];
            valY += in[(y-1)*Width+x]*0;
            valY += in[(y-1)*Width+x+1]*(-1);
            //j = 0;
            valY += in[(y)*Width+x-1]*2;
            valY += in[(y)*Width+x]*0;
            valY += in[(y)*Width+x+1]*(-2);
            // j = -1
            valY += in[(y+1)*Width+x-1];
            valY += in[(y+1)*Width+x]*(0);
            valY +=in[(y+1)*Width+x+1]*(-1);
        }
        res = abs(valX)+abs(valY);
        if (res > 200){
          out[y*Width+x] = 255;
        }else{
          out[y*Width+x] = 0;
        }
      }
    }
    return;
}
void tsk_fxn(UArg a0, UArg a1) {
  int status;
  int offset = 0;
  unsigned char* imgBuffer = (unsigned char*) malloc(Height*Width);
  unsigned char* imgBuffer_out = (unsigned char*)
malloc(Height*Width);
```

```
DataMsg dm;
System_printf("In Task Function\n");
MessageQ_Params mqp;
MessageQ_Params_init(&mqp);
MessageQ_Handle localQ = MessageQ_create(localQueueName, &mqp);
if (localQ == NULL) {
 System_printf("Error: create failed\n");
 System_exit(1);
}
System_printf("Create queue succeeded, name: ");
System_printf(localQueueName);
System_printf("\n");

MessageQ_QueueId remoteQ;
System_printf("Size of DataMsg: %i\n", sizeof(structDataMsg));
int b = 1;
int timeoutcount = 0;

while (b) {
 status = MessageQ_get(localQ, (MessageQ_Msg *) &dm, 1000);
 if (status == MessageQ_S_SUCCESS) {
  remoteQ = MessageQ_getReplyQueue(dm);

  if(dm->totalSize>0)
  {
   memcpy(imgBuffer + offset, dm->data, dm->size);
   offset += dm->size;
   if(offset == dm->totalSize) {
    offset = 0;
    edge(imgBuffer,imgBuffer_out);
   }
  }
  else
  {
   memcpy(dm->data, imgBuffer_out + offset, dm->size);
   offset += dm->size;
  }
  MessageQ_setReplyQueue(localQ, (MessageQ_Msg) dm);
  do {
   status = MessageQ_put(remoteQ, (MessageQ_Msg) dm); // send
message back as ack
  } while (status < 0);
 }else if (status == MessageQ_E_TIMEOUT) {
  timeoutcount++;
 } else {
 }
}
MessageQ_free((MessageQ_Msg) dm);
BIOS_exit(0);
}
```

9.10.1.4 Compile and run the program

Import the ARM program into Eclipse; see Figure 9.42. Select **Existing Projects into Workspace** and press next as shown in Figure 9.43. Select the root directory as shown in Figure 9.44. Build the project as shown in Figure 9.45, and observe the output shown in Figure 9.46.

Import the DSP project into CCS as shown in Figure 9.47 and Figure 9.48. If CCS is not installed, install it from within VMware. Build the project as shown in Figure 9.49, and observe the console output. The console should be as shown in Figure 9.50.

Upload both of the executable files to the server as shown in Figure 9.51 and Figure 9.52. Upload the DSP loading script **load_all.sh** file to the server as shown in Figure 9.53.

In **PuTTY**, type '**chmod + x** *' to change the permission of all the files to allow executing. These can also be done within FileZilla. Change the file permission using FileZilla as shown in Figure 9.54 to allow execution. Right click on **load.sh** file, and set the permissions. These could also have been done earlier in **PuTTY** by typing **chmod + x**.

Type **./load_all.sh MessageQDSP.out** to load the program on DSP cores. Type **./ArmSide** to run the program. The program detects edges in the input image and writes the output image to **output.pgm**. The console output should be as shown in Figure 9.55. Access the output file **output.pgm** using FileZilla as shown in Figure 9.56. Select **output.pgm** and right click. See output in Figure 9.57.

Figure 9.42 Importing a file.

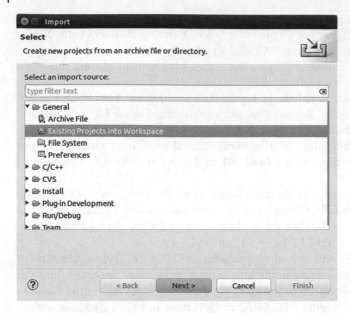

Figure 9.43 Select **Existing Projects into Workspace**.

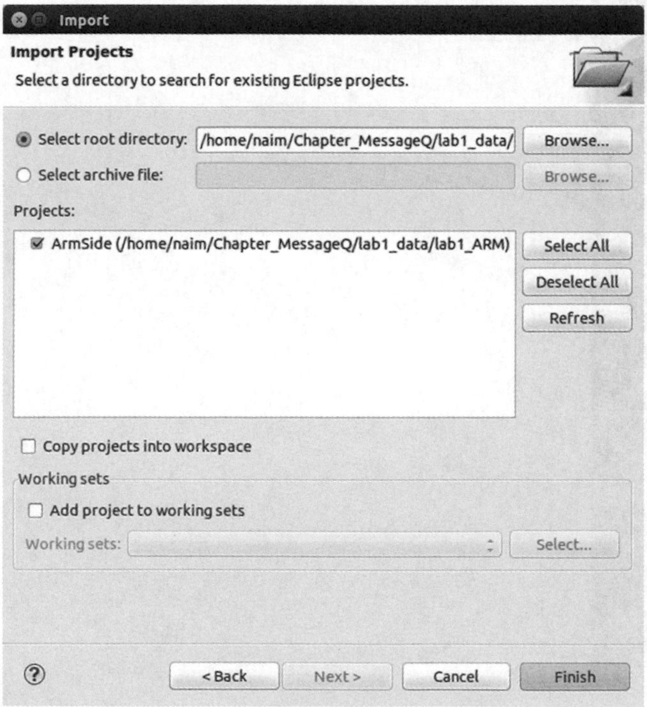

Figure 9.44 Selecting the root directory.

Figure 9.45 Building a project.

```
Tasks  Console 23
CDT Build Console [ArmSide]

Building target: ArmSide
Invoking: Cross G++ Linker
arm-linux-gnueabihf-g++  -o "ArmSide"  ./main.o   -l:/home/naim/ti/ipc_3_00_04_29/linux/src/api/.libs/libtiipc.so -l:/home/naim/ti/ipc_3_00_04_29/linux/src/utils/.libs/libtiipcutils.so
Finished building target: ArmSide

13:29:37 Build Finished (took 1s.280ms)
```

Figure 9.46 Console output.

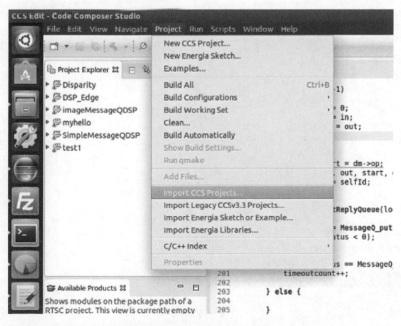

Figure 9.47 Importing a Code Composer Studio project.

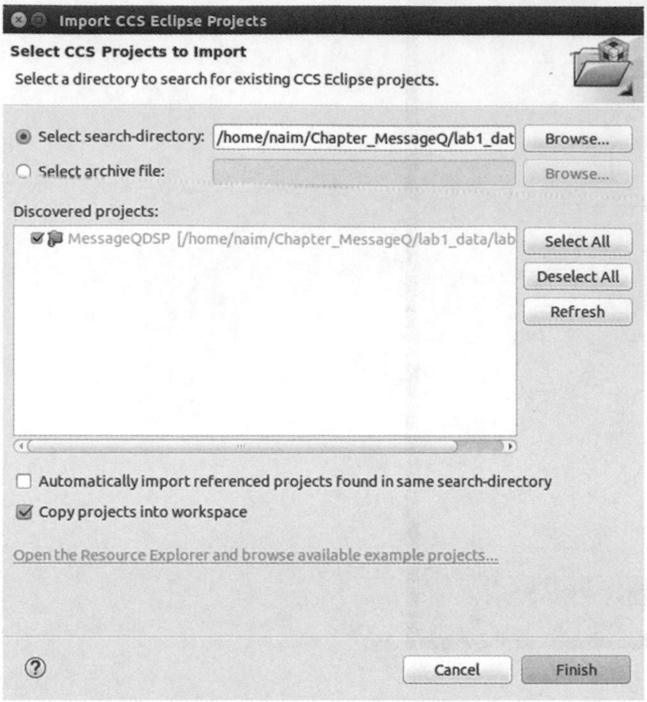

Figure 9.48 Selecting the directory.

Figure 9.49 Building a project.

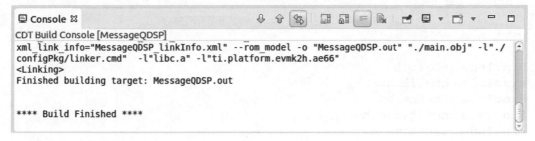

Figure 9.50 Console output.

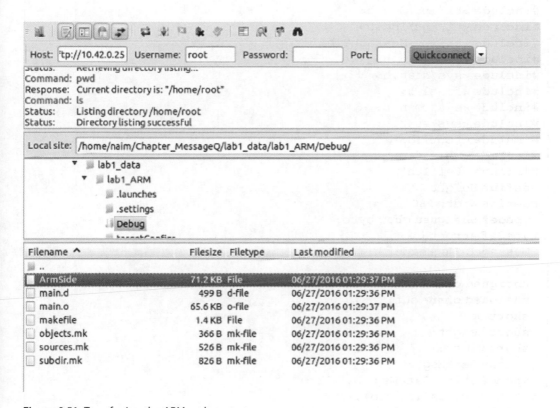

Figure 9.51 Transferring the ARM code.

9.10.2 Laboratory experiment 2: Transferring a pointer

This example is similar to Laboratory experiment 1, except that instead of sending data only a pointer is sent.

9.10.2.1 Explore the ARM code
The files are located in:

/home/Chapter_MessageQ/lab2_pointer/lab2_ARM/main.cpp

```c
/*
 * main.c
 */
#include <stdio.h>
#include <stdlib.h>
#include <string.h>
#include <sys/types.h>
#include <sys/time.h>
#include <unistd.h>
#include <time.h>
#include <sys/time.h>
#include >ti/ipc/Std.h<
#include <ti/ipc/Ipc.h>
#include <ti/ipc/MessageQ.h>
#include <sys/mman.h>
#include <sys/stat.h>
#include <fcntl.h>
#include <sys/mman.h>
#include <sys/stat.h>
#include <fcntl.h>
#include <unistd.h>
#include "Left1.h"
#define Height 223
#define Width 280
typedef unsigned char byte;
typedef struct {
 MessageQ_MsgHeader h;

  unsigned char* in;
  unsigned char* out;
  short op;
  short height;
  short width;
} structDataMsg;
typedef structDataMsg *DataMsg;
int calc_nrows(int cid);

#define DATA_LEN 256
#define DSP_NUM_CORES 1
#define DSP_OFFSET 1
char *localQueueName = "arm";
MessageQ_QueueId remoteQ[DSP_NUM_CORES];
struct timeval tv1, tv2;
unsigned char IN[Height*Width] = {Left};
unsigned char *IN_DSP;
unsigned char *OUT_DSP;

int main(int argc, char **argv) {
```

```c
int i;
int status;
printf("Starting IPC\n");
status = Ipc_start();
if(status !=0 )
{
 printf("IPC fail\n");
 return 1;
}
MessageQ_Params mqp;
printf("init\n");
MessageQ_Params_init(&mqp);
MessageQ_Handle localQ = MessageQ_create(localQueueName, &mqp);
char remoteQName[14];
i=0;
//send request to core 0 for memory address
snprintf(remoteQName, 14, "SLAVE_CORE_%02i", i+DSP_OFFSET);
do {
 //printf("Opening Queue %i\n", i + DSP_OFFSET);
 status = MessageQ_open(remoteQName, &(remoteQ[i]));
 if (status < 0) {
  printf("Error opening queue %s: %i\n", remoteQName, status);
 } else {
  //printf("Succeeded opening queue %s\n", remoteQName);
 }
} while (status < 0);

DataMsg dm;
dm = (DataMsg) MessageQ_alloc(0, (sizeof(structDataMsg) + 3*sizeof
(short) +2*sizeof(char*) ));
dm->op = -1;

//receive pointer of shared memory
MessageQ_setReplyQueue(localQ, (MessageQ_Msg) dm);
status = MessageQ_put(remoteQ[i], (MessageQ_Msg) dm);
if (status == MessageQ_S_SUCCESS) {
 do {
  status = MessageQ_get(localQ, (MessageQ_Msg *) &dm, 1000);
 } while (status < 0);
 IN_DSP = dm->in;
 OUT_DSP = dm->out;
} else {
 printf("Message not sent for some reason (err %i)\n", status);
}

DataMsg dm1[DSP_NUM_CORES];
//memory map
int size = Height*Width;
int g_devmem = open("/dev/mem", O_RDWR | O_SYNC);
```

```c
  if(g_devmem < 0 )
  {
   printf("Error opening /dev/mem\n");
   return -1;
  }
   unsigned char* IN_v = mmap(NULL, size, PROT_READ | PROT_WRITE,
MAP_SHARED, g_devmem, (int)IN_DSP);
   if(IN_v == (void *) -1 )
    printf("mmap failed\n");
   printf("start\n");
 gettimeofday(&tv1, NULL);
 //copy data into shared memory
 memcpy(IN_v,IN,Height*Width);
 int h=0;
 //inform DSPs to start work
 for (i = 0; i < DSP_NUM_CORES; i++) {
  int nrow = calc_nrows(i);
  snprintf(remoteQName, 14, "SLAVE_CORE_%02i", i+DSP_OFFSET);
  do {
   status = MessageQ_open(remoteQName, &(remoteQ[i]));
   if (status < 0) {
    printf("Error opening queue %s: %i\n", remoteQName, status);
   }
  } while (status < 0);
  dm1[i] = (DataMsg) MessageQ_alloc(0, (sizeof(structDataMsg) +
3*sizeof(short) +2*sizeof(char*)));
  dm1[i]->op = h; //
  dm1[i]->height = nrow;
  dm1[i]->width = Width;

  MessageQ_setReplyQueue(localQ, (MessageQ_Msg) dm1[i]);
  status = MessageQ_put(remoteQ[i], (MessageQ_Msg) dm1[i]);
  if (status != MessageQ_S_SUCCESS) {
   printf("Message not sent for some reason (err %i)\n", status);
  }
  h+=nrow;
 }
 //DSPs finish
 int count = 0;
 while(count<DSP_NUM_CORES)
 {
  MessageQ_get(localQ, (MessageQ_Msg *) &dm, 1000);
  count++;
 }
 //memory map of output
   unsigned char* OUT_v = mmap(NULL, size, PROT_READ | PROT_WRITE,
MAP_SHARED, g_devmem, (int)OUT_DSP);
   if(OUT_v == (void *) -1)
    printf("mmap out failed\n");
```

```
gettimeofday(&tv2, NULL);
printf ("Total time = %f seconds\n", (double) (tv2.tv_usec - tv1.
tv_usec) / 1000000 + (double) (tv2.tv_sec - tv1.tv_sec));
FILE *f = fopen("output.pgm", "wb");
fprintf(f, "P2\n%i %i %i\n", Width, Height,60);
for (int y=0; y<Height; y++)
{
  for (int x=0; x<Width; x++)
  {
    fprintf(f,"%d ",OUT_v[y*Width+x]);
  }
}
fclose(f);
Ipc_stop();
return 0;
}

int calc_nrows(int cid)
{
  int nrows = Height / DSP_NUM_CORES;
  int r =  Height % DSP_NUM_CORES;
  if (r != 0 &&#38; cid < r) {
    nrows ++;
  }
  return nrows;
}
```

9.10.2.2 Explore the DSP code
The files are located in:

/home/Chapter_MessageQ/lab2_pointer/lab2_DSP/main.cpp

```
#include <xdc/std.h>
#include <xdc/cfg/global.h>
#include <xdc/runtime/Types.h>
#include <xdc/runtime/System.h>
#include <xdc/runtime/Memory.h>
#include <xdc/runtime/IHeap.h>
#include <xdc/runtime/Timestamp.h>
#include <xdc/runtime/Startup.h>
#include <ti/sysbios/BIOS.h>
#include <ti/sysbios/knl/Semaphore.h>
#include <ti/sysbios/knl/Task.h>
#include <ti/ipc/MultiProc.h>
#include <ti/ipc/Ipc.h>
#include <ti/ipc/Notify.h>
```

```
#include <ti/ipc/MessageQ.h>
#include <ti/ipc/SharedRegion.h>
#include <ti/ipc/transports/TransportRpmsg.h>
#include <ti/ipc/HeapBufMP.h>
#include <ti/ipc/GateMP.h>
#include <c6x.h>
#include <stdio.h>
#include <stdlib.h>
#include <string.h>
#pragma DATA_SECTION(in, ".MSMC0")
#pragma DATA_SECTION(out, ".MSMC1")
#pragma DATA_ALIGN(in,4096)
#pragma DATA_ALIGN(out,4096)
#define Height 223
#define Width 280
unsigned char in[Height*Width];
unsigned char out[Height*Width];
void tsk_fxn(UArg a0, UArg a1);
typedef unsigned char byte;
typedef struct {
 MessageQ_MsgHeader h;

  unsigned char* in;
  unsigned char* out;
  short op;
  short height;
  short width;
} structDataMsg;
typedef structDataMsg *DataMsg;

int selfId;
int numCores;
char localQueueName[14];
char *remoteQueueName = "arm";

//extern "C" {
Void myStartup() {
 int dsp_number = DNUM;
 MultiProc_setLocalId(dsp_number + 1);
}
//}
void main() {
 selfId = MultiProc_self();
 numCores = MultiProc_getNumProcessors();
 snprintf(localQueueName, 14, "SLAVE_CORE_%02i", selfId);
 Task_Params tp;
 Task_Params_init(&tp);
 tp.vitalTaskFlag = TRUE;
```

```c
   tp.priority = 1;
   tp.affinity = selfId;
   Task_create(&tsk_fxn, &tp, NULL);
   BIOS_start();
   volatile int b = 1;
   while (b)
     ;
   return;
}
void edge(unsigned char *in, unsigned char *out, short startrow,
short nrow)
{
  int valX, valY = 0;
  int x, y = 0;
  int res = 0;

  for(y = startrow; y < startrow+nrow; y++)
  {
   for(x = 0; x < Width; x++)
   {
    valX = 0;
    valY = 0;
    if((x != 0)&&(x != Width-1)&&(y != 0)&&(y != Height-1)){

        // j = -1
        valX += in[(y-1)*Width+x-1];
        valX += in[(y-1)*Width+x]*2;
        valX += in[(y-1)*Width+x+1];
        //j = 0;
        valX += in[(y)*Width+x-1]*0;
        valX += in[(y)*Width+x]*0;
        valX += in[(y)*Width+x+1]*0;
        // j = 1
        valX += in[(y+1)*Width+x-1]*(-1);
        valX += in[(y+1)*Width+x]*(-2);
        valX += in[(y+1)*Width+x+1]*(-1);
         // j = -1
        valY += in[(y-1)*Width+x-1];
        valY += in[(y-1)*Width+x]*0;
        valY += in[(y-1)*Width+x+1]*(-1);
        //j = 0;
        valY += in[(y)*Width+x-1]*2;
        valY += in[(y)*Width+x]*0;
        valY += in[(y)*Width+x+1]*(-2);
        // j = -1
        valY += in[(y+1)*Width+x-1];
        valY += in[(y+1)*Width+x]*(0);
        valY +=in[(y+1)*Width+x+1]*(-1);
```

```
      }
      res = abs(valX)+abs(valY);
      if (res > 200){
        out[y*Width+x] = 255;
      }else{
        out[y*Width+x] = 0;
      }
    }
  }
 return;
}
void tsk_fxn(UArg a0, UArg a1) {
 int status;
 //int receiving = 0;
 //int offset = 0;
 //char *imgBuffer;

 System_printf("In Task Function\n");
 MessageQ_Params mqp;
 MessageQ_Params_init(&mqp);
 MessageQ_Handle localQ = MessageQ_create(localQueueName, &mqp);
 if (localQ == NULL) {
  System_printf("Error: create failed\n");
  System_exit(1);
 }
 System_printf("Create queue succeeded, name: ");
 System_printf(localQueueName);
 System_printf("\n");

 MessageQ_QueueId remoteQ;

 System_printf("Size of DataMsg: %i\n", sizeof(structDataMsg));
 volatile int b = 1;
 int timeoutcount = 0;

 while (b) {
  DataMsg dm;

  status = MessageQ_get(localQ, (MessageQ_Msg *) &dm, 1000);
  if (status == MessageQ_S_SUCCESS) {
   remoteQ = MessageQ_getReplyQueue(dm);
   if(dm->op==-1)
   {
    dm->op = 0;
    dm->in = in;
    dm->out = out;
   }
   else
```

```
  {
    int start = dm->op;
    edge(in, out, start, dm->height);
    dm->op = selfId;
  }
  MessageQ_setReplyQueue(localQ, (MessageQ_Msg) dm);
  do {
    status = MessageQ_put(remoteQ, (MessageQ_Msg) dm); // send
message back as ack
  } while (status < 0);
  } else if (status == MessageQ_E_TIMEOUT) {
    timeoutcount++;
  } else {
  }
  }
  BIOS_exit(0);
}
```

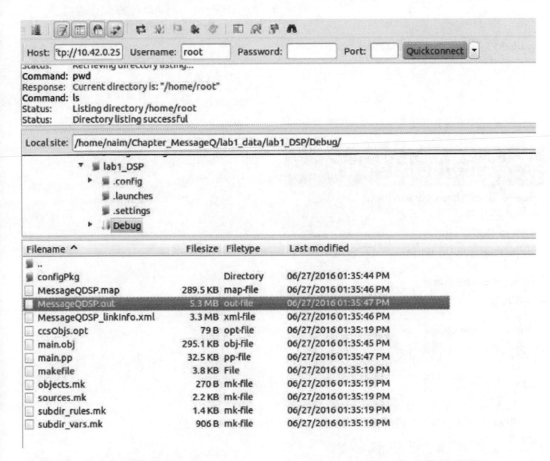

Figure 9.52 Transferring the DSP code.

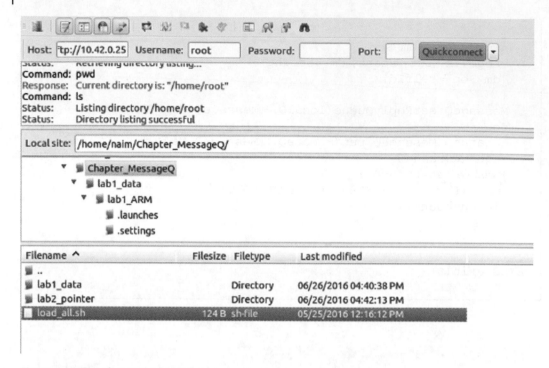

Figure 9.53 Transferring the **load.sh**.

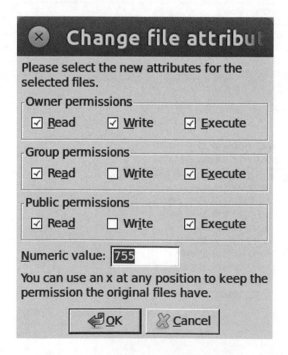

Figure 9.54 Changing the permission of a file.

Figure 9.55 Console output.

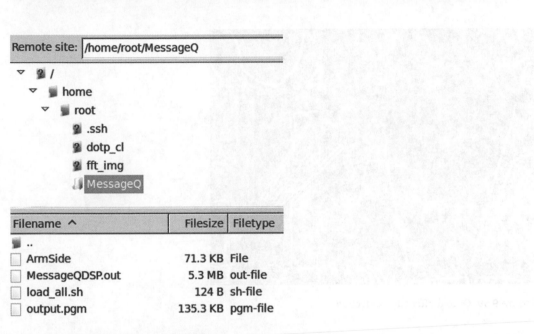

Figure 9.56 Accessing files using FileZilla.

Figure 9.57 Output after edge detection.

```
COM8 - PuTTY                                           —   □   ✕
root@k2hnode1:~# ./ARM_Edge
Starting IPC
init
start
Total time = 0.002114 seconds
root@k2hnode1:~#
```

Figure 9.58 Console output.

Figure 9.59 Output after edge detection.

9.10.2.3 Compile and run the program

Compile and upload the program in the same way as in Laboratory experiment 1. Type **./load_all.sh DSP_Edge.out** to load the program on DSP cores. Type **./ARM_Edge** to run the program (see Figure 9.58). The program detects edges in the input image and writes the output image to **output.pgm** (see Figure 9.59).

9.11 Conclusion

In this book, the communication between cores has been emphasised in many chapters. In this chapter, a comprehensive review of selected inter-processor communications modules and practical examples with both the KeyStone I and KeyStone II have been given.

Please refer to the new IPC releases that can be accessed from Ref. [9].

References

1 Texas Instruments, IPC product releases, [Online]. Available: http://software-dl.ti.com/dsps/ dsps_public_sw/sdo_sb/targetcontent/ipc/.
2 Texas Instruments, IPC benchmarking, February 2015. [Online]. Available: http://processors.wiki. ti.com/index.php/IPC_BenchMarking#Linux.

3 Texas Instruments, xdc.runtime.knl.ISync, [Online]. Available: http://rtsc.eclipse.org/cdoc-tip/xdc/runtime/knl/ISync.html#xdoc-desc.

4 Texas Instruments, SYS/BIOS inter-processor communication (IPC) 1.25 User's Guide, September 2012. [Online]. Available: http://www.ti.com/lit/ug/sprugo6e/sprugo6e.pdf.

5 Texas Instruments, Multiple fixed size buffer heap manager, [Online]. Available: http://software-dl.ti.com/dsps/dsps_public_sw/sdo_sb/targetcontent/sysbios/6_42_01_20/exports/bios_6_42_01_20/docs/cdoc/ti/sysbios/heaps/HeapMultiBuf.html.

6 Texas Instruments, IPC API 3.40.00.06, 2015. [Online]. Available: http://software-dl.ti.com/dsps/dsps_public9_sw/sdo_sb/targetcontent/ipc/latest/docs/doxygen/html/index.html.

7 Texas Instruments, Static and run-time memory manager, February 2015. [Online]. Available: http://rtsc.eclipse.org/cdoc-tip/xdc/runtime/Memory.html.

8 Texas Instruments, Developing with MCSDK: transports, December 2015. [Online]. Available: http://processors.wiki.ti.com/index.php/MCSDK_UG_Chapter_Developing_Transports#IPC_Transports.

9 Texas Instruments, IPC product releases, 2015. [Online]. Available: http://software-dl.ti.com/dsps/dsps_public_sw/sdo_sb/targetcontent/ipc/index.html.

10

Single and multicore debugging

Multicore DSP: From Algorithms to Real-time Implementation on the TMS320C66x SoC, First Edition. Naim Dahnoun.
© 2018 John Wiley & Sons Ltd. Published 2018 by John Wiley & Sons Ltd.
Companion website: www.wiley.com/go/dahnoun/multicoredsp

10.1 Introduction

The debugging phase during the development and/or testing of an application is very important. Depending on the application, debugging can take much more than 50% of the development time. Good development and debugging (software and hardware) tools are a good investment because not only do they reduce the debugging time and therefore decrease the time-to-market, but also they help to produce systems with minimum bugs which are therefore more reliable, robust and low maintenance. The extra instrument embedded on chip, also known as the *embedded debug components* or *embedded emulation components* (EECs) [1], may be insignificant compared to the rest of the system in terms of silicon area and power consumption, but it requires extra development time from the chip designers and therefore an initial extra investment. However, it provides an indispensable tool to the programmer as shown later in this chapter. Unfortunately, bugs cannot be eliminated completely from complex systems, especially for dynamic systems that are built to change, despite good programming practice and style, good software quality (e.g. stability, determinism, robustness and thread safety) and good testing procedures, as these bugs can be intermittent and hard to reproduce during the development or test phases. As a consequence, a constant release of patches is produced to fix these bugs. These can be seen with Microsoft and Apple software, for instance, for which regular patches are released in order to fix and/or upgrade applications.

As diagnosis is a major component in medicine, and without a good diagnosis a patient can suffer irreversible consequences, debugging is also becoming a major complement in software development. Without it, applications such as in medical, aerospace and automotive fields can have dire consequences for the end user and/or may be the end of the manufacturer if bugs are not detected and emerge during a critical time.

Profiling a code to find bottlenecks and optimising code to improve the performance are also considered as part of debugging, as will be shown later in the chapter.

So, what do we expect from the debugging tools? The answer may vary, but good visibility of what software was doing before, during or after reaching a certain location in memory or just after crashing could be very helpful in determining the cause of skulking bugs.

Debugging becomes exorbitant when we move from programming single-core to parallel programming using system-on-chip (SoC) homogeneous multicores, or, even worse, when moving to heterogeneous SoCs. This is because the complexity of systems increases with additional components, and therefore makes signals and buses inaccessible outside the cores.

Debugging tools can be very sophisticated and complicated, and may require a dedicated book to cover every aspect. However, in this chapter, debugging is made easy by giving an overview of the tools available, showing what they can offer and giving step-by-step procedures for each mode of debugging. In general, there are two types of debugging methods, software and hardware, as shown here. By understanding which debugging tools are available and how they operate, the programmer can select the right method(s) for a particular application code to debug.

It is important at this stage to know that optimisation can make debugging difficult. Ref. [2] deals with the trade-off between the ease of debugging and the effectiveness of the compiler optimisation.

Finally, bugs discovered and fixed before production and distribution are less costly than ones discovered after production or distribution.

10.2 Software and hardware debugging

To find bugs, there are two main approaches: the software debugging approach and the hardware debugging approach. Software debugging is intrusive in the sense that:

1) Extra code needs to be added in the application, like a **printf** statement. That, by itself, can cause some timing issues and therefore result in a non-functional code or, even worse, code with some intermittent faults (bugs).
2) Using breakpoints, halting and stepping through code can affect the system functionality, especially for real-time applications when timing issues are the main culprits.

However, that said, software debugging still plays an important role, especially when instrumented software is available, as shown in this chapter. On the other hand, hardware debugging is not intrusive as it uses the on-chip hardware debugging components as shown in Section 10.3.

Debugging can be done at a core level or system level.

10.3 Debug architecture

High-performance SoCs are complex and need some advanced debugging tools that can provide event triggers such as breakpoints, watchpoints, events, counters, bus monitoring and state sequencers. These are now being designed into the chip to provide visibility into the chip at core and system levels. These tools are known as the EECs [1].

It is important to note that events can be specified as a code execution address, or as a data read or write access to an address with a specified value.

The KeyStone debug architecture can be broken into three main parts:

- Trace (standard, event and system)
- Unified Breakpoint Manager (UBM)
- Unified Instrumentation Architecture (UIA).

All three types of debugging make use of the Advanced Event Trigging (AET) logic.

10.3.1 Trace

There are three types of hardware traces: the standard trace, event trace and system trace.

10.3.1.1 Standard trace

The standard trace provides the Program Counter Address (PC), the processor data accesses (read/write for addresses/data) and the timing at which these events occur, as shown in Figure 10.1. These captured data can be combined by selecting the appropriate functions. However, the user should select the minimum data to avoid a clustered output that may be difficult to decipher. The standard trace also provides the possibility of selecting when these data are to be collected as shown in Figure 10.2. For instance, data can be collected only when the program executes a specific part that is specified by the start address and the end address, as shown in Figure 10.3 and Ref. [3].

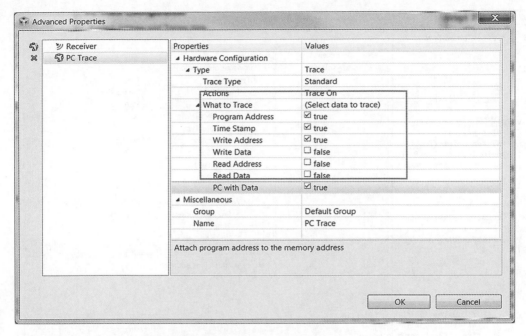

Figure 10.1 Trace functions available with the standard trace.

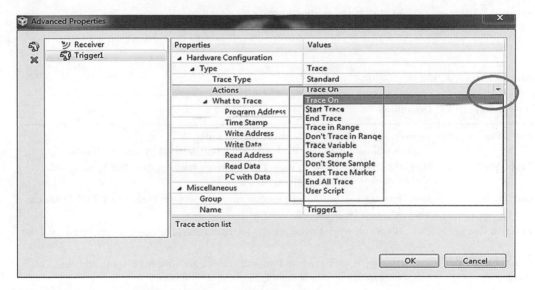

Figure 10.2 How to select when to start and/or stop tracing.

Once it has been decided what to trace, the user has the option to decide when to start tracing by selecting the **Trace On** option, as shown in Figure 10.2 and described in Table 10.1.

10.3.1.2 Event trace

Event traces provide stall events for the CPU, **L1P** and **L1D**, memory system events (**L1D, L2** and External) or **CPU-IDMA/EDMA** bank conflict; see Figure 10.10 and Figure 10.11 [5].

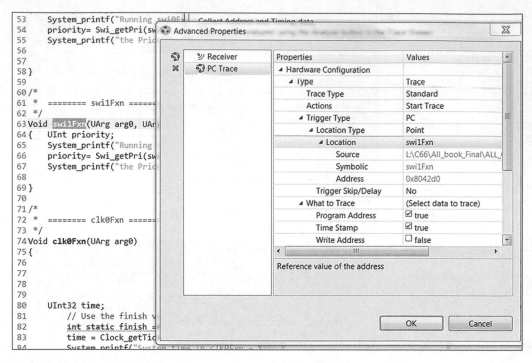

Figure 10.3 Using a function name as the starting location for a capture.

Table 10.1 Trace actions description

Trace action	Description
Trace On	Begins tracing as soon as the target starts running. Capture will continue until the buffer is full or turned off by **End All Trace**.
Trace Start	Starts the trigger only when some conditions are met. For example: Specific address is read from or written to, as shown in Figure 10.3.
End Trace	Ends the trace capture. Can be used in conjunction with **Trace Start**. **End Trace** can allow the choice of which trace to end.
Trace in Range	Traces data only when the PC reaches a certain range. For example: See Figure 10.4 and Figure 10.5.
Don't Trace in Range	This is the opposite of **Trace in Range**. It allows trace outside the specified range. This is useful for capturing code branching to unexpected location(s).
Trace Variable	Traces the **Read** or **Write** to a **Variable**. See Figure 10.6 and Figure 10.7.
Store Sample	Traces on PC range, memory range or events. See Figure 10.8 and Figure 10.9.
Don't Store Sample	This is the reverse of **Store Sample**. Useful for capturing code in an invalid location.
Insert Trace Marker	Inserts a point or range trace marker
End All Trace	Stops all tracing (one cannot choose which trace to end as can be done by **End Trace**)
User Script	Captures trace data and then uses post-processing scripts [4]

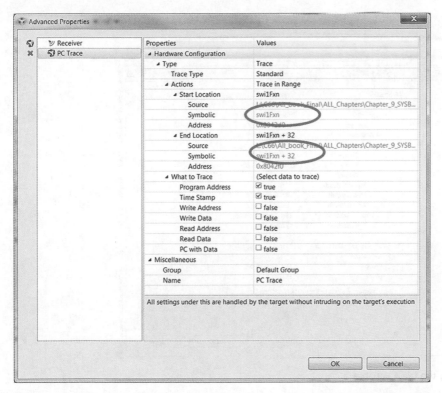

Figure 10.4 Trace range specified with a function name and range (32 bytes) for this example.

	Filename	Function	Program Address	Load Address	Read ...	W...	Disassembly		Source	Cycle	Delta Cycles	Trace Status
1										0		Start of trace
2	L:\C66\All...	swi1Fxn(...	0x008042D4	0x008042D4			018F2128	MVK.S1 ...	System_...	0	1	Stall timing collecti...
3	L:\C66\All...	swi1Fxn(...	0x008042D4	0x008042D4			018F2128	MVK.01 ...		1	7	Pipeline stall
4	L:\C66\All...	swi1Fxn(...	0x008042D8	0x008042D8			018040E8	MVKH.S1 ...		8	1	
5	L:\C66\All...	swi1Fxn(...	0x008042D8	0x008042D8			018040E8	MVKH.S1 ...		9	2	Pipeline stall
6	L:\C66\All...	swi1Fxn(...	0x008042E0	0x008042E0			100EA813	CALLP.S2...		11	6	
7	L:\C66\All...	swi1Fxn(...	0x008042E4	0x008042E4			01BC22F4 ||	STW.D...		11		
8	L:\C66\All...	xdc_runt...	0x0080B820	0x0080B820			07BC1FD9	MV.L1X ...	{	17	1	Stall timing collecti...
9	L:\C66\All...	xdc_runt...	0x0080B824	0x0080B824			27F7 ||	STW.D2T1 ...		17		
10	L:\C66\All...	swi1Fxn(...	0x008042E8	0x008042E8			100F7C13	CALLP.S2...	priorit...	18	6	Stall timing collecti...
11	L:\C66\All...	swi1Fxn(...	0x008042EC	0x008042EC			0200056C ||	LDW.D...		18		
12	C:/TI/bios...	ti_sysbio...	0x0080BEC0	0x0080BEC0			0210A264	LDW.D1T1...	return ...	24	1	Stall timing collecti...
13										24		End of trace

Figure 10.5 Captured data when using **Trace in Range**.

10.3.1.3 System trace

The system trace (STM) allows the user to find system-wide problems by capturing system events in real-time. These events can be, for instance, the memory throughput or power domain status. The STM will also attach a timestamp to each event.

STMs capture system events via modules known as *CP tracers* (CPTs) as shown in Figure 10.12. Figure 10.13 shows how to use the CPTs using the Code Composer Studio (CCS). To use the STM, one can use either the on-chip ETB or the XDS560v2 STM external emulator connected to the pin trace of the device (see videos on TI emulators [6] and how to use traces [7]).

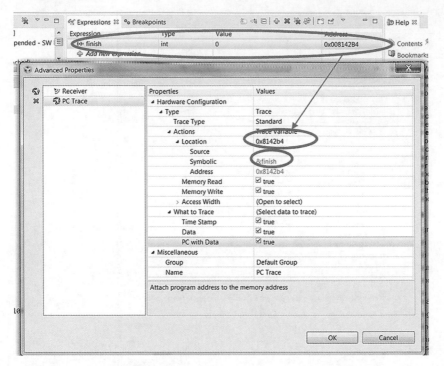

Figure 10.6 Setting the trace variable.

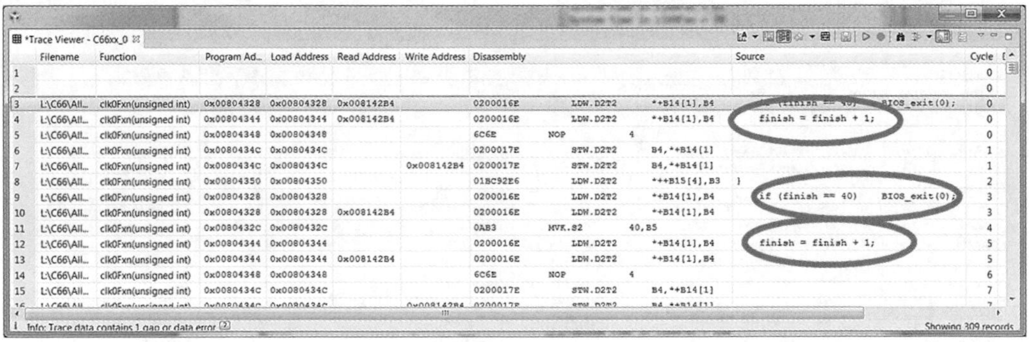

Figure 10.7 Trace output after setting the trace variable.

In general, a combination of core trace and system trace will be required for advanced debugging. However, due to the large amount of data to be analysed and due to the fact that the data collected contain no information about what the operating system was doing at the time of data collection, debugging with traces should be used as a last resort.

10.4 Advanced Event Triggering

The KeyStone debug architecture also contains AET hardware that can be configured to generate triggers based on events that are associated with each core; see Figure 10.15.

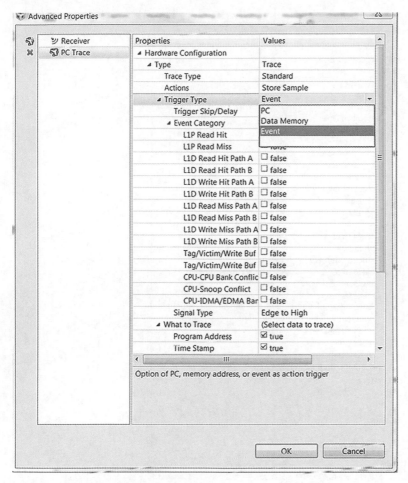

Figure 10.8 Store sample configuration.

The AET can also handle complex event state machines and event counters. The AET is not intrusive, requires no detailed knowledge of underlying on-chip hardware and can offer the following features:

1) Runtime control. Debug tools can start and stop any core, modify registers and give single-step machine instructions.
2) Hardware program breakpoints. By specifying the program addresses or address ranges (by the user) and when these addresses are reached, the AET can generate events such as halting the processor or triggering the trace capture.
3) Data watchpoints. By specifying data variable addresses, address ranges or data values and when these addresses are reached or data value matched, the AET can generate events such as halting the processor or triggering the trace capture.
4) Counters. The counters count the occurrence of an event or cycles for performance monitoring. This is very useful as sometimes we do not want to stop the processor, but we do want to know how many times this event occurred. For example:

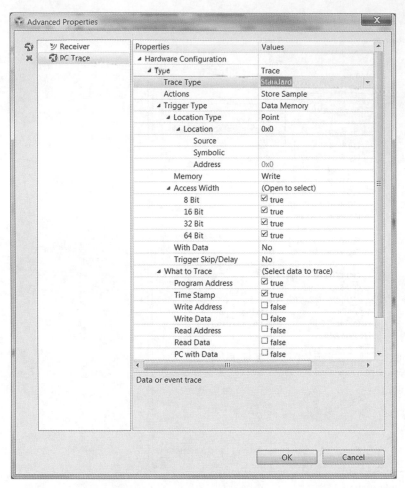

Figure 10.9 Store sample configuration example.

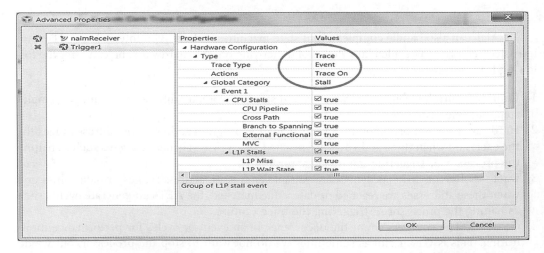

Figure 10.10 Event trace with stall.

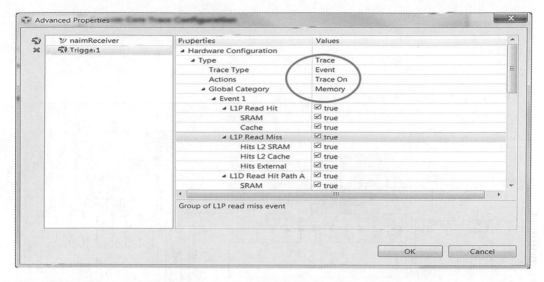

Figure 10.11 Event trace with memory.

Counts the number of times a program address is executed

Counts the number of times a function is entered

Counts the number of cycles that an interrupt service routine may take.

5) State sequencing. This allows combinations of hardware program breakpoints and data watchpoints to precisely generate events for complex sequences.

Each DSP core has a dedicated Trace Buffer (a TI Embedded Trace Buffer (TETB or ETB)) that is used for storing data collected by the AET hardware. These buffers are 4 KB in size each.

The KeyStone architecture also has one debug sub-system that has a system trace for exporting software messages through **printf**-like statements embedded in the application code, exporting bus statistics and events, and full-system events. A dedicated ETB with 16 KB in size is also available for collecting trace data. The ETBs are on-chip circular buffers that store compressed trace information. The compression is used for saving memory; in fact, the ETBs can store up to 30,000 lines of trace data.

All cores and the system trace can be connected to either the ETBs or the External Trace Receiver as shown in Figure 10.14. Figure 10.15 shows the complete debug architecture.

10.4.1 Advanced Event Triggering logic

In order to make good use of AET, it is worth examining its architecture. As can be seen from Figure 10.17, the AET is composed of two main parts: Event Generation Logic and Event Combination Logic.

Event Generation Logic provides events from the processor address and data buses, from external events and some from itself. The main events which are produced by the processor address and data are first passed through a pipeline flatter in order to simplify the event detection and produce events in the order they arrive. This is followed by bus comparators in order to detect the state or states of the buses. Event Generation Logic also contains state resources that implement a state sequencing in order to debug complex code. For instance, if we consider the example code shown in Figure 10.16, where a subroutine starts at **Start_SB** and ends at **End_SB**, and if there is a bad code that is rarely executed (say, between cycle $N + 4$ and $N + 7$), then it

KeyStone CP tracer modules

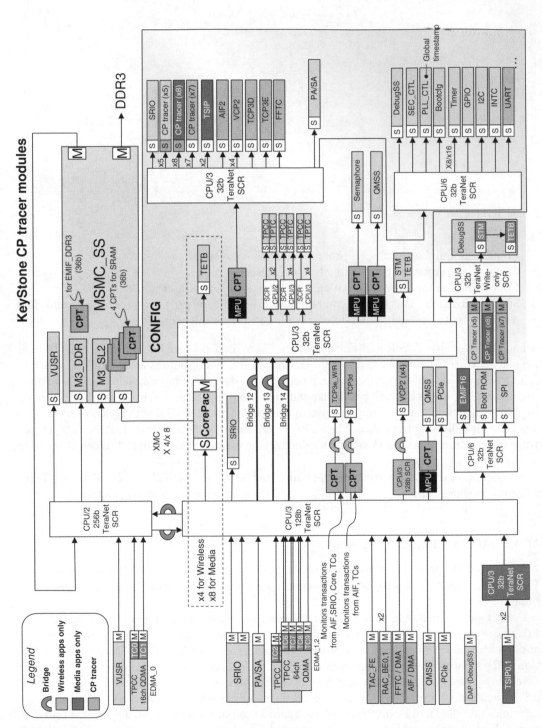

Figure 10.12 KeyStone CP tracer modules [8]. *Source*: Courtesy of Texas Instruments.

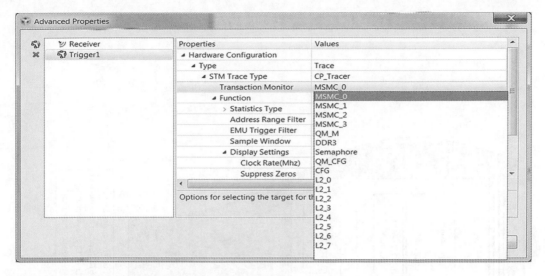

Figure 10.13 Using the CP tracer.

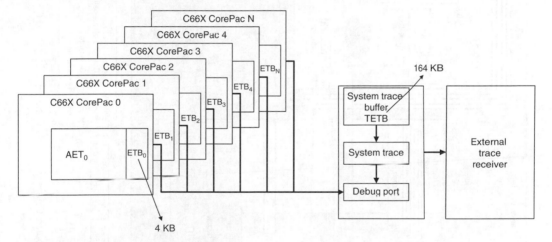

Figure 10.14 Debug sub-system.

would be very difficult to find it. However, the state sequencing allows us to specify that the bad code should run first before examining the data. In other words, the intention is to have a sequence of events happening before taking action to stop the CPU or something else.

Events are combined and triggers are accordingly generated by Event Combination Logic. Trace data are captured based on triggers generated by the AET Unit.

In addition to the debug hardware available, there is also a software library that can be included in the application for either the core traces or the system trace; see Figure 10.18 [9]. This library is useful for configuration and control of the debug and profiling modules.

This library is referred to as the **CToolsLib**. It is a collection of embedded target APIs/libraries (DSP Trace Library, AET Library, STM Library, Tracer Library and ETB Library) that are used by the chip tools that provide the system-level visibility. The chip tools are referred to as the **CTools**. The **CTools** utilise the system trace capability with MIPI System Trace Protocol [9]. Documentation and examples can be found in Ref. [9].

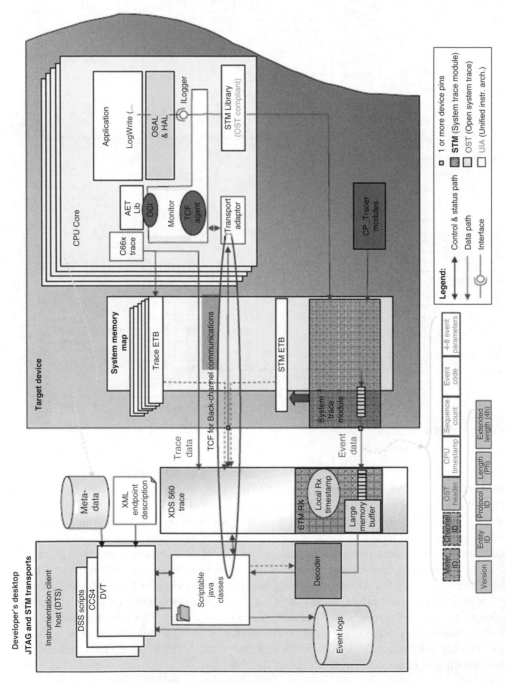

Figure 10.15 TMS320C66x debug architecture [8]. *Source:* Courtesy of Texas Instruments.

cycle *N* + 1	**Start_SB**	**inst**	Good code
cycle *N* + 2		**inst**	Good code
cycle *N* + 3		**inst**	Good code
cycle *N* + 4		**[B0] B loop**	Condition true most of the time
cycle *N* + 5		**inst**	Bad code
cycle *N* + 6		**inst**	Bad code
cycle *N* + 7		**STW**	Bad code
cycle *N* + 8	**loop**	**STW**	Good code
cycle *N* + 8		**inst**	Good code
cycle *N* + 10	**End_SB**	**inst**	Good code

Figure 10.16 Bugged code example that can be detected using state sequencing.

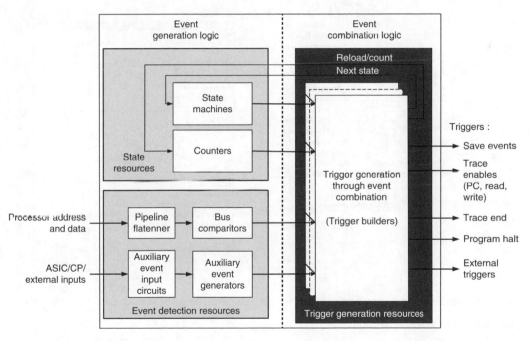

Figure 10.17 AET logic.

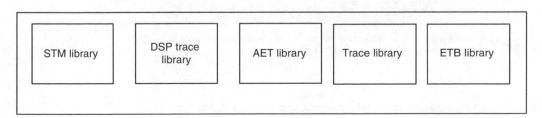

Figure 10.18 System libraries.

10.4.2 Unified Breakpoint Manager

Debugging through the UBM is fairly simple. With the UBM, not only software breakpoints, hardware breakpoints and hardware watchpoints but also counts, trace and so on can be used, as shown in Figure 10.19. The breakpoints stop the execution of the program at a certain address, halt the processor whenever a particular data variable is overwritten with a special value, halt the processor after a certain number of times a particular address is reached, toggle the EMU1 pin for external use or generate an RTOS interrupt. The hardware breakpoints use the AET and therefore are not intrusive while evaluating a condition. However, since they are implemented in hardware, their number is limited to four for the KeyStone. If the user asks for more, an error is generated as shown in Figure 10.20. In contrast, the software interrupts are implemented by software and there is no limit to the number used, but they are intrusive.

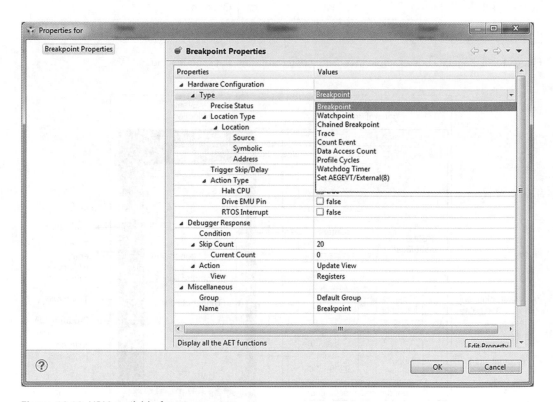

Figure 10.19 UBM available functions.

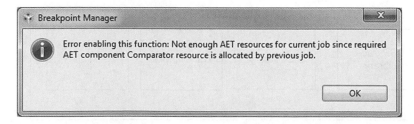

Figure 10.20 Error generated when asking for more resources.

10.5   Unified Instrumentation Architecture

To ease debugging, TI also introduced the UIA which is a combination of components that includes a set of tools, APIs, transports and interfaces that can be used to instrument an application. These UIA components can be found in many products on the side of the host or the target.

10.5.1   Host-side tooling

1) The Data Visualization Technology (DVT) features provide a common platform for:
   • Analysis
   • Display.
2) CCS: An eclipse-based Integrated Development Environment.

10.5.2   Target-side tooling

On the target side, a set of software packages provides additional debugging capabilities:

UIA: Unified Instrumentation Architecture APIs
XDC: eXtenDed C tools and software
IPC: Inter-Processor Communications library
NDK: Network Developers Kit
SYS/BIOS: SYS/BIOS or TI-RTOS.

UIA provides target content that aids in the creation and gathering of instrumentation data (e.g. log data) that can be used with the System Analyzer.
The rest of this chapter deals only with the UIA.
The UIA Target Software Package shown in Figure 10.21 provides the following features:

1) Software instrumentation APIs
2) Predefined software events
3) Event loggers
4) Transports
5) SYS/BIOS event capture and transport
6) Multicore support.

The available events and their descriptions are shown in Figure 10.22 and Figure 10.23.

Figure 10.21 UIA components.

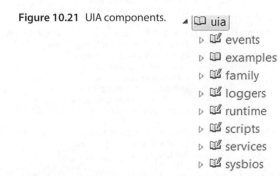

Figure 10.22 Events functions provided by the UIA.

- ▲ 📖 uia
  - ▲ 📗 events
    - 📄 DvtTypes
    - 📄 IUIACtx
    - 📄 IUIAEvent
    - 📄 UIAAppCtx
    - 📄 UIABenchmark
    - 📄 UIAChanCtx
    - 📄 UIAErr
    - 📄 UIAEvt
    - 📄 UIAFrameCtx
    - 📄 UIAHWICtx
    - 📄 UIAMessage
    - 📄 UIAProfile
    - 📄 UIARoundtrip
    - 📄 UIASnapshot
    - 📄 UIAStatistic
    - 📄 UIASWICtx
    - 📄 UIASync
    - 📄 UIAThreadCtx
    - 📄 UIAUserCtx

Module	Events	Diagnostics Control Bit	Comments
UIABenchmark	Start and stop events	diags_ANALYSIS	UIABenchmark reports time-elapsed exclusive of time spent in other threads that preempt or otherwise take control from the thread being benchmarked. This module's events are used by the Duration feature described in Section 4.14
UIAErr	Numerous error events used to identify common errors in a consistent way	diags_STATUS (ALWAYS_ON by default)	These events have an EventLevel of EMERGENCY, CRITICAL or ERROR. Special formatting specifiers let you send the file and line at which an error occurred
UIAEvt	Events with detail, info, and warning priority levels	diags_STATUS or diags_INFO depending on level	An event code or string can be used with each event type
UIAMessage	Events for msgReceived, msgSent, replyReceived, and replySent	diags_INFO	Uses UIA and other tools and services to report the number of messages sent and received between tasks and CPUs
UIAProfile	Start and stop events... Functions can be identified by name or address. These events are designed to be used in hook functions identified by the compiler's --entry_hook and --exit_hook command-line options	diags_ENTRY and diags_EXIT	UIAProfile reports time-elapsed exclusive of time spent in other threads that preempt or otherwise take control from the thread being profiled. This module's events are used by the Context Aware Profile feature described in Section 4.15
UIAStatistic	Reports bytes processed, CPU load, words processed, and free bytes	diags_ANALYSIS	Special formatting specifiers let you send the file and line at which the statistic was recorded

Figure 10.23 **Log_Event** description [10]. *Source*: Courtesy of Texas Instruments.

10.5.2.1 Software instrumentation APIs
The **xdc.runtime.Log** module provides basic instrumentation APIs to log errors, warnings, events and generic instrumentation statements. A key advantage of these APIs is that they are designed for real-time instrumentation. The processing and decoding format strings are left to the host in order to not burden the target; see Section 10.6.

10.5.2.2 Predefined software events and metadata
The **ti.uia.events** package includes software event definitions (Figure 10.22) that have metadata associated with them to enable the RTOS Analyzer and System Analyzer to provide performance analysis, statistical analysis, graphing and real-time debugging capabilities. For example, using the logger and UIABenchmark:

```
Log_write1(UIABenchmark_start, (xdc_IArg)"running");
//insert here the code to benchmark
Log_write1(UIABenchmark_stop, (xdc_IArg)"running");
```

10.5.2.3 Event loggers
A number of event-logging modules are provided to allow instrumentation events to be captured and uploaded to the host over both JTAG and non-JTAG transports. Different logger modules can implement a host-to-target connection. These can be found in:

- **ti.uia.sysbios.LoggingSetup**
- **ti.uia.services.Rta**
- **ti.uia.runtime.ServiceMgr**
- **ti.uia.loggers.LoggerStopMode**
- **ti.uia.runtime.LoggerSM**
- **ti.uia.sysbios.LoggerIdle**.

For instance, for an application to use the UIA instrumentation, the configuration file of this application should include the following command:

var LoggingSetup = xdc.useModule('ti.uia.sysbios.LoggingSetup');

10.5.2.4 Transports
Both JTAG-based and non-JTAG transports can be used for communication between the target and the host. Non-JTAG transports include Ethernet, with UDP used to upload events to the host and TCP used for bidirectional communication between the target and the host.

10.5.2.5 SYS/BIOS event capture and transport
For example, when the UIA is enabled, SYS/BIOS uses the UIA to transfer data about the CPU load, task load and task execution to the host.

10.5.2.6 Multicore support
The UIA supports routing events and messages across a central master core. It also supports logging synchronisation information to enable correlation of events from multiple cores so that they can be viewed on a common timeline.

It addition to these, the UIA also provides the following:

1) Scalable solutions. UIA allows different solutions to be used for different devices.
2) Examples provided. UIA includes working examples for the supported boards.
3) Source code included. UIA modules can be modified and rebuilt to facilitate porting and customisation.

The data provided by the UIA can be analysed by the pre-instrumented SYS/BIOS threads, and therefore no extra programming is required. However, this requires a minimum configuration, as shown in this chapter. Additional target-side code can be used to collect specific data and therefore provide extra instrumentation.

The UIA is a component that the TI-RTOS provides, which means that the UIA is installed automatically when the TI-RTOS is installed. The UIA is included in the Multicore Software Development Kit (MCSDK); see Figure 10.24.

10.6 Debugging with the System Analyzer tools

The System Analyzer is a real-time tool for analysing, visualising and profiling applications running on single-core or multicore systems. Data are collected using the UIA software instrumentation on the target and transported via Ethernet, run-mode JTAG, stop-mode JTAG, USB or UART to the host PC for analysis and visualisation in the CCS. In a multicore system, data from all cores are correlated to a single timeline.

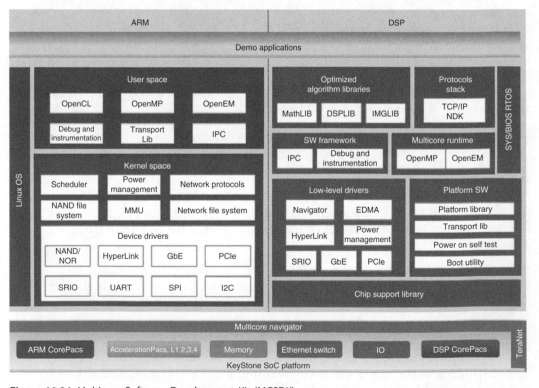

Figure 10.24 Multicore Software Development Kit (MCSDK).

10.6.1 Target-side coding with UIA APIs and the XDCtools

If more instrumentation is required for an application, the UIA provides some APIs that can be added on the target side. These APIs can be used for the following functions:

- Logging events with **Log_write()** functions
- Enabling event output with the diagnostics mask
- **LogSnapshot** APIs for logging state information:
  - **LogSnapshot_getSnapshotId()**
  - **LogSnapshot_writeMemoryBlock()**
  - **LogSnapshot_writeNameOfReference()**
  - **LogSnapshot_writeString()**.
- **LogSync** APIs for multicore timestamps
- **LogCtxChg** APIs for logging context switches
- Module APIs for controlling loggers
- Custom transport functions for use with **ServiceMgr**.

In this chapter, we make use of the service provided by the **xdc.runtime** package [11] that is composed of various modules, as shown in Figure 10.25 and Table 10.2. Some important modules will be studied, and examples will be shown.

The modules can be partitioned into three groups: modules that can generate events, modules that can control which events are generated and when to turn them on or off and modules that generate outputs like a print function does. These three modes are described in more detail in Ref. [12].

1) Modules that generate events. **Assert**, **Error** and **Log** provide methods that are added to the source code and generate events.
2) Modules that allow precise control over when (or if) various events are generated. The **Diags** module provides both configuration and runtime methods to selectively enable or disable different types of events on a per-module basis.
3) Modules that manage the output or display of the events. **LoggerBuf** and **LoggerSys** are simple alternative implementations of the **ILogger** event 'handler' interface.

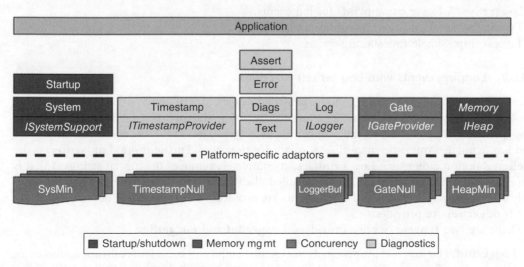

Figure 10.25 The **xdc.runtime** package and its modules [12].

Table 10.2 xdc.runtime package

(a)

Category	Modules	User's Guide Documentation
memory management	Memory⬚	Memory Management
concurrency support	Gate⬚	Mutual Exclusion Gates
real-time diagnostics	Log⬚, Assert⬚, Error⬚, Timestamp⬚, Diags⬚	Event Logging, Error Handling, Timestamp Services
system services	System⬚, Startup⬚	Basic Services

xdc.runtime package

(b)

Diagnostics	Log	Allows events to be logged and then passes those events to a Log handler
	Assert	Provides for configurable assertion diagnostics
	Error	Allows raising, checking and handling errors defined by any modules
	Timestamp	Provides time-stamping APIs that forward calls to a platform-specific time stamper (or one provided by CCS)
	Diags	Allows diagnostics to be enabled/disabled at either configuration or run time on a per-module basis

Diagnostic module

The diagnostics and logger modules used in this chapter are described here:

- **Assert**: Add integrity checks to the code.
- **Diags**: Manage a module's diagnostics mask.
- **Error**: Raise error events.
- **Log**: Generate log events in real time.
- **LoggerBuf**: A logger using a buffer for log events.
- **LoggerSys**: A logger using **printf** for log events.
- **Types**: Define diagnostics configuration parameters.
- **Timestamp**: Simple timestamp service.

10.6.2 Logging events with Log_write() functions

The UIA events can use the log module provided as part of XDCtools to log events. From an application, it is possible to generate events by using the **printf** type functions such as **Log_print()** or **Log_write()** as they consume fewer cycles, less code space and less data space and are made deterministic since their formats are restricted. Furthermore, **Log_write()** is the preferred option rather than **Log_print()**, as it completely removes the format strings [13]. It is important to note that when events are disabled, the time overhead for these two statements is only a few instruction cycles; and when events are enabled, the overhead will depend mainly on the **ILogger** service provider.

There are two **ILogger** Service Providers, **LoggerBuf** and **LoggerSys**:

1) **LoggerBuf**. This logger captures events in a circular buffer and is used for real-time applications.
2) **Logger_Sys**. This logger outputs events as they occur. Not suitable for real-time applications.

The remainder of this section will show how these tools are configured and used.

10.6.3 Advance debugging using the diagnostic feature

Memory space is an important resource for an embedded system, and the debugging software can absorb some of this resource for logging data. Bandwidth is also consumed as these data are then transferred to the host for analysis. The debugging tools for the KeyStone allow the programmer to turn on or off the selected **log** statement(s) and therefore reduce not only memory space but also the bandwidth required for transmitting data from the target to the host.

In order to do this efficiently, every **log** event is defined with a certain category, so that it can be turned on or off separately. There are 16 predefined categories, as shown in Table 10.3. To have further control, each category has four levels of control: Level 1, Level 2, Level 3 and Level 4. Level 1 is the highest priority, and Level 4 is the lowest. By enabling one level, subsequent higher levels will be automatically enabled. For instance, if Level 3 is enabled, automatically the higher levels (Level 2 and Level 1) will be enabled.

There are three main steps for enabling or disabling a **log** event. The following steps show the java script needed for configuring the configuration file.

Step 1. Enable or disable the event. Example:

```
/* Enable USER1 events of all levels for Main. */
Main.common$.diags_USER1 = Diags.RUNTIME_ON;
```

Step 2. Enable or disable filtering.

```
/* enable filtering by level. */
LoggerBuf.filterByLevel = true;
```

Table 10.3 Different categories available

Control character	Category	Meaning
E	ENTRY	Function entry
X	EXIT	Function exit
L	LIFECYCLE	Object life-cycle
I	INTERNAL	Internal diagnostics
A	ASSERT	Assert checking
S	STATUS	Warning or error events
1	USER1	User-defined diagnostics
2	USER2	User-defined diagnostics
3	USER3	User-defined diagnostics
4	USER4	User-defined diagnostics
5	USER5	User-defined diagnostics
6	USER6	User-defined diagnostics
7	USER7	User-defined diagnostics
F	INFO	Informational event
8	USER8	User-defined diagnostics
Z	ANALYSIS	Analysis event

Step 3. Set the filter level.

```
/* Filter out all USER1 events below LEVEL2. */
LoggerBuf.level3Mask = Diags.USER1;
/* the next line is to make sure that USER1 is only enabled once*/
/*if a mask is enabled twice an error will be generated.*/
LoggerBuf.level4Mask = Diags.ALL_LOGGING & (~Diags.USER1);
```

On the target side, to be able to use the filter, the following code can be used:

```
Log_print0(Diags_USER1 | Diags_LEVEL1, "A USER1 category, LEVEL1 event.");
Log_print0(Diags_USER1 | Diags_LEVEL2, "A USER1 category, LEVEL2 event.");
Log_print0(Diags_USER1 | Diags_LEVEL3, "A USER1 category, LEVEL3 event.");
Log_print0(Diags_USER1 | Diags_LEVEL4, "A USER1 category, LEVEL4 event.");
```

In this example, if the **Diags_USERN** and the **Diags_LEVELN** (*N*: 1–4) are selected, then all print statements corresponding to a level below *N* will be enabled. See Laboratory experiment 5 for a practical example (Section 10.8.5).

10.6.4 LogSnapshot APIs for logging state information

The **LogSnapshot** module APIs allow logging of memory values, register values and stack contents [10]. The **LogSnapshot** module provides four functions (examples are provided in Ref. [10]):

1) **LogSnapshot_getSnapshotId()**. Group snapshot events.
2) **LogSnapshot_writeMemoryBlock()**. Generate a snapshot event when a block of memory is accessed.
3) **LogSnapshot_writeNameOfReference()**. This will log a function name; for instance, if a function is created dynamically, this API can retrieve the name of the function that might be used by another function.
4) **LogSnapshot_writeString()**. Retrieve the actual content of a memory location.

10.7 Instrumentation with TI-RTOS and CCS

With CCS, it is easy to view Log messages. The CCS uses the TI-RTOS (that is based on the UIA), the RTOS Object Viewer (ROV), the RTOS Analyzer and the System Analyzer tools.

10.7.1 Using RTOS Object Viewer

The ROV, which is part of the CCS, is the simplest debugging tool as it requires no configuration or extra debugging code for the application. The ROV automatically provides state information about all modules in the RTSC when the target is halted (known as *stop-mode debugging*). The limitation of the ROV is that it only captures data when the target is halted and therefore does not cumulate records like the other debugging instruments discussed earlier [10, 14–16]. A laboratory experiment is shown in Laboratory experiment 1 in Section 10.8.1.

10.7.2 Using the RTOS Analyzer and the System Analyzer

In order to log live data (in real-time) and then view it later, one can use the RTOS Analyzer or the System Analyzer to take advantage of the built-in instrumented tools for debugging.

The XDC debugging functions added to an application can be disabled at runtime or completely removed at compile time by using a single command as shown in Laboratory experiment 2 (in Section 10.8.2). The features available are shown in Figure 10.26.

10.7.2.1 RTOS Analyzer

The RTOS Analyzer is very useful and easy to use as it does not require any setting except a minimum configuration of the UIA as shown above. The RTOS Analyzer functions available are shown in Table 10.4.

See Laboratory experiment 2 (Section 10.8.2) for a practical example showing how to configure the UIA and use the RTOS Analyzer.

10.7.2.2 System Analyzer

The System Analyzer allows additional debugging functions that can be set in the application code in addition to the UIA configuration. The available commands for this mode are shown in Figure 10.27 and Table 10.5.

Duration analysis: This command provides the time elapsed between two execution points specified by the two functions' commands:

UIABenchmark_start and **UIABenchmark_stop**

As an example, one can use the following code to benchmark a code:

```
Log_write1(UIABenchmark_start, (xdc_IArg)"running");
//insert here the code to benchmark
Log_write1(UIABenchmark_stop, (xdc_IArg)"running");
```

The second argument in this example, which is "running", should be the same in both **UIABenchmark_start** and **UIABenchmark_stop**; otherwise, there will be no output. See Laboratory experiment 3 in Section 10.8.3.

Figure 10.26 RTOS Analyzer features available.

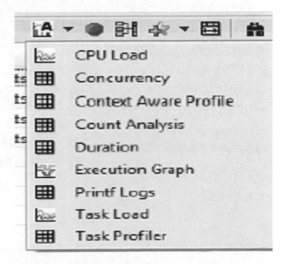

Table 10.4 RTOS Analyzer functions

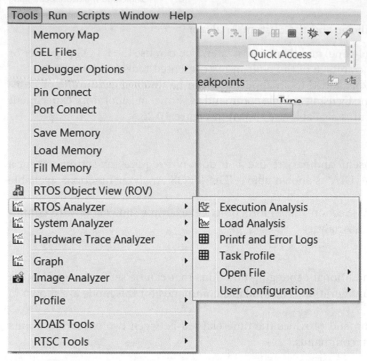

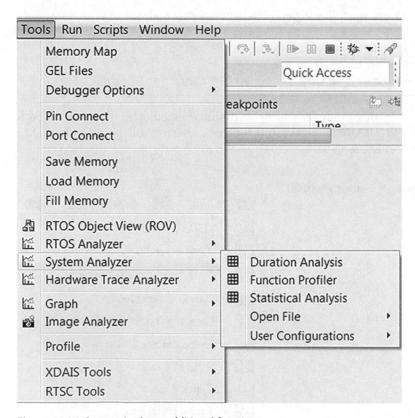

Figure 10.27 System Analyzer additional functions.

Table 10.5 System Analyzer commands

System Analyzer menu command	Definition
Duration analysis	Time elapsed between two execution points
Function profiler	Context-aware profile (calculates duration while considering context switches, interruptions and execution of other functions). Both inclusive and exclusive time can be shown.
Statistical analysis	Count analysis can be used to find the maximum and minimum values reached by some variable or the number of times a variable is changed.

10.8 Laboratory sessions

Four laboratory experiments are introduced to cover the main debugging types available with the KeyStone. The laboratories are independent, and therefore you can start with any laboratory session. However, it is recommended to go through them in order.

10.8.1 Laboratory experiment 1: Using the RTOS ROV

Project location:

\Chapter_10_Code\RTOS_ROV_Example_1

In this laboratory experiment, the program shown in Figure 10.28, the **main()** function starts the operating system by using the **BIOS_start()** function. There are also one task function **taskFxn()** and two software functions **swi0_Fxn()** and **swi1_Fxn()** with priority 6 and 7, respectively.

Note that no instrumentation code is included in either the application or the configuration files. Since we have used **printf**, tasks and SWIs, let's open the ROV (in CCS Debug mode) (**Tools - > RTOS Object View**, ROV), select **SysMin**, run the code, then pause and observe the output (see Figure 10.29).

Note: To minimise the memory footprint of the application, use **SysMin** instead of **SysStd**.

Select **sysbios - > knl - > Task**, and observe the output (see Figure 10.30).

Select **Swi**, and observe the output (see Figure 10.31).

10.8.2 Laboratory experiment 2: Using the RTOS Analyzer

Files location:

\Chapter_10_Code\RTOS_ROV_Example

In the laboratory experiment, the program shown in Figure 10.33 is used in conjunction with the UIA configuration in order to use the built-in instrumentation for debugging. Figure 10.35 shows various built-in functions that can be used with the RTOS Analyzer.

The following program is stand-alone and does not require any modification. Please follow these steps to get an understanding of this debugging mode:

Step 1. Within the CCS, import the project from the following folder to your workspace: ...\ **\Chapter_Debugging\RTOS_Analyzer_Example**. Once the project is imported, check that

```
/*
 *  ======== main.c ========
 */
#include <xdc/std.h>
#include <xdc/runtime/System.h>
#include <ti/sysbios/BIOS.h>
#include <ti/sysbios/knl/Task.h>
#include <ti/sysbios/knl/Swi.h>
#include <xdc/cfg/global.h>
/*
 *  ======== taskFxn ========
 */
void taskFxn(UArg a0, UArg a1)
{
    System_printf("enter taskFxn()\n");

    Task_sleep(10);
    Swi_post(swi0);

    System_printf("exit taskFxn()\n");
}


/*
 *  ======== wsi0 ========
 */
void swi0_Fxn(UArg a0, UArg a1)
{
    System_printf("enter SWI_Fxn0()\n");
    Swi_post(swi1);
    System_printf("exit SWI_Fxn0()\n");

}
/*
 *  ======== wsi1 ========
 */
void swi1_Fxn(UArg a0, UArg a1)
{
    System_printf("enter SWI_Fxn1()\n");
    System_printf("exit SWI_Fxn1()\n");
}
/*
 *  ======== main ========
 */
void main()
{
    /*
     * use ROV->SysMin to view the characters in the circular buffer
     * this is set in the configuration file
     */
    System_printf("enter main()\n");
    BIOS_start();          /* enable interrupts and start SYS/BIOS */
}
```

Figure 10.28 A test code with a **main()**, a **task()** and two **Swi**s functions.

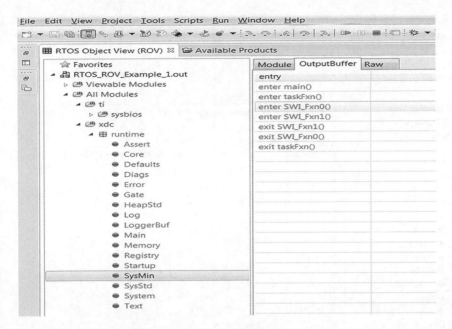

Figure 10.29 Using **SysMin** to display **System_printf** outputs.

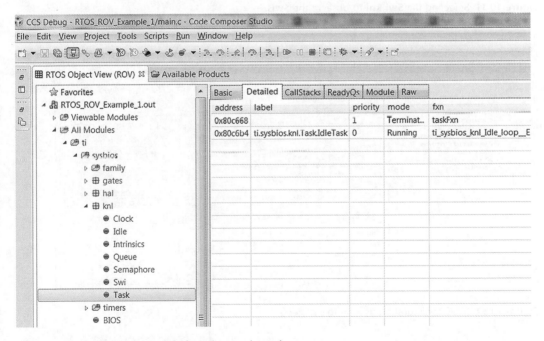

Figure 10.30 Selecting the **task knl** to observe the tasks.

you have the right versions of the SYS/BIOS and the UIA installed in your PC. Figure 10.32 shows the RTSC configuration used in this project.

Step 2. Connect and power up the EVM module.

Step 3. In the CCS, press on the bug to build and load the application. (You should observe no errors or warnings.)

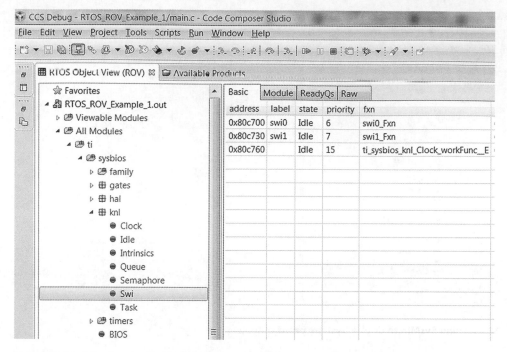

Figure 10.31 Selecting the **Swi knl** to observe the **Swi**s.

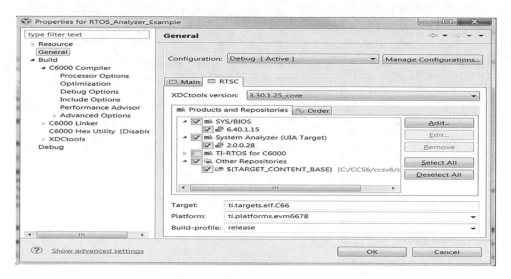

Figure 10.32 RTSC configuration used in this project.

Step 4. Open the configuration file **RTOS_config.cfg**, select **SYS/BIOS – System Overview** and press the arrow circled as shown in Figure 10.34a. Then, select **Logging_Setup** and tick the box for **Add LoggingSetup to my Configuration**. Different options are shown depending on the threads that need to be debugged. For instance, if your application is using only **Hwi** threads and not **Swi** threads, then do not tick **Swi** to the save memory. Close the **RTOS_config.cfg** window to make sure the file is saved.

```
/*
 *  ========= main.c =========
 */
#include <xdc/std.h>
#include <xdc/runtime/Log.h>      // use this for both Log_info and System_printf
#include <xdc/runtime/System.h>  // use this for printf

Int main (Void) {
int arg1 = 0xDAD1;
int arg2 = 0xDAD2;
int arg3 = 0xDAD3;
int arg4 = 0xDAD4;
int arg5 = 0xDAD5;

int i = 0;
int y = 0;
Log_info0( "Zero argument!" ); // this generates a log "info event" with 0 argument.
Log_info1( "One argument!" , arg1); // this generates a log "info event" with 1 arguments.
Log_info2( "Two arguments!" , arg1,arg2);  // this generates a log "info event" with 2 arguments.
Log_info3( "Three arguments!" , arg1,arg2,arg3); // this generates a log "info event" with 3
arguments.
Log_info4( "Four arguments!" , arg1,arg2,arg3,arg4); // this generates a log "info event" with 4
arguments.
Log_info5( "Five arguments!" , arg1,arg2,arg3,arg4,arg5); // this generates a log "info event"
with 5 arguments.

for (i = 0; i < 1000; i++) {
        y = i + 1;
}

System_printf( "y  = %d \n" , y);

return (0);
}
```

Figure 10.33 Test code.

(a)   (b)

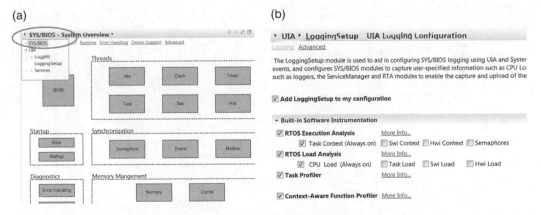

Figure 10.34 (a) Opening the UIA dialogue box. (b) Configuring the UIA.

Step 5. If your program is not loaded, load it again as shown in Step 3, and select **Tools > RTOS Analyzer > Printf and Error Logs** as shown in Figure 10.35. Once the mode is selected, the dialogue box shown in Figure 10.36 appears. Complete the box as shown in the figure and press **START**. Run your code, and after a few seconds the data are collected and displayed as shown in Figure 10.38. If the debugging window cannot be seen, then resize the windows.

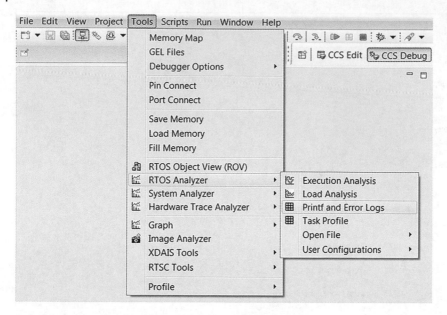

Figure 10.35 Available commands used for the RTOS Analyzer.

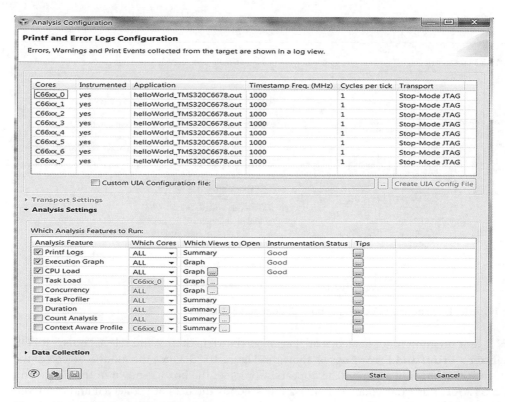

Figure 10.36 Analysis configuration.

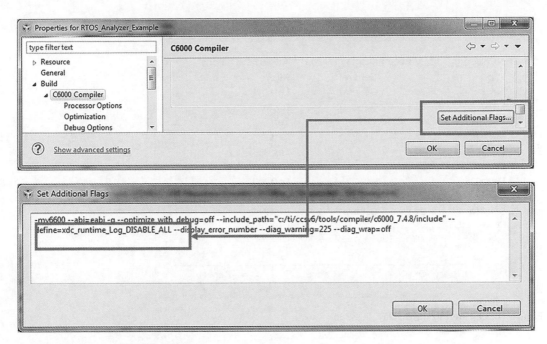

Figure 10.37 Adding compiler options.

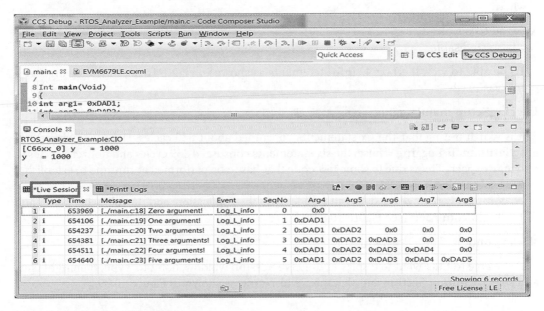

Figure 10.38 Output of the logged data.

To be able to enable or disable logging for **Tasks**, **Swi**s, **Hwi**s or the main function, the following statements can be used:

LoggingSetup.sysbiosTaskLogging = X;
LoggingSetup.sysbiosSwiLogging = X;

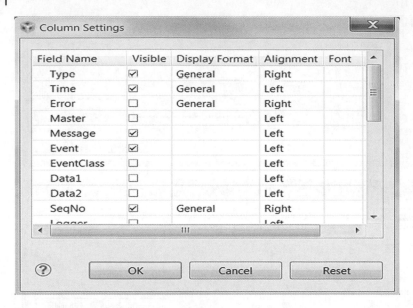

Figure 10.39 Filtering the display message.

LoggingSetup.sysbiosHwiLogging = X;
LoggingSetup.mainLogging = X;

where:

X = true to enable
X = false to disable.

To be able to view the **Log_L_info** statements as shown in Figure 10.38, the following state-ment was used:

LoggingSetup.mainLogging = true

It is worth noting that the six statements consumed about 836 cycles. However, when **Log-gingSetup.mainLogging = false**, the six statements consumed 369 cycles and not zero cycles as one may expect. This is due to some test code that has to be performed at runtime to establish if logging is enabled or not.

To completely remove the debugging code from the application, add the following command to the build command (see Figure 10.37):

-Dxdc_runtime_Log_DISABLE_ALL

To be able to filter what to display in the **Live Session** window, point to **Type**, then right click and select **Column Setting** and tick the **Field Name** as shown in Figure 10.39.

10.8.3 Laboratory experiment 3: Using the System Analyzer

Files location:

\Chapter_10_Code\System_Analyzer_Example

In the laboratory experiment, the program shown in Figure 10.41 is used in conjunction with the UIA configuration in order to use the built-in instrumentation for debugging. Figure 10.35 shows various built-in functions that can be used with the RTOS Analyzer.

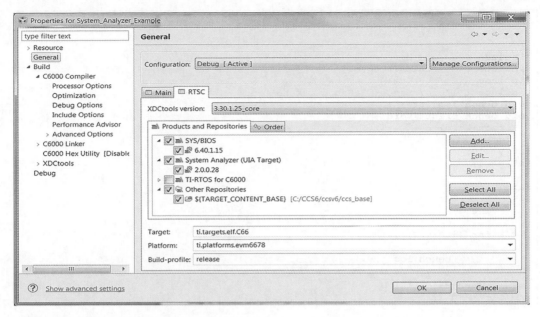

Figure 10.40 Configuration used in this project.

The program shown in Figure 10.41 is stand-alone and does not require any modification. Follow these steps to get an understanding of this debugging mode:

Step 1. Within the CCS, import the project from the following folder to your workspace: ...\ \ **Chapter_Debugging\System_Analyzer_Example**. Once the project is imported, check that you have the right versions of the SYS/BIOS and UIA installed in your PC. Figure 10.40 shows the RTSC configuration used in this project.

Step 2. Connect and power up the EVM module.

Step 3. In the CCS, press on the bug ![bug icon] to build and load the application. (You should observe no errors and no warnings.)

Step 4. Open the **System_config.cfg** file, and navigate to UIA Logging Configuration. Notice in the section **User-written Software Instrumentation**, the **Duration Analysis (Benchmarking)** is selected; see Figure 10.43.

Step 5. Open the configuration file **System_config.cfg**, select **SYS/BIOS – System Overview** and press the arrow circled shown in Figure 10.34. Then, select **Logging_Setup** and tick the box for **Add LoggingSetup to my Configuration**. Different options are shown depending on the threads that need to be debugged. For instance, if your application is using only **Hwi** threads and not **Swi** threads, then do not tick **Swi** to save memory; see Figure 10.43.

Step 6. If your program is not loaded, load it again as shown in Step 3, and select **Tools > System Analyzer > Duration Analysis**. Once this mode is selected, the dialogue box shown in Figure 10.44 will appear. Complete the box as shown in the figure, and press **START**. Run your code, and after a few seconds the data are collected and displayed as shown in Figure 10.45 and Figure 10.46. Notice that **Loginfo0()** took only 238 cycles, whereas **System_- printf** took 2885 cycles. Now, turn the optimisation on (**-03**) and notice the **Loginfo()** function is taking 180 cycles and **Sytem_printf()** is taking 2965 cycles, even higher than the debug mode; see Figure 10.47.

The configuration used is shown in Figure 10.42.

```
/*
 *  ======== main.c ========
 */
#include <xdc/std.h>
#include <xdc/runtime/Log.h>            // use this for the Log_info
#include <xdc/runtime/System.h>         // use this for printf
#include <ti/uia/events/UIABenchmark.h>
void bench_log(int);
void bench_printf(void);
Int main(Void);

int y =10;
int N= 3;

main (){

    int j;

    for (j=0;j<N;j++)
    {
         bench_log(j);
    }
    bench_printf();
    return (0);
}

void bench_log(int i)
{
    Log_write2(UIABenchmark_start, (xdc_IArg)"Log_info running: %d", i);

    Log_info0("the Log_info has been used!");

    Log_write2(UIABenchmark_stop, (xdc_IArg)"Log_info running: %d", i);

    return;
}

void bench_printf(void)

{
    int t =1;

    Log_write2(UIABenchmark_start, (xdc_IArg)"System_printf running: %d", t);
// t has been forced again to 2 in order to show that if the parameters
//2, 3      ..8 for
    // Log_write are not identical therefor there will be no output.
    //t=1;
    System_printf("y   = %d \n",  y);
    Log_write2(UIABenchmark_stop, (xdc_IArg)"System_printf running: %d", t);
    return;
}
```

Figure 10.41 Source code for Laboratory experiment 2.

10.8.4 Laboratory experiment 4: Using diagnosis features

Controlling which events will be enabled or disabled.
  Project location:

Chapter_10_Code\Conditional_Debugging

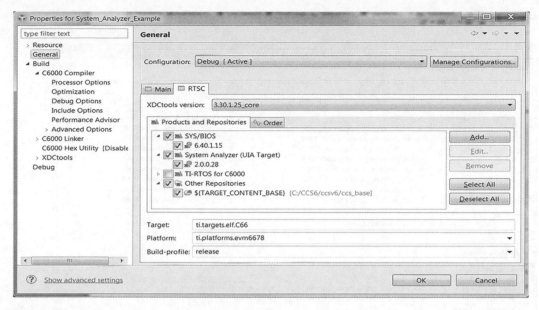

Figure 10.42 RTSC configuration used in this project.

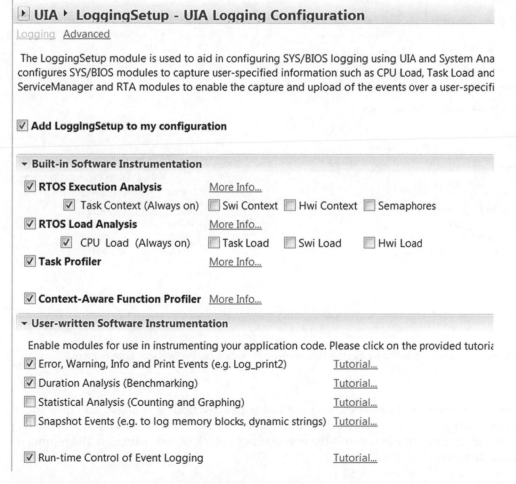

Figure 10.43 Selecting options for user-written software.

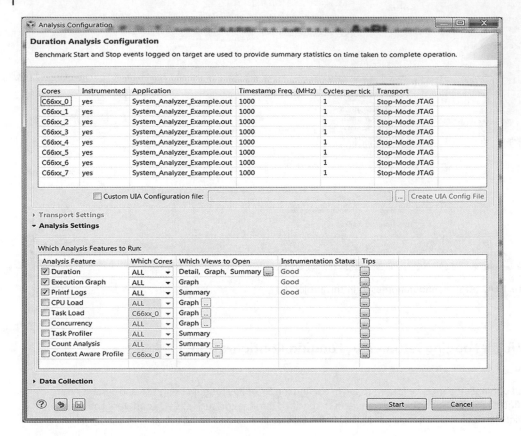

Figure 10.44 Analysis configuration.

	Source	START	STOP	Duration	
1	C66xx_0, Log_info running: 0	281365	281649	284	
2	C66xx_0, Log_info running: 1	281804	282042	238	
3	C66xx_0, Log_info running: 2	282197	282435	238	
4	C66xx_0, System_printf running: 1	282590	285475	2885	

Showing 4 records

Figure 10.45 Duration graph showing the time to execute each function (without optimisation).

The XDCtools allow diagnostics to be enabled or disabled both at configuration time and at runtime. Each XDC module has a diagnostics mask that controls both the **Assert** and the **Log** statements or conditionally controls the execution of a block of code using the **Diags_query**() runtime function [17].

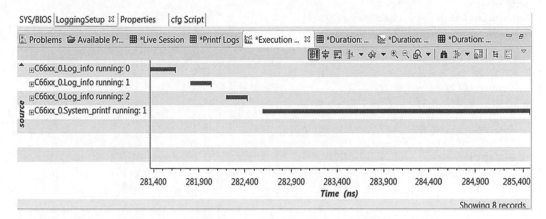

Figure 10.46 Execution graph showing the order in which functions were executed.

	Source	START	STOP	Duration
1	C66xx_0, Log_info running: 0	281504	281717	213
2	C66xx_0, Log_info running: 1	281846	282026	180
3	C66xx_0, Log_info running: 2	282155	282335	180
4	C66xx_0, System_printf running: 1	282467	285432	2965

Showing 4 records

Figure 10.47 Duration graph showing the time to execute each function (with **-o3** optimisation).

By turning off the diagnostic at configuration time (**Mode.ALWAYS_OFF**), the optimising compiler will remove the diagnostics code; see Figure 10.51. This is very convenient as there will be no need to modify the source code. However, by turning on the diagnostics at configuration time (**Mode.ALWAYS_ON**), the optimising compiler eliminates all runtime conditional checking and leaves the diagnostic code; see Figure 10.50. Configuring the flag as either **RUNTIME_OFF** or **RUNTIME_ON** allows the user to determine at runtime whether to log the event or not; see Figure 10.52. Figure 10.48 and Figure 10.49 show the source code and the configuration code used in this example. Figure 10.50, Figure 10.51 and Figure 10.52 show various outputs depending on the setting used. In this example, the control character 'F' (informational event) has been used, but other control characters can be used as shown in Table 10.3.

10.8.5   Laboratory experiment 5: Using a diagnostic feature with filtering

This laboratory session was introduced in Section 10.6.3. In this example, two categories are used, USER1 and USER2, and LEVEL3 is enabled in the configuration file; see the application code in Figure 10.53, configuration code in Figure 10.54 and output in Figure 10.55.

```
void test (void);
int y =10;
int main(int argc, char *argv[])
{
    /* disable info events associated with Main */
    Diags_setMask("xdc.runtime.Main-F");

    Log_info0("hello world");    /* DISABLED */

    /* enable info events associated with Main */
    Diags_setMask("xdc.runtime.Main+F");
    Log_info0("hello again");   /* ENABLED */
    test();
    System_printf(" the code is now completed    = %d \n",  y);

    return (0);
}

void test()
{
        int i= 10; // this is just an example

        Log_write2(UIABenchmark_start, (xdc_IArg)"Log_info running: %d", i);
        Log_info0("the Log_info has been used!");
        Log_write2(UIABenchmark_stop, (xdc_IArg)"Log_info running: %d", i);
}
```

Figure 10.48 Source code.

```
var Log = xdc.useModule("xdc.runtime.Log");
var Diags = xdc.useModule("xdc.runtime.Diags");

/* enable runtime control of "info" events for non -modules */
var Main = xdc.useModule("xdc.runtime.Main");
//Main.common$.diags_INFO = Diags.RUNTIME_OFF
//Main.common$.diags_INFO = Diags.ALWAYS_OFF
Main.common$.diags_INFO = Diags.ALWAYS_ON

/* use the LoggerSys ILogger service provider */
var Logger = xdc.useModule("xdc.runtime.LoggerSys");

/* create and bind a logger for all non -module code */
//Main.common$.logger = Logger.create();

// adding from here
var BIOS = xdc.useModule('ti.sysbios.BIOS');
BIOS.libType = BIOS.LibType_Custom;
BIOS.taskEnabled = false;
var LoggingSetup = xdc.useModule('ti.uia.sysbios.LoggingSetup');
LoggingSetup.sysbiosTaskLogging = true;
LoggingSetup.benchmarkLogging = true;
LoggingSetup.enableTaskProfiler = true;
LoggingSetup.enableContextAwareFunctionProfiler = true;
```

Figure 10.49 Configuration code.

	Type	Time	Error	Master	Message	Event	EventClass	Data1	Data2	SeqNo	Lc
1	i	283756		C66xx...	[../main.c:17] hello world	Log_L_info	Info			0	M
2	i	286211		C66xx...	[../main.c:22] hello again	Log_L_info	Info			1	M
3	📈	286351		C66xx...	Start: Log_info running: 10	Start		Log_info running: 10		2	M
4	i	286473		C66xx...	[../main.c:39] the Log_info ...	Log_L_info	Info			3	M
5	📈	286593		C66xx...	Stop: Log_info running: 10	Stop		Log_info running: 10		4	M

Figure 10.50 Using the **Main.common$.diags_INFO = Diags.ALWAYS_ON**.

	Type	Time	Error	Master	Message	Event	EventClass	Data1	Data2	SeqNo	Logger
1	📈	286176		C66xx_0	Start: Log_info running: 10	Start		Log_info running: 10		0	Main Logger
2	📈	286343		C66xx_0	Stop: Log_info running: 10	Stop		Log_info running: 10		1	Main Logger

Figure 10.51 Using the **Main.common$.diags_INFO = Diags.ALWAYS_OFF**.

	Type	Time	Error	Master	Message	Event	EventClass	Data1	Data2	SeqNo	Logg
1	i	286151		C66xx...	[../main.c:22] hello again	Log_L_info	Info			0	Main
2	📈	286316		C66xx...	Start: Log_info running: 10	Start		Log_...		1	Main
3	i	286451		C66xx...	[../main.c:39] the Log_info ...	Log_L_info	Info			2	Main
4	📈	286571		C66xx...	Stop: Log_info running: 10	Stop		Log_...		3	Main

Figure 10.52 Using the **Main.common$.diags_INFO = Diags.RUNTIME_OFF** or **Main.common$.
diags_INFO = Diags.RUNTIME_ON**.

```
#include <xdc/std.h>
#include <xdc/runtime/Log.h>
#include <xdc/runtime/LoggerBuf.h>
#include <xdc/runtime/Diags.h>
#include <xdc/cfg/global.h>

extern const LoggerBuf_Handle logger;

int main(int argc, String argv[])
{
    /* Print USER1 category events with each of the four levels. */
    Log_print0(Diags_USER1 | Diags_LEVEL1,"A USER1 category, LEVEL1 event.");
    Log_print0(Diags_USER1 | Diags_LEVEL2,"A USER1 category, LEVEL2 event.");
    Log_print0(Diags_USER1 | Diags_LEVEL3,"A USER1 category, LEVEL3 event.");
    Log_print0(Diags_USER1 | Diags_LEVEL4,"A USER1 category, LEVEL4 event.");

    /* Print a USER2 category event. */

    Log_print0(Diags_USER2 | Diags_LEVEL1,"A USER2 category, LEVEL1 event.");
    Log_print0(Diags_USER2 | Diags_LEVEL2,"A USER2 category, LEVEL2 event.");
    Log_print0(Diags_USER2 | Diags_LEVEL3,"A USER2 category, LEVEL3 event.");
    Log_print0(Diags_USER2 | Diags_LEVEL4,"A USER2 category, LEVEL4 event.");

    /* Flush the logger to see the records printed out. */
    LoggerBuf_flush(logger);

    return (0);
}
```

Figure 10.53 Application code.

```
var Log = xdc.useModule("xdc.runtime.Log");
var Diags = xdc.useModule("xdc.runtime.Diags");

/* Configure logging for xdc.runtime.Main */
var Main = xdc.useModule("xdc.runtime.Main");

/* Use the LoggerBuf ILogger service provider */
var LoggerBuf = xdc.useModule("xdc.runtime.LoggerBuf");

/* Create and bind a logger for Main */
Program.global.logger = LoggerBuf.create();

Main.common$.logger = Program.global.logger;

/* Enable the USER1 event category for Main. */
Main.common$.diags_USER2 = Diags.RUNTIME_ON;

/* Disable USER2 events of all levels for Main. */
Main.common$.diags_USER1 = Diags.RUNTIME_OFF;

/* We must enable filtering by level. */
LoggerBuf.filterByLevel = true;

/* Filter out all USER2 events below LEVEL3. */

LoggerBuf.level3Mask = Diags.USER2;

LoggerBuf.level4Mask = Diags.ALL_LOGGING & (~Diags.USER2);
```

Figure 10.54 Configuration file.

Project location:

\Chapter_10_Code\Conditional_debugg_2A

By changing the configuration file and not the application code, a different output can be obtained. This shows how filtering can be done without the need to change the application code. This is demonstrated by using the code shown in Figure 10.53 and the new configuration shown in Figure 10.56. In this example, USER1 and LEVEL2 are selected. The output is shown in Figure 10.57.

Project location:

\Chapter_10_Code\Conditional_debugg_1

```
# [C66xx_0]
#0000000001 xdc.runtime.Main: A USER2 category, LEVEL1 event.
#0000000002 xdc.runtime.Main: A USER2 category, LEVEL2 event.
#0000000003 xdc.runtime.Main: A USER2 category, LEVEL3 event.
```

Figure 10.55 Console output showing USER2 and all levels above three.

```
var Log = xdc.useModule("xdc.runtime.Log");
var Diags = xdc.useModule("xdc.runtime.Diags");

/* Configure logging for xdc.runtime.Main */
var Main = xdc.useModule("xdc.runtime.Main");

/* Use the LoggerBuf ILogger service provider */
var LoggerBuf = xdc.useModule("xdc.runtime.LoggerBuf");

/* Create and bind a logger for Main */
Program.global.logger = LoggerBuf.create();

Main.common$.logger = Program.global.logger;

/* Enable the USER1 event category for Main. */
Main.common$.diags_USER1 = Diags.RUNTIME_ON;

/* Disable USER2 events of all levels for Main. */
Main.common$.diags_USER2 = Diags.RUNTIME_OFF;

/* We must enable filtering by level. */
LoggerBuf.filterByLevel = true;

/* Filter out all USER1 events below LEVEL2. */

LoggerBuf.level2Mask = Diags.USER1;

LoggerBuf.level4Mask = Diags.ALL_LOGGING & (~Diags.USER1);
```

Figure 10.56 Configuration file using USER1 and LEVEL2.

```
[C66xx_0]
#0000000001 xdc.runtime.Main: A USER1 category, LEVEL1 event.
#0000000002 xdc.runtime.Main: A USER1 category, LEVEL2 event.
```

Figure 10.57 Output file.

10.9 Conclusion

Before debugging, the programmer should know what he or she is looking for and what the tools can offer. In this chapter, different types of debugging for the TMS320C66x SoC have been shown. The main element of all the debugging techniques is the Advanced Event Triggering (AET) hardware. It has been shown that AET is used for detecting some of the application activities and generating some events accordingly. These events or triggers can be used to CPU Halt (with Debugger Connected), to generate an interrupt, to start/increment a counter, to stop/reload a counter and so on. AET can be used as a CCS plug-in or be programmed using the AET target library. However, using any library will have an impact on the application code size

Table 10.6 Comparison between a few debugging techniques

Debugging/ profiling methods	Intrusiveness	Changes to the application code	Additional configuration required	Core visibility	System visibility
Printf like	Yes	Yes	No	Yes	No
RTOS Object View (ROV)	Yes	No	No	Yes	No
RTOS Analyzer	Yes			Yes	Yes
System Analyzer	Yes	Possible but not required	Possible but not required	Yes	Yes
Trace	No	No	No	Yes	No
System trace	No	No	Possible but not required	Yes	Yes

and performance. Trace uses the AET hardware and provides a historical account of application code execution, timing and data accesses. The system trace also uses the AET hardware but monitors the whole device and is used to collect and export bus transition of various resources and events from the on-chip CP tracers.

The Universal Instrumentation Architecture (UIA) can be very beneficial in many applications. However, this may affect some applications when code size and performance are critical.

Using the XDCtools can also have some advantages as shown in the examples.

Which technique to use for debugging will depend on the application and the programmer's experience. Table 10.6 shows a comparison between a few techniques.

References

1 Texas Instruments, Using Advanced Event Triggering to debug real-time problems in high speed embedded microprocessor systems, [Online]. Available: http://www.ti.com/lit/an/spra387/spra387.pdf.

2 Texas Instruments, Debug versus optimization tradeoff, [Online]. Available: http://processors.wiki.ti.com/index.php/Debug_versus_Optimization_Tradeoff.

3 Texas Instruments, Trace functionality and capabilities, [Online]. Available: http://processors.wiki.ti.com/index.php/Trace_Functionality_And_Capabilities.

4 Texas Instruments, Trace scripts, [Online]. Available: http://processors.wiki.ti.com/index.php/Trace_Scripts.

5 Texas Instruments, TraceDeviceFeatures, [Online]. Available: http://processors.wiki.ti.com/index.php/TraceDeviceFeatures.

6 Texas Instruments, TI Emulator Showcase 2013, [Online]. Available: https://www.youtube.com/watch?v=OzPRvigznP0&list=PL3NIKJ0FKtw4w_bK7FASz6RrTZb8PD3j5&index=9.

7 Texas Instruments, Trace - embedded development tools, [Online]. Available: https://www.youtube.com/watch?v=G3noymHTvGI&index=10&list=PL3NIKJ0FKtw4w_bK7FASz6RrTZb8PD3j5.

8 Texas Instruments, KeyStone advanced debug, [Online]. Available: http://processors.wiki.ti.com/images/5/52/Keystone_Advanced_Debug.pdf.

9 Texas Instruments, CToolsLib, [Online]. Available: http://processors.wiki.ti.com/index.php/CToolsLib.

10 Texas Instruments, System Analyzer user's guide, [Online]. Available: http://www.ti.com/lit/ug/spruh43f/spruh43f.pdf.

11 RTSC, XDCtools user's guide, [Online]. Available: http://rtsc.eclipse.org/docs-tip/XDCtools_User%27s_Guide.

12 RTSC, Overview of xdc.runtime, [Online]. Available: http://rtsc.eclipse.org/docs-tip/Overview_of_xdc.runtime.

13 RTSC, Using xdc.runtime logging, [Online]. Available: http://rtsc.eclipse.org/docs-tip/Using_xdc.runtime_Logging.

14 RTSC, Runtime Object Viewer, [Online]. Available: http://rtsc.eclipse.org/docs-tip/RTSC_Object_Viewer.

15 RTSC, Using xdc.runtime Logging, [Online]. Available: http://rtsc.eclipse.org/docs-tip/Using_xdc.runtime_Logging.

16 RTSC, Using xdc.runtime logging/example 5, [Online]. Available: http://rtsc.eclipse.org/docs-tip/Using_xdc.runtime_Logging/Example_5.

17 RTSC, Diagnostics manager, [Online]. Available: http://rtsc.eclipse.org/cdoc-tip/xdc/runtime/Diags.html#xdoc-desc.

Further reading

18 Texas Instruments, KeyStone II debug tools, [Online]. Available: http://processors.wiki.ti.com/index.php/Keystone_II_Debug_Tools.

19 Texas Instruments, CToolsLib, [Online]. Available: https://gforge.ti.com/gf/project/ctoolslib/frs/.

20 Texas Instruments, Common platform tracer examples, [Online]. Available: http://processors.wiki.ti.com/index.php/Common_Platform_Tracer_Examples.

21 Texas Instruments, Using system trace (STM), [Online]. Available: http://processors.wiki.ti.com/index.php/Using_System_Trace_%28STM%29.

22 Texas Instruments, CP tracer details, [Online]. Available: http://processors.wiki.ti.com/index.php/CP_Tracer_Details.

23 Texas Instruments, Visualizing common platform tracer data on the KeyStone architecture, [Online]. Available: http://processors.wiki.ti.com/index.php/Visualizing_Common_Platform_Tracer_Data_On_The_Keystone_Architecture.

24 Texas Instruments, Adding An STM node to CCSv5 target configuration, [Online]. Available: http://processors.wiki.ti.com/index.php/Adding_An_STM_Node_to_CCSv5_Target_Configuration.

25 Texas Instruments, XDS560v2 system trace, [Online]. Available: http://processors.wiki.ti.com/index.php/XDS560v2_System_Trace.

26 Texas Instruments, MCSDK UG chapter developing debug trace, [Online]. Available: http://processors.wiki.ti.com/index.php/MCSDK_UG_Chapter_Developing_Debug_Trace.

27 RTSC, Basic runtime support for RTSC programs, [Online]. Available: http://rtsc.eclipse.org/cdoc-tip/xdc/runtime/package.html.

28 Texas Instruments, ITM: instrumentation trace module for Cortex M, [Online]. Available: https://www.youtube.com/watch?v=HG_i_uln6Es.

29 Texas Instruments, 64x + advanced emulation techniques and technologies, [Online]. Available: http://wiki.tiprocessors.com/images/a/a7/AdvancedEmulation07.pdf.

30 Texas Instruments, XDS target connection guide, [Online]. Available: http://processors.wiki.ti.com/index.php/XDS_Target_Connection_Guide.

31 Texas Instruments, TI development tools information, [Online]. Available: http://processors.wiki.ti.com/index.php/TI_Development_Tools_Information.

32 Texas Instruments, Category:Emulation, [Online]. Available: http://processors.wiki.ti.com/index.php/Category:Emulation.

11

Bootloader for KeyStone I and KeyStone II

11.1 Introduction

A bootloader is a program that transfers a code (program and/or data) from one memory location to another. The code which is transferred is called *boot code*. The bootloader and the boot code are indispensable in any microprocessor-based system. When a microprocessor, which is connected to some peripherals, is powered up, it will not function properly since these peripherals need to be configured and the application, which may reside in non-volatile memory, should be transferred in the internal memory where it is run. For instance, if the KeyStone II is connected to double data rate (DDR) memory and flash memory, the DDR needs to be configured and the clock signals need to be set up before the bootloader transfers the code from the flash memory to DDR. The boot code can be another bootloader or an image (an application, for instance), depending on the application.

In this chapter, the bootloaders for both the KeyStone I and the KeyStone II will be discussed.

11.2   How to start the boot process

The boot process only happens after a reset. There are, however, different types of resets (see Table 11.1 and Table 11.2 for the KeyStone I and the KeyStone II, respectively):

- Power-on reset (POR) (higher priority)
- Hard reset
- Soft reset (lower priority)
- Local reset (only affects the local core that has been reset).

For the local reset, the boot process is not triggered and, therefore, no boot process will take place.

After power-up, the POR puts the complete system to a reset state until the initialisation (to the default values) of internal registers is completed, after the power supply voltage and the internal clock of the device reach a stable state. However, the RESETFULL (see Table 11.1 and Table 11.2) which can be asserted by the host is also asserted during POR, and it will have the same effect as the POR and therefore reset the whole device. More information on the reset can be found in Refs. [1, 2], and some useful information is shown in Table 11.1 and Table 11.2 and summarized in Table 11.3.

11.3   The boot process

Both KeyStone I and II support various boot processes. These processes start executing the bootloader (after a reset, as shown) located at the beginning of the **BOOT ROM** address (see Section 11.4). There are two **BOOT ROM** addresses depending on whether the DSP boot mode or ARM boot mode is selected.

Since the KeyStone II has both DSPs and ARMs, it has four hardware modes supported on top of the various software-driven boot processes [1]:

A) Public ROM boot for non-secure devices
   1) When the **DSP CorePac0** is the boot master (DSP boot mode):
      - Both the **DSP CorePac0** and the **ARM CorePac0** are released at the same time.
      - Both the ARM and the DSP start executing.
      - The bootloader reads the value of these pins as latched into the **bootCFG** register at device POR reset.
      - Both the ARM and the DSP read the boot mode inside the **bootCFG** register in order to determine which the boot master is.
      - If the ARM reads that the DSP is the master, then all ARM cores will enter a low-power standby state by executing the WFI (wait for interrupt) instruction and waiting for a DSP interrupt.

Table 11.1 Reset types for the KeyStone I [3]

Reset type	Initiator	Effect on device when reset occurs	RESETSTAT pin status
Power-on reset (POR)	• POR pin active low • RESETFULL pin active low	• Total reset of the chip • Everything on the device is reset to its default state in response to this. • Activates the POR signal on the chip, which is used to reset test/emu logic • Boot configurations are latched. • ROM boot process is initiated.	Toggles RESETSTAT pin
Hard reset	• RESET pin active low • Emulation • PLL Control Register (PLLCTL) (or Reset Control Register (RSCTRL)) • Watchdog timers	• Resets everything except for test/emu logic and reset isolation modules • Emulator and reset isolation modules stay alive during this reset. • This reset is also different from POR in that the PLLCTL assumes power and clocks are stable when device rest is asserted. • Boot configurations are not latched. • The ROM boot process is initiated.	Toggles RESETSTAT pin
Soft reset	• RESET pin active low • PLLCTL register (RSCTRL) • Watchdog timers	• Software can program these initiators to be hard or soft. • Hard reset is the default, but it can be programmed to be soft reset. • Soft reset will behave like hard reset except that external memory contents, EMIF16 memory map registers (MMRs), DDR3 EMIF MMRs and the sticky bits in PCIe MMRs are retained. • Boot configurations are not latched. • ROM boot process is initiated.	Toggles RESETSTAT pin
C66x CorePac local reset	• Software (through LPSC MMR) • Watchdog timers • LRESET pin	• MMR bit in LPSC controls the C66x CorePac local reset. • Used by watchdog timers (in the event of a timeout) to reset the C66x CorePac • Can also be initiated by LRESET device pin. The C66x CorePac memory system and slave DMA port are still alive when the C66x CorePac is in local reset. • Provides a local reset of the C66x CorePac, without destroying clock alignment or memory contents • Does not initiate ROM boot process	Does not toggle RESETSTAT pin

Source: Courtesy of Texas Instruments.

- If the DSP is the boot master, then **CorePac0** will execute the bootloader (after being downloaded) and the other DSP cores stay idling by executing the **IDLE** instruction.
- The downloaded code inside the bootloader may contain code for either the other DSP cores or the ARM cores that need to be run. If this is the case, then the downloaded

Table 11.2 Reset types for the KeyStone II [1]

Type	Initiator	Effect(s)
Power-on reset (POR)	• POR pin • RESETFULL pin	• Resets the entire chip, including the test and emulation logic • The device configurations are latched only during POR.
Hard reset	• RESET pin • PLL Control* Register (RSCTRL) • Watchdog timers • Emulation	• Hard reset resets everything except for test, emulation logic and reset isolation modules. • This reset is different from POR in that the PLL Controller assumes power and clocks are stable when a hard reset is asserted. • The device configuration pins are not re-latched. • Emulation-initiated reset is always a hard reset. • By default, these initiators are configured as hard reset, but they can be configured (except emulation) as soft reset in the RSCFG Register of the PLL Controller. • Contents of the DDR3 SDRAM memory can be retained during a hard reset if the SDRAM is placed in self-refresh mode.
Soft reset	RESET pin PLL Control Register (RSCTRL) Watchdog timers	• Soft reset behaves like hard reset except that the PCIe MMRs (memory-mapped registers) and DDR3 EMIF MMRs contents are retained. • By default, these initiators are configured as hard reset, but can be configured as soft reset in the RSCFG Register of the PLL Controller. • Contents of the DDR3 synchronous dynamic random-access memory (SDRAM) can be retained during a soft reset if the SDRAM is placed in self-refresh mode.
Local reset	LRESET pin Watchdog timer timeout LPSC MMRs	• Resets the C66x CorePac without disturbing clock alignment or memory contents • The device configuration pins are not re-latched.

*All masters in the device have access to the PLL Control Registers.
Source: Courtesy of Texas Instruments.

code should contain control code that specifies the execution addresses for each core. These addresses are written to each core's boot address register. The **CorePac0** then issues interrupts to the relevant cores through the Interrupt Control Generation Registers (**IPCGRx**) to execution code.

2) When the **ARM CorePac0** is the boot master (ARM boot mode):
   • This mode is similar to the previous mode except that the **DSP CorePac0** and the **ARM CorePac0** roles are swapped.
   • This is the preferred boot mode when using Linux.

B) Secure ROM boot for secure devices. Secure boot allows the user bootloader to be authenticated before use, ensuring that the customer software is not modified from the original customer signed image. Secure devices always boot from the DSP internal secure ROM, and the bootloader is authenticated before use.

   1) When the TMS320C66x **CorePac0** is the boot master. This mode operates in the same way as public ROM boot, when the **DSP CorePac0** is the boot master.

Table 11.3 Reset types summary for both KeyStone I and II

Reset types	Initiator	What is not reset	Boot pins	Boot process
Power-on reset (POR)	• POR active low pin • RESETFULL active low	None (reset everything on the system)	Latched and updated	Yes
Hard reset	• RESET active low pin • Emulation • PLLCTL register • Watchdog timers	• Test/emu logic • Reset isolation modules. • Clocking is unaffected within the device.	No	Yes
Soft reset	• RESET active low • PLL Control Register (RSCTRL) • Watchdog timers	• Test/emu logic • Reset isolation modules. • EMIF16 MMRs, DDR3 EMIF MMRs and the sticky bits in PCIe MMRs	No	Yes
C66x CorePac local reset	• Software (through LPSC MMR) • Watchdog timers • LRESET pin	• Only CorePac reset that does not destroy memory.	No	No

2) When the **ARM CorePac0** is the boot master. This mode operates in the same way as public ROM boot, when the **ARM CorePac0** is the boot master.

Bit 8 in Figure 11.9 determines whether the boot is a C66x CorePac boot or ARM CorePac boot (0 = ARM is boot master; 1 = C66x is boot master).

11.4 ROM Bootloader (RBL)

The KeyStone I and II devices contain RBL software that is designed to transfer code to initialize the platform and branch to the start of a user application or to a second bootloader such as U-Boot, as shown in Section 11.5.2. The RBL, which is burned onto the on-chip ROM at production time, cannot be modified, and it is the first piece of code to run and start the device.

The RBL is located in the ROM at the following addresses:

- 0x20B00000 (128 K for the KeyStone I)
- 0x00000000 (256 K for the KeyStone II).

The task of the RBL is twofold: one is to configure the device resources (which are determined by the boot mode) at the beginning of the boot process (initialisation stage), and the second is to

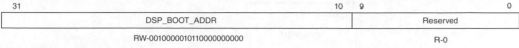

	31			10	9		0

DSP_BOOT_ADDR	Reserved
RW-0010000010110000000000	R-0

Legend: R = Read only; RW = Read/Write; -*n* = value after reset

Bit	Field	Description
31–10	DSP_BOOT_ADDR	Boot address of CorePac. CorePac will boot from that address when a reset is performed. The reset value is 22 MSBs of ROM base address = 0×20B00000.
9–0	Reserved	Reserved

Figure 11.1 Boot address of a CorePac, **DSP_BOOT_ADDR**, for the KeyStone I.

0x02620040	0x02620043	4B	DSP_BOOT_ADDR0	The boot address for C66x CorePac0
0x02620044	0x02620047	4B	DSP_BOOT_ADDR1	The boot address for C66x CorePac1
0x02620048	0x0262004B	4B	DSP_BOOT_ADDR2	The boot address for C66x CorePac2
0x0262004C	0x0262004F	4B	DSP_BOOT_ADDR3	The boot address for C66x CorePac3
0x02620050	0x02620053	4B	DSP_BOOT_ADDR4	The boot address for C66x CorePac4 (66AK2H14/12 only)
0x02620054	0x02620057	4B	DSP_BOOT_ADDR5	The boot address for C66x CorePac5 (66AK2H14/12 only)
0x02620058	0x0262005B	4B	DSP_BOOT_ADDR6	The boot address for C66x CorePac6 (66AK2H14/12 only)
0x0262005C	0x0262005F	4B	DSP_BOOT_ADDR7	The boot address for C66x CorePac7 (66AK2H14/12 only)

Figure 11.2 Boot address of a CorePac, **DSP_BOOT_ADDR**, for the KeyStone II.

load the image into the device and execute this image (boot process stage); see Figure 11.3 and Figure 11.4 for the KeyStone I and II, respectively.

The DSP **CorePacN** will boot from the address specified in the **DSP_BOOT_ADDRn** register (n: CorePac number) which has a default value of 0x20B00000; see Figure 11.1 and Figure 11.2.

It is important not to confuse the **DSP_BOOT_ADDR** with the *boot magic address*. The **DSP_BOOT_ADDR** is used by the RBL to branch to the boot code, and the boot magic address is used by the DSP core to get the execution address of the image and start executing.

During the POST or RESETFULL assertion, the general-purpose pins (see Figure 11.5 and Figure 11.6) that are used for selecting a boot mode are sampled, and their corresponding boot configuration values are latched into the Device Status Register (DEVSTAT) located at address 0x2620020 (for all devices) (see Figure 11.7 for the C6678 and Figure 11.8 for the 66AK2H14/12/06).

The corresponding boot configuration values (stored in the Boot Parameter Table) are then copied by the RBL to a reserved L2 section of the **CorePac0** and also modified according to the bootstrap pins (see Figure 11.12 and Figure 11.13).

The Boot Parameter Table used by the RBL to determine the boot flow (see Figure 11.12) is composed of some parameters that are common to all boot modes and are referred to as the *Boot Parameter Table Common Parameters*. It is also composed of other boot parameter tables, each one corresponding to a specific boot mode. Table 11.4 shows the C6678's Boot Parameter Table Common Parameters, Table 11.5 is its EMIF16 Boot Parameter Table and Table 11.6 and Table 11.7 are its SPI (Serial Peripheral Interconnect) Boot Parameter Tables. For the other boot tables, please refer to the user guide [3]. Table 11.8, Table 11.9 and Table 11.10 show the 66AK2H14/12/06's Boot Parameter Table Common Parameters, the EMIF16 Boot Parameter Table and the SPI Boot Parameter Table, respectively. Figure 11.9 shows the boot mode pins for the TMS320C66AK2H14/12/06. For the other boot tables, please refer to the user guide [1].

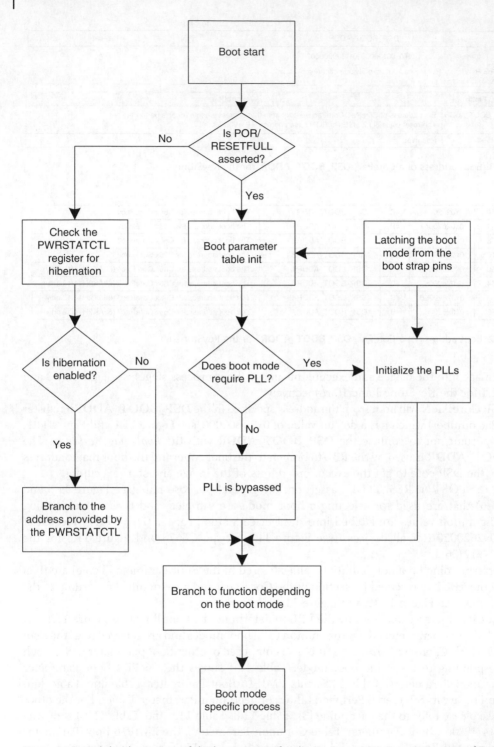

Figure 11.3 High-level overview of the boot process for the KeyStone I [4]. *Source*: Courtesy of Texas Instruments.

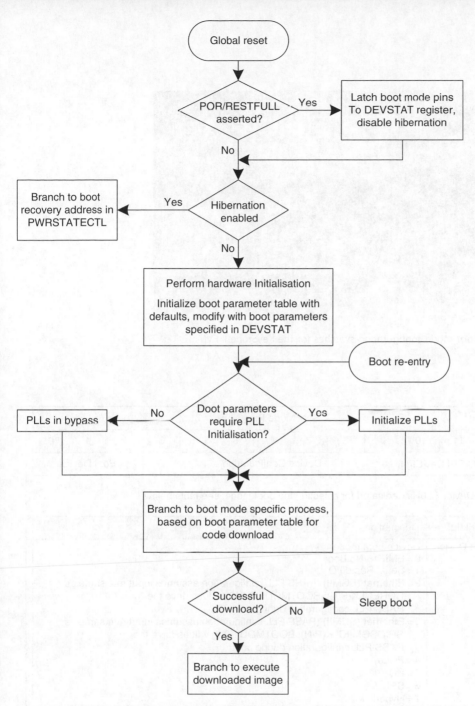

Figure 11.4 High-level overview of the boot process for the KeyStone II [2]. *Source*: Courtesy of Texas Instruments.

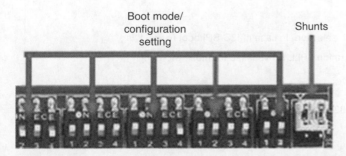

Figure 11.5 Boot mode configuration switches for the TMS320C6678 EVM.

Figure 11.6 Boot mode configuration switches for the KeyStone II EVM.

Boot Mode Pins												
12	11	10	9	8	7	6	5	4	3	2	1	0
PLL Mult I²C/SPI Dev Cfg			Device Configuration							Boot Device		

- Boot Device [2:0] is dedicated for selecting the boot mode; see table below.

Bit	Field	Description
2-0	Boot Device	Device boot mode 0 = EMIF16/No Boot 1 = Serial Rapid I/O 2 = Ethernet (SGMII) (PASS PLL configuration assumes input rate same as CORECLK(P/N); BOOTMODE[12:10] values drive the PASS PLL configuration during boot) 3 = Ethernet (SGMII) (PASS PLL configuration assumes input rate same as SRIOSGMIICLK(P/N); BOOTMODE[9:8] values drive the PASS PLL configuration during boot) 4 = PCLe 5 = I²C 6 = SPI 7 = Hyperlink

- Device Configuration [9:3] is used to specify the boot mode specific configurations; please refer to the user guide.

- PLL Multi [12:10] are used for PLL selection. In case of I2C/SPI boot mode, it is used for extended device configuration. (PLL is bypassed for these two boot modes.)

Figure 11.7 Boot mode pin for the TMS320C6678 (DEVSTAT) [3]. *Source*: Courtesy of Texas Instruments.

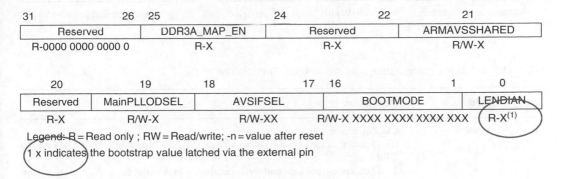

Figure 11.8 Device status register for the TMS320C66AK2H14/12/06 [1]. *Source*: Courtesy of Texas Instruments.

Table 11.4 Boot Parameter Table Common Parameters (TMS320C6678) [3]

Byte offset	Name	Description
0	Length	The length of the table, including the length field, in bytes
2	Checksum	The 16-bit ones complement of the ones complement of the entire table. A value of 0 will disable checksum verification of the table by the boot ROM.
4	Boot Mode	Internal values used by RBL for different boot modes
6	Port Num	Identifies the device port number to boot from, if applicable
8	SW PLL, MSW	PLL configuration, MSW
10	SW PLL, LSW	PLL configuration, LSW

Source: Courtesy of Texas Instruments.

Table 11.5 EMIF16 boot mode parameter table (TMS320C6678) [3]

Byte offset	Name	Description	Configured through boot configuration pins
12	Options	Option for EMIF16 boot (currently none)	–
14	Type	Boot only from NOR flash is supported for TMS320C6678.	–
16	Branch Address MSW	Most significant bit for branch address (depends on chip select)	–
18	Branch Address LSW	Least significant bit for branch address (depends on chip select)	–
20	Chip Select	Chip select for the NOR flash	–
22	Memory Width	Memory width of the EMIF16 bus (16 bits)	–
24	Wait Enable	Extended wait mode enabled: 0 = Wait enable is disabled. 1 = Wait enable is enabled.	Yes

Source: Courtesy of Texas Instruments.

Table 11.6 SPI device configuration bit fields

12	11	10	9	8	7	6	5	4	3
Mode		4, 5 Pin	Addr Width	Chip Select		Parameter Table Index			

Table 11.7 SPI device configuration field descriptions (TMS320CC6678) [3]

Bit	Field	Description
12–11	Mode	Clk Pol/Phase 0 = Data are output on the rising edge of SPICLK. Input data are latched on the falling edge. 1 = Data are output one half-cycle before the first rising edge of SPICLK and on subsequent falling edges. Input data are latched on the rising edge of SPICLK. 2 = Data are output on the falling edge of SPICLK. Input data are latched on the rising edge. 3 = Data are output one half-cycle before the first falling edge of SPICLK and on subsequent rising edges. Input data are latched on the falling edge of SPICLK.
10	4, 5 Pin	SPI operation mode configuration: 0 = 4-pin mode is used. 1 = 5-pin mode is used.
9	Addr Width	SPI address width configuration: 0 = 16-bit address values are used. 1 = 24-bit address values are used.
8–7	Chip Select	The chip select field value: 00b = CS0 and CS1 are both active (not used). 01b = CS1 is active. 10b = CS0 is active. 11b = None is active.
6–3	Parameter Table Index	Specifies which parameter table is loaded from SPI. The boot ROM reads the parameter table (each table is 0x80 bytes) from the SPI starting at SPI address (0x80*parameter index). The value can range from 0 to 15.

Source: Courtesy of Texas Instruments.

Table 11.8 EMIF16 boot parameter table common parameters (66AK2H14/12/06)

Byte offset	Name	Description
0	Length	The length of the table, including the length field, in bytes
2	Checksum	The 16-bit ones complement of the ones complement of the entire table. A value of 0 will disable Checksum verification of the table by the boot ROM.
4	Boot Mode	Internal values used by RBL for different boot modes
6	Port Num	Identifies the device port number to boot from, if applicable
8	SW PLL, MSW	PLL configuration, MSW
10	SW PLL, LSW	PLL configuration, LSW
12	Sec PLL Config, MSW	ARM PLL configuration, MSW
14	Sec PLL Config, LSW	ARM PLL configuration, LSW
16	System Freq	The frequency of the system clock in MHz
18	Core Freq	The frequency of the core clock in MHz
20	Boot Master	Set to TRUE if C66x is the master core

Table 11.9 EMIF16 boot parameter table (TMS320C66AK2H14/12/06)

Byte offset	Name	Description	Configuration through boot configuration pins
22	Options	Async Config Parameters are used. 0 = Values in the sync config parameters are not used to program async config registers. 1 = Values in the async config parameters are used to program async config registers.	NO
24	Type	Set to 0 for EMIF16 (NOR) boot	NO
26	Branch Address MSW	Most significant bit for branch address (depends on chip select)	YES
28	Branch Address LSW	Least significant bit for branch address (depends on chip select)	YES
30	Chip Select	Chip Select for the NOR flash	YES
32	Memory Width	Memory width of the EMIF16 bus (16 bits)	YES
34	Wait Enable	Extended wait mode enabled: 0 = Wait enable is disabled. 1 = Wait enable is enabled.	YES
36	Async Config MSW	Async Config Register MSW	NO
38	Async Config LSW	Async Config Register LSW	NO

Table 11.10 SPI boot parameter table (TMS32066AK2H14/12/06) [1]

Byte offset	Name	Description	Configuration through boot configuration pins
22	Options	Bits 01 & 00 modes: 00 = Load a boot parameter from the SPI (default mode). 01 = Load boot records from the SPI (boot tables). 10 = Load boot config records from the SPI (boot config tables). 11 = Load GP header blob. Bits 15-02 = Reserved.	NO
24	Address Width	The number of bytes in the SPI device address. Can be 16 or 24 bit.	YES
26	NPin	The operational mode, 4 or 5 pin	YES
28	Chipsel	The chip select used (valid in 4-pin mode only). Can be 0–3.	YES
30	Mode	Standard SPI mode (0–3)	YES
32	C2Delay	Setup time between chip assert and transaction	NO
34	Bus Freq, 100 kHz	The SPI bus frequency in kilohertz	NO
36	Read Addr MSW	The first address to read from, MSW (valid for 24-bit address width only)	YES
38	Read Addr LSW	The first address to read from, LSW	YES
40	Next Chip Select	Next Chip Select to be used (used only in boot config mode)	No
42	Next Read Addr MSW	The next read address (used in boot config only)	NO
44	Next Read Addr LSW	The next read address (used in boot config only)	NO

Source: Courtesy of Texas Instruments.

16	15	14	13	12	11	10	9	8	7	6	5	4	3	2	1	Mode
X	X	0	ARMEN	SYSEN	ARM PLL config				SYS PLL config				0	0	0	Sleep
Slaveaddr		1	Port		ARM PLL config				SYS PLL config			Min	0	0	0	I²C slave
X	X	X	Bus addr		Param Idx				X	Port			0	0	1	I²C master
Width	Csel		Mode		Param Idx				Npin				0	1	0	SPI
0	Base addr		Wait	Width	ARM PLL config							0	0	1	1	EMIF (ARM master)
					X	Chip set										EMIF (DSP master)
1	First block			Clear	ARM PLL config											NAND (ARM master)
					X	Chip set										NAND (DSP master)
Lane	Ref clock		Data rate		ARM PLL config			Boot master	Min				1	0	0	SRIO (ARM master)
X					Lane setup											SRIO (DSP master)
PA clk	Ref clk		Ext Con		ARM PLL config				SYS PLL config				1	0	1	Ethernet (ARM master)
					Rsvd	Lane setup										Ethernet (DSP master)
Ref clk	Bar config				ARM PLL config							0	1	1	0	PCLe (ARM master)
					SerDes Cfg											PCLe (DSP master)
Port	Ref clk		Data rate		ARM PLL config							1	1	1	0	Hyperlink (ARM master)
					SerDes Cfg											Hyperlink (DSP master)
X	X	X	X	Port	ARM PLL config				Min				1	1	1	UART (ARM master)
X	X	X	X		X	X	X									UART (DSP master)

Figure 11.9 Boot mode pins for the TMS320C66AK2H14/12/06 (DEVSTAT) [1].

DEVSTAT boot mode pins ROM mapping																
16	15	14	13	12	11	10	9	8	7	6	5	4	3	2	1	0
Width	Csel		Mode		Param Idx/offset			Boot master	Npin	Port		Min		010		Lendian

Figure 11.10 DEVSTAT boot mode pins ROM mapping.

The RBL is very flexible, and it operates differently depending on the boot mode selected. The RBL initialisation, the loading process and the handover process when the RBL has completed are different. Please refer to Chapter 3 in Ref. [4] for the description of each boot mode and the action taken by the RBL.

Figure 11.10 and Figure 11.11 show SPI device configuration fields and descriptions, respectively. Please refer to the data sheet [1] for the other modes.

11.4.1 The boot configuration format

In addition to the boot parameter tables shown in Section 11.4, there are also two other tables that RBL uses. Therefore, three files referred to as *tables* have to be defined:

1) Boot parameter tables seen in Section 11.4 (determine the boot flow)
2) Boot table (the image itself)
3) Boot configuration table (needed for peripherals configuration).

11.4.1.1 Creating the boot parameter table

See Section 11.4.

Bit	Field	Description
16	Width	SPI address width configuration 0 = 16-bit address values are used 1 = 24-bit address values are used (default)
15-14	Csel	The chip select field value 0-3[default = 0]
13-12	Mode	Clk Polarity/ Phase 0 = Data is output on the rising edge of SPICLK. Input data is latched on the falling edge. 1 = Data is output one half-cycle before the first rising edge of SPICLK and on subsequent falling edges. Input data is latched on the edge of SPICLK. 2 = Data is output on the falling edge of SPICLK. Input data is latched on the rising edge (default). 3 = Data is output one half-cycle before the first falling edge of SPICLK and on subsequent rising edges. Input data is latched on the falling edge of SPICLK.
11-9	Param Idx/Offset	Parameter Table Index: 0-7 This value specifies the parameter table index when the C66x is the boot master This value specifies the start read address at 8k time this value when the ARM is the boot master
8	Boot master	Boot master select 0 = ARM is boot master (default) 1 = C66x is boot master
7	Npin	Selected Chip Select driven 0 = C50 to the selected chip select is driven 1 = C50–CS4 to the selected chip select are driven (default)
6-5	Port	Specify SPI port 0 = SPI0 used (default) 1 = SPI1 used 2 = SPI2 used 3 = Reserved
4	Min	Minimum boot configuration select bit. 0 = Minimum boot pin select disabled 1 = Minimum boot pin select enabled. When Min = 1 , a predetermined set of values is configured (see the Device Configuration Field Descriptions table for configuration bits with a "(default)" tag added in the description column). When Min = 0, all fields must be independently configured.
3-1	Boot devices	Boot Devices[3:1] 010 = SPI boot mode Others = other boot modes
0	Lendian	Endianess 0 = Big endian 1 = Little endian

Figure 11.11 SPI device configuration field descriptions.

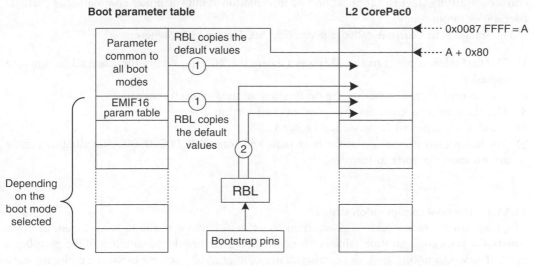

Figure 11.12 The RBL boot process.

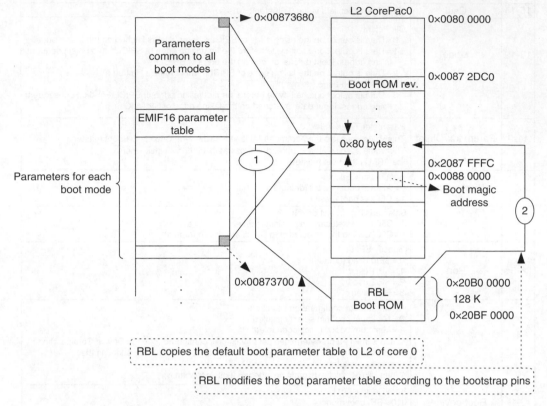

Figure 11.13 Detailed RBL boot process for the TMS320C6678.

11.4.1.2 Creating the boot table

Before booting an application, this application has to be converted to an image. The hexadecimal conversion utility takes the application and information about the linker command file (*.**RMD file**; see example below).

The image is in a format called a *boot table*; see Figure 11.14, where:

1) The first address contains the address to where the RBL should branch when all sections are copied.
2) The second address contains the destination address of each section.
3) The third address contains the count of each section.
4) 2 and 3 are repeated *N* times; see Figure 11.14.
5) The last memory locations of the boot table will contain **0x00000000**, indicating that there are no more sections to transfer.

11.4.1.3 The boot configuration table

The boot configuration table is used when some default value needs to be changed; in other words, the boot configuration table is where the user specifies the parameters of the peripheral used. It is worth noting that all peripherals are configured by setting or clearing bits for each

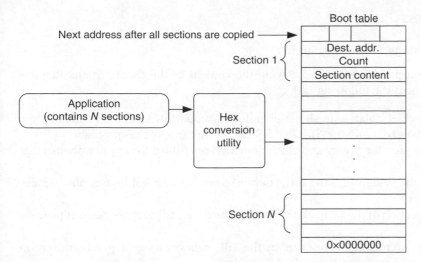

Figure 11.14 The boot table.

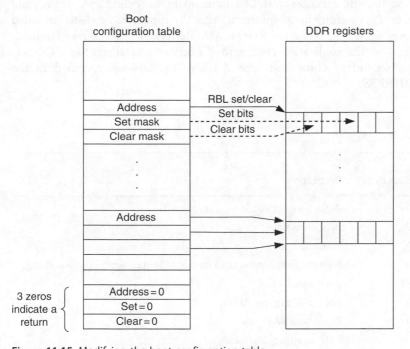

Figure 11.15 Modifying the boot configuration table.

specific register. For instance, if the image needs to be loaded in the DDR memory, this DDR needs to be configured properly.

To configure a peripheral, therefore, the address of the register and the set and clear marks will be required; see Figure 11.15. When the address and the set and clear registers are set to zero, this indicates to the RBL that all registers are set.

11.5   Boot process

11.5.1   Initialisation stage for the KeyStone I

During the initialisation phase, the RBL, knowing the content of the Device Status Register (**DEVSTAT**), will perform the following tasks:

- The RBL enables the reset isolation in all peripherals that support it. The power state of these peripherals is not changed. Only the SRIO and **SmartReflex** support reset isolation [3].
- The RBL also ensures that the power and clock domains are enabled for any peripherals that are required for boot.
- The RBL configures the system PLL to set the device speed. See user guides for more details: DSP [4] and ARM [2].
- The RBL reserves a portion of the L2 in all the cores in the device to perform the boot process; see Table 11.11.
- For the EMIF16 boot, no memory is reserved by the RBL; memory usage depends entirely on the image stored in, and executed from, the NOR flash.
- All the interrupts are disabled, except inter-processor communication (IPC) interrupts and the host interrupts that are needed for the **PCIe**, SRIO (DirectIO) and HyperLink boot modes.
- During the boot process, the RBL executes an **IDLE** command on the secondary CorePacs and keeps the secondary CorePacs waiting for an interrupt. After the application code to be loaded in these secondary CorePacs is loaded and the **BOOT_MAGIC_ADDRESS** values in individual CorePacs are populated, the application code in the **CorePac0** can trigger the IPC interrupt to wake up the secondary cores and branch up to the address specified in the **BOOT_MAGIC_ADDRESS**.

Table 11.11 Bootloader Section in L2 SRAM

Start address (hex)	Size (hex bytes)	Description
0x00872DC0	0x40	ROM boot version string (unreserved)
0x00872E00	0x400	Boot code stack
0x00873200	0xE0	Boot log
0x008732E0	0x20	Boot progress register stack (copies of boot program on mode change)
0x00873300	0x100	Boot internal stats
0x00873400	0x20	Boot table arguments
0x00873420	0xE0	ROM boot FAR data
0x00873500	0x100	DDR configuration table
0x00873600	0x80	RAM table
0x00873680	0x80	Boot parameter table
0x00873700	0x4900	Clear text packet scratch
0x00878000	0x7F80	Ethernet/SRIO packet/message/descriptor memory
0x0087FF80	0x40	Small stack
0x0087FFC0	0x3C	Not used
0x0087FFFC	0x4	Boot magic address

31		8	7	6	5	4	3	2	1	0
Reserved			BC7	BC6	BC5	BC4	BC3	BC2	BC1	BC0
R-0000 0000 0000 0000 0000 0000			RW-0	RW-0	RW-0	RW-0	RW-0	RW-0	RW-0	RW-0

Legend: R = Read only; RW = Read/write; n = value after rest.

Note: BCx CorePacx boot status:
      0 = CorePacx boot NOT complete.
      1 = CorePacx boot complete.

Figure 11.16 Boot complete register, **BCx**. *Note*: **BCx CorePacx** boot status: 0 = **CorePacx** boot NOT complete, 1 = **CorePacx** boot complete.

- All L1D and L1P memory is configured by the boot code as cache memory. L2 memory, however, is configured as addressable memory.
- The RBL also provides the ability for the user to configure the DDR EMIF before loading the image into the external memory during the boot process using a DDR structure that is reserved in the L2. For every section that the RBL reads, it verifies if the **DDR enable magic word** is set. If the **magic word** is set, then the DDR structure is used to initialize the DDR. For the KeyStone I, the DDR structure is located at address **0x00873500**.
- The RBL uses the pin-strapped boot mode pins (available through the **DEVSTAT** register) to set up the initial configuration structure, which is called the *boot parameter table*. This table is stored in the reserved section of L2 in **CorePac0**. Even though the boot parameter table format varies based on the boot mode selected, there are a few offsets that are common across all boot modes for a specific device. As an example, see Table 11.4 and Table 11.5 for the common ones, and EMIF16 configurations in Refs. [1] and [3] for complete lists.
- The RBL uses the **BOOTCOMPLETE** register (see Figure 11.16), which controls the **BOOT-COMPLETE** pin status, to indicate the completion of the RBL boot process. The **BOOT-COMPLETE** pin goes high when the boot complete bits in the **BOOTCOMPLETE** register for all the cores are set. The RBL sets the bits for each CorePac once it completes the boot process in the CorePac and just before it exits the process.

11.5.2 Second-level bootloader

In order to support other boot modes, customize an available boot mode or extend the capability of a bootloader, a second-level boot mode can be used. In this chapter, two second-level bootloaders are used. On the TMS320C6678 this bootloader is called the *intermediate bootloader* (IBL), and on the KeyStone II it is called *U-Boot*.

The RBL is flexible and allows the downloaded code to modify the boot parameter table in order, for instance, to select a new boot mode and/or new parameter table. This can be very useful, for instance when one would like to have a fast booting mode. In this case, use the RBL to download a small piece of code that will modify the device clock (from the default and slow clock to a faster clock), and re-enter the boot process to resume using a faster clock.

11.5.2.1 Intermediate bootloader
For the C6678, the IBL is used as the second-level bootloader to allow:

- Boot from the NAND/NOR flash on TMS320C667x.
- Boot from an FTP server.

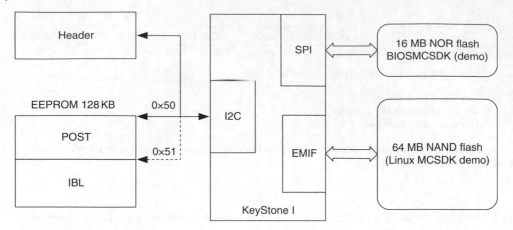

Figure 11.17 IBL location. *Note*: In Rev 1.0 of the C6670 EVM, the FPGA is programmed to invoke the IBL in order to execute the PLL fix, and then jump right back to RBL which continues the process. See the reference for the IBL update [6].

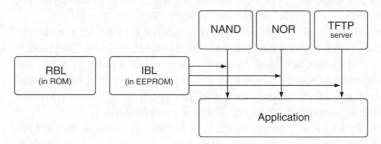

Figure 11.18 IBL boot modes.

- Boot from images with different formats (ELF [5] or BBLOB).
- Boot from multiple images.
- Extended functions before boot.

The IBL is flashed into the EEPROM that is connected to the DSP via the I2C bus at address **0x51**, as shown in Figure 11.17.

Boot mode dip switch settings. The EVM supports an IBL booting an image from the NAND, NOR or Ethernet; see Figure 11.18. Table 11.12 shows the boot mode dip settings for the different boot modes that the EVM supports.

11.5.2.2 How to use the IBL

The following steps show how to use the IBL to boot from the NAND/NOR or upload to an FTP server.

Step 1. Compile the IBL source code, which is available in the MCSDK directory:
C:\TI\mcsdk_2_01_02_06\tools\boot_loader\ibl
Step 2. Burn the IBL and parameter set to I2C EEPROM (using eepromwriter_evm6678.out within the Code Composer Studio (CCS)).

Table 11.12 Boot mode for the C6678 EVM [7]

Boot mode	DIP SW3 (Pin1, 2, 3, 4)	DIP SW4 (Pin1, 2, 3, 4)	DIP SW5 (Pin1, 2, 3, 4)	DIP SW6 (Pin1, 2, 3, 4)
IBL NOR boot on image 0 (default)	(off, off, on, off)	(on, on, on, on)	(on, on, on, off)	(on, on, on, on)
IBL NOR boot on image 1	(off, off, on, off)	(off, on, on, on)	(on, on, on, off)	(on, on, on, on)
IBL NAND boot on image 0	(off, off, on, off)	(on, off, on, on)	(on, on, on, off)	(on, on, on, on)
IBL NAND boot on image 1	(off, off, on, off)	(off, off, on, on)	(on, on, on, off)	(on, on, on, on)
IBL TFTP boot	(off, off, on, off)	(on, on, off, on)	(on, on, on, off)	(on, on, on, on)
I2C POST boot	(off, off, on, off)	(on, on, on, on)	(on, on, on, on)	(on, on, on, on)
ROM SPI boot	(off, on, off, off)	(on, on, on, on)	(on, on, off, on)	(on, on, on, on)
ROM SRIO boot	(off, off, on, on)	(on, on, on, off)	(on, off, on, off)	(off, on, on, on)
ROM Ethernet boot	(off, on, off, on)	(on, on, on, off)	(on, on, off, off)	(off, on, on, on)
ROM PCIe boot	(off, on, on, off)	(on, on, on, on)	(on, on, on, off)	(off, on, on, on)
No boot	(off, on, on, on)	(on, on, on, on)	(on, on, on, on)	(on, on, on, on)

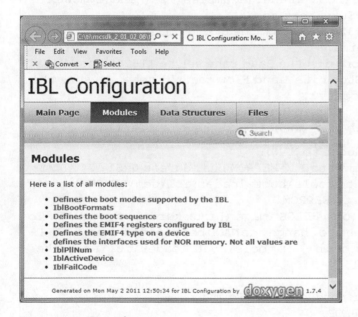

Figure 11.19 IBL configuration.

Step 3. Set the IBL parameter to I2C EEPROM.
Step 4. Get the user application and generate a user image.
Step 5. Burn the user image to NAND/NOR or upload to an FTP server.
Step 6. Set the boot mode pins, see refs [7,8] and Table 11.12.
Step 7. Power on the DSP.

For more details on the IBL configuration, see Figure 11.19 and Ref. [9].

Examples. A few examples can be found in: C:\TI\mcsdk_2_01_02_06\tools\boot_loader \examples. Please read the README.txt file for each boot mode. A NAND boot over I2C example is shown here:

NAND boot over I2C example

A simple Hello World example demonstrating NAND boot over I2C.

Steps to build the example:
1. Import the i2cnandboot CCS project from tools\boot_loader \examples\i2c\nand\evmc66xxl directory (in CCSv5, Project->Import Existing CCS/CCE Eclipse Projects).
2. Clean the i2cnandboot project and rebuild the project. After build is completed, i2cnandboot_evm66xxl.out and i2cnandboot_evm66xxl.map will be generated under the tools \boot_loader\examples\i2c\nand\evmc66xxl\bin directory.

Steps to run i2cnandboot in CCSv5:
1. Be sure to set the boot mode dip switch to no boot/EMIF16 boot mode on the EVM.
2. Load the program tools\boot_loader\examples\i2c\nand\evmc66xxl \bin\i2cnandboot_evm66xxl.out to CCS.
3. Connect the 3-pin RS-232 cable from the EVM to the serial port of the PC, and start Hyper Terminal.
4. Create a new connection with the Baud rate set to 115200 bps, Data bits 8, Parity none, Stop bits 1 and Flow control none. Be sure the COM port # is set correctly.
5. Run the program in CCS. i2cnandboot will send the hello world booting info to both the CCS console and the Hyper Terminal.

Steps to program i2cnandboot to NAND:
1. Be sure IBL is programmed to I2C EEPROM bus address 0x51. If IBL is not programmed, refer to tools\boot_loader\ibl\doc\README.txt on how to program the IBL to EEPROM.
2. By default, IBL will boot a BBLOB image (Linux kernel) from NAND. To run this example, we need to change the NAND boot image format to ELF:
   a. In setConfig_c66xx_main() of tools\boot_loader\ibl\src\make \bin\i2cConfig.gel,
      replace
      ibl.bootModes[1].u.nandBoot.bootFormat   = ibl_BOOT_FORMAT_BBLOB;
      with
      ibl.bootModes[1].u.nandBoot.bootFormat = ibl_BOOT_FORMAT_ELF;
   b. Reprogram the boot configuration table. Refer to tools \boot_loader\ibl\doc\README.txt on how to program the boot configuration table to EEPROM.
3. Copy tools\boot_loader\examples\i2c\nand\evmc66xxl\bin \i2cnandboot_evm66xxl.out to tools\writer\nand\evmc66xxl\bin, rename it app.bin and refer to tools\writer\nand\docs\README.txt on how to program the app.bin to NAND flash.

4. Once the programming is completed successfully, set the boot dip
   switches to I2C master mode, bus address 0x51 and boot parameter
   index to be 2.
5. After POR, IBL will boot the hello world image from NAND.
   Please refer to C6678L/C6670L/C6657L EVM boot mode dip switch
   settings:
   http://processors.wiki.ti.com/index.php/TMDXEVM6678L_EVM_Hardware
   _Setup#Boot_Mode_Dip_Switch_Settings
   http://processors.wiki.ti.com/index.php/TMDXEVM6670L_EVM
   _Hardware_Setup#Boot_Mode_Dip_Switch_Settings
   http://processors.wiki.ti.com/index.php/TMDSEVM6657L_EVM_
   Hardware_Setup#Boot_Mode_Dip_Switch_Settings
   And please refer to the User's Guide for more details:
   http://processors.wiki.ti.com/index.php/BIOS_MCSDK
   _2.0_User_Guide

11.6 Laboratory experiment 1

Booting with the KeyStone I. In this laboratory experiment, the booting is from an SPI (TMS320C6678 EVM). The TMS320C6678 EVM layout is shown in Figure 11.20.

In this laboratory experiment, the user will create an application that blinks one of the EVM LEDs using the CCS and flashes it to an NOR flash. The user will also do the appropriate setting for booting from the SPI NOR flash. The procedure is as follows:

1) Create a project for the image you would like to boot.
   A) Open the project: **BlinkLED**

   Project location:

   \Chapter_11_Code\NOR_Booting\NORbootWS\BlinkLED

   B) Explore the project.

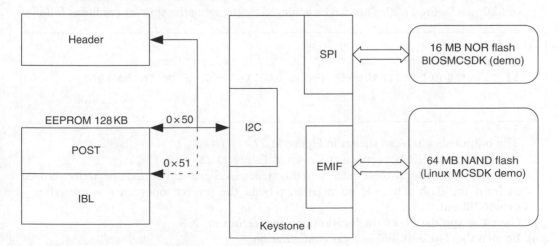

Figure 11.20 TMS320C6678 EVM memory layout.

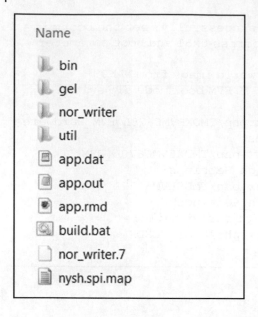

Figure 11.21 File locations.

2) Build the project and rename the **BlinkLED.out** program 'app.out'. The name is changed in order to make use of an existing batch file.
3) Place **app.out** in the same directory as **build.bat** as shown in Figure 11.21.
4) Run the **build.bat** file (select **build.bat**, right click and run). The file generated, **app.dat**, will be burned into the flash. Explore the **built.bat** file. The **built.bat** file contains four executable files:
   1) **hex6x.exe app.rmd**: Converts the application code into a boot table format.
   2) **b2i2c.exe app.btbl app.btbl.i2c**: Converts the boot table into an SPI format table, as the image is to be transferred using the SPI.
   3) **b2ccs.exe app.btbl.i2c app.i2c.ccs**: Converts the previous file to a format that CCS recognizes since the CCS will be used to burn the flash.
   4) **romparse.exe nysh.spi.map**: Appends the boot parameter to the boot table.

   Note: The address **0x01F40051** needs to be changed to **0x01F40000** in order to use **bank0**; see Figure 11.20. This can be achieved by changing the **spirom.ccs** file as follows:

```
>"spirom.ccs" (
 for /f "usebackq delims=" %%A in ("i2crom.ccs") do (
  if "%%A" equ "0x01f40051" (echo 0x01f40000) else (echo %%A)
 )
)
```

   The output should be as shown in Figure 11.22.
5) Set the EVM to NO boot mode as shown in Figure 11.23.
6) Load the **nor_writer** project in CCS. In this step, a CCS project entitled **nor_writer** is used to burn the flash. There is no need to rebuild the project; one can use **norwriter_-evm6678l.out**.
7) Copy the **app.dat** file to the location **nor_writer/bin** in CCS.
8) Launch the TMS320C6678 target configuration.

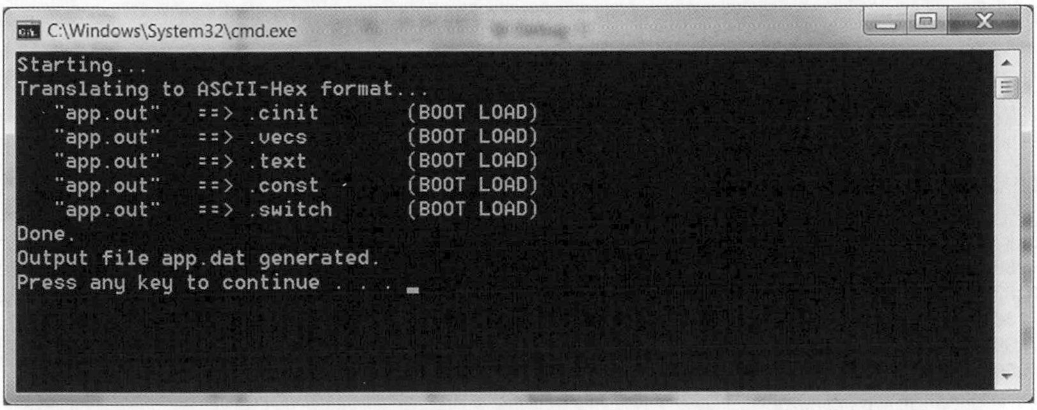

Figure 11.22 Output after running the **build.bat** file.

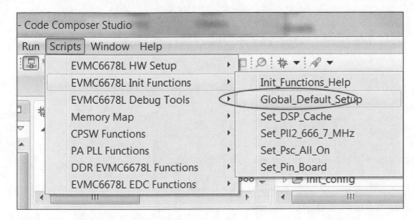

Figure 11.23 No boot mode switches.

9) **Run- > Load** the file, and leave suspended at **main()**:
   \Chapter_11_Code\NOR_Booting\NORbootWS\norwriter_evmc6678l\bin\norwriter_evm6678l.out.
10) Load **gel\ evmc6678l.gel** in CCS, and execute Scripts- > **EVMC6678L Init Functions- >
    Global_Default_Setup** as shown in Figure 11.24.

Figure 11.24 Loading the GEL file.

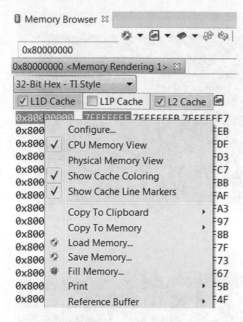

Figure 11.25 Loading the image.

11) Open the Memory Browser, go to address **0x80000000**, right click and pick **Load Memory** as shown in Figure 11.25.

12) Pick the **app.dat** file, choose TI Data from File type, tick the Use file header information checkbox and press Next, as shown in Figure 11.26.

13) Fill out **0x80000000** in the Start Address field, and leave Length as it is; see Figure 11.27.

14) Press Finish.

15) Unsuspend the **nor_writer** process by pressing the Run button, and wait for completion. If all the steps are followed correctly, the NOR will be flashed; see Figure 11.28.

16) Turn the power of the EVM off, change the bootmode switches to the SPI boot mode as shown in Figure 11.29 and power on the EVM. The LED should now blink.

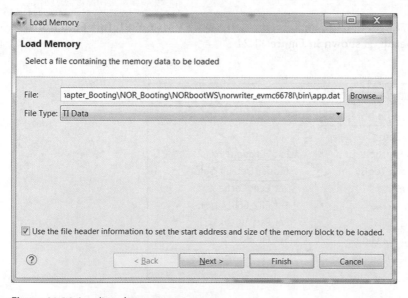

Figure 11.26 Loading the memory.

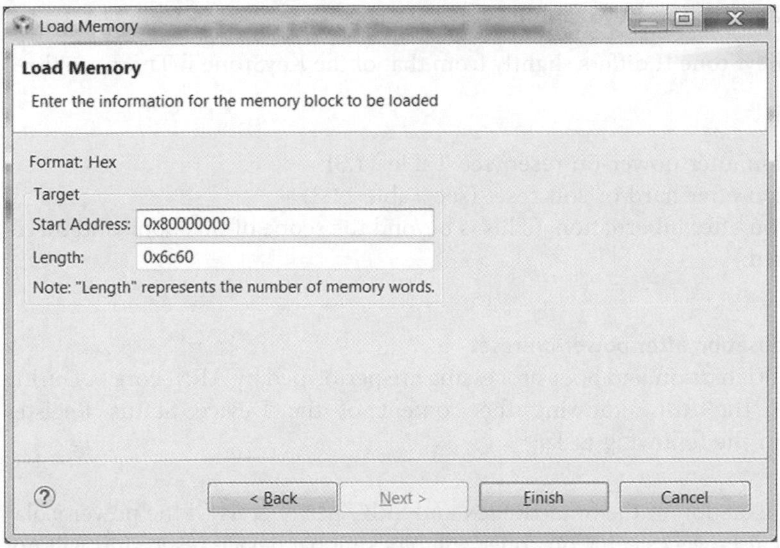

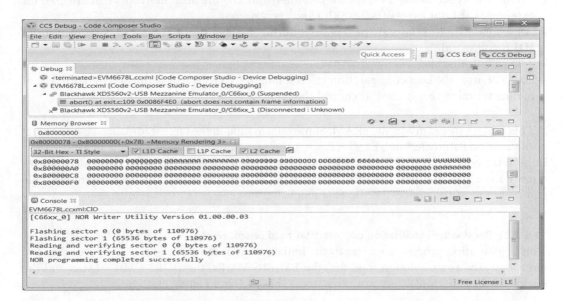

Figure 11.27 Entering the information for the memory block to be loaded.

Figure 11.28 Output when the NOR is flashed properly.

Figure 11.29 ROM SPI boot mode.

11.6.1 Initialisation stage for the KeyStone II

The initialisation of the KeyStone II differs slightly from that of the KeyStone I. There are three initialisation types:

1) Bootloader initialisation after power-on reset (see Table 11.3)
2) Bootloader initialisation after hard or soft reset (see Table 11.3)
3) Bootloader initialisation after hibernation. (This is beyond the scope of this book and therefore will not be covered.)

11.6.1.1 Bootloader initialisation after power-on reset

For the KeyStone II, all initialisation and boot processing are performed by ARM core 0. During this initialisation phase, the RBL, knowing the content of the Device Status Register (DEVSTAT), will perform the following tasks:

- The RBL enables reset isolation in the SmartReflex and SRIO peripherals (if the power isolation is enabled, the Local Power/Sleep Controller will block global device resets and will not pause the clocks during reset transitions).
- All interrupts are disabled except for IPC interrupts and the host interrupts that are used for external host boot modes (PCIe, SRIO and HyperLink).
- All secondary ARM cores are held in reset during the boot process. All DSP cores execute an **IDLE** command.
- All cache is disabled.
- The RBL uses the boot configuration information in DEVSTAT to set up and initialize a boot parameter table that is used to control the boot process.

This table is stored in MSMC SRAM. Some information in the table is initialized based on the configuration parameters in DEVSTAT, while the remaining information is default values based only on the boot mode. The format of the table varies depending on the boot mode. All start with a few entries that are common to all boot modes. Information about the boot parameter table can be found in the device-specific data manual.

11.6.1.2 Bootloader initialisation process after hard or soft reset

This initialisation process is similar to the initialisation of the power-on reset except the test/emu logic, the reset isolation modules, the EMIF16 MMRs, the DDR3 EMIF MMRs and the sticky bits in PCIe MMRs are not reset (see Table 11.1 and Figure 11.4).

11.6.2 Second bootloader for the KeyStone II

The ARM subsystem runs the following software components:

- U-Boot: Bootloader
- Boot monitor: Monitor and other secure functions
- SMP Linux: ARM A15 port of SMP Linux.

Please refer to Ref. [10] for guidelines on how to build the Linux kernel and how to use the naming convention for Linux, U-Boot and the boot monitor.

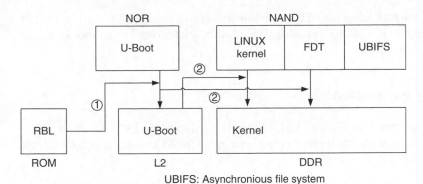

UBIFS: Asynchronious file system

Figure 11.30 The boot sequence for the KeyStone II EVM.

Figure 11.30 shows the memory layout of the KeyStone II EVM. The NOR contains the U-Boot code, and the NAND contains the Linux kernel, the FDT[1] (flat device tree) and the file system (UBIFS: Asynchronous Filesystem). The ROM contains the RBL as described in this chapter.

The SPI boot process is as follows:

1) The RBL loads U-Boot from the NOR to the L2 Memory. Note the RBL cannot boot the Linux kernel as the RBL is small and has limited functionality.
2) The U-Boot boots the Linux kernel and the FDT to the DDR.

11.6.2.1 U-Boot

U-Boot is an open-source, cross-platform bootloader that provides out-of-box support for a large number of embedded platforms including the KeyStone II. The main advantage is that U-Boot is easy to customize, has a rich feature set, is very well documented and has a small binary foot print which is very critical for embedded systems. It is worth looking at the following definition if the user is not familiar with U-Boot:

1) U-Boot environment. The U-Boot environment is a block of memory that is kept on persistent storage and copied to RAM when U-Boot starts.
2) Variables. U-Boot uses environment variables that can be used to configure the boot process.
3) Terminal. The user can access these variables via a terminal during the boot process.
4) Ethenet and USB. U-Boot can download a kernel image via either an Ethernet or a USB port.

Secondary program loader (SPL)   The RBL will only load the U-Boot SPL, which will first initialize hardware and then look for **u-boot.img**. An SPL is added to U-Boot in version 2012-10. The SPL is supported in U-Boot for KeyStone II devices. This feature is configured using the config option **CONFIG_SPL**. It allows the creation of a small first-stage bootloader (SPL) that can be loaded by ROM Bootloader (RBL) which would then load and run the second-stage bootloader (a full version of U-Boot) from NOR or NAND. The required config options are added to **tci6638_evm.h**. The user may refer to the **README** text file in the U-Boot root source directory for more details.

---

1 The FDT is a specific database that represents the hardware components on a given board. The FDT is the default mechanism to pass low-level hardware information from the bootloader to the kernel, and it is also referred to as the Device Tree Blob, Device Tree Binary or simply Device Tree. The Device Tree is expected to be loaded from the same media as the kernel, and from the same relative path.

For the KeyStone II, the RBL loads the SPL image from offset 0 of the SPI NOR flash in the SPI boot mode. The first 64 K of the SPI NOR flash is flashed with SPL followed by **u-boot.img**. The initial 64 K is padded with zeros.

11.7 Laboratory experiment 2

1) Connect the EVM as shown in Figure 11.31. Figure 11.32 shows the EVM hardware.
2) Set the Boot Mode Switches to 0010 [11]. Set the boot mode to SPI mode. Various modes are shown in Table 11.13.

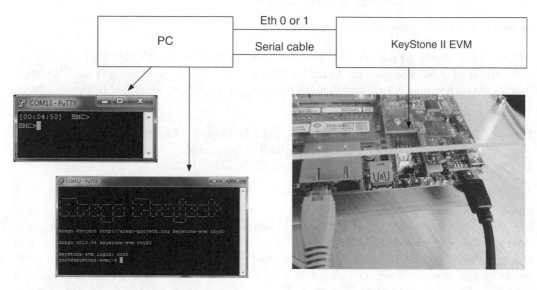

Figure 11.31 EVM connection to the PC.

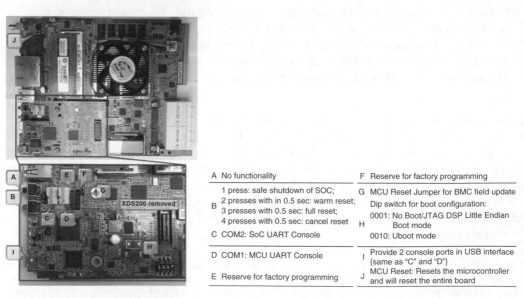

A	No functionality
B	1 press: safe shutdown of SOC; 2 presses with in 0.5 sec: warm reset; 3 presses with 0.5 sec: full reset; 4 presses with 0.5 sec: cancel reset
C	COM2: SoC UART Console
D	COM1: MCU UART Console
E	Reserve for factory programming

F	Reserve for factory programming
G	MCU Reset Jumper for BMC field update
H	Dip switch for boot configuration: 0001: No Boot/JTAG DSP Little Endian Boot mode 0010: Uboot mode
I	Provide 2 console ports in USB interface {same as "C" and "D"}
J	MCU Reset: Resets the microcontroller and will reset the entire board

Figure 11.32 EVMK2H hardware.

Table 11.13 Boot mode switches

Boot mode	SW1 (pin1, pin2, pin3, pin4)
No boot	off, off, off, on
UART	off, on, off, off
NAND	off, off, off, off
SPI	*off, off, on, off*
I2C	off, off, on, on
Ethernet	off, on, off, on

3) Determine the port address. The port addresses on the PC will change every time a serial device is connected. Open the Device Manager and find out the port addresses. In Figure 11.33, the port addresses found for the serial ports are COM12 and COM13.

4) Open two terminals. In Figure 11.34, **PuTTY** has been used as a terminal emulator. Open two terminals with COM12 and COM13 (that will depend on your settings, as it is very likely that you will have different ports); see Figure 11.34.

5) Set the IP addresses. Set addresses as shown in Figure 11.35.
   a) VMware. Start VMware and configure VMware to Bridged as shown in Figure 11.36 and Figure 11.37, then set the IP address of VMware as shown in Figure 11.38.
   b) Setting the host PC. See Figure 11.39.

6) Power cycle the EVM, and set up the IP address of the EVM.

Figure 11.33 Device manager for identifying the COM ports.

Figure 11.34 Setting up the COM ports.

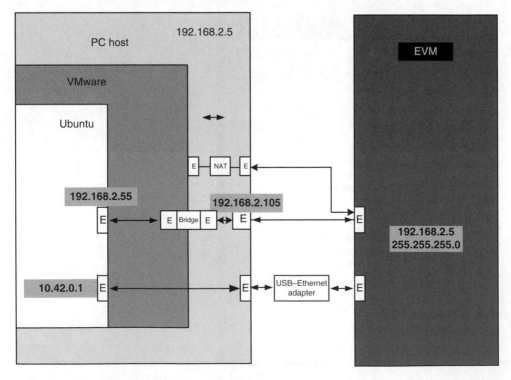

Figure 11.35 IP addresses used in this experiment.

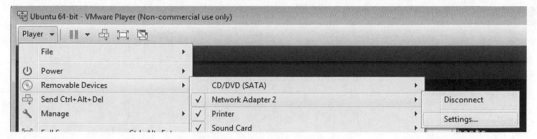

Figure 11.36 Accessing the network settings.

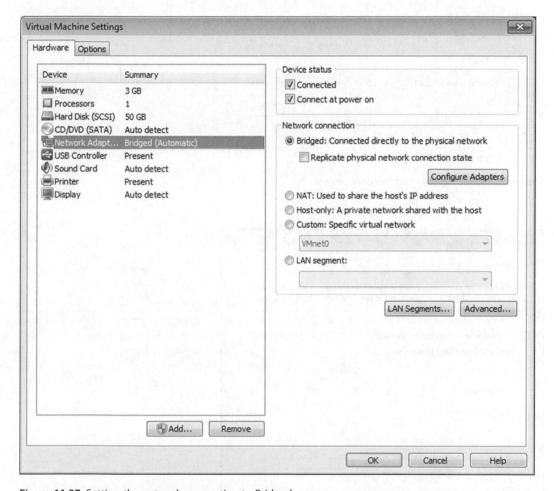

Figure 11.37 Setting the network connection to Bridged.

Open the COM ports, power cycle the EVM and follow instructions 1 to 7 as shown in Table 11.14.

If the EVM cannot be booted, check the EVM boot mode using the BMC as shown in Figure 11.43. Type **bootmode #N** as shown by the arrows pointing to the left, and observe the outputs as shown by the arrows pointing to the right as shown in Figure 11.43. Select **bootmode 0** and type **reboot**. Further details on the EVM hardware setup can be found in Ref. [12].

Figure 11.38 Setting the IP address of VMware.

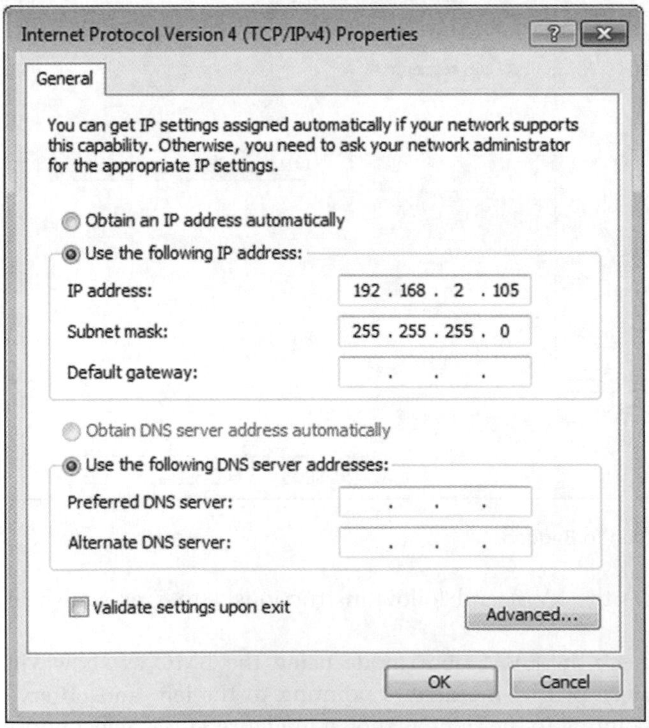

Figure 11.39 Setting the host PC IP address.

Table 11.14 Steps required for booting the EVM

KS2 EVM side	PC side
1) Power the EVM to boot, and wait for a few seconds. 2) Type **root** as shown in Figure 11.40. 3) Type **ifconfig** to check the EVM IP address. 4) You can change the IP address by typing **ifconfig eth0 192.168.2.5**; see Figure 11.41. Please note that your IP address that you selected will be erased every time you reboot your EVM.	5) Log in to the EVM from your virtual machine via ssh. Type: **ssh root@192.168.2.5**; when prompted for a password, type **root**. Then, type '**ls/**' to explore the root directory of the EVM. See Figure 11.42. 6) Now that you are connected to the EVM, you can perform tasks on the ARM or DSP. 7) Explore the files.

Figure 11.40 Console output after booting Linux.

Figure 11.41 Setting **eth0**'s IP address to 192.168.2.5.

```
⊗ ⊝ ⊡   naim@ubuntu: ~
naim@ubuntu:~$ ssh root@192.168.2.5
root@192.168.2.5's password:
root@keystone-evm:~# ls /
bin      debug    etc      ipaddr   media    proc     srv      tmp      var
boot     dev      home     lib      mnt      sbin     sys      usr      www
root@keystone-evm:~#
```

Figure 11.42 Accessing the EVM from VMware.

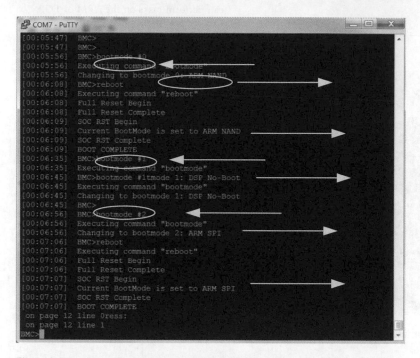

Figure 11.43 Using the BMC to verify the EVM boot mode selected.

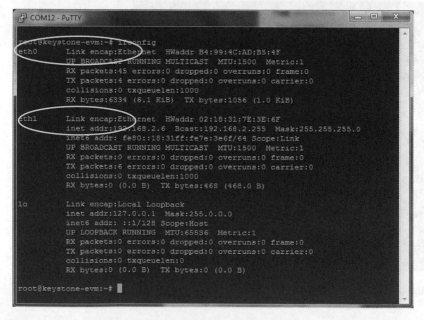

Figure 11.44 Using **ipconfig** to check the IP addresses.

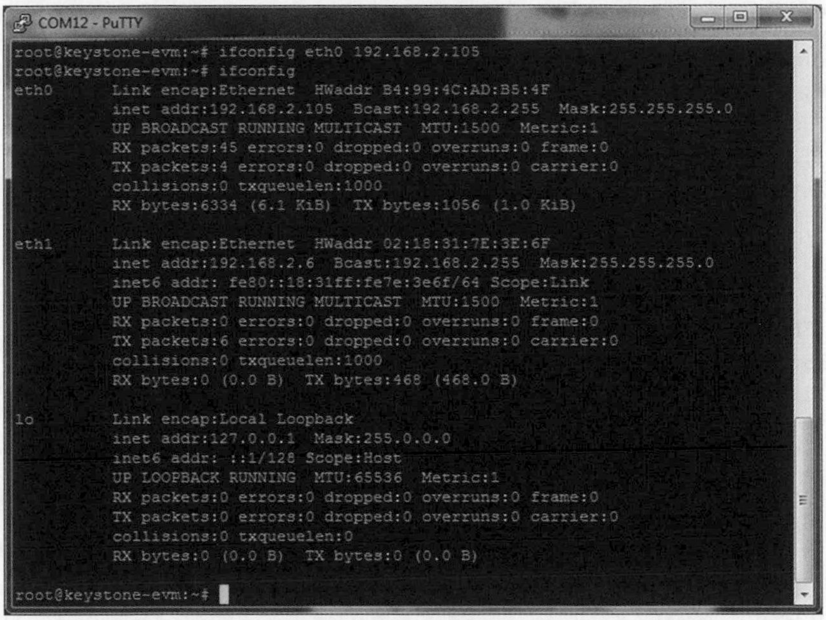

Figure 11.45 Setting the **eth0** to IP address 192.168.2.105.

When the boot ends, log in as **root** as shown in Figure 11.40, then type **ifconfig** to check the EVM IP addresses as shown in Figure 11.44.

You can change the IP address of **eth0** or **eth1** by typing **ifconfig eth0 192.168.2.105**, for example, as shown in Figure 11.45.

Now that Linux has completed booting, explore the file system as shown in Figure 11.46.

Figure 11.46 Monitor showing the file system.

11.7.1 Printing the U-Boot environment

The U-Boot environment stores some important configuration parameters. You can read and write these values when you are connected to the U-Boot console via the serial port. Type **printenv** to display the U-Boot envirionement variables as shown here:

```
U-Boot 2013.01-00004-g0c2f8a2 (Aug 16 2013 - 19:04:15)

I2C:  ready
DRAM:  2 GiB
NAND:  512 MiB
Net:  TCI6638_EMAC
Warning: TCI6638_EMAC using MAC address from net device
, TCI6638_EMAC1
Hit any key to stop autoboot:  0
TCI6638 EVM # printenv
addr_fdt=0x87000000
addr_fs=0x82000000
addr_kern=0x88000000
addr_mon=0x0c5f0000
addr_ubi=0x82000000
addr_uboot=0x87000000
args_all=setenv bootargs console=ttyS0,115200n8 rootwait=1
args_net=setenv bootargs ${bootargs} rootfstype=nfs root=/dev/nfs
rw nfsroot=${serverip}:${nfs_root},${nfs_options} ip=dhcp
args_ramfs=setenv bootargs ${bootargs} earlyprintk rdinit=/sbin/
init rw root=/dev/ram0 initrd=0x802000000,9M
args_ubi=setenv bootargs ${bootargs} rootfstype=ubifs root=ubi0:
rootfs rootflags=sync rw ubi.mtd=2,2048
args_uinitrd=setenv bootargs ${bootargs} earlyprintk rdinit=/sbin/
init rw root=/dev/ram0
baudrate=115200
boot=ubi
bootcmd=run init_${boot} get_fdt_${boot} get_mon_${boot} get_kern_
${boot} run_mon run_kern
bootdelay=3
bootfile=uImage
burn_ubi=nand erase.part ubifs; nand write ${addr_ubi} ubifs ${filesize}
burn_uboot=sf probe; sf erase 0 0x100000; sf write ${addr_uboot} 0 ${filesize}
ethact=TCI6638_EMAC
ethaddr=b4:99:4c:9d:ba:f2
fdt_high=0xffffffff
get_fdt_net=dhcp ${addr_fdt} ${tftp_root}/${name_fdt}
get_fdt_ramfs=dhcp ${addr_fdt} ${tftp_root}/${name_fdt}
get_fdt_ubi=ubifsload ${addr_fdt} ${name_fdt}
get_fdt_uinitrd=dhcp ${addr_fdt} ${tftp_root}/${name_fdt}
get_fs_ramfs=dhcp ${addr_fs} ${tftp_root}/${name_fs}
```

```
get_fs_uinitrd=dhcp ${addr_fs} ${tftp_root}/${name_uinitrd}
get_kern_net=dhcp ${addr_kern} ${tftp_root}/${name_kern}
get_kern_ramfs=dhcp ${addr_kern} ${tftp_root}/${name_kern}
get_kern_ubi=ubifsload ${addr_kern} ${name_kern}
get_kern_uinitrd=dhcp ${addr_kern} ${tftp_root}/${name_kern}
get_mon_net=dhcp ${addr_mon} ${tftp_root}/${name_mon}
get_mon_ramfs=dhcp ${addr_mon} ${tftp_root}/${name_mon}
get_mon_ubi=ubifsload ${addr_mon} ${name_mon}
get_mon_uinitrd=dhcp ${addr_mon} ${tftp_root}/${name_mon}
get_ubi_net=dhcp ${addr_ubi} ${tftp_root}/${name_ubi}
get_uboot_net=dhcp ${addr_uboot} ${tftp_root}/${name_uboot}
has_mdio=0
init_net=run set_fs_none args_all args_net
init_ramfs=run set_fs_none args_all args_ramfs get_fs_ramfs
init_ubi=run set_fs_none args_all args_ubi; ubi part ubifs;
ubifsmount boot
init_uinitrd=run set_fs_uinitrd args_all args_uinitrd
get_fs_uinitrd
initrd_high=0xffffffff
mem_lpae=1
mem_reserve=512M
mtdparts=mtdparts=davinci_nand.0:1024k(bootloader)ro,512k
(params)ro,129536k(ubifs)
name_fdt=uImage-k2hk-evm.dtb
name_fs=arago-console-image.cpio.gz
name_kern=uImage-KeyStone-evm.bin
name_mon=skern-KeyStone-evm.bin
name_ubi=KeyStone evm-ubifs.ubi
name_uboot=u-boot-spi-KeyStone-evm.gph
name_uinitrd=uinitrd.bin
nfs_options=v3,tcp,rsize=4096,wsize=4096
nfs_root=/export
no_post=1
run_kern=bootm ${addr_kern} ${addr_uinitrd} ${addr_fdt}
run_mon=mon_install ${addr_mon}
serverip=192.168.1.195
set_fs_none=setenv addr_uinitrd -
set_fs_uinitrd=setenv addr_uinitrd ${addr_fs}
stderr=serial
stdin=serial
stdout=serial
ver=U-Boot 2013.01-00004-g0c2f8a2 (Aug 16 2013 - 19:04:15)
Environment size: 2960/262140 bytes
TCI6638 EVM #
```

11.7.2 Using the help for U-Boot

To access the commands available, type **help** as shown here:

```
Environment size: 2960/262140 bytes
TCI6638 EVM # help
?       - alias for 'help'
askenv  - get environment variables from stdin
base    - print or set address offset
boot    - boot default, i.e. run 'bootcmd'
bootd   - boot default, i.e. run 'bootcmd'
bootm   - boot application image from memory
bootp   - boot image via network using BOOTP/TFTP protocol
chpart  - change active partition
cmp     - memory compare
coninfo - print console devices and information
cp      - memory copy
crc32   - checksum calculation
ddr     - DDR3 test
dhcp    - boot image via network using DHCP/TFTP protocol
echo    - echo args to console
editenv - edit environment variable
eeprom  - EEPROM sub-system
env     - environment handling commands
exit    - exit script
facimg  - Read nand page data and oob data in raw format to memory
false   - do nothing, unsuccessfully
fatinfo - print information about filesystem
fatload - load binary file from a dos filesystem
fatls   - list files in a directory (default /)
fdt     - flattened device tree utility commands
fmtimg  - Format image file into TI's KeyStone boot mode format
getclk  - get clock rate
go      - start application at address 'addr'
help    - print command description/usage
highmem - highmem test
i2c     - I2C sub-system
iminfo  - print header information for application image
imxtract- extract a part of a multi-image
itest   - return true/false on integer compare
loadb   - load binary file over serial line (kermit mode)
loads   - load S-Record file over serial line
loady   - load binary file over serial line (ymodem mode)
loop    - infinite loop on address range
md      - memory display
mdc     - memory display cyclic
mm      - memory modify (auto-incrementing address)
mon_install - Install boot kernel at 'addr'
mon_power - power on/off secondary core
```

```
mtdparts - define flash/nand partitions
mtest  - simple RAM read/write test
mw     - memory write (fill)
mwc    - memory write cyclic
nand   - NAND sub-system
nboot  - boot from NAND device
nfs    - boot image via network using NFS protocol
nm     - memory modify (constant address)
oob   - reformat the oob data from the U-boot layout to the RBL readable layout
ping   - send ICMP ECHO_REQUEST to network host
pllset - set pll multiplier and pre divider
printenv - print environment variables
psc    - <enable/disable psc module os disable domain>
reset  - Perform RESET of the CPU
run    - run commands in an environment variable
saveenv - save environment variables to persistent storage
saves  - save S-record file over serial line
setenv - set environment variables
setmpax - set mpax ses for ARM privid
sf     - SPI flash sub-system
showvar - print local hushshell variables
sleep  - delay execution for some time
smtest  - simple RAM read/write test
source - run script from memory
test   - minimal test like /bin/sh
tftpboot - boot image via network using TFTP protocol
true   - do nothing, successfully
ubi    - ubi commands
ubifsload - load file from an UBIFS filesystem
ubifsls - list files in a directory
ubifsmount - mount UBIFS volume
ubifsumount - unmount UBIFS volume
usb    - USB sub-system
usbboot - boot from USB device
version - print monitor, compiler and linker version
TCI6638 EVM #
```

11.8 TFTP boot with a host-mounted Network File System (NFS) server – NFS booting

In order to boot the root file system over NFS, the following four steps have to be performed:

1) Install the TFTP server. A TFTP server needs to be installed as it is required to host the kernel image.
2) NFS server. To boot a system over NFS, an NFS server must be available on the local network. This is often the same machine that is being used for software development.

An NFS file server can provide a variety of services to Linux machines on a network:

1) It can provide a place to store files for a machine to use or write to.
2) It can be used to allow machines to boot a root file system image stored on the NFS server.
3) It can provide a place to store file system images when they are captured from a flash like a NAND.
4) It can be connected to a desktop system to provide a common file store.

It is possible to boot using NFS as the root file system. This method can have two advantages:

A) Save time during development where the root file system is modified frequently.
B) Reduce the wear on the on-board flash device as the flash devices have only a finite number of reprograming cycles.

Figure 11.47 shows the Linux kernel running on a target. Linux is able to mount the root file system from the host.

For the KeyStone II, there are various examples, as shown in Table 11.15 [11] where the content is shown, and a complete workshop (KeyStone II Multicore Workshop) is included in Ref. [13].

To build and run U-Boot and the Linux kernel, please refer to Ref. [14].

11.8.1 Laboratory experiment 3

This laboratory experiment shows how to bring up Linux on the EVMK2H using the NFS file system.

1) Connect the EVM as shown in Figure 11.48.
2) Set the boot mode to 0010.

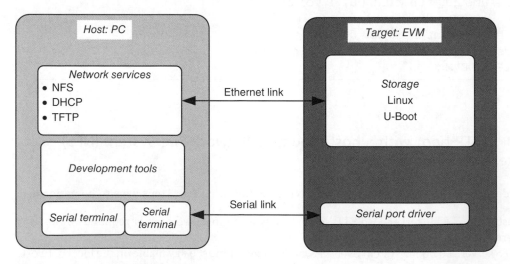

Figure 11.47 Host-mounted NFS server.

Table 11.15 KeyStone II boot examples [11]

Source: Courtesy of Texas Instruments.

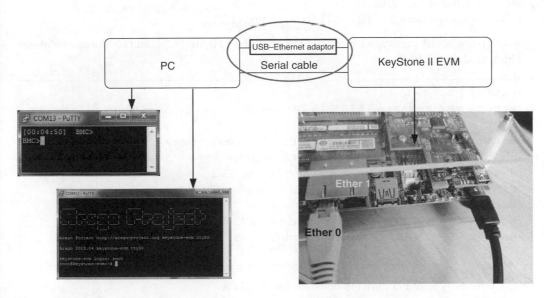

Figure 11.48 EVM setup.

3) Copy the Linux kernel.
   a) Create a director to hold the TI SDK Linux root file system (e.g. **mynfs2**) as shown in Figure 11.49.
      To do so, type:

```
cd /
sudo mkdir mynfs2
```

   b) Locate the file + **tisdk-rootfs.tar.gz**, and extract it in the created directory:
      cd/home/naim/ti/mcsdk_linux_3_00_04_18/images
      sudo tar xvf tisdk-rootfs.tar.gz -C/mynfs2 (x: extract, v: verbose, f: File, C: change to directory DIR).
4) Set the environment variables.
   Follow the steps shown in Figure 11.50 to Figure 11.52.

   Power up the board, press any key to hold the board (see Figure 11.52) and set the environment variables as shown here:

```
1.setenv boot net
2.setenv mem_reserve 1536 M [A larger size can be used when using more
  than 2 GB DIMM.]
3.setenv gatewayip 10.42.0.1 [This is the gateway IP of the subnet on
  which the host PC and the board are present.]
4.setenv serverip 10.42.0.1 [This is the IP of the host Linux machine.]
5.setenv tftp_root/tftpboot [Path to the TFTP server on your host
  machine]
6.setenv nfs_root/mynfs [Path to the NFS on your host machine]
7.saveenv [This saves the environment variables to the flash.]
```

   Note: Do not type the comments between the brackets.

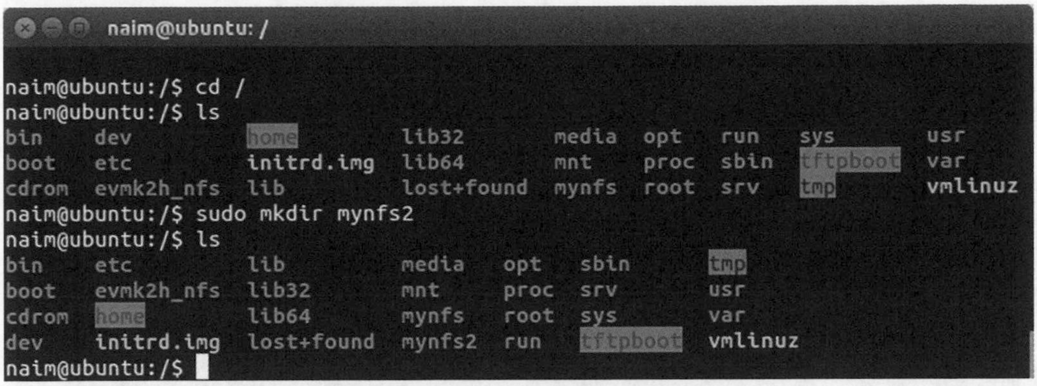

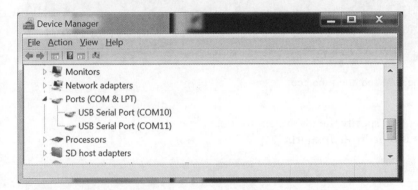

Figure 11.49 Creating directory to hold the TI SDK Linux root file system.

Figure 11.50 Check the COM ports.

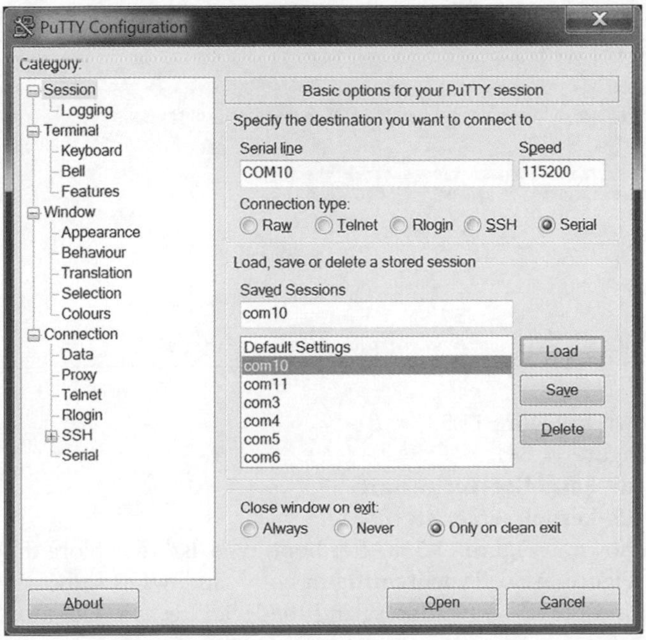

Figure 11.51 Configure the COM ports.

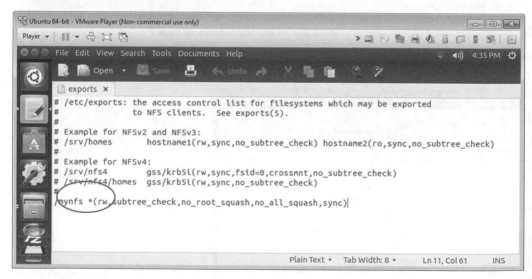

Figure 11.52 Power up the EVM and abort the boot.

5) Edit the exports file, and specify the file system to use:
   naim@ubuntu:/$ sudo gedit/etc/exports

```
# /etc/exports: the access control list for filesystems which may be exported
#               to NFS clients.  See exports(5).
#
# Example for NFSv2 and NFSv3:
# /srv/homes       hostname1(rw,sync,no_subtree_check) hostname2(ro,sync,no_subtree_check)
#
# Example for NFSv4:
# /srv/nfs4        gss/krb5i(rw,sync,fsid=0,crossmnt,no_subtree_check)
# /srv/nfs4/homes  gss/krb5i(rw,sync,no_subtree_check)
#
/mynfs *(rw,subtree_check,no_root_squash,no_all_squash,sync)
```

6) Set the Ethernet to **shared** as shown in Figure 11.53.
7) Restart the server by typing (see Figure 11.54):
   naim@ubuntu:~$ sudo service nfs-kernel-server restart
   naim@ubuntu:~$ sudo service nfs-kernel-server status
8) Booting the board. Type boot as shown in Figure 11.55. After boot, type '**ls/**' to explore the 'main' root directory as shown in Figure 11.56. Type **ifconfig** in both windows as shown in Figure 11.57 and Figure 11.58 to find the EVM IP address and the IP address for Ubuntu.

Figure 11.53 Setting the Ubuntu Ethernet connection.

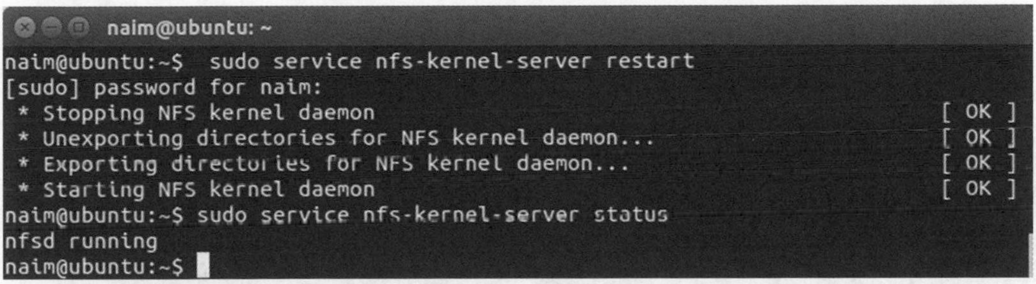

Figure 11.54 Restarting the server.

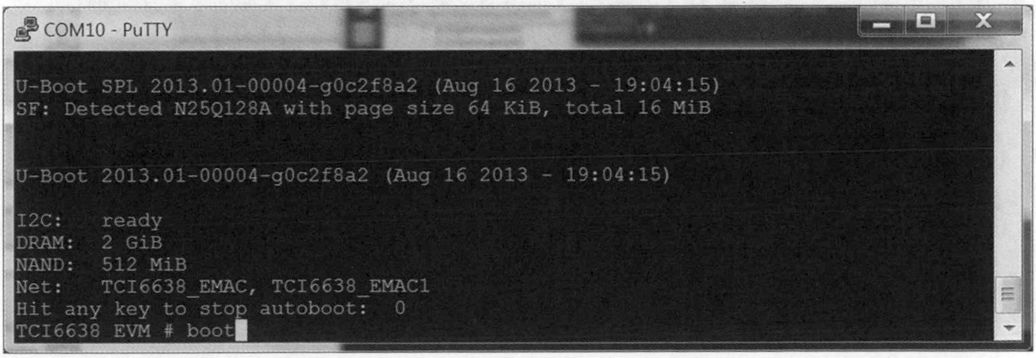

Figure 11.55 Booting the EVM.

Figure 11.56 EVM booted.

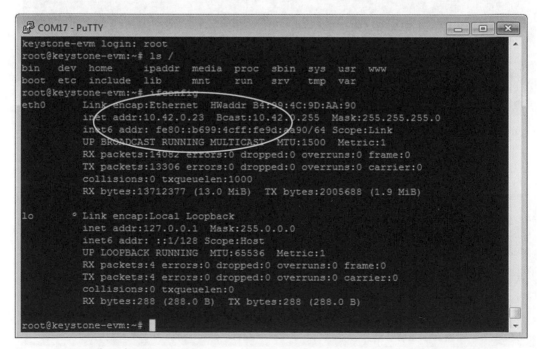

Figure 11.57 Finding the EVM IP address.

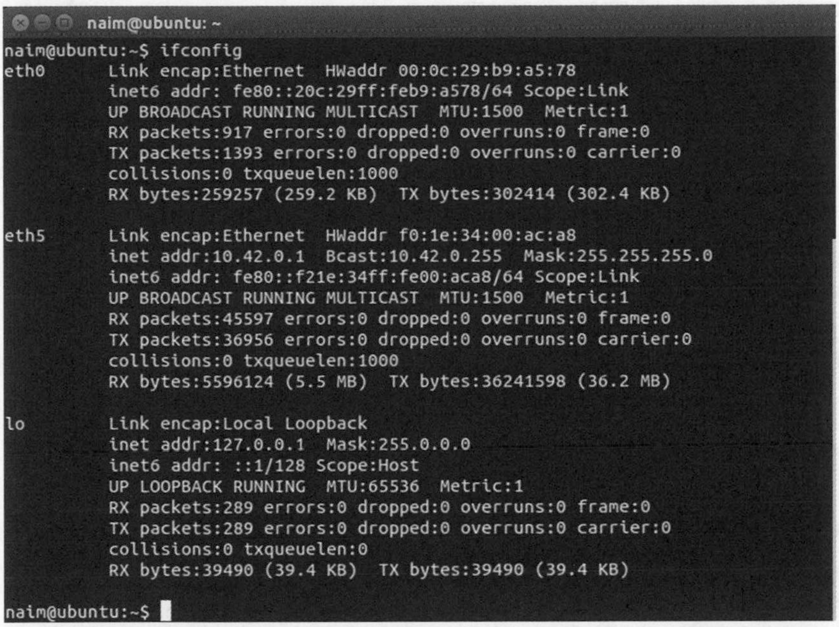

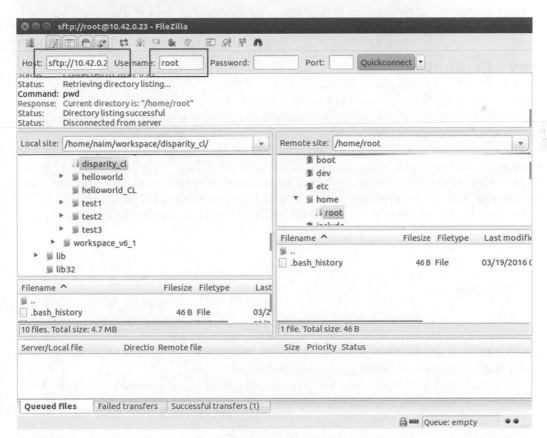

Figure 11.58 Finding the IP address for Ubuntu.

Figure 11.59 Connect to the EVM and Ubuntu using FileZilla.

Figure 11.60 Creating a directory.

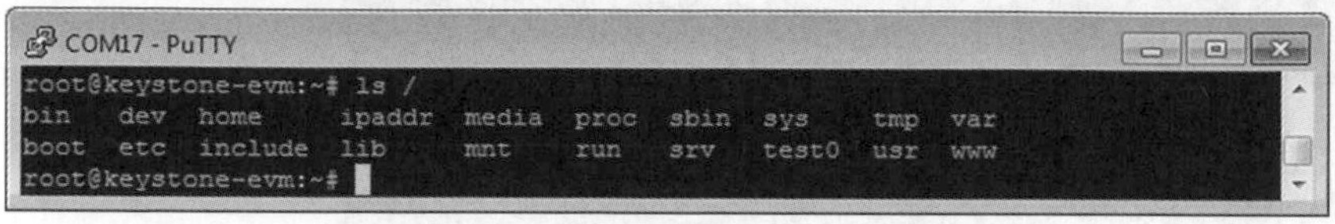

Figure 11.61 Terminal showing the created directory **test0**.

Open FileZilla and type the address. In this case, it is **10.42.0.23** as shown in Figure 11.59, and the username is **root**.

Test if the file system is really on Ubuntu. To do so, navigate to **mynfs** and create a directory **test0** for example, as shown in Figure 11.60.

Now open the terminal for the EVM and check if the file is available; see Figure 11.61.

More examples can be found in Ref. [11], and the content is shown in Table 11.15.

11.9 Conclusion

In this chapter, the bootloader for the KeyStone I and II devices has been introduced. It has been shown the boot is driven only on the device reset, and the boot configuration is determined by the boot pins that can be read by the device from the DEVSTAT register.

Before booting, the ROM code has no information about the peripheral(s) connected and therefore uses the default parameter table that can be modified by the DEVSTAT. Then the PLLs are initialized if need be and the boot mode specified is performed. Once the download is complete, the specific processor starts executing the downloaded image.

Two practical examples, one for the TMS320C6678 EVM and one for the KeyStone II, have been given.

References

1 Texas Instruments, Multicore DSP + ARM KeyStone II System-on-Chip (SoC), November 2013. [Online]. Available: http://www.ti.com/lit/ds/symlink/66ak2h12.pdf.
2 Texas Instruments, KeyStone II Architecture ARM Bootloader user guide, July 2013. [Online]. Available: http://www.ti.com/lit/ug/spruhj3/spruhj3.pdf.
3 Texas Instruments, Multicore fixed and floating-point digital signal processor, March 2014. [Online]. Available: http://www.ti.com/lit/ds/symlink/tms320c6678.pdf.
4 Texas Instruments, KeyStone Architecture DSP Bootloader user guide, July 2013. [Online]. Available: http://www.ti.com/lit/ug/sprugy5c/sprugy5c.pdf.

5 The Santa Cruz Operation, Inc, System V Application Binary Interface edition 4.1, March 18 1997. [Online]. Available: http://www.sco.com/developers/devspecs/gabi41.pdf.

6 Texas Instruments, Bios MCSDK 2.0.2 IBL update, 14 September 2011. [Online]. Available: http://processors.wiki.ti.com/index.php/Bios_MCSDK_2.0.2_IBL_Update.

7 Texas Instruments, TMDXEVM6678L EVM hardware setup, 12 January 2016. [Online]. Available: http://processors.wiki.ti.com/index.php/TMDXEVM6678L_EVM_Hardware_Setup.

8 Processor SDK RTOS BOOT C66x: http://processors.wiki.ti.com/index.php/Processor_SDK_RTOS_BOOT_C66x. [Online].

9 Texas Instruments, IBL Configuration: C:\ti\mcsdk_2_01_02_06\tools\boot_loader\ibl\doc, [installation of the mcsdk is required].

10 Texas Instruments, MCSDK user guide: exploring the MCSDK, 11 March 2016. [Online]. Available: http://processors.wiki.ti.com/index.php/MCSDK_UG_Chapter_Exploring#Running_U-Boot.2C_Boot_Monitor_and_Linux_Kernel_on_EVM.

11 Texas Instruments, KeyStone II boot examples, [Online]. Available: http://processors.wiki.ti.com/index.php/KeystoneII_Boot_Examples?keyMatch=KeyStoneII%20Boot%20Examples&tisearch=Search-EN. [Accessed January 2017].

12 Texas Instruments, EVMK2H hardware setup, 1 December 2016. [Online]. Available: http://processors.wiki.ti.com/index.php/EVMK2H_Hardware_Setup#DIP_Switch_and_Bootmode_Configurations.

13 T. Instruments, Keystone multicore workshop: lab manual-SPRP820, April 2014. [Online]. Available: http://www.ti.com/lit/ml/sprp820/sprp820.pdf. [Accessed December 2016].

14 Build and run U-boot and Linux kernel on TCI6638 EVM, [Online]. Available: http://www.deyisupport.com/cfs-file.ashx/__key/telligent-evolution-components-attachments/00-53-00-00-00-02-38-12/Build-and-Run-U_2D00_boot-and-Linux-Kernel-on-TCI6638-EVM.pdf.

12

Introduction to OpenMP

Multicore DSP: From Algorithms to Real-time Implementation on the TMS320C66x SoC, First Edition. Naim Dahnoun.
© 2018 John Wiley & Sons Ltd. Published 2018 by John Wiley & Sons Ltd.
Companion website: www.wiley.com/go/dahnoun/multicoredsp

Chapter 1 showed that continually increasing the clock frequency of a processor in order to increase the performance is not an option anymore, and the way forward is to increase the number of cores. However, this introduces other problems; for instance, was the application written for parallel processing? Can this application be parallelised, and has its performance increased as a consequence of this parallelism? How much effort has to be put for parallelising it? And so on.

To solve this, software parallelisation is not an easy task for a compiler alone to accomplish. Software parallelisation is a very important subject as it improves the performance of applications that can take advantage of high-performance computing (HPC). Many programming models like Message Passing Interface (MPI) and Open specifications for Multi-Processing (OpenMP) have been introduced.

OpenMP is supported on Texas Instruments' (TI) KeyStone family of Multicore TMS320C66x digital signal processor (DSP) System-on-Chips (SoCs) using the Multicore Software Development Kit MCSDK-C66. A list of other vendors' compilers can be found in Ref. [1].

In this chapter, OpenMP which is the de facto industry standard for shared memory parallel programming will be introduced, and a few examples on both the KeyStone I and II will be given. The aim of this chapter is to give the reader a quick start to implement OpenMP on the KeyStone devices. The complete specification, documentation, tutorials and examples can be found in Ref. [2].

12.1 Introduction to OpenMP

OpenMP was first introduced in 1997 [3], and it is now gaining more popularity with the emergence of multicore. OpenMP is a set of standard directives and tools that are inserted into serial code to help the compiler in parallelising it. OpenMP works well for applications running on a multicore platform with shared memory which is the case for the KeyStone processors. However, if the memory is not shared, it will not be accessible. The only way to pass information to an OpenMP-compatible compiler is for the programmer to identify parallel regions in the code to be parallelised and insert the appropriate directives. It is important to note at this stage that, as in general directives are optional compiler hints and may be ignored, the programmer can always test the original serial code without having to remove the directive. It is also important to understand the meaning of the **#pragma** directives before proceeding.

OpenMP was designed to offer the following features:

- Standardisation (OpenMP-compatible compilers)
- Ease of use (minimum modification of the serial code)
- Portability (APIs are specified for C/C++ and Fortran).

In this book, only OpenMP for C/C++ is discussed as it is supported for the KeyStone. OpenMP is simple and consists of three elements:

1) Compiler directives
2) Runtime library support
3) Environment variables.

The programmer inserts the directives which in turn are implemented by the runtime library support. OpenMP is based on three main components: work sharing, synchronisation and data sharing (see Figure 12.1). A serial code is parallelised using a work-sharing construct, the data attribute is specified and the method of synchronisation is decided in order to show how data are shared (see Figure 12.2).

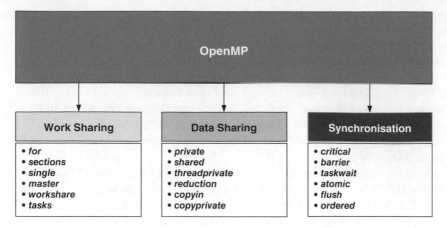

Figure 12.1 The three main components of OpenMP.

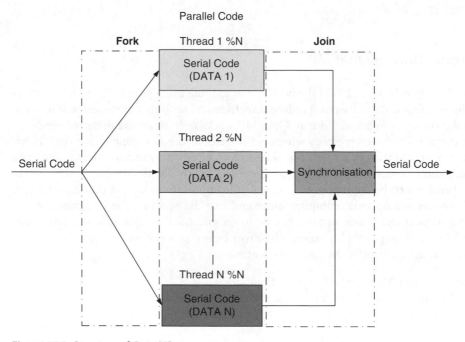

Figure 12.2 Structure of OpenMP.

12.2 Directive formats

The syntax of an OpenMP directive is as follows:

```
#pragma omp directive-name optional_clauses…
```

All directives start with **#pragma omp** to avoid confusion with other **#pragma**.

12.3   Forking region

The forking region starts with a single thread and converts it to parallel threads as illustrated in Figure 12.3.

12.3.1   omp parallel – parallel region construct

The **parallel** construct defines a parallel region of the program that is to be executed by multiple threads in parallel. The **parallel** construct is always used at the beginning of a shared region.

Syntax:

```
#pragma omp parallel [clause[ [, ]clause] …] new-line
{
//structured-block
}
```

A structured-block can be a single statement or a list of statements.

Example:

```
int count = 0;
#pragma omp parallel num_threads(8)
        {
                count++;
                printf("thread %d: count = %d\n", omp_get_thread_num
                (), count);
        }
        printf("thread %d: nb of threads= %d\n", omp_get_thread_num
        (), count);
```

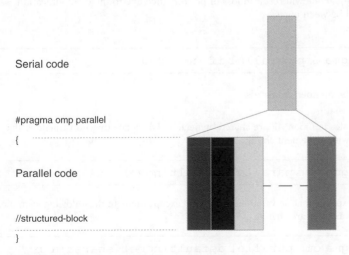

Serial code

#pragma omp parallel

{

Parallel code

//structured-block

}

Figure 12.3 Illustration of forking.

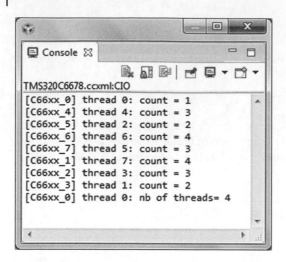

Figure 12.4 Output console.

See the output in Figure 12.4 and Laboratory experiment 1 in Section 12.8.1.

12.3.1.1 Clause descriptions

A single clause or a combination of clauses can be used as shown in Table 12.1, and these clauses may also be used by other pragmas as shown in Table 12.2.

Table 12.1 Clauses and descriptions

Clause	Description
if(scalar-expression)	The **if** statement makes the parallel region directive conditional. It only executes the parallel region if the condition is true. This is very useful; when **n** is small, parallelising code will result in loss of performance due to the communication overhead between cores. Example:
	```#pragma omp parallel if (n>10000)```
num_threads(integer-expression)	Setting the number of threads.
default(shared | none)	All variables in OpenMP are shared by default. To specify that no variable is shared, the following statement should be used:
	```#pragma omp parallel default(none)```
	One can specify that no variables are shared except some. In the following example, on▶ **a** and **b** variables are shared:
	```#pragma omp parallel default(none) shared(a,b)```

Table 12.1 (Continued)

Clause	Description
private(a list of variables)	If a variable is private, it will be declared once in every thread, and if it is initialised before declared as private, its value will be lost as shown in Figure 12.5. This is very important for work sharing, as each thread should have its own variable.

```
#pragma omp parallel default(none)  shared(a,b)
private(i)
{
// if i is used here, thread will have its own copy of i.
// if not initialised here, the initial value is random
    and shared
//by all thread.
}
```

See project:

Chapter_12_Code\OpenMPBasic2

Clause	Description
firstprivate(a list of variables)	**firstprivate()** is similar to **private()**, but the variable inside each thread will be initialised as shown in Figure 12.5 and Figure 12.6. Notice in both cases that the return value has been preserved. See project: Chapter_12_Code\OpenMPBasic2

```
// test for private and firstPrivate   --start--
int i = 10;
#pragma omp parallel private(i)
{    printf("thread %d: i = %d\n", omp_get_thread_num(), i);
i = i+ omp_get_thread_num();
printf("thread %d: i modified= %d\n", omp_get_thread_num(), i);
}

printf("i after private = %d\n\n", i);

#pragma omp parallel firstprivate(i)
{
        printf("thread %d: i = %d\n", omp_get_thread_num(), i);
        i = i + omp_get_thread_num();
        printf("thread %d: i modified= %d\n", omp_get_thread_num(),
i);
}

printf("i after firstprivate= %d\n", i);

for(;;);
// test for private and firstPrivate -- end--
```

Figure 12.5 private() and **firstprivate()** examples.

Table 12.1 (Continued)

Clause	Description

```
[C66xx_6] thread 6: i = 201547120
thread 6: i modified= 201547126
[C66xx_0] thread 0: i = 0
[C66xx_1] thread 1: i = 201547120
[C66xx_2] thread 2: i = 201547120
[C66xx_3] thread 3: i = 201547120
[C66xx_4] thread 4: i = 201547120
[C66xx_5] thread 5: i = 201547120
[C66xx_7] thread 7: i = 201547120
[C66xx_0] thread 0: i modified= 0
[C66xx_1] thread 1: i modified= 201547121
[C66xx_2] thread 2: i modified= 201547122
[C66xx_3] thread 3: i modified= 201547123
[C66xx_4] thread 4: i modified= 201547124
[C66xx_5] thread 5: i modified= 201547125
[C66xx_7] thread 7: i modified= 201547127
[C66xx_0] i after private = 10

thread 0: i = 10
thread 0: i modified= 10
[C66xx_1] thread 1: i = 10
[C66xx_2] thread 2: i = 10
[C66xx_3] thread 3: i = 10
[C66xx_4] thread 4: i = 10
[C66xx_5] thread 5: i = 10
[C66xx_6] thread 6: i = 10
[C66xx_7] thread 7: i = 10
[C66xx_1] thread 1: i modified= 11
[C66xx_2] thread 2: i modified= 12
[C66xx_3] thread 3: i modified= 13
[C66xx_4] thread 4: i modified= 14
[C66xx_5] thread 5: i modified= 15
[C66xx_6] thread 6: i modified= 16
[C66xx_7] thread 7: i modified= 17
[C66xx_0] i after firstprivate= 10
```

Figure 12.6 Console output (using **private** and **firstprivate**).

Table 12.1 (Continued)

Clause	Description
shared(list)	Specifies variables that are shared among all the threads. Can also be used with the **default()** clause; see the **default** section.

Example:

```
#pragma omp parallel default(none) shared(a)
```

In this example, all variables are not shared except the variable **a**.

A variable (global, local static or namespace) can be made private to a thread. Example:

```
#pragma omp threadprivate(var)
```

Clause	Description
copyin(list)	The **copyin** clause can be used to copy a variable to a thread on entering the parallel region. Example:

```
int x,y,z;
#pragma omp threadprivate(x,y,z)
// sepecifies that x,y and z are private to a thread.
#pragma omp parallel copyin(x,y)
{
}
#pragma omp parallel copyin(z)
{

}
```

Clause	Description
reduction(operator: list of variables)	Makes the specified variable private, and specifies that the variable is going to be operated on with the specified operator at the end of the parallel region. See the example in Figure 12.7. The operator can be one of the following: +, *, -, /, &, ^, |, && or ||. See project:

Chapter_12_Code\OpenMPBasic2

```
long long dotp_omp(int count) {

        int   i;

        long long acc = 0;

#pragma omp parallel for reduction(+:acc) private(i) num_threads (8)

                for (i = 0; i < count; i++) {

                        acc += a[i] * x[i];

                                                }

        return acc;

}
```

Figure 12.7 Using **reduction**.

Table 12.2 Pragmas where the clauses can be used

Clause	PARALLEL	FOR	SECTIONS	SINGLE	PARALLEL FOR	PARALLEL SECTIONS
IF	•				•	•
PRIVATE	•	•	•	•	•	•
SHARED	•				•	•
FIRSTPRIVATE	•	•	•	•	•	•
LASTPRIVATE		•	•		•	•
DEFAULT	•				•	•
REDUCTION	•	•	•		•	•
COPYIN	•				•	•
ORDERED		•			•	
SCHEDULE		•			•	
NOWAIT		•	•	•		
NUM_THREADS	•				•	•

12.4 Work-sharing constructs

Work-sharing constructs divide the code into threads.

12.4.1 omp for

omp for can be combined with **parallel** as **#pragma omp parallel for**.
omp for looks inside the loop and divides it into threads.

Syntax:

```
#pragma omp parallel for [clauses]
{
// for loop to be executed in parallel
}
```

Clauses:

- **private(list)**
- **shared(list)**

- **default(shared | none)**
- **firstprivate(list)**
- **lastprivate(list)**
- **reduction(operator: list)**
- **copyin(list)**
- **if(scalar_expression)**
- **ordered**
- **schedule (kind[, chunk]).**

12.4.1.1 OpenMP loop scheduling

In this section, how loops are scheduled is demonstrated, in other words how loops are divided among cores.

With OpenMP, there are three types of scheduling that are referred to in the literature as *kinds*:

1) Static. This scheduling type distributes a set number of iterations to each core in a round-robin fashion. Consider Figure 12.8, where each iteration **i** contains a delay **i** (to create an imbalance loading). It can be seen from the output (**LoopStatic**, Figure 12.11) that the first iteration was sent to core 2 (the smallest delay), then the next iterations were sent to core 3, core 4, core 5, core 6, core 7, core 0 and so on; the last iteration was run in core 7.
2) Dynamic. In this scheduling type, a chunk of loop iterations is taken from an internal queue at runtime and sent to a core. Once a core finishes, it retrieves the next chunk; see Figure 12.11 **LoopDynamic**. Figure 12.9 shows code for a dynamic scheduling of chunk size of five, and Figure 12.12 shows the output. In this case, a chunk of five consecutive iterations is sent to the same core.
3) Guided. This scheduling type starts by sending a large chunk of iterations to each available core (as dynamic) and gradually reduces the chunk size to the specified chunk size (**CHUNK_SIZE**); see Figure 12.10. Notice in Figure 12.12 that the last chunk size is effectively five.

```
#define count 18

#pragma omp parallel for reduction(+:acc) private(i)  schedule(static, 1)
      for (i = 0; i < count; i++) {

              Log_write1(UIABenchmark_start,  (xdc_IArg)"LoopStatic");

              Task_sleep(i);

              Log_write1(UIABenchmark_stop,  (xdc_IArg)"LoopStatic");


      }
```

Figure 12.8 Using static scheduling.

```
#define CHUNK_SIZE 5
#define count92

#pragma omp parallel for reduction(+:acc) private(i) schedule(dynamic,CHUNK_SIZE)
      for (i = 0; i < count; i++) {
            int tid - omp_get_thread_num();
            Log_write1(UIABenchmark_start, (xdc_IArg)"LoopDynamic");
            Task_sleep(i);

            Log_write1(UIABenchmark_stop, (xdc_IArg)"LoopDynamic");


      }
```

Figure 12.9 Dynamic scheduling.

```
#pragma omp parallel for reduction(+:acc) private(i)  schedule(guided,CHUNK_SIZE)
      for (i = 0; i < count; i++) {
            int tid = omp_get_thread_num();
            Log_write1(UIABenchmark_start, (xdc_IArg)"LoopGuided");
            Task_sleep(i);

            Log_write1(UIABenchmark_stop, (xdc_IArg)"LoopGuided");
      }
```

Figure 12.10 Guided scheduling.

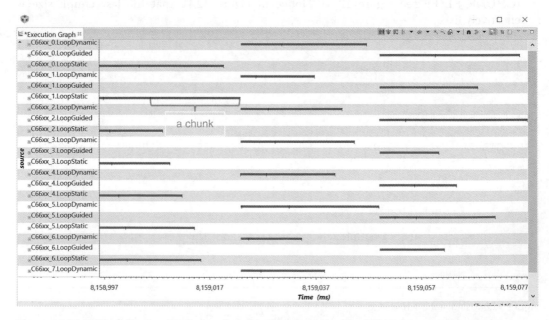

Figure 12.11 Output showing the three scheduling kinds (types) with small iteration number.

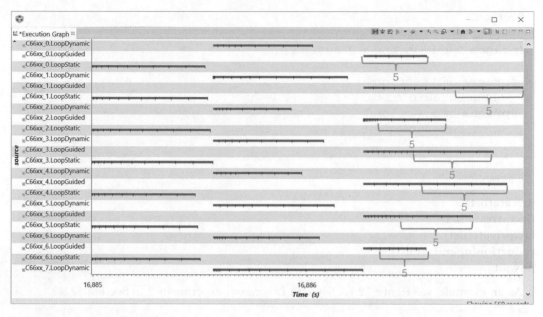

Figure 12.12 Output showing the three scheduling kinds (types) with large iteration number.

See Laboratory files location for experiment 3 in Section 12.8.3.

12.4.2 omp sections

omp sections are very useful for functional-level parallelism. Assuming that some tasks need to be executed in a certain order, running them in parallel will produce the wrong results.

As an example, consider the code shown in Figure 12.13 where Funct_2 may depend on the result of Funct_1.

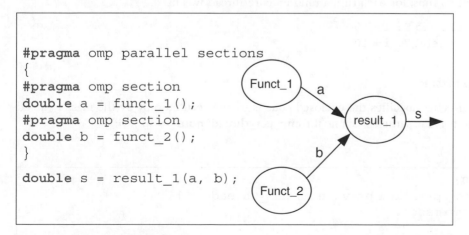

Figure 12.13 Using **omp sections**.

Syntax:

```
#pragma omp sections [ Clauses... ]
{
  #pragma omp section
    structured_block
  #pragma omp section
    structured_block
}
```

Clauses:

private(list)
firstprivate(list)
lastprivate(list)
reduction(operator: list)
nowait

As an example, see Figure 12.13. See also Laboratory experiment 2 in Section 12.8.2.

12.4.3 omp single

The single directive specifies that the enclosed code is to be executed only by one thread. It is similar to **master** shown below, but **single** can be used with clauses shown in Table 12.2.

Example:

```
int i = 10;
#pragma omp parallel
#pragma omp single
{
    printf("thread %d: i = %d\n", omp_get_thread_num(), i);
}
```

Output for two runs; for each run, a random core/thread was used:

```
[C66xx_5] thread 5: i = 10
[C66xx_7] thread 7: i = 10
```

12.4.4 omp master

The master directive specifies the enclosed code is to be executed only by the master thread that is core 0. It is the same as saying 'if (**omp_get_thread_num** == 0) {...};'

Example:

```
int i = 10;
#pragma omp parallel private(i) num_threads (8)
#pragma omp master
{
    printf("thread %d: i = %d\n", omp_get_thread_num(), i);
}
```

The output is:

```
[C66xx_0] thread 0: i = 0
```

12.4.5   omp task

The task (tasking) has been introduced in version 3.0 of the OpenMP specification. The task operation is very useful as tasks (functions) can be automatically scheduled to run on the available threads (cores). When a task construct is reached, the code will be executed in one available core, and subsequent code will be executed on any other available core.

Consider the example shown in Figure 12.17, where ten tasks take different times to execute and two cores have been selected. In this case, Task 1 will run in Core 1 and Task 2 will be dispatched to Core 2. Core 1 finishes Task 1 before Core 2 finishes Task 2, and therefore it will run Task 3 and so on. In this example, by using **task** directive the performance can be double (since two cores are used). See Laboratory experiment 4 in Section 12.8.4 (Example 1). In Example 1, 16 tasks with different execution times running on two threads are programmed with **task** and without **task** directives as shown in Figure 12.14. The output when using task directive is shown in Figure 12.15, and the output when not using task directive is shown in Figure 12.16; the time taking in this case is doubled.

```c
#include <xdc/std.h>
#include <xdc/runtime/Log.h>
#include <xdc/runtime/System.h>
#include <xdc/runtime/Diags.h>
#include <xdc/runtime/Types.h>
#include <xdc/runtime/Timestamp.h>

#include <ti/omp/omp.h>
#include <stdio.h>
#include <time.h>

#define N_TASKS 16

int times[N_TASKS] = { 10000000, 10000000, 20000000, 10000000, 10000000,
                       10000000, 20000000, 10000000, 10000000, 10000000,
                       40000000, 10000000, 10000000, 10000000, 10000000, 0 };

int count = 0;
omp_lock_t lock;

void main() {
omp_set_num_threads(2);
omp_init_lock(&lock);
#pragma omp parallel
{
Types_Timestamp64 tsstart, tsend;
Types_FreqHz fhfreq;
long long start, end, diff, freq;
Timestamp_getFreq(&fhfreq);
freq = ((long long) (fhfreq.hi) * (long long) 1000000)
                    + ((long long) (fhfreq.lo));
Timestamp_get64(&tsstart);
```

Figure 12.14 Example with and without **task** directives.

```c
#pragma omp single
{
    int i;

    for (i = 0; i < N_TASKS; i++) {
#pragma omp task
        {
            printf("Core %i for task %i: Starting\n",
                   omp_get_thread_num(), i);
            volatile int j = times[i];
            while (j--)
                ;
            printf("Core %i for task %i: Done\n", omp_get_thread_num(),
                   i);
        }
    }

}
#pragma omp taskwait
Timestamp_get64(&tsend);
start = ((long long) (tsstart.hi) << (long long) 32)
                    + ((long long) (tsstart.lo));
end = ((long long) (tsend.hi) << (long long) 32)
                    + ((long long) (tsend.lo));
diff = end - start;
double time = diff / (double) freq;
printf("Time with tasks: %lf\n", time);

} // End of parallel region

#pragma omp parallel
{
Types_Timestamp64 tsstart, tsend;
Types_FreqHz fhfreq;
long long start, end, diff, freq;
Timestamp_getFreq(&fhfreq);
freq = ((long long) (fhfreq.hi) * (long long) 1000000)
                    + ((long long) (fhfreq.lo));
Timestamp_get64(&tsstart);
#pragma omp single
{
    int i;

    for (i = 0; i < N_TASKS; i++) {
        //#pragma omp task
        {
            printf("Core %i for task %i: Starting\n",
                   omp_get_thread_num(), i);
            volatile int j = times[i];
            while (j--)
                ;
        printf("Core %i for task %i: Done\n", omp_get_thread_num(),i);
        }
    }

}
#pragma omp taskwait
Timestamp_get64(&tsend);
start = ((long long) (tsstart.hi) << (long long) 32)
                    + ((long long) (tsstart.lo));
end = ((long long) (tsend.hi) << (long long) 32)
                    + ((long long) (tsend.lo));
diff = end - start;
double time = diff / (double) freq;
printf("Time without tasks: %lf\n", time);

} // End of parallel region

omp_destroy_lock(&lock);
}
```

Figure 12.14 (Continued)

Figure 12.15 Output when using **task** directive.

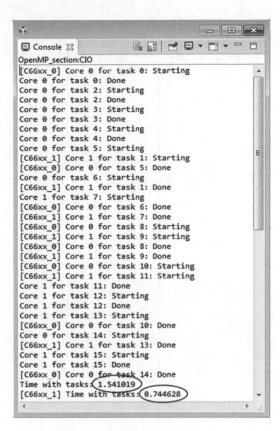

Figure 12.16 Output when not using **task** directive.

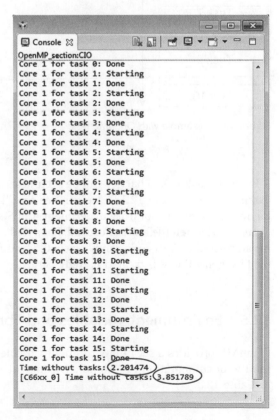

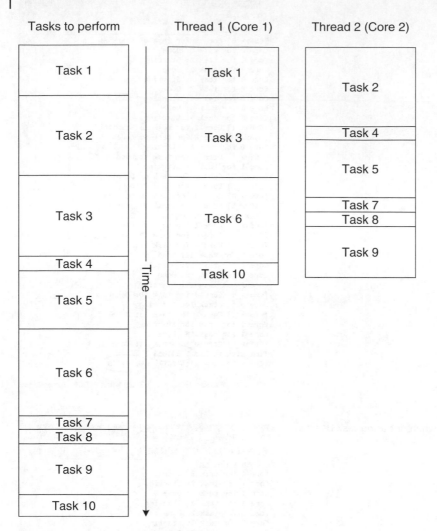

Figure 12.17 Example with OpenMP task.

By using **taskwait**, a barrier will be set and all tasks will wait at this barrier. For the following example code (Figure 12.18), we can see that 'Four' is always printed last; see the output in Figure 12.19. See Laboratory experiment 4 in Section 12.8.4 (Example 2).

task can be **tied** (default) or **untied**. For a **tied** task its code is executed by the same thread, and for an **untied** task its code can be executed by different threads (this is currently not supported by TI OpenMP and is treated is **tied**).

12.5 Environment variables and library functions

OpenMP provides a set of environment variables and library functions to ease programming; see the documentation [4]. Some environment variables used in the chapter are summarised in Table 12.3.

```
int four_times =4;

void task()
{
      while (four_times!=0)
      {
#pragma omp parallel

#pragma omp single // block will be executed only once by a single thread (core).

            {
                  printf("One\n ");
#pragma omp task
                  {
                        printf("Two \n");
                  }

#pragma omp task

                  {
                        printf("Three \n");
                  }

#pragma omp taskwait // wait here tor threads to finish
                  printf("Four \n\n ");

            }// End of parallel region
            four_times = four_times -1;     } // end of while


}
```

Figure 12.18 Using **task** and **taskwait**.

Figure 12.19 Output console for four runs.

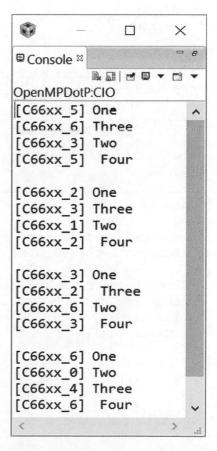

Table 12.3 Environment variables used

Environment variables	Brief description
OMP_NUM_THREADS	Sets the number of threads to be used
OMP_DYNAMIC	Specifies if the OpenMP runtime can adjust the number of threads in a parallel region
OMP_SCHEDULE	Controls the scheduling of the **for** loop construct

12.6 Synchronisation constructs

So far we have been using implicit barriers in **parallel, for, sections** and **single** constructs. However, adding an explicit barrier can be very useful for maintaining correct operation.

Implicit barriers example:

```
#pragma omp parallel
{
 #pragma omp for
 for(int i=0; i<10; ++n) function1();

 //implicit barrier is here due to the for pragma.
 //function2() can only be reached when all threads finish the above
for loop.
 function2();
}

// another implicit barrier is here due to the parallel pragma
//function3() can only be reached when fuction2() finishes.

 function3();
```

The implicit barrier of the **for** pragma can be ignored by using the **nowait** clause as shown here:

```
#pragma omp parallel
{
  #pragma omp for nowait    // removing the barrier
  for(int i=0; i<10; ++n) function1();

//implicit barrier has been removed by using nowait.
//function2() can be executed while other thread are still executing
  function1().
  function2();
}
// another implicit barrier is here due to the parallel pragma
//function3() can only be reached when fuction2() finishes.

 function3();
```

Using the **nowait** clause can improve the performance of a program as it removes the barrier. OpenMP has the following constructs to perform synchronisation:

- **atomic**
- **critical section**
- **barrier**
- **flush**
- **ordered**
- **single**
- **master**
- **taskwait**.

12.6.1 atomic

atomic guarantees that only one thread can update a variable at one time; this is known as *fine-grained synchronisation*.

Syntax:

```
#pragma omp atomic [read|write|update|capture] [ seq_cst]
expression_statement
```

For more details, refer to Ref. [4].

12.6.1.1 Clauses

For a complete description, please refer to Ref. [4].

update	Updates the value of a variable atomically
read	Reads the value of a variable atomically
write	Writes the value of a variable atomically
capture	This clause is useful if one would like to save the old value of the variable before updating it.

Example:

```
int count = 0;
  #pragma omp parallel num_threads(8)
  {
    #pragma omp atomic
    count++;
    printf("thread %d: count = %d\n", omp_get_thread_num(), count);
  }
  printf("thread %d: nb of threads= %d\n", omp_get_thread_num
(), count);
```

The output is shown in Figure 12.20.

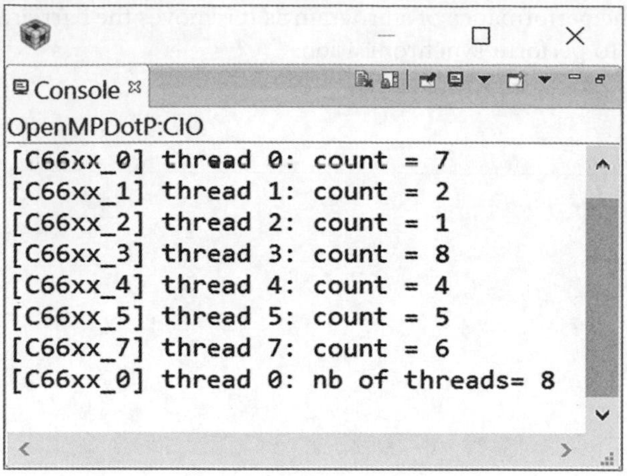

Figure 12.20 Console output.

The **atomic** construct only takes care of the count update. However, **function_4_all()** can be executed by all threads as shown here:

```
int count = 0;
#pragma omp parallel num_threads(8)
{
#pragma omp atomic
    count+= function_4_all();
    printf("thread %d: count = %d\n", omp_get_thread_num(), count);
}
printf("thread %d: nb of threads= %d\n", omp_get_thread_num
(), count);
int function_4_all()
{
    int i;
    i = omp_get_thread_num();
    return i;
}
```

The output is shown in Figure 12.21.

If **#pragma omp atomic** is commented out, as shown here:

```
int count = 0;
#pragma omp parallel num_threads(8)
{
    //#pragma omp atomic
    count+= function_4_all();
    printf("thread %d: count = %d\n", omp_get_thread_num(), count);
}
printf("thread %d: nb of threads= %d\n", omp_get_thread_num
(), count);
```

Figure 12.21 Console output.

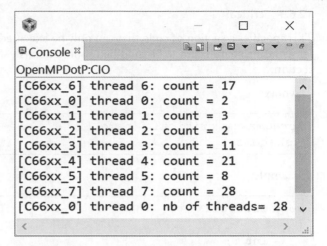

Figure 12.22 Console output.

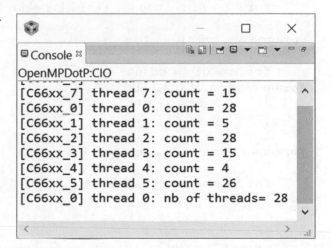

The output could be as shown in Figure 12.22.

12.6.2 barrier

This specifies a location in the program that all threads in the parallel region must arrive at, before any thread can proceed past it.

Example:

```
#pragma omp parallel
{
    // All threads execute code_executed_by_all();

    code_executed_by_all();

#pragma omp barrier // the barrier is here.

    // when all threads before the barrier finish, the next function
will be executed by all threads
    code_executed_by_all();
}
```

12.6.3 critical

This specifies a section of code that can only be executed by one thread at a time. This is useful for avoiding shared data to be modified simultaneously. No branch is allowed out of the critical section.

Syntax:

```
#pragma omp critical [(name)]
  statement_block
```

Example:

```
void critical()
{
    int i;
    for ( i = 0; i < count; ++i) b[i] = i;

#pragma omp parallel for

    for (i = 0; i < count; i++) {
#pragma omp critical (lock1)
            acc1 += b[i] ;
    }

#pragma omp parallel for

    for (i = 0; i < count; i++) {

            acc2 += b[i] ;
    }
    printf("\nThe correct value is 4950 (sum: 0- 99)\n", acc1);
    printf("acc with omp critical : %d\n", acc1);
    printf("acc WITHOUT omp critical : %d\n", acc2);
}
```

The output is shown in Figure 12.23.

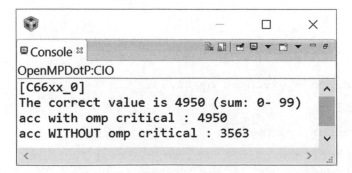

Figure 12.23 Console output.

12.7 OpenMP accelerator model

The OpenMP 4.0 specification enables the use of OpenMP on heterogeneous systems by adding support for a set of device constructs. The OpenMP community uses the term 'OpenMP Accelerator Model' to refer to this set. OpenMP 4.0 defines a host device on which the OpenMP programs begin execution, and target devices onto which regions of code can be offloaded.

12.7.1 Supported OpenMP device constructs

- #pragma omp target
- #pragma omp declare target
- #pragma omp target data
- #pragma omp target update
- omp_set_default_device()
- omp_get_default_device()
- omp_get_num_devices()
- omp_is_initial_device()

Descriptions and examples with **#pragma omp target, #pragma omp target data, #pragma omp target update** and **#pragma omp declare target** are given in this section.

12.7.1.1 #pragma omp target
#pragma omp target specifies the code to be offloaded to the targets as shown in Figure 12.24.

Syntax:

```
#pragma omp target [clause[, clause, ...]]
```

Arguments:

Clause	Can be any of the following clauses:
	- **device**(integer-expression)
	- **map**([map-type:] list)
	- **if**(scalar-expression).

There is also a **map** clause for the **target** construct that specifies which variable(s)/memory(s) should be passed from the host to the device or from the device to the host; see Figure 12.25. There are also two other clauses that can be used: **alloc** and **tofrom**.

The *map-type* may be any of the following:

- **alloc**: A new variable with an undefined value that corresponds to each list item created on the device.
- **to**: The data are copied from the host to the target.
- **from**: On exit from the target, the data are copied from the target to the host.
- **tofrom(default)**: The data are copied from the host to the target on entry and from the target to the host on exit.

Example:

```
#pragma omp target device(int) map(map-type: list if (scalar expression)
{
//structured code (single entry and single exit) executing in the target device.

}
```

Example:

```
int data[N];
// loop 1
#pragma omp target
#pragma omp for
for (int i = 0; i < N; i++)
    data[i] *= 2;
```

Figure 12.24 Using **#pragma omp target**.

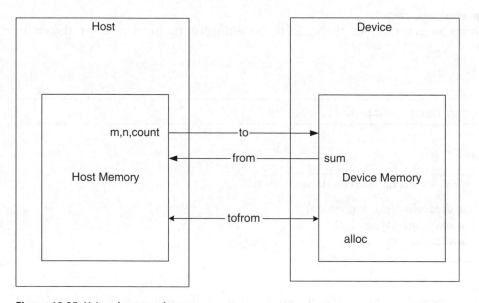

Figure 12.25 Using the **map** clause.

- **if(scalar-expression)**: When an **if** clause is used and the **if** expression evaluates to false, the device will be the host.

An issue of speed can happen here if the data are not used by the host but copied from the device to the host and copied back to the device. Let's have a look at the example shown in Figure 12.26.

```
int data[N];
// loop 1
#pragma omp target
#pragma omp for
for (int i = 0; i < N; i++)
      data[i] *= 2;

// Do something else on the Host without using the data.
//loop 2 continue using the same data
#pragma omp target
#pragma omp for
for (int i = 0; i < N; i++)


      data[i] += 5;
```

Figure 12.26 Example when data do need to be processed by the host.

The **target** constructs will create two data environments, one for loop 1 and one for loop 2. The following sequence will happen:

1) Data are copied from the host to the device and processed in loop 1.
2) Data are copied from the device to the host.
3) The host does some processing and does not use the same data.
4) The data are copied from the host to the device (again!).
5) loop 2 processes the data.

Steps 2 and 4 are unnecessary and consume a large number of cycles.
To solve this problem, one can use **#pragma omp target data** as shown in Section 12.7.1.2.

12.7.1.2   #pragma omp target data
To avoid unnecessary data movements, from the host to the target and from the target to the host, the **#pragma omp target data** has been introduced.

Syntax:

```
#pragma omp target data [clause, clause, ...]
```

Arguments:

Clause	Can be zero or more of the following clauses: **device**(integer-expression) **if**(scalar-expression) **map**([map-type:] list).

Example:

```
int data[N];

#pragma omp target data map(tofrom: data)
{
#pragma omp target
#pragma omp for
        for (int i = 0; i < N; i++)
                data[i] *= 2;

#pragma omp target update from(data)

        // Do something else

#pragma omp target
#pragma omp for
        for (int i = 0; i < N; i++)
                data[i] += 5;
}
```

1) Data are copied from the host to the device and processed in loop 1.
2) The data are explicitly copied from the device to the host.
3) The host does some processing and uses the data.
4) loop 2 processes the data.
5) The data are copied at the end of the target region.

In this example, only explicit data movement is used and therefore unnecessary data movement is avoided.

12.7.1.3 #pragma omp target update
```
#pragma omp target update [clause[, clause, …]]
```

Arguments:

Clause	Can be any of the following clauses:
	• **device**(integer-expression)
	• **if**(scalar-expression)
	• **from**(list)
	• **to**(list).

The **from** clause causes data to be copied from the device to the host.
The **to** clause causes data to be copied from the host to the device.
If an **if** expression evaluates to false, then no copies will occur.

```
#pragma omp target update to(a,b) from(c,d)
```

Use this pragma to copy the values of *a* and *b* (located on the host) to the device, then copy the values of *c* and *d* (located on the device) to the host.

Example:

```
int data[N];
#pragma omp target data map(tofrom: data)
{
#pragma omp target
#pragma omp for
        for (int i = 0; i < N; i++)
                data[i] *= 2;

#pragma omp target update from(data)

        // Do something else

#pragma omp target
#pragma omp for
        for (int i = 0; i < N; i++)
                data[i] += 5;
```

12.7.1.4   #pragma omp declare target

Any function or variable that appears between the **omp declare target** pragma and the **omp end declare target** pragma statements that is specified inside a target region but executes on the device.

Example:

```
#pragma omp declare target
#include <omp.h>
extern int printf(const char *_format, …);
#pragma omp end declare target

int dotp(short* m, short* n, int count, int ncore)
{

        int sum = 0;
#pragma omp target data map(to:m[0:count],n[0:count],count)
map(from: sum)
        {

#pragma omp target map(to:m[0:count],n[0:count],count)
map(from: sum)
                {
                        int i;
                        sum = 0;
                        double t0 = omp_get_wtime();
#pragma omp parallel for private(i) shared(m,n,count)
                        reduction(+:sum) num_threads(ncore)
                                for(i=0;i<count;i++)
```

```
                        {
                            sum += m[i]*n[i];
                        }
                    double t1 = omp_get_wtime();
                    printf("time from target side %f ns using %d
                    cores\n",
                            (t1-t0)*1e9, ncore);
                }
            }
        return sum;
}
```

12.8 Laboratory experiments

12.8.1 Laboratory experiment 1

Project location:

\Chapter_12_Code\OpenMP_Parallel

In this experiment, the structured code contains a counter **count** which is incremented, and a thread number representing the core it is running on is printed. By using the **parallel** construct, the structured code will run on a number of cores specified by **num_threads()**.

1) Open the project **OpenMPDotP**.
2) Explore the source code **open_parallel.c** and notice that the header file < **ti/omp/omp.h** > has been added.
3) Open the project properties and enable the support for OpenMP by ticking the box as shown in Figure 12.27.
4) Connect the TMS320C6678 EVM, build the project and load the code. If no errors are reported, the debug window would look like that shown in Figure 12.28 (left). Select all cores, right click and select **Group cores**; see Figure 12.28 (right). Select Group 1 and press **run** to run all the cores together. The output is shown in Figure 12.29.
5) Sometimes when copying or renaming and loading the new project, some cores will start running automatically before starting a manual run, and this will lead to the project not running properly. To avoid this, select each core individually (see Figure 12.28), select **Tools > Debug options > Auto run** and **Load options**, and right click and untick the **On a program load or restart** box as shown in Figure 12.30. Make sure you press **Remember my settings**.

12.8.2 Laboratory experiment 2

Project location:

\Chapter_12_Code\openMP_Sections

Explore the **omp_section.c** file shown in Figure 12.31, then build and run the project. The output is shown in Figure 12.32.

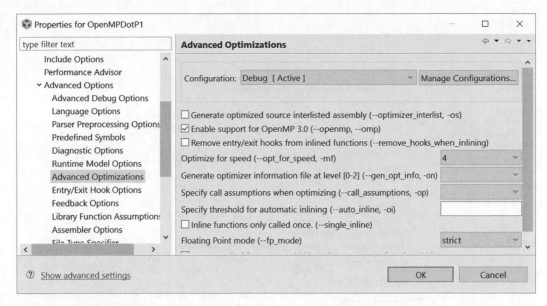

Figure 12.27 Enabling OpenMP.

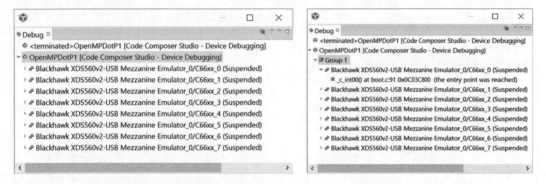

Figure 12.28 Group the cores before running the code.

Figure 12.29 Output console.

```
OpenMPDotP1:CIO
[C66xx_0] thread 0: count = 5
[C66xx_6] thread 6: count = 8
[C66xx_1] thread 1: count = 6
[C66xx_2] thread 2: count = 7
[C66xx_3] thread 3: count = 4
[C66xx_4] thread 4: count = 5
[C66xx_5] thread 5: count = 1
[C66xx_7] thread 7: count = 3
[C66xx_0] thread 0: nb of threads= 8
```

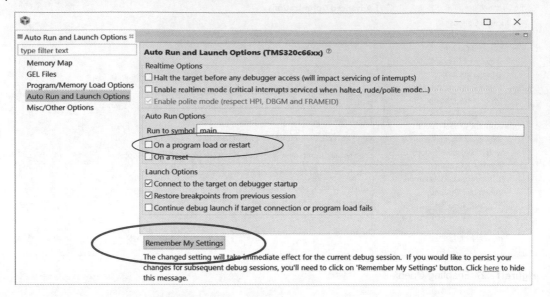

Figure 12.30 Remove the **Auto run option**.

```
#pragma omp sections
{
    #pragma omp section
    {
        printf ("\ncore id = %d, \n", omp_get_thread_num());
    }

    #pragma omp section
    {
        printf ("core id = %d, \n", omp_get_thread_num());
    }

        #pragma omp section
    {
        printf ("core id = %d, \n", omp_get_thread_num());
    }

}
```

Figure 12.31 Code using **omp section**.

12.8.3   Laboratory experiment 3

Project location:

\Chapter_12_Code\OpenMP_scheduling

Two threads   Three threads

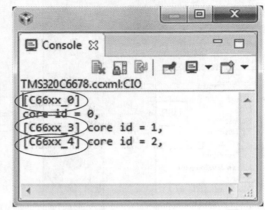

Figure 12.32 Console outputs.

12.8.4 Laboratory experiment 4

Project location:

Example 1: \Chapter_12_Code\OpenMP_task1
Example 2: \Chapter_12_Code\OpenMP_task2.

12.8.5 Laboratory experiment 5

This laboratory experiment is to be tested with the KeyStone II EVM.
Copy all files from:

\Chapter_12_Code\KS2 to /home/dotp2_Openmp

Project location:

/home/dotp2_Openmp/dotp_ompacc/ (OpenMP with target)

In this laboratory, the **dotp** product will be implemented on the ARM and on DSPs using OpenMP with the accelerator model. The hardware setup is shown in Figure 12.33.
Software setup:

1) Identify the serial ports available on the host PC. To do this, connect the USB cable to the PC and power up the EVM; see Label 1 in Figure 12.33. Then open the device manager and identify which serial ports are connected to the EVM as shown in Figure 12.34; in this example, COM14 and COM16 are used. Note that every time a PC is restarted, new COM ports can be used.
2) Connect the EVM to a terminal. One can use the **PuTTY** as a terminal [5]. Set two ports as shown in Figure 12.35 and Figure 12.36.
3) Start Ubuntu as shown in Figure 12.37 and Figure 12.38.

Establish the connection between the VMware and the EVM by following the steps illustrated in Figure 12.39 to Figure 12.45. Once powered up, the EVM should boot as shown in Figure 12.46. Type **root** as shown in Figure 12.47. Type **ifconfig** to check the IP addresses; see Figure 12.48.

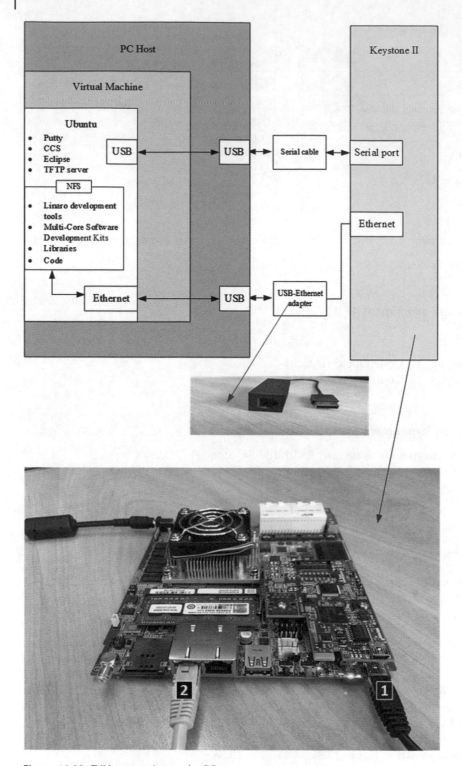

Figure 12.33 EVM connection to the PC.

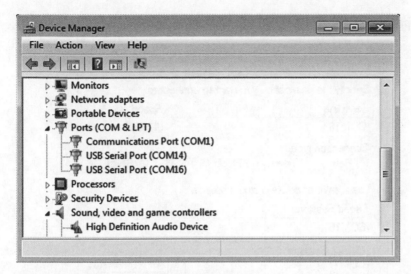

Figure 12.34 Identifying ports used by the EVM.

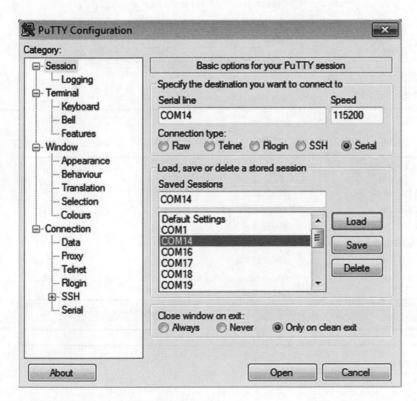

Figure 12.35 Setting up the COM port 14.

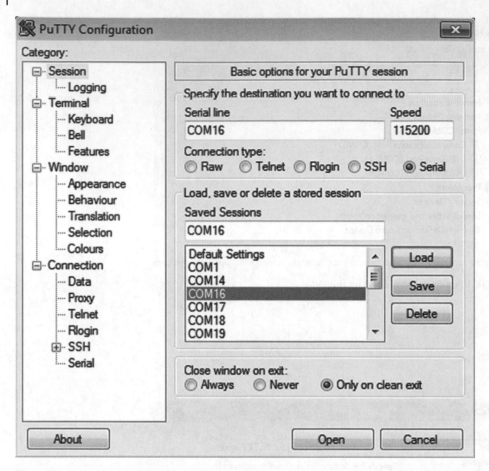

Figure 12.36 Setting up the COM port 16.

Name	Type	Size
Ubuntu 64-bit-5d21de91.vmem.lck	File folder	
Ubuntu 64-bit-b1098b90.vmem.lck	File folder	
VMWARE_for_Windows	File folder	
564d86ca-a77c-9017-5ca3-f49d56f65cff.vmem	VMEM File	3,145,728 KB
Ubuntu 64-bit.nvram	NVRAM File	9 KB
Ubuntu 64-bit.vmdk	VMware virtual disk file	40,972,992 ...
Ubuntu 64-bit.vmsd	VMSD File	0 KB
Ubuntu 64-bit.vmx	VMware virtual machine configuration	4 KB
Ubuntu 64-bit.vmxf	VMXF File	1 KB

Figure 12.37 Evoking the VMware.

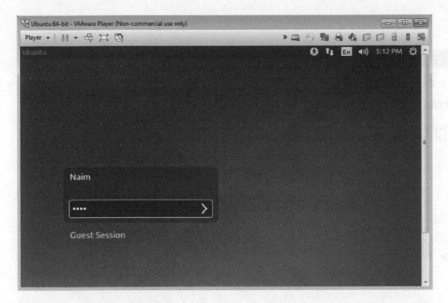

Figure 12.38 The VMware.

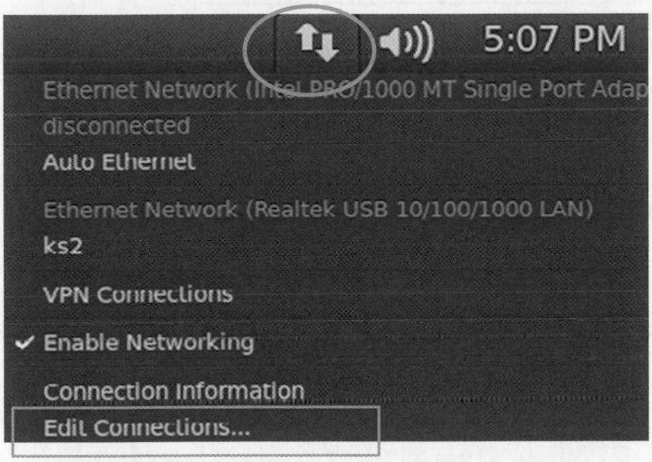

Figure 12.39 Edit connections.

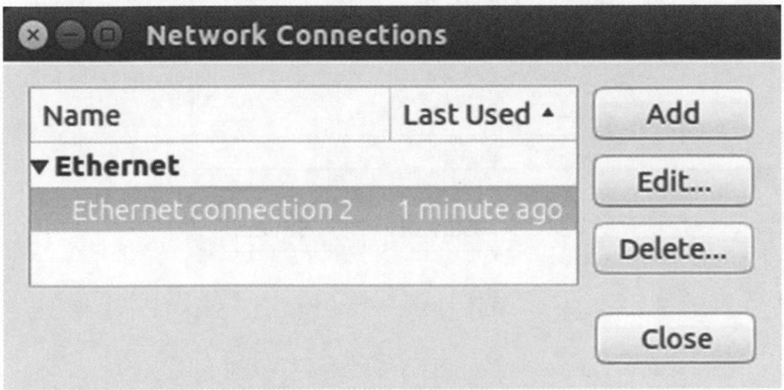

Figure 12.40 Select an Ethernet connection or add one.

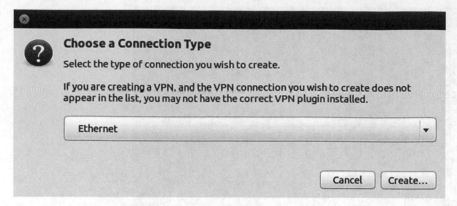

Figure 12.41 Create an Ethernet connection.

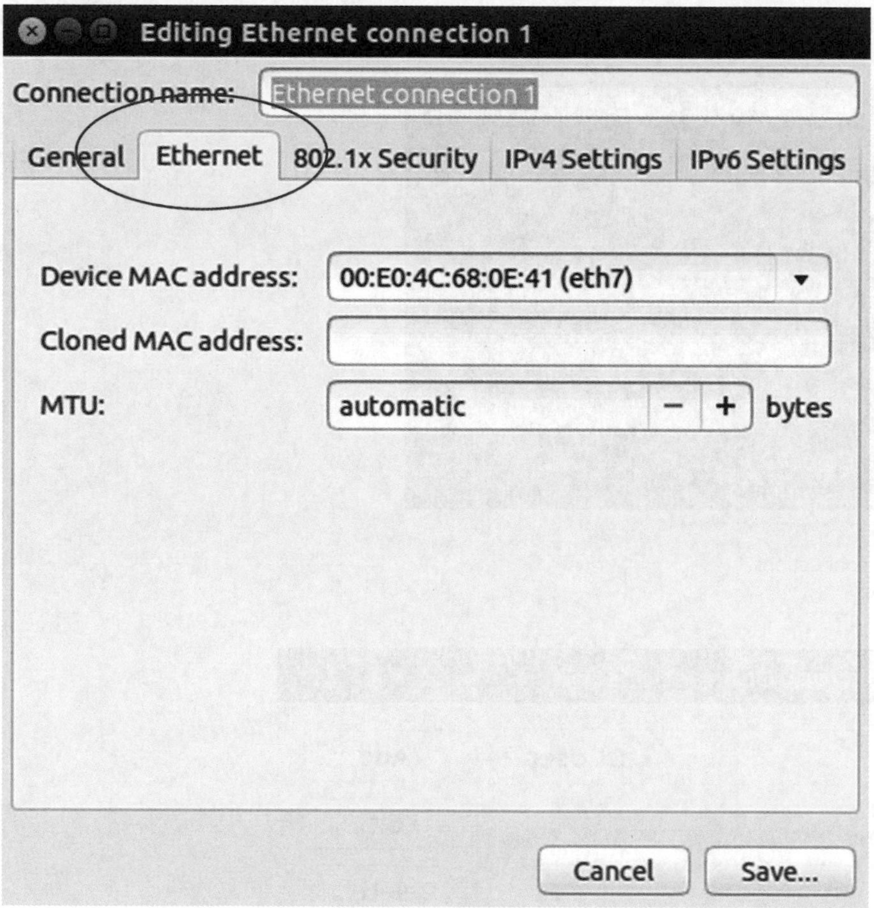

Editing Ethernet connection 1

Connection name: Ethernet connection 1

General | **Ethernet** | 802.1x Security | IPv4 Settings | IPv6 Settings

Device MAC address: 00:E0:4C:68:0E:41 (eth7)

Cloned MAC address:

MTU: automatic − + bytes

Cancel | Save...

Figure 12.42 Select Ethernet.

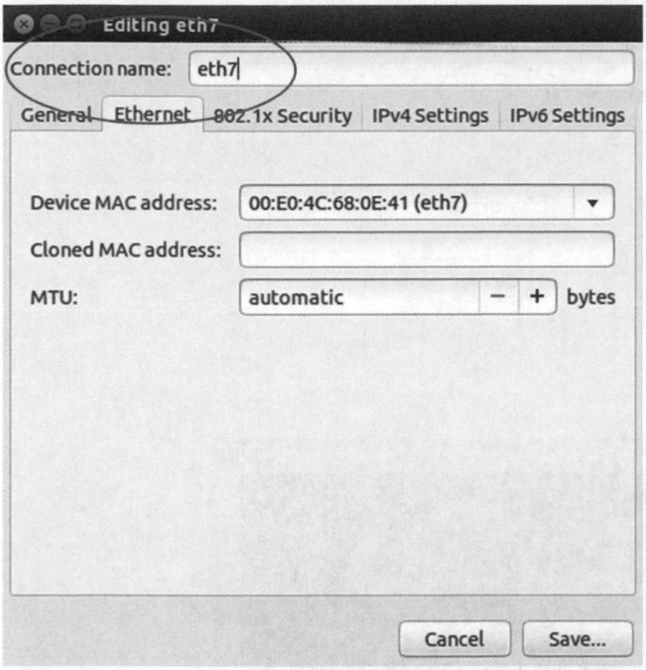

Figure 12.43 Edit the **Connection name**.

Figure 12.44 Select **IPv4 settings**.

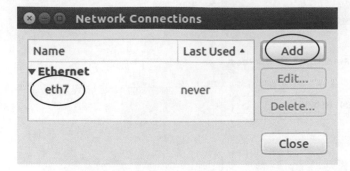

Figure 12.45 Add the Ethernet created.

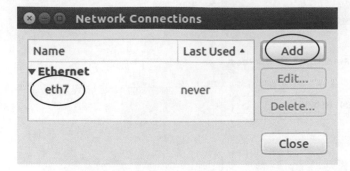

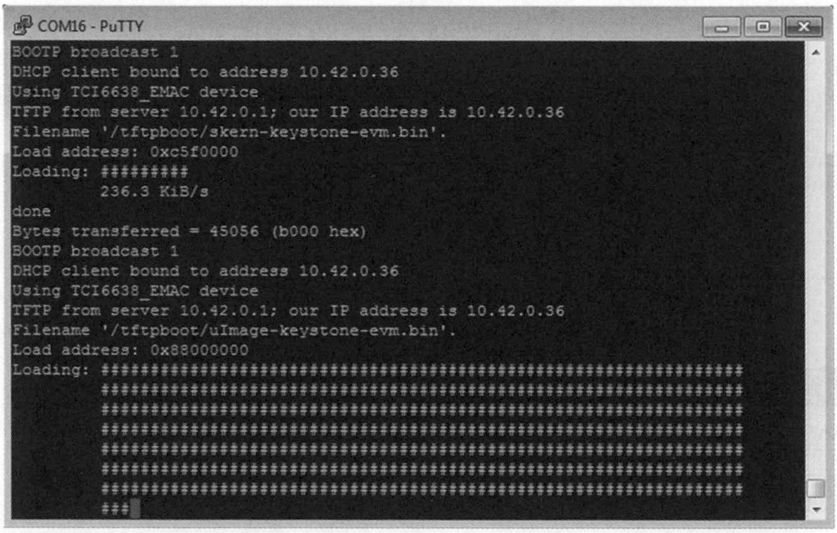

Figure 12.46 KeyStone II EVM booting.

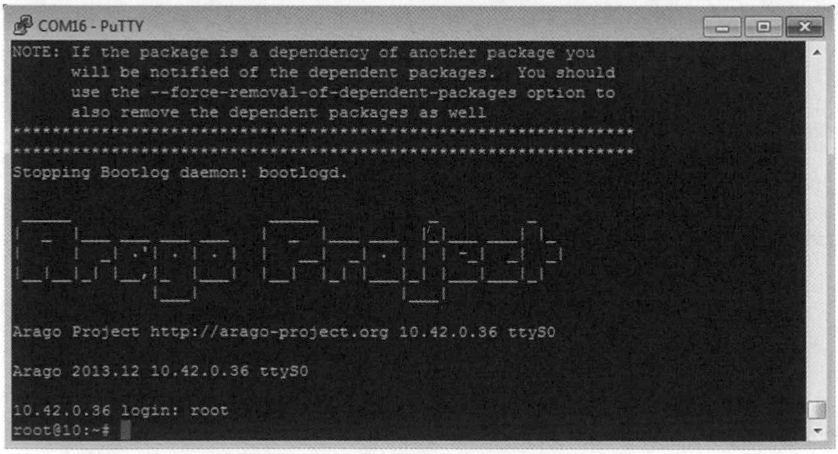

Figure 12.47 Login as **root**.

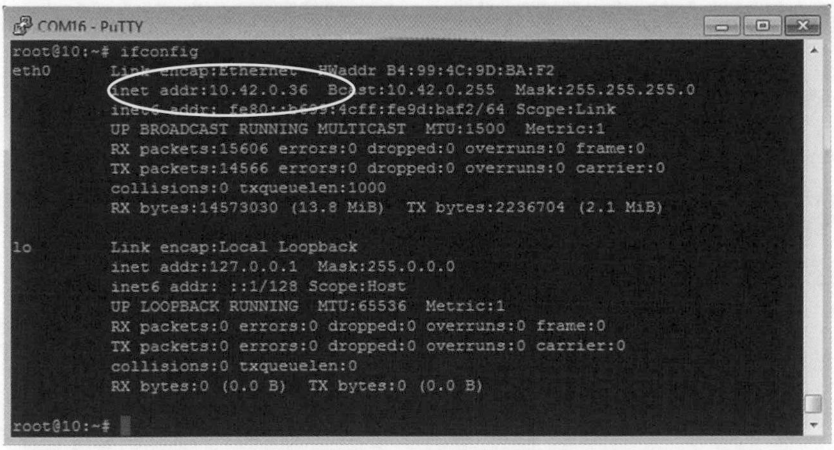

Figure 12.48 Use **ifconfig** to check the IP addresses.

Figure 12.49 Starting FileZilla.

Start FileZilla for transferring files; see Figure 12.49. FileZilla is an easy tool for transferring files between the host and the EVM. Evoke FileZilla, and type the EVM address as shown in Figure 12.50. Right click on the project and select upload.

Select the files to transfer as shown in Figure 12.51. Change the files' properties as shown in Figure 12.52. Run the **dotp** file by typing **./dotp** as shown in Figure 12.53. The output should be as shown in Figure 12.54. Explore the code running on the host (Figure 12.55) and the code running on the target (Figure 12.56).

Explore the makefile named **Makefile**; see Figure 12.57. Locate the **dotp** file on the EVM, run it by typing **./dotp** and observe the output file; see Figure 12.58.

File location:

/home/dotp2_Openmp/dotp_ompacc/ (OpenMP with target)

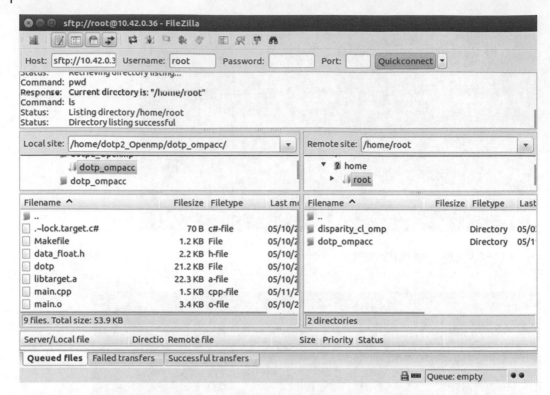

Figure 12.50 Using FileZilla to transfer files from the host to the EVM.

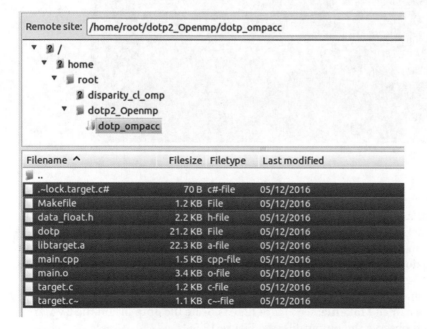

Figure 12.51 Selecting the files to be transferred to the EVM.

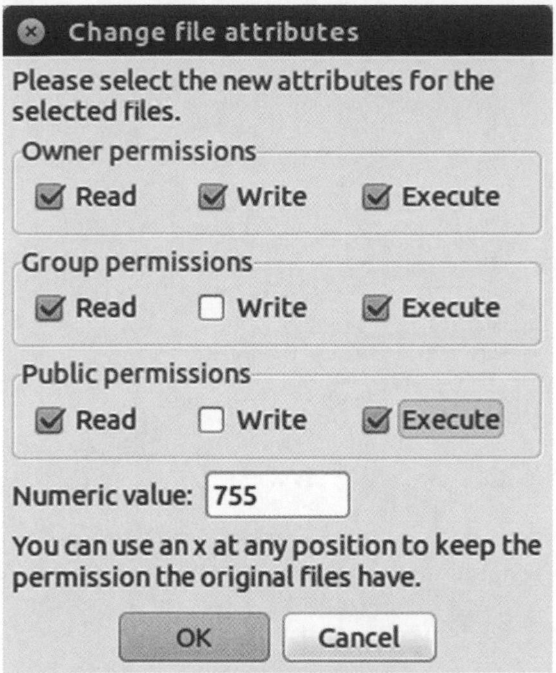

Figure 12.52 Changing the file permission.

Figure 12.53 Running the **dotp** project on the EVM.

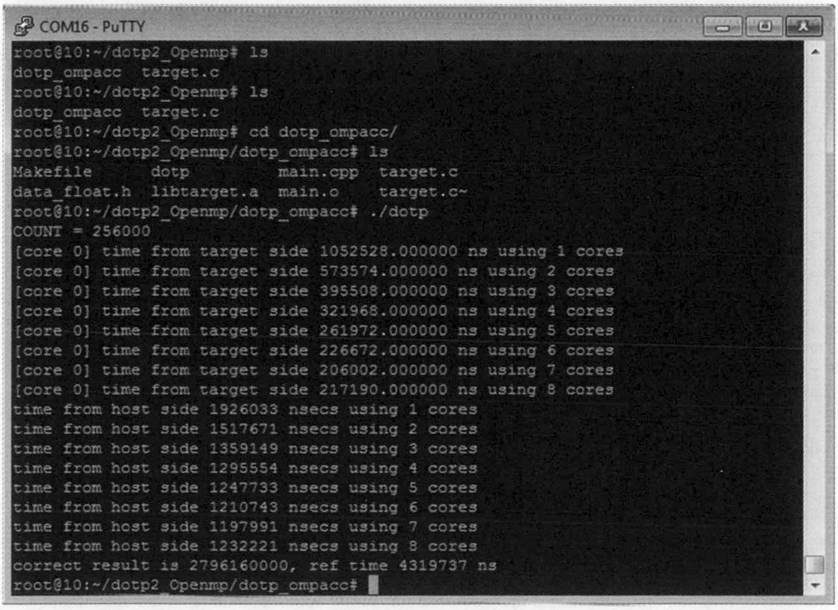

Figure 12.54 Output console showing time consumed when code is running on the target or host.

```cpp
#include <stdio.h>
#include <stdlib.h>
#include <iostream>
#include <fstream>
#include <cassert>
#include <cstdlib>
#include <signal.h>
#include <math.h>


#define COUNT 256000
//#include "data_float.h"
//short a[COUNT] = {A_ARRAY_F};
//short x[COUNT] = {X_ARRAY_F};

extern "C" {
int dotp(short* m, short* n, int count, int ncore);
int dotp1(short* m, short* n, int count, int ncore);
}
static unsigned ns_diff (struct timespec &t1, struct timespec &t2)
{ return (t2.tv_sec - t1.tv_sec) * 1e9 + (t2.tv_nsec - t1.tv_nsec); }


/**********************************************************************
* main
**********************************************************************/
int main()
{

       struct timespec t0, t1;


       double time;
       int result;

       short *a = new short[COUNT];
       short *x = new short[COUNT];

for(int i=0;i<COUNT;i++){
           a[i] = i%256;
           x[i] = 256-a[i];
       }
printf("COUNT = %d\n",COUNT);
dotp1 (a, x, COUNT, 8);


for(int n = 1; n <= 8; n++)
{
       result = 0;
       result = dotp (a, x, COUNT, n);
}


for(int n = 1; n <= 8; n++)
{
       result = 0;
       clock_gettime(CLOCK_MONOTONIC, &t0);
       result = dotp1 (a, x, COUNT, n);
       clock_gettime(CLOCK_MONOTONIC, &t1);
printf("time from host side %d nsecs using %d cores\n",ns_diff(t0, t1),n);
}

       long long sum = 0;
       clock_gettime(CLOCK_MONOTONIC, &t0);
       for(int i=0;i<COUNT;i++){
           sum += a[i]*x[i];
       }
       clock_gettime(CLOCK_MONOTONIC, &t1);
       printf("correct result is %lld, ref time %d ns\n", sum, ns_diff(t0, t1));


       return 0;

}
```

Figure 12.55 Code running on the host.

```
#pragma omp declare target
#include<omp.h>
extern int printf(const char *_format, ...);
#pragma omp end declare target


double dotp(short* m, short* n, int count, int ncore)
{
      double sum2 = 0;

#pragma omp target data map(to:m[0:count],n[0:count],count) map(from: sum2)
      {

#pragma omp target map(to:m[0:count],n[0:count],count) map(from: sum2)
              {
                    int i;
                    int sum = 0;
                    double t0 = omp_get_wtime();
#pragma omp parallel for private(i) shared(m,n,count) reduction(+:sum)
num_threads(ncore)
                    for(i=0;i<count;i++)
                    {
                          sum += m[i]*n[i];
                    }
                    double t1 = omp_get_wtime();
                    sum2 = t1-t0;
      printf("time from target side %f ns using %d cores\n",(t1-t0)*1e9, ncore);
                    printf("dsp sum result  %d\n",sum);
              }
      }
      return sum2;

}


int dotp1(short* m, short* n, int count, int ncore)
{
      int sum = 0;
#pragma omp target data map(to:m[0:count],n[0:count],count) map(from: sum)
      {

#pragma omp target map(to:m[0:count],n[0:count],count) map(from: sum)
              {
                    int i;
                    sum = 0;
#pragma omp parallel for private(i) shared(m,n,count) reduction(+:sum)
num_threads(ncore)
                    for(i=0;i<count;i++)
                    {
                          sum += m[i]*n[i];
                    }
              }
      }
      return sum;

}
```

Figure 12.56 OpenMP code.

12.9 Conclusion

This chapter has shown how to parallelise programs using OpenMP, which is very suitable for multicore architectures like the KeyStone I and II. Many examples have been provided that can be used as foundations for future projects. Although OpenMP has been shown to provide good scalability and performance, this is not the case for all types of code or algorithms; for instance,

```
CL                      = cl6x
OA_SHELL                = clacc
CPP                     = arm-linux-gnueabihf-g++
CC                      = arm-linux-gnueabihf-gcc
OA_SHELL_TMP_FILES      = *.out __TI_CLACC_KERNEL.c *.cl *.asm *.dsp_h *.bc *.objc *.if *.map
*.opt *.int.c *.o *.obj

UNAME_M :=$(shell uname -m)
ifneq (,$(findstring x86, $(UNAME_M)))
 ifeq ($(TARGET_ROOTDIR),)
 $(call crror,ERROR - TARGET_ROOTDIR must be set for cross compiling)
 endif
endif

ifeq ($(TI_OCL_CGT_INSTALL),)
   TI_OCL_CGT_INSTALL = $(TARGET_ROOTDIR)/usr/share/ti/cgt-c6x
endif



OA_TC_OPTS              = -O3 -pm
OA_HC_OPTS              = -O3 -Wall -Wextra -fopenmp -lrt
OA_SHELL_OPTS       = -v --hc="$(OA_HC_OPTS)" --tc="$(OA_TC_OPTS)"
OA_SHELL_LIB_OPTS   = $(OA_SHELL_OPTS) --make_lib

EXE                 = dotp
HOST_CODE           = main.cpp
OBJS                = $(patsubst %.cpp,%.o,$(HOST_CODE))

TARGET_CODE   = target.c
CLACC_LIBS          = $(patsubst %.c,%.a,$(TARGET_CODE))

$(EXE): $(OBJS) $(CLACC_LIBS)
       $(OA_SHELL) $(OA_SHELL_OPTS) $(OBJS) $(addprefix lib, $(CLACC_LIBS)) -o $@


%.a: %.c
       $(OA_SHELL) $(OA_SHELL_LIB_OPTS) $< -o lib$@

%.o: %.cpp
       $(CPP) $(CPP_OPTS) -c $<

clean:
             @rm -f $(EXE) $(addprefix lib, $(CLACC_LIBS)) $(OA_SHELL_TMP_FILES) *.log
```

Figure 12.57 The makefile used **Makefile**.

Figure 12.58 Console output showing the results.

an adaptive filter is not very suitable to be optimised using OpenMP. Also, when the number of iterations is small, the overhead introduced for the communication between cores dominates. Different compilers' vendors implement OpenMP differently, and therefore the same code compiled with different compilers will be very likely to provide different performance.

In terms of implementation, the software setup can be very time consuming. However, a large number of detailed examples have been given for both the KeyStone I and II in order to remedy this. It has also been shown that the KeyStone II is supported with OpenMP device constructs that can make programming very easy by hiding the ARM–DSP communication from the programmer. Combination of OpenMP and OpenCL is also shown in Chapter 13.

References

1 OpenMP, OpenMP Compilers, 2016. [Online]. Available: http://www.openmp.org/resources/openmp-compilers/. [Accessed 16 January 2016].
2 Open MP, The OpenMP API specification for parallel programming, 2016. [Online]. Available: http://www.openmp.org/. [Accessed 16 January 2016].
3 OpenMP in the era of low power devices and accelerators, in *Proceedings of 9th International Workshop on OpenMP (IWOMP)*, Canberra, Australia, 2013.
4 OpenMP, OpenMP Application Program Interface version 4.0, July 2013. [Online]. Available: http://www.openmp.org/wp-content/uploads/OpenMP4.0.0.pdf. [Accessed 16 January 2016].
5 PuTTY, Download PuTTY, [Online]. Available: http://www.putty.org/. [Accessed 16 January 2016].

13

Introduction to OpenCL for the KeyStone II

Multicore DSP: From Algorithms to Real-time Implementation on the TMS320C66x SoC, First Edition. Naim Dahnoun.
© 2018 John Wiley & Sons Ltd. Published 2018 by John Wiley & Sons Ltd.
Companion website: www.wiley.com/go/dahnoun/multicoredsp

13.1 Introduction

In the previous chapters of this book, various programming models (IPC/MessageQ and OpenMP) have been introduced, and it has been shown how they leverage the communications between cores. In this chapter, another programming model called Open Computing Language (OpenCL) is introduced. This chapter will emphasize OpenCL for the KeyStone rather than other devices.

OpenCL has been around for barely a decade (2008); it was first submitted by Apple to the non-profit member-funded Khronos Group [1], a consortium of which the University of Bristol, Texas Instruments (TI), ARM, Xilinx, Altera, AMD, IBM, Qualcomm, Intel, Nvidia and others are part. The complete list of compliant devices from different vendors can be found in Ref. [2].

The main idea for writing an OpenCL application is to be able to execute kernels without having to deal with the details of communication between different cores within a heterogeneous device. OpenCL is the open and royalty-free programming standard for general-purpose computations on heterogeneous systems like the KeyStone II. OpenCL offers an ease-of-use programming model that can improve the performance while maintaining portability by hiding architecture details. However, by knowing the architecture details more performance can be extracted, as demonstrated later in this chapter.

The simplicity of OpenCL comes from the fact that it provides a standard using task-based and data-based parallelisation. In other words, the program to be parallelized is divided into tasks, and the data are divided between tasks. This concept is reiterated in the remaining sections of this chapter.

13.2 Operation of OpenCL

The heterogeneous model used by OpenCL consists of a control processor called a **host processor** and multiple compute processors called **compute devices**; see Figure 13.1. Each compute device may contain one or more **compute unit(s)**. For the KeyStone II device, one ARM core, for example, could be used as the host processor, one compute device could contain the

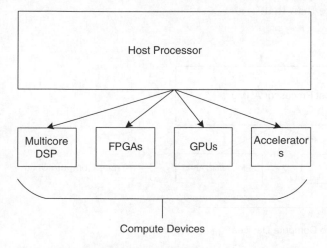

Figure 13.1 OpenCL platform model.

eight DSP cores and the other compute device could contain the other three ARM cores; see Figure 13.2. However, this mode is not supported by TI, but the model shown in Figure 13.4 is supported.

To write an application using OpenCL, two distinct programs have to be written, one for the host (this is the main application) and one for the compute devices (this is the program that one wants to run in parallel on the DSPs or other compute units to accelerate the execution). The program or function to be computed in a compute unit is called a **kernel** and is declared in the OpenCL C program. The application runs on the host and submits kernels to the compute units via command queues, as shown in Figure 13.3.

In order to introduce OpenCL in a simple and concise manner, clear definitions and examples will be necessary. The *host code* is responsible for setting up and managing the execution of kernels on OpenCL devices and performing other tasks that are required by the application.

Figure 13.5 illustrates how an application using OpenCL is executed. In this example, the application starts running **task_1** on the ARM core, and when completed, it initializes some buffers that the DSP cores will be using. This is achieved by the ARM using the command **enqueueWriteBuffer()**. When the memory is initialized, the ARM core starts issuing work items to the DSPs by using the command **enqueueNDRangeKernel()**. The DSPs then start executing the code (DSP tasks). When the ARM finishes issuing the work items, the ARM starts processing **task_2** while the DSP may still be processing the data. As shown in Figure 13.5, the ARM and the DSP in this case are running in parallel. When the ARM finishes **task_2**, it will be instructed to wait for the DSP to complete. When the DSP completes, it will write the data to memory, and the ARM is instructed to read the data using the command **enqueueReadBuffer()**. Finally, when the ARM completes reading the data, it will proceed to executing **task_3**.

Before proceeding, it is important to understand what is meant by work items, workgroups, workgroup IDs, local work-item IDs and global work-item IDs. Figure 13.6, which shows an example that performs a **dotp** function as a kernel, illustrates these definitions:

Work item. When creating a kernel, a number of work items are created. Each *work item* executes the same kernel, and each kernel may or may not use the same data. In other words, a work item is a code to be executed by a processing unit (a DSP core). Figure 13.6 shows that a work item will perform a simple multiplication of two variables ($ai \times xi$). Of course, a work item can be programmed to do more than that.

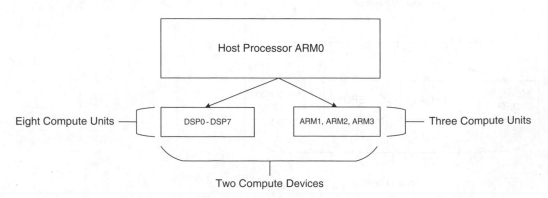

Figure 13.2 Host, compute devices and compute units.

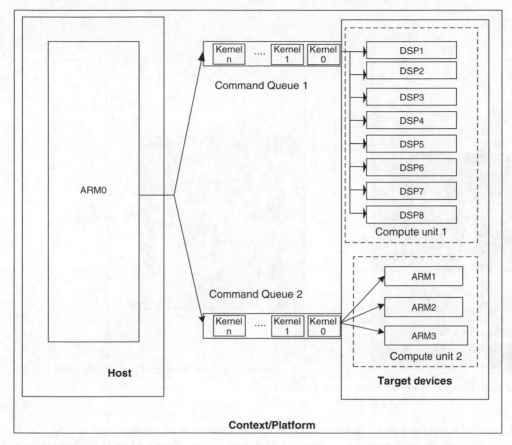

Figure 13.3 Operation of OpenCL.

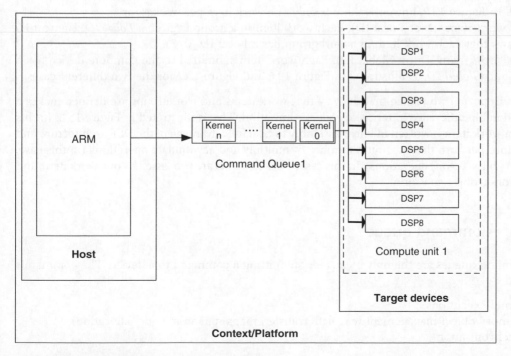

Figure 13.4 Context/platform used in this chapter.

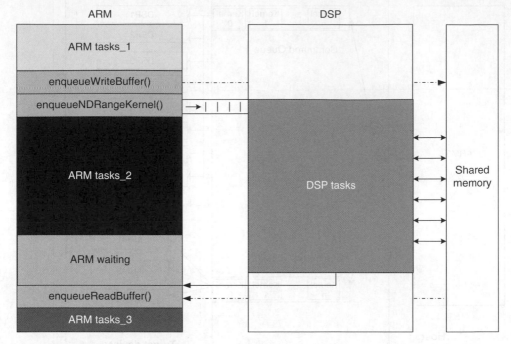

Figure 13.5 Example of and application using OpenCL.

Workgroups. A *workgroup* is a group of work items executing in a compute unit (a DSP core). Figure 13.6 shows workgroups consisting of four work items. The number of work items in a group will depend on the application. Figure 13.6 shows that four work items are executed in one workgroup on the same DSP core, but the execution of these four work items is sequential. In Figure 13.7, however, there is only one work item per workgroup.

Local work-item ID (or Local ID). Each work item in a group is given a *Local ID*. Figure 13.6 shows that each work item in a workgroup has a Local ID (0–3).

Global work-item ID (or Global ID). Each work item submitted to the compute device has a unique *Global ID*, as illustrated in Figure 13.6 and Figure 13.7 for the two different cases.

It is important to realize that the way that work items and workgroups are defined makes a considerable difference in terms of code optimisation. This is illustrated in Figure 13.8. In this case, a work item is written in linear assembly to take full advantage of the DSP architecture and therefore perform the maximum number of multiply and accumulate operations (in this case, $4 \times$ (16-bit $\times$ 16-bit) operations). In this case, 4×4 elements are processed by one work item, and the workgroup size is one.

13.3 Command queue

Command queues are the only means for submitting a command to a device. The commands include:

- Kernel executions
- Memory object management (e.g. data transfer, mapping and memory allocation)
- Synchronisation.

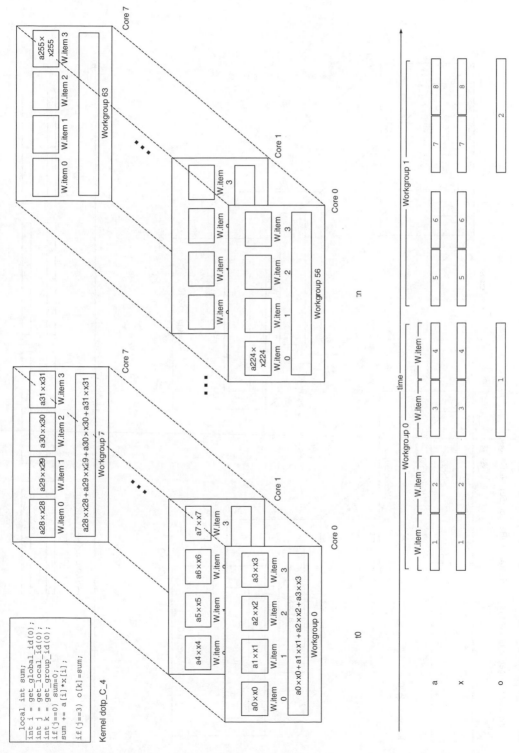

Figure 13.6 Example 1: Illustration of work item and workgroups.

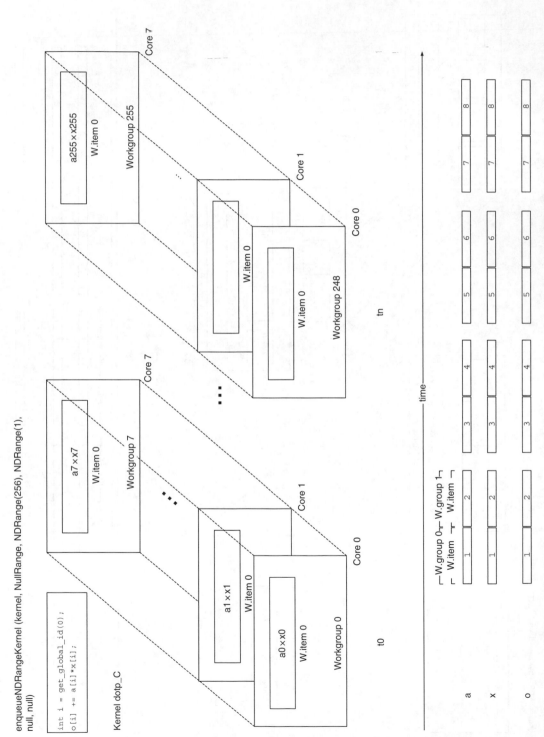

Figure 13.7 Example 2: Illustration of work item and workgroups.

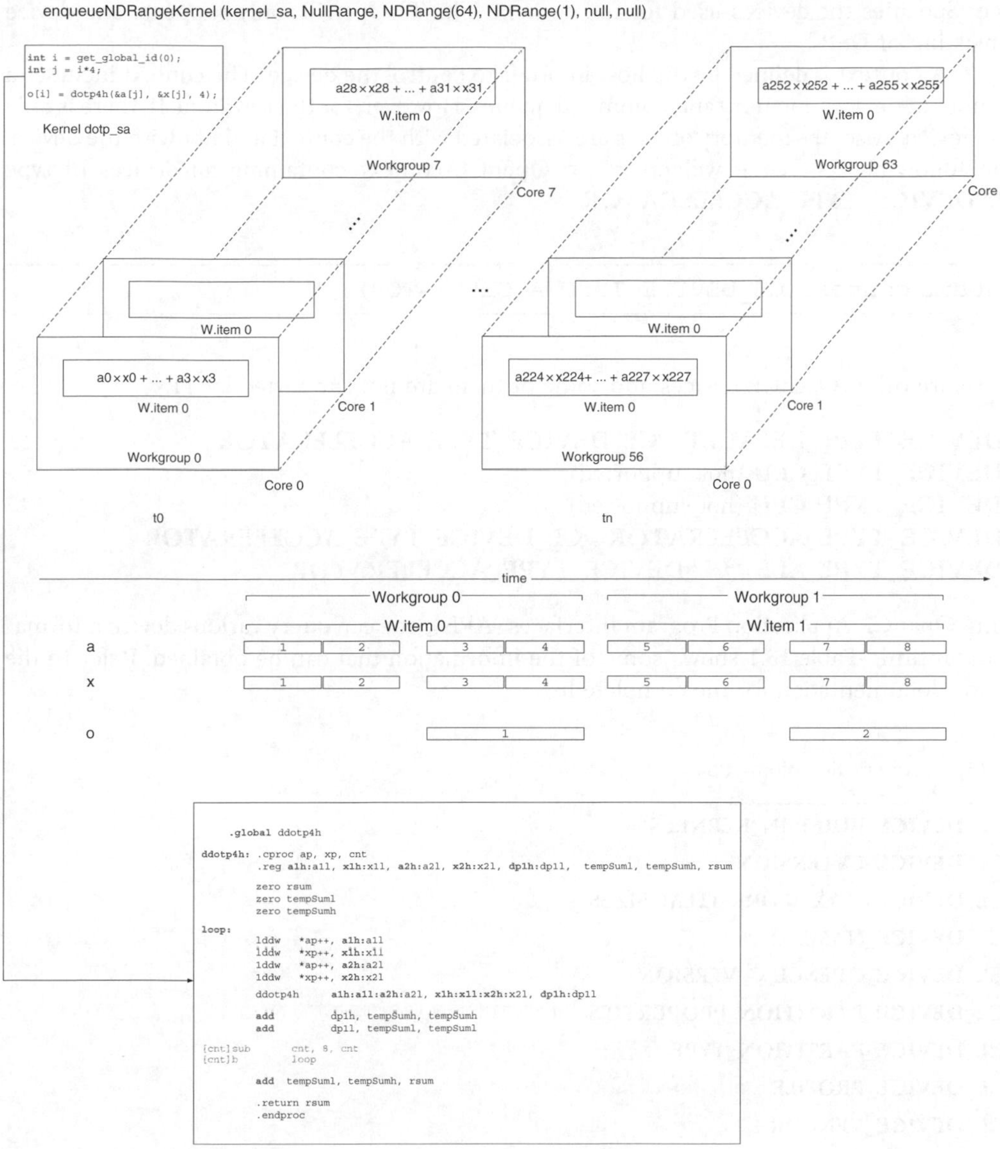

enqueueNDRangeKernel (kernel_sa, NullRange, NDRange(64), NDRange(1), null, null)

Figure 13.8 Example 3: Illustration of work item and workgroups.

13.3.1 Creating a command queue

Commands sent to a command queue are queued in order, but may be set up for out-of-order execution depending on the attributes specified when the command queue is created. The syntax for creating a command queue is shown here:

```
CreateCommandQueue(cl_context context,
                   cl_device_id device,
                   cl_command_queue_properties properties,
                   cl_int *errcode_ret)
```

device. Specifies the devices used for the context. The TI OpenCL only supports one device consisting of DSPs.

context. A context is defined by the host in order to control the device. The context includes a number of devices, memory and command queues. However, for the KeyStone II, there is only one device. Also, the memory objects are associated with the context and not with the device. The following C++ code will create an OpenCL context containing all devices of type **CL_DEVICE_TYPE_ACCELERATOR**.

```
Context context(CL_DEVICE_TYPE_ACCELERATOR);
```

There are other OpenCL devices and some of them are not supported by TI:

CL_DEVICE_TYPE_DEFAULT = **CL_DEVICE_TYPE_ACCELERATOR**
CL_DEVICE_TYPE_CPU (not supported)
CL_DEVICE_TYPE_GPU (not supported)
CL_DEVICE_TYPE_ACCELERATOR = **CL_DEVICE_TYPE_ACCELERATOR**
CL_DEVICE_TYPE_ALL = **CL_DEVICE_TYPE_ACCELERATOR**

Using OpenCL Application Program Interfaces (APIs), one can query various device information at runtime. Table 13.1 shows some of the information that can be obtained. Refer to the Khronos documentation for the complete list.

Table 13.1 Some device information

1. **CL_DEVICE_BUILT_IN_KERNELS**
2. **CL_DEVICE_EXTENSIONS**
3. **CL_DEVICE_MAX_WORK_ITEM_SIZES**
4. **CL_DEVICE_NAME**
5. **CL_DEVICE_OPENCL_C_VERSION**
6. **CL_DEVICE_PARTITION_PROPERTIES**
7. **CL_DEVICE_PARTITION_TYPE**
8. **CL_DEVICE_PROFILE**
9. **CL_DEVICE_VENDOR**
10. **CL_DEVICE_VERSION**

The following code queries the created context and prints various context information:

```
// using C++
Context context(CL_DEVICE_TYPE_ACCELERATOR);
std::vector<Device> devices = context.
getInfo<CL_CONTEXT_DEVICES>();
int  nDev  = devices.size();
printf("Number of Devices: %d\n",nDev);
char s[100];
int m;
devices[0].getInfo(CL_DEVICE_NAME,&s);
```

```
printf("DEVICE_NAME: %s\n",s);
devices[0].getInfo(CL_DEVICE_VENDOR,&s);
printf("DEVICE_VENDOR: %s\n",s);
devices[0].getInfo(CL_DEVICE_MAX_COMPUTE_UNITS,&m);
printf("DEVICE_MAX_COMPUTE_UNITS: %d\n",m);
devices[0].getInfo(CL_DEVICE_MAX_CLOCK_FREQUENCY,&m);
printf("DEVICE_MAX_CLOCK_FREQUENCY: %d\n",m);
```

The output is:

```
Number of Devices: 1
DEVICE_NAME: TI Multicore C66 DSP
DEVICE_VENDOR: Texas Instruments, Inc.
DEVICE_MAX_COMPUTE_UNITS: 8
DEVICE_MAX_CLOCK_FREQUENCY: 1228
```

13.3.1.1 Command-queue properties

There are three command-queue properties:

1) **In-order** is the default, and the switch docs not need to be specified. In this case, the commands execute in the order they are sent. The next command can execute only when the previous command has completed and the memory written.
2) **CL_QUEUE_OUT_OF_ORDER_EXEC_MODE_ENABLE** specifies that commands in the command queue will be executed out of order. In this case, there is no guaranteed order for the completion of the command; see Example 13.1.
3) **CL_QUEUE_PROFILING_ENABLE** specifies that commands in the command queue will enable profiling.

Example 13.1

```
// Create command queue using the context, device and Out-of-order
// property
CommandQueue Q (context, devices[0],
CL_QUEUE_OUT_OF_ORDER_EXEC_MODE_ENABLE);
```

Example 13.2

```
// Create command queue using the context, device and profiling
// Property
CommandQueue Q (context, devices[0], CL_QUEUE_PROFILING_ENABLE);
```

Example 13.3

```
// Create command queue using the context, device and both Properties
CommandQueue Q (context, devices[0],
CL_QUEUE_OUT_OF_ORDER_EXEC_MODE_ENABLE) |
CL_QUEUE_PROFILING_ENABLE);
```

Example 13.4

```
// Create command queue using the context and device
CommandQueue Q (context, devices[0]);
```

If a property is not specified, the operation will be ignored. For instance, in Example 13.1 the profiling will be ignored, whereas in Example 13.4 the queue will execute in order and no profiling will be used.

Each command queue should point to only a single device; see Figure 13.3. For the case of the KeyStone II, where there is only one device (DSP), there will be only one queue. The command sent to the command queue will specify how many work items (kernels) to be executed before it proceeds to the next instruction or command in the host.

13.3.2   Enqueueing a kernel

To execute the kernel, a command has to be sent to a queue (i.e. enqueueing a kernel). Two APIs are available:

1) **enqueueNDRangeKernel()**
2) **enqueueTask()**.

The **enqueueTask()** is similar to **enqueueNDRangeKernel()** but always has:

Offset = 0
Global size = local size = 1.

For the **enqueueNDRangeKernel()**, the syntax is:

```
Q.enqueueNDRangeKernel (
const Kernel& kernel,
const NDRange& offset,
const NDRange& global,
const NDRange& local,
const VECTOR_CLASS<Event> * events = NULL, // See Section 13.7
Event * event = NULL)  // See Section 13.7
)
```

```
Q.enqueueNDRangeKernel(kernel_C, NullRange, NDRange(), NDRange(),
NULL, NULL);
```

where the **commandqueue** is defined as:

```
CommandQueue    Q    (context, devices[0]);
```

and the parameters are defined as:

kernel_C. The kernel object
NullRange. The data offset (NullRange = 0)
NDRange() (third parameter). The number of global items (i.e. the total number of work items)

NDRange() (fourth parameter). The number of work items in a group
NULL (fifth parameter). Used for synchronisation (see Section 13.7)
NULL (sixth parameter). Used for synchronisation (see Section 13.7).

For example:

```
Q.enqueueNDRangeKernel(kernel_C, NullRange, NDRange(COUNT),
NDRange(1), NULL, NULL);
```

13.4 Kernel declaration

A kernel has to be written in the following formats.
  On the device, the kernel should be declared as:

```
__kernel void dotp_C()
{
//do something eg. Dotp() , FIR() etc.
}
```

On the host, the kernel should be declared as:

```
Kernel kernel_C(program, "dotp_C");
```

13.5 How do the kernels access data?

To illustrate how data are accessed by a kernel, an example with a *dot product* is shown here.
Within a kernel, one can get the global work-item ID to retrieve the appropriate data to process.
This can be achieved by using the **get_global_id(0)** instruction. This global ID is used to access
data. In general, data are divided as shown in Figure 13.9, and the work item can be used as
shown here:

```
// kernel for performing a multiplication of 16-bit by 16-bit
  __kernel void dotp_C(__global short* a, __global short* x, __global
    int* o)
{
    int i = get_global_id(0); // reading the global id of this kernel.
    // 0 :dimension 1, 1 is dimension 2 and 2 is dimension 3.
    o[i] = a[i]*x[i];
}
```

  A complete application is developed in the laboratory experiment shown in Sections 13.11.1
and 13.11.2.

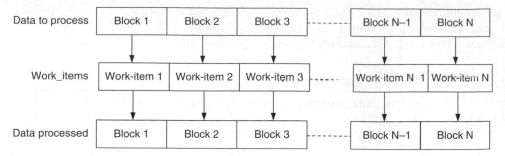

Figure 13.9 How data are divided amongst work items.

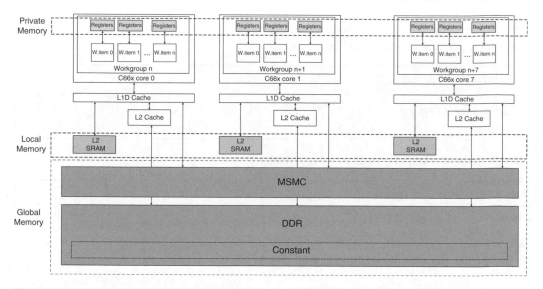

Figure 13.10 KeyStone II memory map definition.

13.6 OpenCL memory model for the KeyStone

There are four memory regions defined for OpenCL implementation on the KeyStone; see Figure 13.10. These memory regions are:

Global memory. This memory region, which is located in the DDR or the MSMC memory, is where the program and the kernel(s) will reside. As shown in Figure 13.10, this memory is visible to all cores (compute units). The address space qualifier __global is used to specify that data reference by a pointer residing in the global memory. For example, the following OpenCL code declares three pointers (**inputx**, **inputy** and **output**) that point to locations that reside in the global memory.

```
__kernel product(
    __global float *inputx,
    __global float *inputy,
    __global float *output,
    const unsigned int count)
```

```
{
    int i = get_global_id(0);
    if (i < count)
        output[i] = inputx[i] * inputy[i];
}
```

Constant memory. Data that never change (i.e. data that remain unchanged during the execution of a kernel and the program code) are declared in this memory region. The address space qualifier __**constant** is used to map data to the constant memory; see previous OpenCL code and Figure 13.10.

Local memory. This memory is located in the L2 memory. When performance is very critical, one may want the kernel to use the local memory. To achieve high performance, whenever possible keep the data in the local memory. The address space qualifier __**local** is used to map data to the local memory; see the OpenCL code here and Figure 13.10:

```
__kernel void dotp_C_4(__global short* a, __global short* x, __global
    int* o)
{
        __local int sum;
        int i = get_global_id(0);
        int j = get_local_id(0);
        int k = get_group_id(0);
        if(j==0) sum=0;
        sum += a[i]*x[i];
        if(j==3) o[k]=sum;
}
```

Private memory. Private memory is the memory that is accessible only to the specific work items. For the KeyStone II, it is the core registers as shown in Figure 13.10. When defining a local variable in a kernel, by default it is declared as private. However, since the private memory is very small (64 registers per core), if arrays or extra registers are declared as local to the kernel, they will be mapped into the private L2 SRAM automatically.

It's worth noting at this stage that the host memory management is explicit, and therefore it is up to the programmer to transfer data from the host and the local memory (the host on its own cannot access local memory).

13.6.1 Creating a buffer

Before creating a buffer, let's define a buffer object in order to remove any ambiguity that may arise later. *Buffer objects* store data that can be accessed by the host using OpenCL calls and accessed by the device using the pointer within the kernel. A buffer object contains the size of the buffer, the data and the properties describing how the buffer is going to be used, as described further here. Data location and sharing between the host and the device are very important when communication speed and memory size need to be optimized.

When creating a buffer, not only the memory location but also the memory operation are specified at the same time (by specifying the **cl_mem_flags**). The **clCreateBuffer** is the main function for allocating buffer objects of a given size.

To specify a buffer in OpenCL, the following prototype is used:

```
Buffer(const Context& context, cl_mem_flags flags, size_t size,
       void* host_ptr = NULL, cl_int* err = NULL);
```

1) **context**. The context was defined in Section 13.3.1.
2) **cl_mem_flags**. This flag defines the property of the memory and specifies how the kernel and the host access this memory. There are seven flags that can be combined except that 1, 2 and 3 are exclusive; for more details, refer to Ref. [3]. See below for a detailed description and usage of the available flags.
3) **host_ptr**. A pointer to the buffer data that have already been allocated by the application.
4) **Err**. if NULL, no error is returned.

13.6.1.1 Cl_mem_flags

1) **CL_MEM_READ_WRITE**. This kernel can read from and write to memory. This is the default flag if no flag is specified.
2) **CL_MEM_WRITE_ONLY**. This kernel can only write to memory.
3) **CL_MEM_READ_ONLY**. This kernel can only read from memory. There is no cache coherency between the ARMs and DSPs, but when using OpenCL, the programmer does not need to deal with coherency as it is maintained by the OpenCL runtime. However, to increase the performance, one should use **CL_MEM_WRITE_ONLY** or **CL_MEM_READ_ONLY** if possible instead of **CL_MEM_READ_WRITE** in order to avoid some unnecessary coherency operations.
4) **CL_MEM_USE_MSMC_TI**. This is a TI extension to OpenCL for the KeyStone. It specifies that the memory will be located in the MSMC; see Figure 13.10.
5) **CL_MEM_USE_HOST_PTR**. The memory allocated by the host using **malloc**, **calloc** or other similar functions will not be accessible by the device as illustrated in Figure 13.11. However, when using the switch **CL_MEM_USE_HOST_PTR**, the buffer will be copied from the host memory to the device memory if the memory was located in the host as shown in Figure 13.12, or use a pointer to access the memory if the buffer was allocated in the device memory as shown in Figure 13.13. See Laboratory experiment 3 in Section 13.11.3.
6) **CL_MEM_ALLOC_HOST_PTR (default)**. This flag specifies that both the host and device will share the same memory, and therefore no memory copy is required. However, the buffer needs to be mapped into the host before the host can access it. From Figure 13.14, we can see that the host will be using the **L** pointer and the kernel will be using the **LBuff** that has been passed by the following instruction:

```
kernel_name.setArg(0, LBuff);
```

See Laboratory experiment 4 in Section 13.11.4.
7) **CL_MEM_COPY_HOST_PTR**. When using **CL_MEM_READ_ONLY|CL_MEM_CO-PY_HOST_PTR**, OpenCL will create a memory buffer in the device that will be read-only by the kernel, and the host memory is copied to this memory that has been created; see

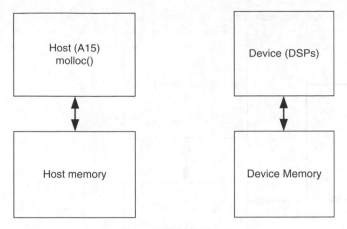

Figure 13.11 Memory not accessible by the device.

Using CL_MEM_USE_HOST_PTR

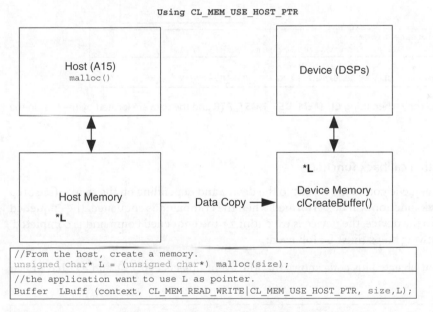

```
//From the host, create a memory.
unsigned char* L = (unsigned char*) malloc(size);
```
```
//the application want to use L as pointer.
Buffer  LBuff (context, CL_MEM_READ_WRITE|CL_MEM_USE_HOST_PTR, size,L);
```

Figure 13.12 Data copy when using **CL_MEM_USE_HOST_PTR** and the data are allocated in the host.

Figure 13.15. **CreateBuffer** and **EnqueueWriteBuffer** can be combined into a single command using the **CL_MEM_COPY_HOST_PTR** flag. See Laboratory experiment 5 in Section 13.11.5.

13.7 Synchronisation

Synchronisation can happen at the host level or device level. Synchronisation between work items is possible only within workgroups, and barriers or memory fences can be used within work items. For the host synchronisation, the following possibilities are available.

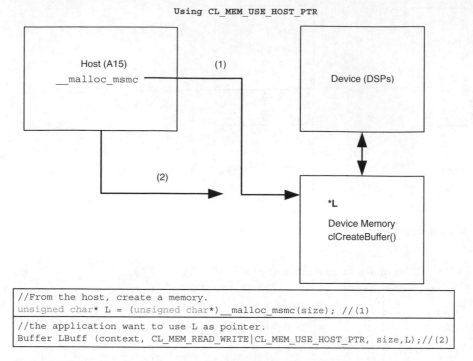

Using **CL_MEM_USE_HOST_PTR**

```
//From the host, create a memory.
unsigned char* L = (unsigned char*)__malloc_msmc(size); //(1)
```
```
//the application want to use L as pointer.
Buffer LBuff (context, CL_MEM_READ_WRITE|CL_MEM_USE_HOST_PTR, size,L);//(2)
```

Figure 13.13 No data copy when using **CL_MEM_USE_HOST_PTR** and the data are located by the host in the device memory.

13.7.1   Event with a callback function

In this case, when a queue command is sent to the device and depending on the switch (see Step 3 below), a **callback** function will be performed when the command is enqueued, the enqueued command is sent to the device, the device is executing or the enqueued command is completed. To do so, the following steps must be followed:

1) Declare an event object. This code shows that the event objects **ev0**, **ev1** and **ev2** have been declared:

```
Event ev0,ev1,ev2;
```

2) Set the event object to be used in the command queue: in this case, **ev0** is used.

```
Q.enqueueNDRangeKernel(kernel_C, NullRange, NDRange(COUNT),
NDRange(1), NULL, &ev0);
```

3) Hook the **callback** function to the event, and specify when the **callback** function is called.

```
ev0.setCallback(CL_COMPLETE,&mycallback); // 0 = when the previous
//command is completed.
```

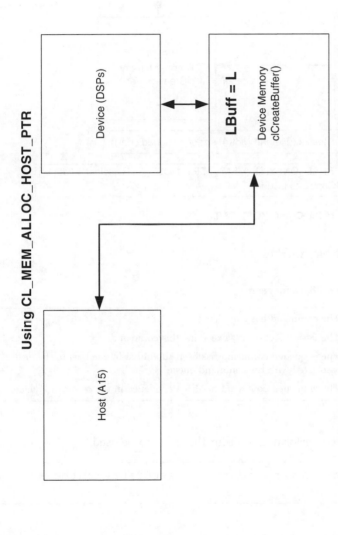

Using CL_MEM_ALLOC_HOST_PTR

Host (A15)

Device (DSPs)

LBuff = L
Device Memory
clCreateBuffer()

```
// the Buffer is created to be shared but data cannot be access with the pointer buffer, so mapping is required.
Buffer LBuff (context,  CL_MEM_READ_ONLY|CL_MEM_USB_MSMC_TI|CL_MEM_ALLOC_HOST_PTR, size);
```

```
// map the region of Lbuff into the host address space and return this mapped region.
unsigned char* L =(unsigned char*)Q.enqueueMapBuffer(LBuff,  CL_TRUE,  CL_MAP_WRITE,  0,  size);
```

Figure 13.14 Using CL_MEM_ALLOC_HOST_PTR.

Using CL_MEM_COPY_HOST_PTR

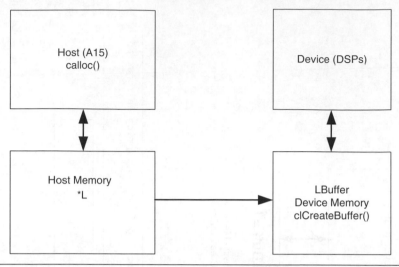

unsigned char* L = (unsigned char*)calloc((Height*Width), sizeof(unsigned char));

Buffer LBuff (context, CL_MEM_READ_ONLY|CL_MEM_COPY_HOST_PTR, Height*Width,L);
//CL_MEM_COPY_HOST_PTR: Copy L to LBuffer

Figure 13.15 Data copy using **CL_MEM_COPY_HOST_PTR**.

Table 13.2 shows the available switches.

Table 13.2 Different levels for the **callback** function

define CL_COMPLETE 0	The command has completed.
define CL_RUNNING 1	The device is currently executing this command.
define CL_SUBMITTED 2	The enqueued command has been submitted by the host to the device associated with the command queue.
define CL_QUEUED 3	The enqueued command has been enqueued in the command queue.

4) Now that the setup has been performed, wait for the event to be used.

```
ev0.wait(); //wait here until the command //associate with the event ev0
//is completed.
```

5) This step is not required but is used for debugging. It returns a value **m** to specify the status of the command queue.

```
ev0.getInfo(CL_EVENT_COMMAND_EXECUTION_STATUS,&m);
//Returns the execution status of the command identified by event. See
//Step 3 for the valid values.
```

6) Writing a **callback** function:

```
void mycallback(cl_event event, cl_int status, void *user_data)
{
printf("[kernel callback] finish\n");
}
```

See Laboratory experiment 6 in Section 13.11.6.

13.7.2   User event

In this case, one can set an event manually before a command queue can start. In this case, there are four steps to follow.

1) Declare the event:

```
evs[0] = UserEvent(context); // in this case we have only one event.
```

2) Specify that the command queue will be waiting for an event before a command queue is enqueued; this is highlighted in the next command queue.

```
Q.enqueueNDRangeKernel(kernel_C, NullRange, NDRange(COUNT),
NDRange(1), (std::vector<cl::Event>*)&evs, &ev);
```

3) Hook the **callback** function to the event, and specify when the **callback** function is called.

```
ev.setCallback(0,&mycallback); // 0 = The callback function is called
//when the command has completed
```

4) Set the event manually to complete.

```
evs[0].setStatus(CL_COMPLETE); // the user set the event to complete
```

See Laboratory experiment 6 in Section 13.11.6.

13.7.3   Waiting for one command or all commands to finish

To wait for one command to complete, use the **wait()** API. For example:

```
ev.wait() // ev: event.
```

To wait for all commands to complete, use the **finish()** API. For example:

```
Q.finish() // Q: commandqueue.
```

See Laboratory experiment 6 in Section 13.11.6.

13.7.4   wait_group_events

On the device side, one can use the **wait_group_events** API to wait for data to be copied. This has to be used in conjunction with the **async_work_group_copy()** API. The **async_work_group_copy()** API is used to transfer data from one memory location to another, and when the transfer is completed an event is generated.

In the following code example, the kernel saves the processed data in a **temp[]** array that is declared with **__local**. When the loop is completed, the data will be transferred from the local memory to the slower global memory by using the **async_work_group_copy()** API that generates an event when the data are completed.

```
kernel void dotp_loop( __global short* a, __global short* x, __global
short* output, __local short* temp)
{
        int i = get_global_id(0);
        int j;
        for(j=0;j<256;j++)
        {
                temp[j] = a[i*256+j]*x[i*256+j];
        }
        event_t ev = async_work_group_copy(&output[i*256],
        &temp[0], 256, 0);
        wait_group_events(1,&ev);
}
```

See Laboratory experiment 7 in Section 13.11.7.

13.7.5   Barrier

The barrier allows synchronisation of work items within a workgroup. Figure 13.16 shows an example of a command queue with four work items and one group that have been sent. On the kernel side, it shows an example without a barrier; and the output is **0,0,0,1,1,1,2,2,2,3,3,3**, as expected, since each work item executes to completion. However, when a barrier is used (see Figure 13.17), when each work item reaches a barrier it will wait until all other work items (three of them in this example) are completed. So, Work item 0 will perform Loop 1, then wait for the other work items to complete. When all the work items are executed (but not completed) to the barrier, then Work item 0 can proceed to Loop 2 and the process continues. The output is shown in Figure 13.17.

See Laboratory experiment 8 in Section 13.11.8.

13.8   Basic debugging profiling

From a programmer's point of view, debugging and profiling can be performed at the host level or device level. On the market, there are very sophisticated OpenCL debugging and profiling tools available, such as the OpenCL Code Builder from Intel (covering only Intel devices), and gDEBugger from AMD. However, for the KeyStone, the tools are not as sophisticated. On the ARM side one can use **gdb**, and on the device side one can debug a kernel with the **gdbc6x** hosted debugger.

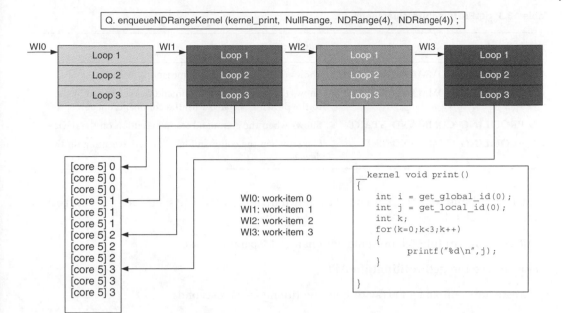

Figure 13.16 No barrier is used.

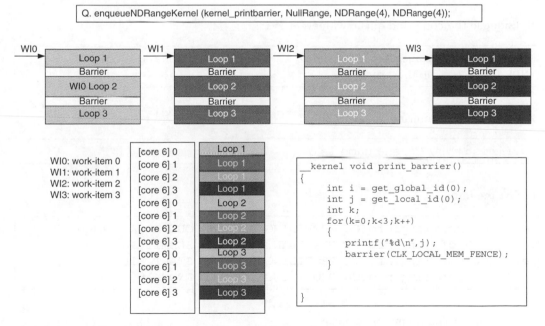

Figure 13.17 A barrier is used.

For basic debugging and profiling, one can use the **printf** or **getProfilingInfo** API. However, as discussed in the debugging chapter (Chapter 10), these will be intrusive. **printf** can be used to debug the application running on the ARM as well as the kernel running on the DSP. When using the **printf** in the kernel, the output of the **printf** will be redirected to the host automatically [4]. By using the **getProfilingInfo** API, one can obtain the time it takes a command queue to be

Table 13.3 getProfilingInfo parameters

Parameter	Description
CL_PROFILING_COMMAND_QUEUED	Shows the time when the command is enqueued
CL_PROFILING_COMMAND_SUBMIT	Shows the time when the command was submitted by the host to the device associated with the command queue
CL_PROFILING_COMMAND_START	Shows when the command starts execution on the device
CL_PROFILING_COMMAND_END	Shown when the command has finished executing on the device

enqueued, submitted, started or ended as shown in Table 13.3. The syntax for the **getProfilingInfo** API is as follows:

cl_int **getProfilingInfo** (cl_profiling_info name, T *param) const

How to use the **getProfilingInfo** API.

1) Create an event and an array to hold the timing in nanoseconds:

```
Event ev1 // creating and event ev1
cl_ulong lt [4];
```

2) Enqueue the command queue with event **ev1**:

```
Q.enqueueNDRangeKernel(kernel_C, NullRange, NDRange(COUNT),
NDRange(1), NULL, &ev1);
```

3) Wait for the command queue to finish:

```
Q.finish(); // queue name is Q
```

4) Get the info:

```
ev1.getProfilingInfo(CL_PROFILING_COMMAND_QUEUED,&lt[0]);
ev1.getProfilingInfo(CL_PROFILING_COMMAND_SUBMIT,&lt[1]);
ev1.getProfilingInfo(CL_PROFILING_COMMAND_START,&lt[2]);
ev1.getProfilingInfo(CL_PROFILING_COMMAND_END,&lt[3]);
```

5) Print the timing (in microseconds):

```
printf("kernel1 queue %0.0f us\n",(int)(lt[0]-lt[0])/1e3);
printf("kernel1 submit %0.0f us\n",(int)(lt[1]-lt[0])/1e3);
printf("kernel1 start %0.0f us\n",(int)(lt[2]-lt[0])/1e3);
printf("kernel1 end %0.0f us\n",(int)(lt[3]-lt[0])/1e3);
```

Note: The timing printed is from the time the command queue is submitted.
See Laboratory experiment 9 in Section 13.11.9.

13.9 OpenMP dispatch from OpenCL

OpenMP code can be used for kernels running on the ARM cores or on the DSPs. Both scenarios are shown in this section.

13.9.1 OpenMP for the kernel code

Writing an OpenMP code with OpenCL is quite simple. The kernel can call an OpenMP code. In addition to what has been covered in this chapter, an OpenMP code has to be written separately (to be compiled separately). The call to an OpenMP code from the kernel is shown here:

```
long long dotp_ompfunction(short* a, short* x, int count)
{
  int i;
  long long sum=0;
  #pragma omp parallel for private(i) reduction(+:sum) shared
  (count,a,x) num_threads(8)
  for(i=0;i<count;i++){
    sum += a[i]*x[i];
  }
  return sum;
}
__kernel void dotp_omp(__global short* a, __global short* x, __global
long *result, int count)
{
        result[0]=dotp_ompfunction(a,x,count);
}
```

The **makefile** should include all compilers as shown here:

```
CPP   = g++ # compiler for the ARM
CLOCL = clocl # compiler for OpenCL
CL6X  = cl6x -mv6600 -abi=eabi $(DSP_INCLUDE) # compiler for the DSP
```

More details can be found in Ref. [5]. See Laboratory experiment 10 in Section 13.11.10.

13.9.2 OpenMP for the ARM code

One can use OpenMP and OpenCL on the ARM. The host code should include **#include "omp. h"** and the **makefile** should include **-fopenmp**. This code shows both the OpenMP code on the ARM and the kernel:

```
#include "omp.h"
#pragma omp parallel default(shared)
{
        int tid = omp_get_thread_num();
        for(int i = 0; i < COUNTPC; i++)
        {
                a[tid*COUNTPC+i]=i;
                x[tid*COUNTPC+i]=255-i;
        }
```

```
#pragma omp critical
{
        Q.enqueueWriteBuffer(aBuff, CL_FALSE, 0, COUNTPC*sizeof
        (short), &a[tid*COUNTPC], NULL, NULL);
        Q.enqueueWriteBuffer(xBuff, CL_FALSE, 0, COUNTPC*sizeof
        (short), &x[tid*COUNTPC], NULL, NULL);
        Q.enqueueNDRangeKernel(kernel_C, NullRange,
        NDRange(COUNTPC), NDRange(1), NULL, NULL);
        Q.enqueueReadBuffer (oBuff, CL_TRUE, 0, COUNTPC*sizeof
        (short), &o[tid*COUNTPC], NULL, NULL);
        printf("core %d finished\n",tid);
}
}
__kernel void dotp_C(__global short* a, __global short* x, __global
  short* o)
{
        int i = get_global_id(0);
        o[i] = a[i]*x[i];
}
```

See Laboratory experiment 11 in Section 13.11.11.

13.10   Building the OpenCL project

To build the project, a **makefile** is written for each application and it is included with source code. The following **makefile** is an example.

/home/Chapter_OpenCL/lab1_helloworld/makefile

```
DSP_INCLUDE = -I$(TI_OCL_CGT_INSTALL)/include
DSP_INCLUDE += -I$(TARGET_ROOTDIR)/usr/share/ti/cgt-c6x/include
DSP_INCLUDE += -I$(TARGET_ROOTDIR)/usr/share/ti/opencl

EXE = helloworld
CPP_FLAGS  = -O3 -fopenmp -Wall

CLOCL_FLAGS = -O3 -Wall
CL6X_FLAGS  = -O3 -omp
CPP  = g++
CLOCL = clocl
```

```
CL6X = cl6x -mv6600 -abi=eabi $(DSP_INCLUDE)

LIBS = -lOpenCL -locl_util

%.o: %.cpp
    @$(CPP) -c $(CPP_FLAGS) $<
    @echo Compiling $<

%.o: %.c
    @$(CPP) -c $(CPP_FLAGS) $<
    @echo Compiling $<

%.obj: %.c
    @$(CL6X) -c $(CL6X_FLAGS) $<
    @echo Compiling $<

%.out: %.cl
    @$(CLOCL) $(CLOCL_FLAGS) $^
    @echo Compiling $<

%.dsp_h: %.cl
    @$(CLOCL) -t $(CLOCL_FLAGS) $^
    @echo Compiling $<

$(EXE):


clean::
    @rm -f $(EXE) *.o *.out *.asm *.if *.opt *.bc *.objc *.map *.bin *.
    dsp_h *.pgm

test: clean $(EXE)
    @echo Running  $(EXE)
    @./$(EXE) >> /dev/null
    @if [ $$? -ne 0 ] ; then echo "FAILED !!!" ; fi


$(EXE): main.o
    @$(CPP) $(CPP_FLAGS) main.o $(LIBS) -lrt -o $@

main.o: hello.dsp_h
```

13.11 Laboratory experiments

Copy all files from /Chapter_13_Code to your Virtual Machine (/home/Chapter_OpenCL).

13.11.1 Laboratory experiment 1: Hello World

Project location:

/home/Chapter_OpenCL/lab1_helloworld1

1) Set the connection between the host (PC) and the KeyStone II EVM. This section applies to all laboratory experiments.
2) Connect the EVM, as shown in Figure 13.18 and Figure 13.19.
3) Determine and set the port addresses. The port addresses on the PC will change every time a serial device is connected. Open the Device Manager and find out the port addresses. In Figure 13.20, the port addresses found for the serial ports are COM7 and COM8 (these will depend on your settings, as it is very likely that you will have different ports). **PuTTY** as a terminal emulator has been used. Open two terminals with COM7 and COM8; see Figure 13.21.
4) Start the VMware. It is assumed that VMware has been built; see Figure 13.22. See Appendix 1 for instructions on how to build the VMWare required for this application. Make sure that the USB adaptor is connected, and then press OK as shown in Figure 13.23.
5) Set an Ethernet connection. Follow the steps shown from Figure 13.24 to Figure 13.29.

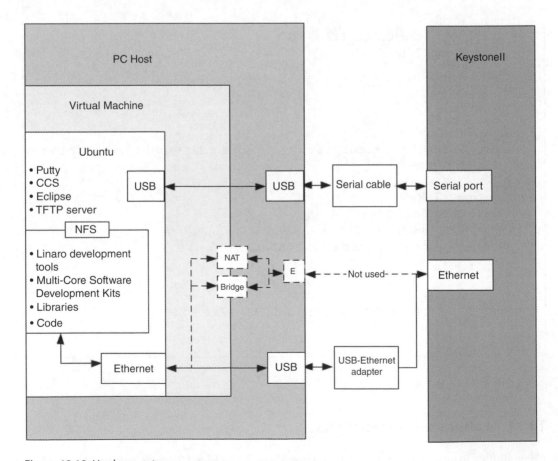

Figure 13.18 Hardware setup.

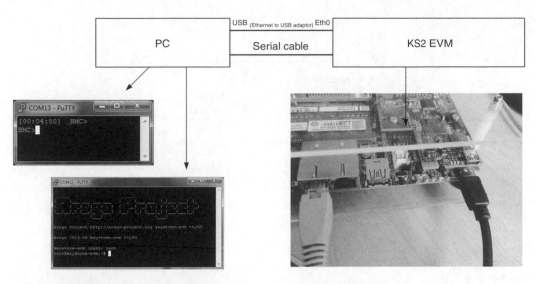

Figure 13.19 EVM connection to the PC.

Figure 13.20 Device manager for identifying the COM ports.

6) Check the IP address. Open a terminal by typing (**ctrl + alt + t**), and type **ifconfig** to check the IP address as shown in Figure 13.29.
7) Check the IP address. Go back to **PuTTY**, wait for the booting process to complete, then type root and press return; see Figure 13.30.
8) Check the IP address on the EVM. Check the IP address of the EVM by typing **ifconfig** as shown in Figure 13.31.
9) Use FileZilla to transfer files between the host and the EVM. FileZilla is an easy tool for transferring files between the host and the EVM. Evoke FileZilla, and type the EVM address as shown in Figure 13.32. Right click on the project, and select upload.

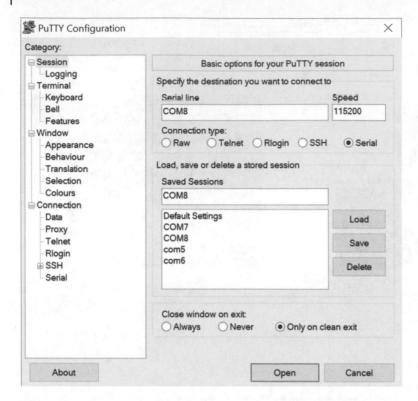

Figure 13.21 Setting up the COM ports.

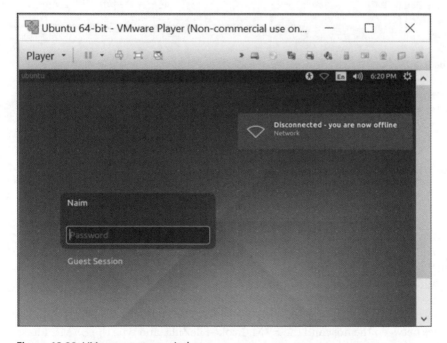

Figure 13.22 VMware start-up window.

Ubuntu 64-bit - VMware Player

A USB device is about to be unplugged from the host and connected to this virtual machine. It will first be stopped to enable safe removal. With some devices, the host may display the message "The device can now safely be removed."

☐ Do not show this message again

OK   Cancel

Figure 13.23 VMware detected the USB device.

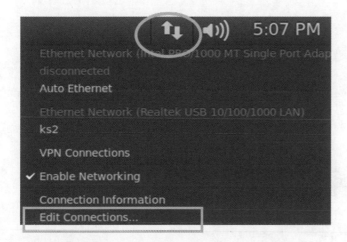

Figure 13.24 Edit connections.

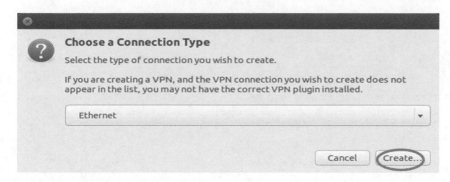

Figure 13.25 Choose a connection type.

Figure 13.26 Select a MAC address and a connection name.

Figure 13.27 Select **Shared to other computers**.

Figure 13.28 Output when the connection is made.

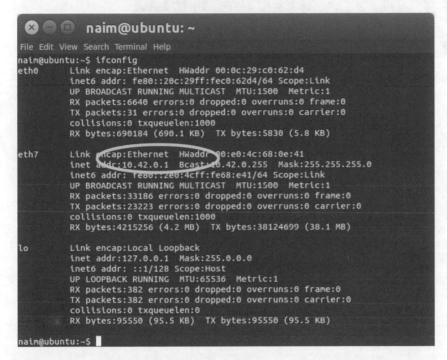

Figure 13.29 Use **ifconfig** to check the IP addresses.

Figure 13.30 Boot process completed successfully.

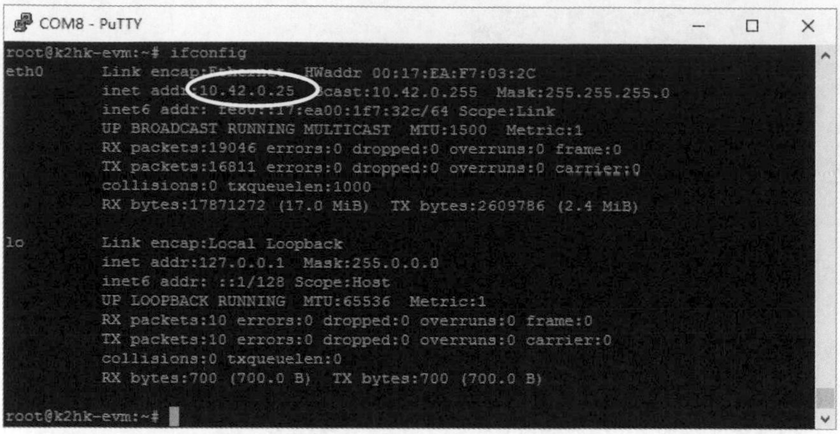

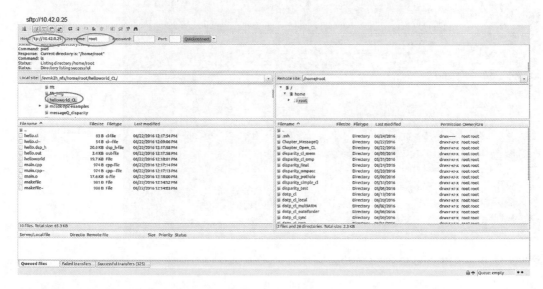

Figure 13.31 Checking the IP address on the EVM.

Figure 13.32 Using FileZilla to transfer files from the host to the EVM.

10) Explore the main function: **main.cpp**. Open and explore the **main.cpp** file shown below and located in:
/home/Chapter_OpenCL/lab1_helloworld/main.cpp

```cpp
#define __CL_ENABLE_EXCEPTIONS
#include <CL/cl.hpp>
#include <iostream>
#include <stdio.h>
#include "ocl_util.h"
#include "CL/cl.hpp"
#include <stdlib.h>
```

```
#include <stdint.h>
#include <omp.h>
#include <string.h>
#include "hello.dsp_h"

using namespace cl;
using namespace std;

int main(int argc, char **argv) {
  try{
    Context context(CL_DEVICE_TYPE_ACCELERATOR);
    std::vector<Device> devices = context.getInfo<CL_CONTEXT
    _DEVICES>();
    CommandQueue Q (context, devices[0]);
    Program::Binaries binary(1,make_pair(hello_dsp_bin,sizeof
    (hello_dsp_bin)));
    Program program = Program(context, devices, binary);
    program.build(devices);

    Kernel kernel(program, "hello");

    printf(" [ARM] start OpenCL task\n");
    Q.enqueueNDRangeKernel(kernel, NullRange, NDRange(8),
    NDRange(1));
    Q.finish();
    printf(" [ARM] finish OpenCL task\n");

  }
  catch(Error err){
    std::cerr << "Error: " << err.what() << "(" << err.err() << ")." <<
    std::endl;
  }
  return 0;
}
```

11) Explore the kernel: **hello.cl.** Open and explore the **hello.cl** file shown below and located in: /home/Chapter_OpenCL/lab1_helloworld/hello.cl

```
__kernel void hello(){
    printf("Hello World from DSP\n");
}
```

12) Run the program. Use the **cd** command to enter the project directory, type **make** to compile the project and then type **./helloworld** to run the program. The output should be similar to this:

```
COM8 - PuTTY                                          —   □   ×
root@k2hnode1:~/Chapter_Open_CL/lab1_helloworld# ./helloworld
[ARM] start OpenCL task
[core 0] Hello World from DSP
[core 4] Hello World from DSP
[core 6] Hello World from DSP
[core 3] Hello World from DSP
[core 5] Hello World from DSP
[core 1] Hello World from DSP
[core 7] Hello World from DSP
[core 2] Hello World from DSP
[ARM] finish OpenCL task
root@k2hnode1:~/Chapter_Open_CL/lab1_helloworld#
```

13.11.2   Laboratory experiment 2: dotp functions

This laboratory experiment uses various **dotp** functions.
   Project location:

/home/Chapter_OpenCL/lab2_dotp

   The code in **main.cpp** performs of the following functions:

1) OpenCL setup: contexts, devices, program and command queues
2) Read the device info.
3) Create some buffer objects and set the kernel arguments.
4) Initialize the output buffer.
5) Initialize the input buffers.
6) Send a command queue.
7) Read the output buffer.
8) Print the output.

13.11.2.1   Explore the *main.cpp* (shown below) function
Explore the **main.cpp** file (see below) which is located at:

/home/Chapter_OpenCL/lab2_dotp/main.cpp

```
#define __CL_ENABLE_EXCEPTIONS
#include <CL/cl.hpp>
#include <SDL/SDL.h>
#include <iostream>
#include <fstream>
#include <cassert>
#include <cstdlib>
#include <signal.h>
#include <math.h>
#include "ocl_util.h"
#include "dotp.dsp_h"
#include "omp.h"
```

```cpp
#define COUNT 256
#define offset 10
#define COUNTMAX 300
short a[COUNTMAX];
short x[COUNTMAX];
int o[COUNTMAX];

using namespace std;
using namespace cl;

static unsigned ns_diff (struct timespec &t1, struct timespec &t2);

int main(int argc, char *argv[])
{

  for(int i = 0; i < COUNTMAX; i++)
  {
    a[i]=(i>COUNT)?0:i%256;
    x[i]=(i>COUNT)?0:256-i%256;
  }
  struct timespec t1, t3;
  std::string str;
  try
  {
    Context context(CL_DEVICE_TYPE_ACCELERATOR);
    std::vector<Device> devices = context.
    getInfo<CL_CONTEXT_DEVICES>();
    int           nDev  = devices.size();
    printf("Number of Devices: %d\n",nDev);
    char s[100];
    int m;
    devices[0].getInfo(CL_DEVICE_NAME,&s);
    printf("DEVICE_NAME: %s\n",s);
    devices[0].getInfo(CL_DEVICE_VENDOR,&s);
    printf("DEVICE_VENDOR: %s\n",s);
    devices[0].getInfo(CL_DEVICE_MAX_COMPUTE_UNITS,&m);
    printf("DEVICE_MAX_COMPUTE_UNITS: %d\n",m);
    devices[0].getInfo(CL_DEVICE_MAX_CLOCK_FREQUENCY,&m);
    printf("DEVICE_MAX_CLOCK_FREQUENCY: %d\n",m);
    devices[0].getInfo(CL_DEVICE_MAX_WORK_ITEM_DIMENSIONS, &m);
    printf("DEVICE_MAX_WORK_ITEM_DIMENSIONS: %d\n", m);
    std::vector<std::size_t> witemsize;
    devices[0].getInfo(CL_DEVICE_MAX_WORK_ITEM_SIZES,
    &witemsize);
    printf("DEVICE_MAX_WORK_ITEM_SIZES: (%d",witemsize[0]);
    for (int i = 1; i < m; i++)
  {
      printf(",%d", witemsize[i]);
```

```
        }
        printf(")\n");
        devices[0].getInfo(CL_DEVICE_PROFILE, &s);
        printf("DEVICE_PROFIL: %s\n",s);
        devices[0].getInfo(CL_DEVICE_VERSION, &s);
        printt("DEVICE_VERSION: %s\n", s);
        devices[0].getInfo(CL_DRIVER_VERSION, &s);
        printf("DRIVER_VERSION: %s\n", s);
        devices[0].getInfo(CL_DEVICE_OPENCL_C_VERSION, &s);
        printf("DEVICE_OPENCL_C_VERSION: %s\n", s);

        printf("Count = %d\n",COUNT);
        CommandQueue    Q    (context, devices[0]);
        Program::Binaries binary(1, make_pair(dotp_dsp_bin,sizeof
        (dotp_dsp_bin)));
        Program        program = Program(context, devices, binary);
        program.build(devices);
        Kernel kernel_sa(program, "dotp_sa");
        Kernel kernel_C(program, "dotp_C");
        Kernel kernel_C_4(program, "dotp_C_4");
        Kernel kernel_C_4_private(program, "dotp_C_4_private");

        long long result1=0,result2=0;
        Buffer aBuff (context, CL_MEM_READ_ONLY|CL_MEM_USE_MSMC_TI,
        COUNTMAX*sizeof(short));
        Buffer xBuff (context, CL_MEM_READ_ONLY|CL_MEM_USE_MSMC_TI,
        COUNTMAX*sizeof(short));
        Buffer oBuff (context, CL_MEM_READ_WRITE|CL_MEM_USE_MSMC_TI,
        COUNTMAX*sizeof(int));
        Buffer ompresult (context, CL_MEM_READ_WRITE|
        CL_MEM_USE_MSMC_TI, sizeof(long long));

        kernel_sa.setArg(0, aBuff );
        kernel_sa.setArg(1, xBuff );
        kernel_sa.setArg(2, oBuff );

        kernel_C.setArg(0, aBuff );
        kernel_C.setArg(1, xBuff );
        kernel_C.setArg(2, oBuff );

        kernel_C_4.setArg(0, aBuff );
        kernel_C_4.setArg(1, xBuff );
        kernel_C_4.setArg(2, oBuff );

        kernel_C_4_private.setArg(0, aBuff );
        kernel_C_4_private.setArg(1, xBuff );
        kernel_C_4_private.setArg(2, oBuff );
```

```
for(int i = 0; i < 5; i++)
{
   result1 = 0;
   clock_gettime(CLOCK_MONOTONIC, &t1);
   Q.enqueueWriteBuffer(aBuff, CL_FALSE, 0, COUNT*sizeof(short),
   a, NULL, NULL);
   Q.enqueueWriteBuffer(xBuff, CL_FALSE, 0, COUNT*sizeof(short),
   x, NULL, NULL);
Q.enqueueNDRangeKernel(kernel_sa, NullRange, NDRange(COUNT/4),
NDRange(1), NULL, NULL);
   Q.enqueueReadBuffer (oBuff, CL_TRUE, 0, COUNT/4*sizeof(int), o,
   NULL, NULL);
   clock_gettime(CLOCK_MONOTONIC, &t3);
   for(int n = 0; n < COUNT/4; n++)
   {
    result1 += o[n];
   }
   if(i==4) printf("sa result = %lld Elapsed: %d ns\n", result1,
   ns_diff(t1, t3));
}
for(int i = 0; i < 5; i++)
{
   result2 = 0;
   clock_gettime(CLOCK_MONOTONIC, &t1);
   Q.enqueueWriteBuffer(aBuff, CL_FALSE, 0, COUNT*sizeof(short),
   a, NULL, NULL);
   Q.enqueueWriteBuffer(xBuff, CL_FALSE, 0, COUNT*sizeof(short),
   x, NULL, NULL);
   Q.enqueueNDRangeKernel(kernel_C, NullRange, NDRange(COUNT),
   NDRange(1), NULL, NULL);
   Q.enqueueReadBuffer (oBuff, CL_TRUE, 0, COUNT*sizeof(int), o,
   NULL, NULL);
   clock_gettime(CLOCK_MONOTONIC, &t3);
   for(int n = 0; n < COUNT; n++)
   {
    result2 += o[n];
   }
   if(i==4) printf("c result = %lld Elapsed: %d ns\n", result2,
   ns_diff(t1, t3));
}
for(int i = 0; i < 5; i++)
{
   result2 = 0;
   clock_gettime(CLOCK_MONOTONIC, &t1);
   Q.enqueueWriteBuffer(aBuff, CL_FALSE, 0, COUNT*sizeof(short),
   a, NULL, NULL);
   Q.enqueueWriteBuffer(xBuff, CL_FALSE, 0, COUNT*sizeof(short),
   x, NULL, NULL);
   Q.enqueueNDRangeKernel(kernel_C_4, NullRange, NDRange(COUNT),
   NDRange(4), NULL, NULL);
```

```
      Q.enqueueReadBuffer (oBuff, CL_TRUE, 0, COUNT/4*sizeof(int), o,
    NULL, NULL);
     clock_gettime(CLOCK_MONOTONIC, &t3);
     for(int n = 0; n < COUNT/4; n++)
     {
      result2 += o[n];
     }
     if(i==4) printf("c_4 result = %lld Elapsed: %d ns\n", result2,
     ns_diff(t1, t3));
    }
    for(int i = 0; i < 5; i++)
    {
      result2 = 0;
      clock_gettime(CLOCK_MONOTONIC, &t1);
      Q.enqueueWriteBuffer(aBuff, CL_FALSE, 0, COUNT*sizeof(short),
      a, NULL, NULL);
      Q.enqueueWriteBuffer(xBuff, CL_FALSE, 0, COUNT*sizeof(short),
      x, NULL, NULL);
      Q.enqueueNDRangeKernel(kernel_C_4_private, NullRange,
      NDRange(COUNT), NDRange(4), NULL, NULL);
     Q.enqueueReadBuffer (oBuff, CL_TRUE, 0, COUNT/4*sizeof(int), o,
    NULL, NULL);
     clock_gettime(CLOCK_MONOTONIC, &t3);
     for(int n = 0; n < COUNT/4; n++)
     {
      result2 += o[n];
     }
     if(i==4) printf("c_4_private result = %lld Elapsed: %d ns\n",
     result2, ns_diff(t1, t3));
    }
    long long sum = 0;
    for(int i=0;i<COUNT;i++){
      sum += a[i]*x[i];
    }
    printf("correct result is %lld\n\n", sum);
    for(int i = 0; i < 5; i++)
    {
      result2 = 0;
      clock_gettime(CLOCK_MONOTONIC, &t1);
      Q.enqueueWriteBuffer(aBuff, CL_FALSE, 0, COUNTMAX*sizeof
      (short), a, NULL, NULL);
      Q.enqueueWriteBuffer(xBuff, CL_FALSE, 0, COUNTMAX*sizeof
      (short), x, NULL, NULL);
      Q.enqueueNDRangeKernel(kernel_C, NDRange(offset),
      NDRange(COUNT), NDRange(1), NULL, NULL);
      Q.enqueueReadBuffer (oBuff, CL_TRUE, 0, COUNTMAX*sizeof(int),
      &o, NULL, NULL);
      clock_gettime(CLOCK_MONOTONIC, &t3);
      for(int n = offset; n < COUNT+offset; n++)
```

```
      {
        result2 += o[n];
      }
      if(i==4) printf("c with offset %d result = %lld Elapsed: %d ns\n",
      offset, result2, ns_diff(t1, t3));
    }
    sum = 0;
    for(int i=0;i<COUNT;i++){
      sum += a[i+offset]*x[i+offset];
    }
    printf("correct result with offset %d is = %lld\n",offset, sum);
  }
  catch (Error err)
  {
    cerr << "ERROR: " << err.what() << "(" << err.err() << ")" << endl;
    exit(-1);
  }
}
static unsigned ns_diff (struct timespec &t1, struct timespec &t2)
{ return (t2.tv_sec - t1.tv_sec) * 1e9 + (t2.tv_nsec - t1.tv_nsec); }
```

13.11.2.2 Explore the kernel *dotp.cl (shown below)*
Explore the **dotp.cl** file (shown below) which is located at:

/home/Chapter_OpenCL/lab2_dotp/dotp.cl

```
extern int ddotp4h2(__global short *m, __global short *n, int count);
extern int dotp4h(__global short *m, __global short *n, int count);
__kernel void dotp_sa(__global short* a, __global short* x, __global
  int* o)
{
  int i = get_global_id(0);
  int j = i*4;
  o[i] = dotp4h(&a[j], &x[j], 4);
}
__kernel void dotp_C(__global short* a, __global short* x, __global
  int* o)
{
  int i = get_global_id(0);
  o[i] = a[i]*x[i];
}
__kernel void dotp_C_4(__global short* a, __global short* x, __global
  int* o)
{
  __local int sum;
  int i = get_global_id(0);
  int j = get_local_id(0);
  int k = get_group_id(0);
```

```
  if(j==0) sum=0;
  sum += a[i]*x[i];
  if(j==3) o[k]=sum;
}
  __kernel void dotp_C_4_private(__global short* a, __global short* x,
  __global int* o)
{
  __local int sum;
  __private int i = get_global_id(0);
  __private int j = get_local_id(0);
  __private int k = get_group_id(0);
  if(j==0) sum=0;
  sum += a[i]*x[i];
  if(j==3) o[k]=sum;
}
```

13.11.2.3 Run the dotp program

Use the **cd** command to enter the project directory, type **make** to compile the project and then type **./dotp** to run the program. The output should be similar to this:

```
COM8 - PuTTY                                                    —  □  ✕
root@k2hnode1:~/Chapter_Open_CL/lab2_dotp# ./dotp
Number of Devices: 1
DEVICE_NAME: TI Multicore C66 DSP
DEVICE_VENDOR: Texas Instruments, Inc.
DEVICE_MAX_COMPUTE_UNITS: 8
DEVICE_MAX_CLOCK_FREQUENCY: 1228
DEVICE_MAX_WORK_ITEM_DIMENSIONS: 3
DEVICE_MAX_WORK_ITEM_SIZES: (1073741824,1073741824,1073741824)
DEVICE_PROFIL: FULL_PROFILE
DEVICE_VERSION: OpenCL 1.1 TI
DRIVER_VERSION:
DEVICE_OPENCL_C_VERSION: OpenCL C 1.1 LLVM 3.3
Count = 256
sa result = 2796160 Elapsed: 137163 ns
c result = 2796160 Elapsed: 162710 ns
c_4 result = 2796160 Elapsed: 114707 ns
c_4_private result = 2796160 Elapsed: 124546 ns
correct result is 2796160

c with offset 10 result = 2784925 Elapsed: 147993 ns
correct result with offset 10 is = 2784925
root@k2hnode1:~/Chapter_Open_CL/lab2_dotp# █
```

13.11.3 Laboratory experiment 3: USE_HOST_PTR

Using **CL_MEM_USE_HOST_PTR**.

Project location:

/home/Chapter_OpenCL/lab3_mallocmsmc

1) Explore the **main.cpp** function (shown below) located in:
   /home/Chapter_OpenCL/lab3_mallocmsmc/main.cpp

```cpp
#define __CL_ENABLE_EXCEPTIONS
#include <CL/cl.hpp>
#include <SDL/SDL.h>
#include <iostream>
#include <fstream>
#include <cassert>
#include <cstdlib>
#include <signal.h>
#include <math.h>
#include "ocl_util.h"
#include "dotp.dsp_h"
#include "omp.h"
#define COUNT 256
using namespace std;
using namespace cl;
static unsigned ns_diff (struct timespec &t1, struct timespec &t2);
int main(int argc, char *argv[])
{
 short* a = (short*) __malloc_msmc(sizeof(short)*COUNT);
 short* x = (short*) __malloc_msmc(sizeof(short)*COUNT);
 short* o = (short*) __malloc_msmc(sizeof(short)*COUNT);
 for(int i = 0; i < COUNT; i++)
 {
  a[i]=i%256;
  x[i]=256-i%256;
 }
 struct timespec t1, t3;
 std::string str;
 try
 {
  Context context(CL_DEVICE_TYPE_ACCELERATOR);
  std::vector<Device> devices = context.
getInfo<CL_CONTEXT_DEVICES>();
  printf("Count = %d\n",COUNT);
  CommandQueue    Q    (context, devices[0]);
  Program::Binaries binary(1, make_pair(dotp_dsp_bin,sizeof
(dotp_dsp_bin)));
  Program       program = Program(context, devices, binary);
  program.build(devices);
  Kernel kernel_C(program, "dotp_C");
  long long result1=0;
  Buffer aBuff (context, CL_MEM_READ_ONLY|CL_MEM_USE_HOST_PTR,
COUNT*sizeof(short), a);
  Buffer xBuff (context, CL_MEM_READ_ONLY|CL_MEM_USE_HOST_PTR,
COUNT*sizeof(short), x);
  Buffer oBuff (context, CL_MEM_WRITE_ONLY|CL_MEM_USE_HOST_PTR,
COUNT*sizeof(short), o);
  kernel_C.setArg(0, aBuff );
  kernel_C.setArg(1, xBuff );
  kernel_C.setArg(2, oBuff );
  for(int i = 0; i < 5; i++)
```

```
    {
      result1 = 0;
      clock_gettime(CLOCK_MONOTONIC, &t1);
      Q.enqueueNDRangeKernel(kernel_C, NullRange, NDRange(COUNT),
  NDRange(1), NULL, NULL);
      Q.finish();
      clock_gettime(CLOCK_MONOTONIC, &t3);
      for(int n = 0; n < COUNT; n++)
      {
        result1 += o[n];
      }
      printf("OpenCL result = %lld Elapsed: %d ns\n", result1, ns_diff
  (t1, t3));
      }
    long long sum = 0;
    for(int i=0;i<COUNT;i++){
      sum += a[i]*x[i];
    }
    printf("correct result is %lld\n\n", sum);
    }
  catch (Error err)
  {
    cerr << "ERROR: " << err.what() << "(" << err.err() << ")" << endl;
    exit(-1);
  }
}
static unsigned ns_diff (struct timespec &t1, struct timespec &t2)
{ return (t2.tv_sec - t1.tv_sec) * 1e9 + (t2.tv_nsec - t1.tv_nsec); }
```

2) Explore the **dotp.cl** kernel shown below and located in:

   /home/Chapter_OpenCL/lab3_mallocmsmc/dotp.cl

```
__kernel void dotp_C(__global short* a, __global short* x, __global
short* o)
{
  int i = get_global_id(0);
  o[i] = a[i]*x[i];
}
```

3) Run the program. Use the command **cd** to enter the project directory, type **make** to compile the project and type **./dotp** to run the program. The output should be similar to this:

13.11.4   Laboratory experiment 4: ALLOC_HOST_PTR

Using **CL_MEM_ALLOC_HOST_PTR** (default).
  Project location:

/home/Chapter_OpenCL/lab4_memmap

1) Explore the **main.cpp** function shown below and located in:
    /home/Chapter_OpenCL/lab4_memmap/main.cpp

```
#define __CL_ENABLE_EXCEPTIONS
#include <CL/cl.hpp>
#include <SDL/SDL.h>
#include <iostream>
#include <fstream>
#include <cassert>
#include <cstdlib>
#include <signal.h>
#include <math.h>
#include "ocl_util.h"
#include "dotp.dsp_h"
#include "omp.h"

#define COUNT 256

using namespace std;
using namespace cl;

static unsigned ns_diff (struct timespec &t1, struct timespec &t2);
int main(int argc, char *argv[])
{

  struct timespec t1, t3;
  std::string str;
  try
  {
   Context context(CL_DEVICE_TYPE_ACCELERATOR);
   std::vector<Device> devices = context.
   getInfo<CL_CONTEXT_DEVICES>();
   printf("Count = %d\n",COUNT);
   CommandQueue    Q    (context, devices[0]);
   Program::Binaries binary(1, make_pair(dotp_dsp_bin,sizeof
(dotp_dsp_bin)));
   Program        program = Program(context, devices, binary);
   program.build(devices);
   Kernel kernel_C(program, "dotp_C");
   long long result1=0;
   Buffer aBuff (context, CL_MEM_READ_ONLY|CL_MEM_USE_MSMC_TI|
CL_MEM_ALLOC_HOST_PTR, COUNT*sizeof(short));
```

```
   Buffer xBuff (context, CL_MEM_READ_ONLY|CL_MEM_USE_MSMC_TI|
CL_MEM_ALLOC_HOST_PTR, COUNT*sizeof(short));
   Buffer oBuff (context, CL_MEM_WRITE_ONLY|CL_MEM_USE_MSMC_TI|
CL_MEM_ALLOC_HOST_PTR, COUNT*sizeof(short));
   short* a = (short*)Q.enqueueMapBuffer(aBuff, CL_TRUE,
CL_MAP_WRITE, 0, COUNT*sizeof(short));
   short* x = (short*)Q.enqueueMapBuffer(xBuff, CL_TRUE,
CL_MAP_WRITE, 0, COUNT*sizeof(short));
   for(int i = 0; i < COUNT; i++)
   {
    a[i]=i%256;
    x[i]=256-i%256;
   }
   Q.enqueueUnmapMemObject(aBuff, a);
   Q.enqueueUnmapMemObject(xBuff, x);

   kernel_C.setArg(0, aBuff );
   kernel_C.setArg(1, xBuff );
   kernel_C.setArg(2, oBuff );

   for(int i = 0; i < 5; i++)
   {
    result1 = 0;
    clock_gettime(CLOCK_MONOTONIC, &t1);
    Q.enqueueNDRangeKernel(kernel_C, NullRange, NDRange(COUNT),
NDRange(1), NULL, NULL);
    Q.finish();
    clock_gettime(CLOCK_MONOTONIC, &t3);
    short* o = (short*)Q.enqueueMapBuffer(oBuff, CL_TRUE,
CL_MAP_READ, 0, COUNT*sizeof(short));
    for(int n = 0; n < COUNT; n++)
    {
     result1 += o[n];
    }
    Q.enqueueUnmapMemObject(oBuff, o);
    printf("OpenCL result = %lld Elapsed: %d ns\n", result1, ns_diff
(t1, t3));
   }

   long long sum = 0;
   for(int i=0;i<COUNT;i++){
    sum += a[i]*x[i];
   }
   printf("correct result is %lld\n\n", sum);
  }
 catch (Error err)
 {
  cerr << "ERROR: " << err.what() << "(" << err.err() << ")" << endl;
```

```
  exit(-1);
  }
}
static unsigned ns_diff (struct timespec &t1, struct timespec &t2)
{ return (t2.tv_sec - t1.tv_sec) * 1e9 + (t2.tv_nsec - t1.tv_nsec); }
```

The kernel file is the same as in Laboratory experiment 3.

2) Run the program. Use the **cd** command to enter the project directory, type **make** to compile the project and then type **./dotp** to run the program. The output should be similar to this:

13.11.5 Laboratory experiment 5: COPY_HOST_PTR

Using **CL_MEM_COPY_HOST_PTR**.
  Project location:

/home/Chapter_OpenCL/lab5_copyptr

1) Explore the **main.cpp** function shown below and located in:
   /home/Chapter_OpenCL/lab5_copyptr/main.cpp

```
#define __CL_ENABLE_EXCEPTIONS
#include <CL/cl.hpp>
#include <SDL/SDL.h>
#include <iostream>
#include <fstream>
#include <cassert>
#include <cstdlib>
#include <signal.h>
#include <math.h>
#include "ocl_util.h"
#include "dotp.dsp_h"
#include "omp.h"

#define COUNT 256

using namespace std;
using namespace cl;
```

```
static unsigned ns_diff (struct timespec &t1, struct timespec &t2);

int main(int argc, char *argv[])
{

  short* a = (short*) malloc(sizeof(short)*COUNT);
  short* x = (short*) malloc(sizeof(short)*COUNT);
  short* o = (short*) malloc(sizeof(short)*COUNT);
  for(int i = 0; i < COUNT; i++)
  {
   a[i]=i%256;
   x[i]=256-i%256;
  }

  struct timespec t1, t3;
  std::string str;
  try
  {
   Context context(CL_DEVICE_TYPE_ACCELERATOR);
   std::vector<Device> devices = context.
getInfo<CL_CONTEXT_DEVICES>();
   printf("Count = %d\n",COUNT);
   CommandQueue    Q    (context, devices[0]);
   Program::Binaries binary(1, make_pair(dotp_dsp_bin,sizeof
(dotp_dsp_bin)));
   Program       program = Program(context, devices, binary);
   program.build(devices);
   Kernel kernel_C(program, "dotp_C");
   long long result1=0;
   Buffer aBuff (context, CL_MEM_READ_ONLY|CL_MEM_USE_MSMC_TI|
CL_MEM_COPY_HOST_PTR, COUNT*sizeof(short), a);
   Buffer xBuff (context, CL_MEM_READ_ONLY|CL_MEM_USE_MSMC_TI|
CL_MEM_COPY_HOST_PTR, COUNT*sizeof(short), x);
   Buffer oBuff (context, CL_MEM_WRITE_ONLY|CL_MEM_USE_MSMC_TI,
COUNT*sizeof(short));
   kernel_C.setArg(0, aBuff );
   kernel_C.setArg(1, xBuff );
   kernel_C.setArg(2, oBuff );

   for(int i = 0; i < 5; i++)
   {
     result1 = 0;
     clock_gettime(CLOCK_MONOTONIC, &t1);
     Q.enqueueNDRangeKernel(kernel_C, NullRange, NDRange(COUNT),
NDRange(1), NULL, NULL);
     Q.enqueueReadBuffer (oBuff, CL_TRUE, 0, COUNT*sizeof(short), o,
NULL, NULL);
     clock_gettime(CLOCK_MONOTONIC, &t3);
     for(int n = 0; n < COUNT; n++)
```

```
    {
      result1 += o[n];
    }
    printf("OpenCL result = %lld Elapsed: %d ns\n", result1, ns_diff
(t1, t3));
  }
  long long sum = 0;
  for(int i=0;i<COUNT;i++){
    sum += a[i]*x[i];
  }
  printf("correct result is %lld\n\n", sum);
}
catch (Error err)
{
  cerr << "ERROR: " << err.what() << "(" << err.err() << ")" << endl;
  exit(-1);
}
}
static unsigned ns_diff (struct timespec &t1, struct timespec &t2)
{ return (t2.tv_sec - t1.tv_sec) * 1e9 + (t2.tv_nsec - t1.tv_nsec); }
```

The kernel file is the same as in Laboratory experiment 3.

2) Run the **dotp** program. Use the **cd** command to enter the project directory, type **make** to compile the project and type **./dotp** to run the program. The output should be similar to this:

13.11.6   Laboratory experiment 6: Synchronisation

Project location:

/home/Chapter_OpenCL/lab6_dotp_sync

1) Explore the **main.cpp** function shown below and located in:
   /home/Chapter_OpenCL/lab6_dotp_sync/main.cpp

```
#define __CL_ENABLE_EXCEPTIONS
#include <CL/cl.hpp>
#include <SDL/SDL.h>
#include <iostream>
```

```cpp
#include <fstream>
#include <cassert>
#include <cstdlib>
#include <signal.h>
#include <math.h>
#include "ocl_util.h"
#include "dotp.dsp_h"
#include "omp.h"
#define COUNT 2560

int o[COUNT];
int o1[COUNT];
short a[COUNT];
short x[COUNT];
using namespace std;
using namespace cl;
static unsigned ns_diff (struct timespec &t1, struct timespec &t2);
void mycallback(cl_event event, cl_int event_command_exec_status,
void *user_data);

int main(int argc, char *argv[])
{
  struct timespec t1, t2, t3;
  std::string str;
  for(int i = 0; i < COUNT; i++)
  {
    a[i]=i%256;
    x[i]=256-a[i];
  }

  try
  {
    Context context(CL_DEVICE_TYPE_ACCELERATOR);
    std::vector<Device> devices = context.
getInfo<CL_CONTEXT_DEVICES>();
    int m;
    printf("Count = %d\n\n",COUNT);
    CommandQueue    Q    (context, devices
[0],CL_QUEUE_PROFILING_ENABLE);
    CommandQueue    Q_ooo   (context, devices
[0],CL_QUEUE_PROFILING_ENABLE|
CL_QUEUE_OUT_OF_ORDER_EXEC_MODE_ENABLE);
    Program::Binaries binary(1, make_pair(dotp_dsp_bin, sizeof
(dotp_dsp_bin)));
    Program         program = Program(context, devices, binary);
    program.build(devices);

    Kernel kernel_sa(program, "dotp_sa");
    Kernel kernel_C(program, "dotp_C");
    long long result1=0;
```

```
  Buffer aBuff (context, CL_MEM_READ_ONLY|CL_MEM_USE_MSMC_TI,
COUNT*sizeof(short));
  Buffer xBuff (context, CL_MEM_READ_ONLY|CL_MEM_USE_MSMC_TI,
COUNT*sizeof(short));
  Buffer oBuff (context, CL_MEM_READ_WRITE|CL_MEM_USE_MSMC_TI,
COUNT*sizeof(int));
  Buffer oBuff1 (context, CL_MEM_READ_WRITE|CL_MEM_USE_MSMC_TI,
COUNT*sizeof(int));

  kernel_sa.setArg(0, aBuff );
  kernel_sa.setArg(1, xBuff );
  kernel_sa.setArg(2, oBuff );
  kernel_C.setArg(0, aBuff );
  kernel_C.setArg(1, xBuff );
  kernel_C.setArg(2, oBuff );

  Q.enqueueWriteBuffer(aBuff, CL_FALSE, 0, COUNT*sizeof(short), a,
NULL, NULL);
  Q.enqueueWriteBuffer(xBuff, CL_FALSE, 0, COUNT*sizeof(short), x,
NULL, NULL);
  Event ev,ev1,ev2;
  std::vector<UserEvent> evs(1);

/***********************event with callback************/
  printf("example 1 event with callback function:\n");

  printf("enqueue kernel\n");
  Q.enqueueNDRangeKernel(kernel_C, NullRange, NDRange(COUNT),
NDRange(1), NULL, &ev);
  ev.setCallback(0,&mycallback);
  ev.wait();
  ev.getInfo(C`L_EVENT_COMMAND_EXECUTION_STATUS,&m);
  printf("finish processing kernel, status = %d (0=CL_COMPLETE)
\n",m);
  Q.enqueueReadBuffer (oBuff, CL_TRUE, 0, COUNT*sizeof(int), o,
NULL, NULL);
  for(int n = 0; n < COUNT; n++)
  {
   result1 += o[n];
  }
  printf("opencl result = %lld\n\n", result1);
/***********************user event************************/
  printf("example 2 userevent:\n");
  result1 = 0;
  evs[0] = UserEvent(context);
  printf("enqueue kernel\n");
  //have to wait for the user event to be finished before processing the
kerenel
```

```
  Q.enqueueNDRangeKernel(kernel_C, NullRange, NDRange(COUNT),
NDRange(1), (std::vector<cl::Event>*)&evs, &ev);
  ev.setCallback(0,&mycallback);
  ev.getInfo(CL_EVENT_COMMAND_EXECUTION_STATUS,&m);
  printf("waiting for user event, kernel status = %d (3=CL_QUEUED)
\n",m);
  //finish the user event
  evs[0].setStatus(CL_COMPLETE);
  //now can start processing kernel
  printf("user event complete, start processing kernel\n");
  ev.wait();
  ev.getInfo(CL_EVENT_COMMAND_EXECUTION_STATUS,&m);
  printf("finish processing kernel, status = %d (0=CL_COMPLETE)
\n",m);
  Q.enqueueReadBuffer (oBuff, CL_TRUE, 0, COUNT*sizeof(int), o,
NULL, NULL);
  for(int n = 0; n < COUNT; n++)
  {
    result1 += o[n];
  }
  printf("opencl result = %lld\n\n", result1);
/*******************event.wait() and Q.finish()***************/
  printf("example 3 event.wait() and Q.finish():\n");
  clock_gettime(CLOCK_MONOTONIC, &t1);
  for(int i = 0; i < 5; i++)
  {
    Q.enqueueNDRangeKernel(kernel_C, NullRange, NDRange(COUNT),
NDRange(1), NULL, &ev);
  }
  //commands submitted but not finished yet
  clock_gettime(CLOCK_MONOTONIC, &t2);
  Q.finish();
  //wait until all commands are finished
  clock_gettime(CLOCK_MONOTONIC, &t3);
  printf("return without wait %d ns\n", ns_diff(t1, t2));
  printf("time of Q.finish() %d ns\n", ns_diff(t2, t3));
  clock_gettime(CLOCK_MONOTONIC, &t1);
  for(int i = 0; i < 5; i++)
  {
    Q.enqueueNDRangeKernel(kernel_C, NullRange, NDRange(COUNT),
NDRange(1), NULL, &ev);
    ev.wait();//wait for the command to be finished
  }
  clock_gettime(CLOCK_MONOTONIC, &t3);
  printf("wait after each command is finished %d ns\n\n", ns_diff
(t1, t3));
/************************block**************************/
  printf("example 4 block writing and reading:\n");
```

```
    for(int n = 0; n < COUNT; n++)
    {
     o[n]=0;
    }
    result1=0;
    Q.enqueueNDRangeKernel(kernel_C, NullRange, NDRange(COUNT),
NDRange(1), NULL, &ev);
    ev.wait();
    clock_gettime(CLOCK_MONOTONIC, &t1);
    //non-block reading the output
    //return immediately but the reading is not finished yet
    Q.enqueueReadBuffer (oBuff, CL_FALSE, 0, COUNT*sizeof(int), o,
NULL, NULL);
    clock_gettime(CLOCK_MONOTONIC, &t3);
    //data is not ready so it is likely to be an error
    for(int n = 0; n < COUNT; n++)
    {
     result1 += o[n];
    }
  printf("non-block reading return time %d ns, result = %lld (likely to
be incorrect)\n",  ns_diff(t1, t3),result1);
    for(int n = 0; n < COUNT; n++)
    {
     o[n]=0;
    }
    result1=0;
    Q.enqueueNDRangeKernel(kernel_C, NullRange, NDRange(COUNT),
NDRange(1), NULL, &ev);
    ev.wait();
    clock_gettime(CLOCK_MONOTONIC, &t1);
    //block reading the output
    //return after the reading is finished
    Q.enqueueReadBuffer (oBuff, CL_TRUE, 0, COUNT*sizeof(int), o,
NULL, NULL);
    clock_gettime(CLOCK_MONOTONIC, &t3);
    //data is ready so the result is correct
    for(int n = 0; n < COUNT; n++)
    {
     result1 += o[n];
    }
  printf("block reading return time %d ns, result = %lld\n\n", ns_diff
(t1, t3),result1);
  }
 catch (Error err)
 {
  cerr << "ERROR: " << err.what() << "(" << err.err() << ")" << endl;
  exit(-1);
 }
}
void mycallback(cl_event event, cl_int status, void *user_data)
```

```
{
  printf("[kernel callback] finish\n");
}
static unsigned ns_diff (struct timespec &t1, struct timespec &t2)
{ return (t2.tv_sec - t1.tv_sec) * 1e9 + (t2.tv_nsec - t1.tv_nsec); }
```

2) Explore the **dotp.cl** kernel shown below and located in:
   /home/Chapter_OpenCL lab6_dotp_sync/dotp.cl

```
extern int ddotp4h2(__global short *m, __global short *n, int count);
extern int dotp4h(__global short *m, __global short *n, int count);
__kernel void dotp_sa(__global short* a, __global short* x, __global
int* o)
{

  int i = get_global_id(0);
  int j = i*4;
  o[i] = dotp4h(&a[j], &x[j], 4);
}

__kernel void dotp_C(__global short* a, __global short* x, __global
int* o)
{

  int i = get_global_id(0);
  o[i] = a[i]*x[i];
}
```

3) Run the **dotp** program. Use **cd** to enter the project directory, type **make** to compile the
   project and type **./dotp** to run the program. The output should be similar to this:

13.11.7   Laboratory experiment 7: Local buffer

Project location:

/home/Chapter_OpenCL/lab7_dotp_local

1) Explore the **main.cpp** function shown below and located in:
   /home/Chapter_OpenCL/lab7_dotp_local/main.cpp

```cpp
#define __CL_ENABLE_EXCEPTIONS
#include <CL/cl.hpp>
#include <SDL/SDL.h>
#include <iostream>
#include <fstream>
#include <cassert>
#include <cstdlib>
#include <signal.h>
#include <math.h>
#include "ocl_util.h"
#include "dotp.dsp_h"
#include "omp.h"
#define COUNT 256*256*10
#define LOOPNUM 256

short output[COUNT];
short a[COUNT];
short x[COUNT];
using namespace std;
using namespace cl;
static unsigned ns_diff (struct timespec &t1, struct timespec &t2);
int main(int argc, char *argv[])
{
  struct timespec t1, t3;
  std::string str;
  for(int i = 0; i < COUNT; i++)
  {
    a[i]=i%256;
    x[i]=256-a[i];
  }
  try
  {
    Context context(CL_DEVICE_TYPE_ACCELERATOR);
    std::vector<Device> devices = context.
getInfo<CL_CONTEXT_DEVICES>();
    printf("\nCount = %d\n",COUNT);
    CommandQueue    Q    (context, devices[0]);
    Program::Binaries binary(1, make_pair(dotp_dsp_bin,sizeof
(dotp_dsp_bin)));
    Program    program = Program(context, devices, binary);
    program.build(devices);
```

```
    Kernel kernel_null(program, "null");
    Kernel kernel_data(program, "data");
    Kernel kernel_dotp(program, "dotp");
    Kernel kernel_local(program, "dotp_local");

    long long result1=0;
    Buffer a_ddr (context, CL_MEM_READ_ONLY, COUNT*sizeof(short));
    Buffer x_ddr (context, CL_MEM_READ_ONLY, COUNT*sizeof(short));
    Buffer o_ddr (context, CL_MEM_WRITE_ONLY, COUNT*sizeof(short));

    kernel_dotp.setArg(0, a_ddr);
    kernel_dotp.setArg(1, x_ddr);
    kernel_dotp.setArg(2, o_ddr);
    kernel_local.setArg(0, a_ddr);
    kernel_local.setArg(1, x_ddr);
    kernel_local.setArg(2, o_ddr);
    kernel_data.setArg(0, a_ddr);
    kernel_data.setArg(1, x_ddr);
    kernel_data.setArg(2, o_ddr);

    int nulltime = 0, datatime=0, time1=0, time2=0;
    Q.enqueueWriteBuffer(a_ddr, CL_TRUE, 0, COUNT*sizeof(short), a,
NULL, NULL);
    Q.enqueueWriteBuffer(x_ddr, CL_TRUE, 0, COUNT*sizeof(short), x,
NULL, NULL);
    Q.finish();

    int nbytes = COUNT*2;
    int esttime = (3+nbytes/8096)*2*1000;
    printf("estimated time of dispatching empty kernel = 60000 ns\n");
    printf("estimated time of data caching = (3 + %d(bytes)/8096) * 2 = %d
ns\n",nbytes,esttime);
    printf("estimated total overhead = %d\n\n",60000+esttime);

    for(int i = 0; i < 4; i++)
    {

      clock_gettime(CLOCK_MONOTONIC, &t1);
      Q.enqueueNDRangeKernel(kernel_null, NullRange, NDRange(COUNT/
LOOPNUM), NDRange(1), NULL, NULL);
      Q.finish();
      clock_gettime(CLOCK_MONOTONIC, &t3);
      nulltime = ns_diff(t1, t3);
      if(i==3) printf("time of empty kernel = %d ns\n", nulltime);
    }

    for(int i = 0; i < 4; i++)
```

```
  {
    clock_gettime(CLOCK_MONOTONIC, &t1);
    Q.enqueueNDRangeKernel(kernel_data, NullRange, NDRange(COUNT/
LOOPNUM), NDRange(1), NULL, NULL);
    Q.finish();
    clock_gettime(CLOCK_MONOTONIC, &t3);
    datatime = ns_diff(t1, t3);
    if(i==3) printf("time of empty kernel with data args %d ns\n",
datatime);
  }
  for(int i = 0; i < 4; i++)
  {
    result1 = 0;
    clock_gettime(CLOCK_MONOTONIC, &t1);
    Q.enqueueNDRangeKernel(kernel_dotp, NullRange, NDRange(COUNT/
LOOPNUM), NDRange(1), NULL, NULL);
    Q.finish();
    clock_gettime(CLOCK_MONOTONIC, &t3);
    Q.enqueueReadBuffer (o_ddr, CL_TRUE, 0, COUNT*sizeof(short),
output, NULL, NULL);
    for(int n = 0; n < COUNT; n++)
    {
      result1 += output[n];
    }
    time1 = ns_diff(t1, t3) - datatime;
    if(i==3) printf("dotp result = %lld kernel time = %d ns\n",
result1, time1);
  }
  for(int i = 0; i < 4; i++)
  {
    result1 = 0;
    clock_gettime(CLOCK_MONOTONIC, &t1);
    Q.enqueueNDRangeKernel(kernel_local, NullRange, NDRange(COUNT/
LOOPNUM), NDRange(1), NULL, NULL);
    Q.finish();
    clock_gettime(CLOCK_MONOTONIC, &t3);
    Q.enqueueReadBuffer (o_ddr, CL_TRUE, 0, COUNT*sizeof(short),
output, NULL, NULL);
    for(int n = 0; n < COUNT; n++)
    {
      result1 += output[n];
    }
    time2 = ns_diff(t1, t3) - datatime;
    if(i==3) printf("dotp_local result = %lld kernel time = %d ns\n",
result1, time2);
  }

 printf("using local buffer reduced the timing by %0.2f%\n", ((float)
(time1-time2)*100.0f/(float)time1));
 }
```

```
  catch (Error err)
  {
   cerr << "ERROR: " << err.what () << " (" << err.err () << ") " << endl;
   exit (-1) ;
  }
}
static unsigned ns_diff (struct timespec &t1, struct timespec &t2)
{ return (t2.tv_sec - t1.tv_sec) * 1e9 + (t2.tv_nsec - t1.tv_nsec) ; }
```

2) Explore the **dotp.cl** kernel shown below and located in:
   /home/Chapter_OpenCL/lab7_dotp_local/dotp.cl

```
#define LOOPNUM 256
__kernel void dotp(__global short* a, __global short* x, __global
short* output)
{
 int i = get_global_id(0);
 int j;
 for(j=0;j<LOOPNUM;j++)
 {
  output[i*LOOPNUM+j] = a[i*LOOPNUM+j]*x[i*LOOPNUM+j];
 }
}
__kernel void dotp_local(__global short* a, __global short* x,
__global short* output)
{
 int i = get_global_id(0);
 int j;
 __local short temp[LOOPNUM];
 for(j=0;j<LOOPNUM;j++)
 {
  temp[j] = a[i*LOOPNUM+j]*x[i*LOOPNUM+j];
 }
 event_t ev = async_work_group_copy(&output[i*LOOPNUM], &temp[0],
LOOPNUM, 0);
 wait_group_events(1,&ev);
}
__kernel void data(__global short* a, __global short* x, __global
short* output)
{
 //empty kernel with data args

}
__kernel void null()
{
 //empty kernel
}
```

3) Run the **dotp** program. Use **cd** to enter the project directory, type **make** to compile the project and type **./dotp** to run the program. The output should be similar to this:

```
root@k2hnode1:~/Chapter_Open_CL/lab7_dotp_local# ./dotp

Count = 655360
estimated time of dispatching empty kernel = 60000 ns
estimated time of data caching = (3 + 1310720(bytes)/8096) * 2 = 328000 ns
estimated total overhead = 388000

time of empty kernel = 417578 ns
time of empty kernel with data args 432266 ns
dotp result = 7158169600 kernel time = 2615552 ns
dotp_local result = 7158169600 kernel time = 1410107 ns
using local buffer reduced the timing by 46.09%
root@k2hnode1:~/Chapter_Open_CL/lab7_dotp_local#
```

13.11.8   Laboratory experiment 8: Barrier

Project location:

/home/Chapter_OpenCL/lab8_barrier

1) Explore the **main.cpp** function shown below and located in:
/home/Chapter_OpenCL/lab8_barrier/main.cpp

```cpp
#define __CL_ENABLE_EXCEPTIONS
#include <CL/cl.hpp>
#include <SDL/SDL.h>
#include <iostream>
#include <fstream>
#include <cassert>
#include <cstdlib>
#include <signal.h>
#include <math.h>
#include "ocl_util.h"
#include "dotp.dsp_h"
#include "omp.h"

using namespace std;
using namespace cl;
static unsigned ns_diff (struct timespec &t1, struct timespec &t2);

int main(int argc, char *argv[])
{
  try
  {
   Context context(CL_DEVICE_TYPE_ACCELERATOR);
   std::vector<Device> devices = context.
getInfo<CL_CONTEXT_DEVICES>();
   CommandQueue    Q    (context, devices[0]);
   Program::Binaries binary(1, make_pair(dotp_dsp_bin, sizeof
(dotp_dsp_bin)));
```

```
   Program         program = Program(context, devices, binary);
   program.build(devices);

   Kernel kernel_print(program, "print");
   Kernel kernel_printbarrier(program, "print_barrier");

   printf("barrier example:\n");
   printf("printing without barrier:\n");
   Q.enqueueNDRangeKernel(kernel_print, NullRange, NDRange(4),
NDRange(4));
   Q.finish();
   printf("\nprinting with barrier:\n");
   Q.enqueueNDRangeKernel(kernel_printbarrier, NullRange,
NDRange(4), NDRange(4));
   Q.finish();
  }
 catch (Error err)
 {
  cerr << "ERROR: " << err.what() << "(" << err.err() << ")" << endl;
  exit(-1);
 }
}

static unsigned ns_diff (struct timespec &t1, struct timespec &t2)
{ return (t2.tv_sec - t1.tv_sec) * 1e9 + (t2.tv_nsec - t1.tv_nsec); }
```

2) Explore the **dotp.cl** kernel (shown below) located in:
   /home/Chapter_OpenCL/lab8_barrier/dotp.cl

```
__kernel void print()
{
 int i = get_global_id(0);
 int j = get_local_id(0);
 int k;
 for(k=0;k<3;k++)
 {
  printf("%d\n",j);
 }
}
__kernel void print_barrier()
{
 int i = get_global_id(0);
 int j = get_local_id(0);
 int k;
 for(k=0;k<3;k++)
 {
  printf("%d\n",j);
```

```
  barrier(CLK_LOCAL_MEM_FENCE);
  }
}
```

3) Run the **dotp** program. Use the command **cd** to enter the project directory, type **make** to compile the project and then type **./dotp** to run the program. The output should be similar to this:

```
COM8 - PuTTY                                         —    □    ×
root@k2hnode1:~/Chapter_Open_CL/lab8_barrier# ./dotp
barrier example:
printing without barrier:
[core 3] 0
[core 3] 0
[core 3] 0
[core 3] 1
[core 3] 1
[core 3] 1
[core 3] 2
[core 3] 2
[core 3] 2
[core 3] 3
[core 3] 3
[core 3] 3

printing with barrier:
[core 2] 0
[core 2] 1
[core 2] 2
[core 2] 3
[core 2] 0
[core 2] 1
[core 2] 2
[core 2] 3
[core 2] 0
[core 2] 1
[core 2] 2
[core 2] 3
root@k2hnode1:~/Chapter_Open_CL/lab8_barrier#
```

13.11.9   Laboratory experiment 9: Profiling

Project location:

/home/Chapter_OpenCL/lab9_dotp_profile

1) Explore the **main.cpp** function shown below and located in:
   /home/Chapter_OpenCL/lab9_dotp_profile/main.cpp

```
#define __CL_ENABLE_EXCEPTIONS
#include <CL/cl.hpp>
#include <SDL/SDL.h>
#include <iostream>
#include <fstream>
#include <cassert>
#include <cstdlib>
```

```
#include <signal.h>
#include <math.h>
#include "ocl_util.h"
#include "dotp.dsp_h"
#include "omp.h"

#define COUNT 2560
int o[COUNT];
int o1[COUNT];
short a[COUNT];
short x[COUNT];
using namespace std;
using namespace cl;

int main(int argc, char *argv[])
{
 std::string str;
 for(int i = 0; i < COUNT; i++)
 {
  a[i]=i%256;
  x[i]=256-a[i];
 }

 try
 {
  Context context(CL_DEVICE_TYPE_ACCELERATOR);
  std::vector<Device> devices = context.
getInfo<CL_CONTEXT_DEVICES>();
  printf("Count = %d\n\n",COUNT);
  CommandQueue    Q    (context, devices
[0],CL_QUEUE_PROFILING_ENABLE);
  CommandQueue    Q_ooo  (context, devices
[0],CL_QUEUE_PROFILING_ENABLE|
CL_QUEUE_OUT_OF_ORDER_EXEC_MODE_ENABLE);
  Program::Binaries binary(1, make_pair(dotp_dsp_bin,sizeof
(dotp_dsp_bin)));
  Program      program = Program(context, devices, binary);
  program.build(devices);

  Kernel kernel_sa(program, "dotp_sa");
  Kernel kernel_C(program, "dotp_C");
  long long result1=0,result2=0;
  Buffer aBuff (context, CL_MEM_READ_ONLY|CL_MEM_USE_MSMC_TI,
COUNT*sizeof(short));
  Buffer xBuff (context, CL_MEM_READ_ONLY|CL_MEM_USE_MSMC_TI,
COUNT*sizeof(short));
  Buffer oBuff (context, CL_MEM_READ_WRITE|CL_MEM_USE_MSMC_TI,
COUNT*sizeof(int));
  Buffer oBuff1 (context, CL_MEM_READ_WRITE|CL_MEM_USE_MSMC_TI,
COUNT*sizeof(int));
```

```
kernel_sa.setArg(0, aBuff );
kernel_sa.setArg(1, xBuff );
kernel_sa.setArg(2, oBuff );
kernel_C.setArg(0, aBuff );
kernel_C.setArg(1, xBuff );
kernel_C.setArg(2, oBuff );

Q.enqueueWriteBuffer(aBuff, CL_FALSE, 0, COUNT*sizeof(short), a,
NULL, NULL);
Q.enqueueWriteBuffer(xBuff, CL_FALSE, 0, COUNT*sizeof(short), x,
NULL, NULL);
Event ev,ev1,ev2;
std::vector<UserEvent> evs(1);

printf("in-order queue vs out-of-order queue:\n");
result1=0;
result2=0;
cl_ulong lt[8];
kernel_C.setArg(2, oBuff );
kernel_sa.setArg(2, oBuff1 );

Q.enqueueNDRangeKernel(kernel_C, NullRange, NDRange(COUNT),
NDRange(1), NULL, &ev1);
Q.enqueueNDRangeKernel(kernel_sa, NullRange, NDRange
(COUNT/32/4), NDRange(1), NULL, &ev2);
Q.finish();
ev1.getProfilingInfo(CL_PROFILING_COMMAND_QUEUED,&lt[0]);
ev1.getProfilingInfo(CL_PROFILING_COMMAND_SUBMIT,&lt[1]);
ev1.getProfilingInfo(CL_PROFILING_COMMAND_START,&lt[2]);
ev1.getProfilingInfo(CL_PROFILING_COMMAND_END,&lt[3]);
ev2.getProfilingInfo(CL_PROFILING_COMMAND_QUEUED,&lt[4]);
ev2.getProfilingInfo(CL_PROFILING_COMMAND_SUBMIT,&lt[5]);
ev2.getProfilingInfo(CL_PROFILING_COMMAND_START,&lt[6]);
ev2.getProfilingInfo(CL_PROFILING_COMMAND_END,&lt[7]);

Q.enqueueReadBuffer (oBuff, CL_TRUE, 0, COUNT*sizeof(int),
o, NULL, NULL);
Q.enqueueReadBuffer (oBuff1, CL_TRUE, 0, COUNT/4/32*sizeof(int),
o1, NULL, NULL);

for(int n = 0; n < COUNT; n++)
{
 result1 += o[n];
}
for(int n = 0; n < COUNT/4/32; n++)
```

```
   {
    result2 += o1[n];
   }
  printf("in-order-queue, kernel 1 has %d numbers, kernel 2 has %d
  numbers:\n", COUNT,COUNT/32);
  printf("(! not time staped) kernel1 result = %lld kernel2 result = %
  lld\n", result1,result2);
  printf("kernel1 queue %0.0f us\n",(int)(lt[0]-lt[0])/1e3);
  printf("kernel1 submit %0.0f us\n",(int)(lt[1]-lt[0])/1e3);
  printf("kernel1 start %0.0f us\n",(int)(lt[2]-lt[0])/1e3);
  printf("kernel1 end %0.0f us\n",(int)(lt[3]-lt[0])/1e3);
  printf("kernel2 queue %0.0f us\n",(int)(lt[4]-lt[0])/1e3);
  printf("kernel2 submit %0.0f us\n",(int)(lt[5]-lt[0])/1e3);
  printf("kernel2 start %0.0f us\n",(int)(lt[6]-lt[0])/1e3);
  printf("kernel2 end %0.0f us\n",(int)(lt[7]-lt[0])/1e3);

  //same task using out-of-order queue Q_ooo
  result1=0;
  result2=0;
  Q_ooo.enqueueNDRangeKernel(kernel_C, NullRange, NDRange(COUNT),
NDRange(1), NULL, &ev1);
  Q_ooo.enqueueNDRangeKernel(kernel_sa, NullRange, NDRange(COUNT/
32/4), NDRange(1), NULL, &ev2);//32 to make the kernel smaller
  Q_ooo.finish();
  ev1.getProfilingInfo(CL_PROFILING_COMMAND_QUEUED,&lt[0]);
  ev1.getProfilingInfo(CL_PROFILING_COMMAND_SUBMIT,&lt[1]);
  ev1.getProfilingInfo(CL_PROFILING_COMMAND_START,&lt[2]);
  ev1.getProfilingInfo(CL_PROFILING_COMMAND_END,&lt[3]);

  ev2.getProfilingInfo(CL_PROFILING_COMMAND_QUEUED,&lt[4]);
  ev2.getProfilingInfo(CL_PROFILING_COMMAND_SUBMIT,&lt[5]);
  ev2.getProfilingInfo(CL_PROFILING_COMMAND_START,&lt[6]);
  ev2.getProfilingInfo(CL_PROFILING_COMMAND_END,&lt[7]);
  Q_ooo.enqueueReadBuffer (oBuff, CL_TRUE, 0, COUNT*sizeof(int),
o, NULL, NULL);
  Q_ooo.enqueueReadBuffer (oBuff1, CL_TRUE, 0, COUNT/4/32*sizeof
(int), o1, NULL, NULL);
  for(int n = 0; n < COUNT; n++)
  {
    result1 += o[n];
  }
  for(int n = 0; n < COUNT/4/32; n++)
  {
    result2 += o1[n];
  }
```

```
printf("out-of-order-queue, kernel 1 has %d numbers, kernel 2 has %d
numbers:\n", COUNT, COUNT/32);
printf("(! not time staped) kernel1 result = %lld kernel2 result = %
lld\n", result1, result2);
printf("kernel1 queue %0.0f us\n", (int)(lt[0]-lt[0])/1e3);
printf("kernel1 submit %0.0f us\n", (int)(lt[1]-lt[0])/1e3);
printf("kernel1 start %0.0f us\n", (int)(lt[2]-lt[0])/1e3);
printf("kernel1 end %0.0f us\n", (int)(lt[3]-lt[0])/1e3);
printf("kernel2 queue %0.0f us\n", (int)(lt[4]-lt[0])/1e3);
printf("kernel2 submit %0.0f us\n", (int)(lt[5]-lt[0])/1e3);
printf("kernel2 start %0.0f us\n", (int)(lt[6]-lt[0])/1e3);
printf("kernel2 end %0.0f us\n\n", (int)(lt[7]-lt[0])/1e3);
}
catch (Error err)
{
 cerr << "ERROR: " << err.what() << "(" << err.err() << ")" << endl;
 exit(-1);
}
}
```

2) Explore the **dotp.cl** kernel shown below and located in:
/home/Chapter_OpenCL/lab9_dotp_profile/dotp.cl

```
extern int ddotp4h2(__global short *m, __global short *n, int count);
extern int dotp4h(__global short *m, __global short *n, int count);

__kernel void dotp_sa(__global short* a, __global short* x, __global
int* o)
{

 int i = get_global_id(0);
 int j = i*4;
 o[i] = dotp4h(&a[j], &x[j], 4);
}
__kernel void dotp_C(__global short* a, __global short* x, __global
int* o)
{
 int i = get_global_id(0);
 o[i] = a[i]*x[i];
}
```

3) Run the **dotp** program. Use the **cd** command to enter the project directory, type **make** to compile the project and then type **./dotp** to run the program. The output should be similar to this:

```
root@k2hnode1:~/Chapter_Open_CL/lab9_dotp_profile# ./dotp
Count = 2560

in-order queue vs out-of-order queue:
in-order-queue, kernel 1 has 2560 numbers, kernel 2 has 80 numbers:
(! not time staped) kernel1 result = 27961600 kernel2 result = 641480
kernel1 queue 0 us
kernel1 submit 2 us
kernel1 start 17 us
kernel1 end 421 us
kernel2 queue 30 us
kernel2 submit 425 us
kernel2 start 434 us
kernel2 end 472 us
out-of-order-queue, kernel 1 has 2560 numbers, kernel 2 has 80 numbers:
(! not time staped)  kernel1 result = 27961600 kernel2 result = 641480
kernel1 queue 0 us
kernel1 submit 2 us
kernel1 start 17 us
kernel1 end 418 us
kernel2 queue 25 us
kernel2 submit 27 us
kernel2 start 31 us
kernel2 end 454 us

root@k2hnode1:~/Chapter_Open_CL/lab9_dotp_profile#
```

13.11.10   Laboratory experiment 10: OpenMP in kernel

Project location:

/home/Chapter_OpenCL/lab10_dotp_cl_omp

1) Explore the **main.cpp** function shown below and located in:
   /home/Chapter_OpenCL/lab10_dotp_cl_omp/main.cpp

```cpp
#define __CL_ENABLE_EXCEPTIONS
#include <CL/cl.hpp>
#include <SDL/SDL.h>
#include <iostream>
#include <fstream>
#include <cassert>
#include <cstdlib>
#include <signal.h>
#include <math.h>
#include "ocl_util.h"
#include "dotp.dsp_h"
#include "omp.h"

#define COUNT 256

short a[COUNT];
short x[COUNT];
int o[COUNT];

using namespace std;
```

```
using namespace cl;

static unsigned ns_diff (struct timespec &MP;t1, struct timespec
&AMP;t2);

int main(int argc, char *argv[])
{
  for(int i = 0; i < COUNT; i++)
  {
   a[i]=i%256;
   x[i]=256-i%256;
  }
  struct timespec t1, t3;
  try
  {
   Context context(CL_DEVICE_TYPE_ACCELERATOR);
   std::vector<Device> devices = context.
getInfo<CL_CONTEXT_DEVICES>();

   printf("Count = %d\n",COUNT);
   CommandQueue    Q    (context, devices[0]);
   Program::Binaries binary(1, make_pair(dotp_dsp_bin,sizeof
(dotp_dsp_bin)));
   Program       program = Program(context, devices, binary);
   program.build(devices);
   Kernel kernel_omp(program, "dotp_omp");

   long long result=0;
   Buffer aBuff (context, CL_MEM_READ_ONLY|CL_MEM_USE_MSMC_TI,
COUNT*sizeof(short));
   Buffer xBuff (context, CL_MEM_READ_ONLY|CL_MEM_USE_MSMC_TI,
COUNT*sizeof(short));
  Buffer ompresult (context, CL_MEM_READ_WRITE|CL_MEM_USE_MSMC_TI,
sizeof(long long));

   kernel_omp.setArg(0, aBuff );
   kernel_omp.setArg(1, xBuff );
   kernel_omp.setArg(2, ompresult);
   kernel_omp.setArg(3, COUNT);

   for(int i = 0; i < 5; i++)
   {
    result = 0;
    clock_gettime(CLOCK_MONOTONIC, &t1);
   Q.enqueueWriteBuffer(aBuff, CL_FALSE, 0, COUNT*sizeof(short), a,
NULL, NULL);
   Q.enqueueWriteBuffer(xBuff, CL_FALSE, 0, COUNT*sizeof(short), x,
NULL, NULL);
    Q.enqueueTask(kernel_omp, NULL, NULL);
```

```
    Q.enqueueReadBuffer (ompresult, CL_TRUE, 0, sizeof(long long),
&result, NULL, NULL);
    clock_gettime(CLOCK_MONOTONIC, &t3);
    if(i==4) printf("omp result = %lld Elapsed: %d ns\n", result,
ns_diff(t1, t3));
    }
  long long sum = 0;
  for(int i=0;i<COUNT;i++){
    sum += a[i]*x[i];
  }
  printf("correct result is %lld\n\n", sum);
}
catch (Error err)
{
  cerr << "ERROR: " << err.what() << "(" << err.err() << ")" << endl;
  exit(-1);
}
}
static unsigned ns_diff (struct timespec &t1, struct timespec &t2)
{ return (t2.tv_sec - t1.tv_sec) * 1e9 + (t2.tv_nsec - t1.tv_nsec); }
```

2) Explore the kernel **dotp.cl** shown below and located.
  /home/Chapter_OpenCL/lab10_dotp_cl_omp/dotp.cl

```
__kernel void dotp_omp(__global short* a, __global short* x, __global
long *result, int count)
{
  result[0]=dotp_ompfunction(a,x,count);
}
```

3) Explore the **dotp_omp.c** function file shown below and located in:
  /home/Chapter_OpenCL/lab10_dotp_cl_omp/dotp_omp.c

```
long long dotp_ompfunction(short* a, short* x, int count)
{
  int i;
  long long sum=0;
  #pragma omp parallel for private(i) reduction(+:sum) shared
(count,a,x) num_threads(8)
  for(i=0;i<count;i++){
    sum += a[i]*x[i];
  }
  return sum;
}
```

4) Run the program. Use the **cd** command to enter the project directory, type **make** to compile the project and then type **./dotp** to run the program. The output should be similar to this:

```
COM8 - PuTTY                                            —   □   ×
root@k2hnode1:~/Chapter_Open_CL/lab10_dotp_cl_omp# ./dotp
Count = 256
omp result = 2796160 Elapsed: 152153 ns
correct result is 2796160

root@k2hnode1:~/Chapter_Open_CL/lab10_dotp_cl_omp#
```

13.11.11   Laboratory experiment 11: OpenMP in ARM

Project location:

/home/Chapter_OpenCL/lab11_dotp_multiARM

1) Explore the **main.cpp** function: **main.cpp** shown below and located in:
   /home/Chapter_OpenCL/lab11_dotp_multiARM/main.cpp

```cpp
#define __CL_ENABLE_EXCEPTIONS
#include <CL/cl.hpp>
#include <SDL/SDL.h>
#include <iostream>
#include <fstream>
#include <cassert>
#include <cstdlib>
#include <signal.h>
#include <math.h>
#include "ocl_util.h"
#include "dotp.dsp_h"
#include "omp.h"
#define COUNT 1024
#define COUNTPC (COUNT/4)

short o[COUNT];
short a[COUNT];
short x[COUNT];
using namespace std;
using namespace cl;
static unsigned ns_diff (struct timespec &t1, struct timespec &t2);

int main(int argc, char *argv[])
{
  std::string str;
  try
  {
```

```
   Context context(CL_DEVICE_TYPE_ACCELERATOR);
   std::vector<Device> devices = context.
getInfo<CL_CONTEXT_DEVICES>();
   printf("Count = %d Count per core = %d\n",COUNT,COUNTPC);
   CommandQueue Q(context, devices[0]);
   Program::Binaries binary(1, make_pair(dotp_dsp_bin,sizeof
(dotp_dsp_bin)));
   Program program = Program(context, devices, binary);
   program.build(devices);
   Kernel kernel_C(program, "dotp_C");
   long long result1=0;
   Buffer aBuff (context, CL_MEM_READ_ONLY|CL_MEM_USE_MSMC_TI,
COUNT/4*sizeof(short));
   Buffer xBuff (context, CL_MEM_READ_ONLY|CL_MEM_USE_MSMC_TI,
COUNT/4*sizeof(short));
   Buffer oBuff (context, CL_MEM_READ_WRITE|CL_MEM_USE_MSMC_TI,
COUNT/4*sizeof(short));
   kernel_C.setArg(0, aBuff );
   kernel_C.setArg(1, xBuff );
   kernel_C.setArg(2, oBuff );
   std::vector<Event> evs;
   #pragma omp parallel default(shared)
   {
    int tid = omp_get_thread_num();
    for(int i = 0; i < COUNTPC; i++)
    {
     a[tid*COUNTPC+i]=i;
     x[tid*COUNTPC+i]=255-i;
    }
    #pragma omp critical
    {
   Q.enqueueWriteBuffer(aBuff, CL_FALSE, 0, COUNTPC*sizeof(short),
   &a[tid*COUNTPC], NULL, NULL);
   Q.enqueueWriteBuffer(xBuff, CL_FALSE, 0, COUNTPC*sizeof(short),
   &x[tid*COUNTPC], NULL, NULL);
   Q.enqueueNDRangeKernel(kernel_C, NullRange, NDRange(COUNTPC),
   NDRange(1), NULL, NULL);
   Q.enqueueReadBuffer (oBuff, CL_TRUE, 0, COUNTPC*sizeof(short), &o
   [tid*COUNTPC], NULL, NULL);
     printf("core %d finished\n",tid);
    }
   }
   for(int n = 0; n < COUNT; n++)
   {
    result1 += o[n];
   }
   printf("total %lld\n",result1);
   long long sum = 0;
   for(int i=0;i<1024;i++){
```

```
    sum += a[i]*x[i];
   }
   printf("correct result is %lld\n", sum);
  }
  catch (Error err)
  {
   cerr << "ERROR: " << err.what() << "(" << err.err() << ")" << endl;
   exit(-1);
  }
 }
 static unsigned ns_diff (struct timespec &t1, struct timespec &t2)
 { return (t2.tv_sec - t1.tv_sec) * 1e9 + (t2.tv_nsec - t1.tv_nsec); }
```

2) Explore the kernel **dotp.cl** shown below and located.
   /home/Chapter_OpenCL/lab11_dotp_multiARM/dotp.cl

```
 __kernel void dotp_C(__global short* a, __global short* x, __global
 int* o)
 {
  int i = get_global_id(0);
  o[i] = a[i]*x[i];
 }
```

3) Run the program. Use the **cd** command to enter the project directory, type **make** to compile the project and type **./dotp** to run the program. The output should be similar to this:

```
COM8 - PuTTY                                              —   □   ×
root@k2hnode1:~/Chapter_Open_CL/lab11_dotp_multiARM# ./dotp
Count = 1024 Count per core = 256
core 0 finished
core 3 finished
core 2 finished
core 1 finished
total 11054080
correct result is 11054080
root@k2hnode1:~/Chapter_Open_CL/lab11_dotp_multiARM#
```

13.12   Conclusion

In this chapter, the OpenCL programming model has been introduced. It has been shown how to write host and kernel programs and use synchronisation at the host or kernel levels. It has also been shown how to use OpenMP (see Chapter 12) with OpenCL. Eleven examples have been given to provide working programs that can be easily modified to implement other applications. The most important feature of OpenCL is that the user does not need to deal with the ARM–DSP communication, which can be difficult to program as seen with the inter-processor communication covered in Chapter 9.

References

1 Khronos Group, The open standard for parallel programming of heterogeneous systems, 2016. [Online]. Available: https://www.khronos.org/opencl/.
2 Khronos Group, Conformant products, 2016. [Online]. Available: https://www.khronos.org/conformance/adopters/conformant-products#topencl.
3 Texas Instruments, Understanding OpenCL memory usage, June 2015. [Online]. Available: http://processors.wiki.ti.com/index.php/Understanding_OpenCL_Memory_Usage.
4 Texas Instruments, Debug with printf, October 2016. [Online]. Available: http://downloads.ti.com/mctools/esd/docs/opencl/debug/debug_printf.html.
5 Texas Instruments, OpenMP dispatch from OpenCL, October 2016. [Online]. Available: http://downloads.ti.com/mctools/esd/docs/opencl/extensions/openmp_dsp_dispatch.html?highlight=queue.

14

Multicore Navigator

14.1   Introduction

The Multicore Navigator, also referred to as simply the Navigator, provides a high-speed packet data transfer to enhance CorePac to accelerator/peripheral data movements, core-to-core data movements, inter-core communication and synchronisation without loading the CorePacs. To achieve this, the Navigator architecture is built with two main components, the hardware Queue Manager Subsystem (QMSS) and the Packet DMA engine (PKDMA) which is similar to the EDMA (Enhanced Direct Memory Access) engine covered in Chapter 8. Figure 14.1

Multicore DSP: From Algorithms to Real-time Implementation on the TMS320C66x SoC, First Edition. Naim Dahnoun.
© 2018 John Wiley & Sons Ltd. Published 2018 by John Wiley & Sons Ltd.
Companion website: www.wiley.com/go/dahnoun/multicoredsp

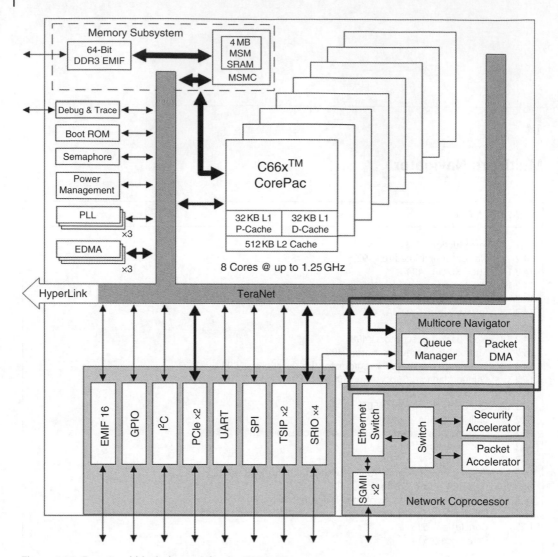

Figure 14.1 Functional block diagram for the C6678 [1].

and Figure 14.2 show the locations of the Navigators for both the KeyStone I and the KeyStone II, respectively.

14.2 Navigator architecture

To use the Navigator efficiently, one needs to understand the following Navigator components:

1) The PKDMA
2) The QMSS
3) The descriptors.

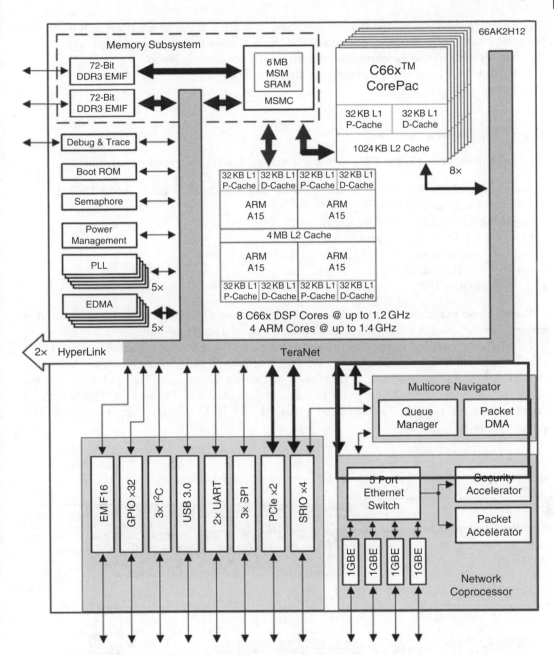

Figure 14.2 Functional block diagram for the 66AK2H14 [2].

Data to be sent are included in structures that also contain other information. These structures are called *descriptors*. To send data, these descriptors are pushed into hardware queues. These queues are managed by the QMSS which is the heart of the Navigator. The PKDMAs then move the descriptors between queues and peripherals. The internal PKDMA, also known as the infrastructure PKDMA, can move data pointed to by one core to another memory pointed to by another core. This method of moving data frees the cores, and therefore no transmitting cores will be blocked by any receiving core or cores.

Once the Navigator is configured and initialised, at runtime the descriptors are only pushed or popped by the application into the queues, and the Navigator hardware dispatcher takes care of the data movement and synchronisation without the cores being involved.

In Section 14.2.1, the main components are introduced.

14.2.1 The PKDMA

As shown in Figure 14.3, the PKDMAs are distributed in the following peripherals:

- QMSS (the PKDMA in QMSS is known as the infrastructure, or core-to-core PKDMA)
- Antenna Interface 2 (AIF2) [3]
- KeyStone Architecture Bit Rate Coprocessor (BCP) [4]
- Fast Fourier Transform Coprocessor (FFTC) [5]
- Network Coprocessor (NETCP) [6,7]
- Serial Rapid IO (SRIO) [8]
- IQ Net #2 (IQN2) [9].

Not all these peripherals are available on a single device. Datasheets should be consulted in order to determine if a peripheral is available on a specific device or not.

The PKDMAs located in the peripherals and the QMSS are composed of a Transmit DMA (Tx DMA) and a Receive DMA (Rx DMA). The PKDMA is instructed to take data using the Tx DMA and pass it to the peripheral engine for processing (e.g. and FFT) via the streaming interface that also provides handshaking as shown in Figure 14.4. The peripheral receives data and provides the processed data to the Rx DMA which in turn retransmits this processed data.

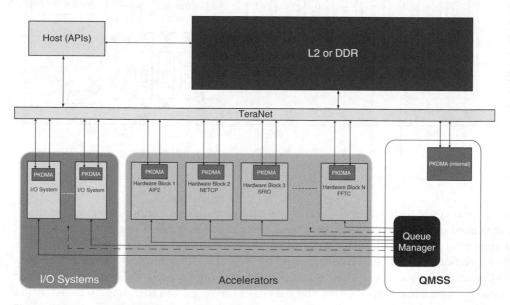

Figure 14.3 The Navigator architecture, simplified.

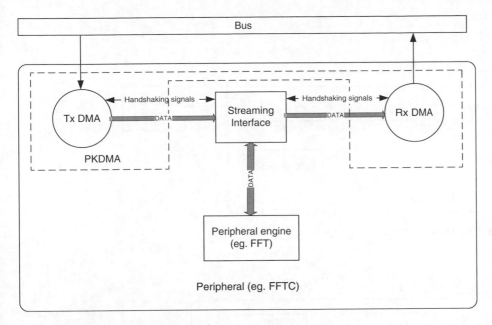

Figure 14.4 PKDMA within a peripheral.

14.2.1.1 PKDMA transmit side

The PKDMA transmit side is composed of:

1) Transmit scheduler that allows prioritisation of queues to channels (the Tx DMA selects the channels using a 4-level round-robin approach). The number of channels is limited and depends on the peripheral; see Table 14.1.
2) Transmit core (Tx Core) that pushes the data from a queue to a transmit channel that includes FIFOs.
3) Transmit channel and a FIFO that buffer the data before sending it to the streaming interface; see Figure 14.5. The transmit channels are programmable and allow features such as channel enable or disable and filtering the extended packet information field included in the descriptors; see Tx Channel N Global Configuration Registers A and B in Ref. [10].

14.2.1.2 PKDMA receive side

The PKDMA receive side (Figure 14.6) is composed of:

1) Rx channel that receives data and control signals from the streaming interface. This Rx channel buffers data coming from the streaming interface to the FIFOs.
2) Rx core pops data from the FIFOs and pushes it into a queue.

Table 14.1 PKDMA channel map

	QMSS	SRIO	NETCP1	NETCP 1.5	AIF	IQN2	BCP	XGE	FFTC
RX channels	32	16	24	91	129	48	8	16	4
TX channels	32	16	9	21	129	48	8	8	4
RX flows	64	20	32	64	129	64	64	32	8

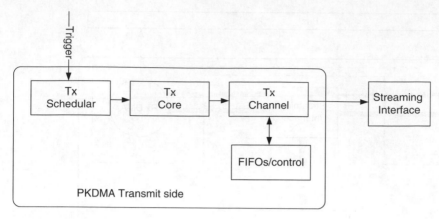

Figure 14.5 PKDMA transmit side.

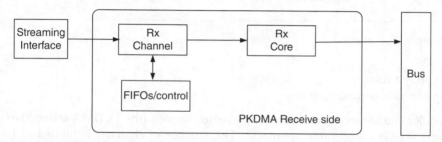

Figure 14.6 PKDMA receive side.

It has been shown that descriptors instruct the PKDMA transmit how data are handled. However, for the PKDMA receive, there are other registers called *Rx flow registers* that need to be programmed.

For each channel (see Table 14.1), there is an Rx DMA channel Config region register to set up; see Figure 14.7. This register is mainly for enabling/disabling, clearing or pausing a channel. These registers are aligned to 32 bytes as shown in Table 14.2.

There are also eight registers (Register A through Register H) for each flow that needs to be set. There are more flows than channels (see Table 14.1) to provide more flexibility. For more details, refer to the user's manual in Ref. [10].

For the infrastructure PKDMA, the Rx flows are specified in the Tx descriptor (Host Packet Descriptor Packet Information Word 1: Source Tag – Lo field) as an index.

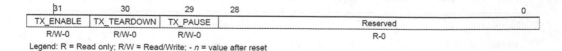

Figure 14.7 Tx Channel N Global Configuration Register A (0x000 + 32 × N) [10].

Table 14.2 Rx DMA channel Config region registers

Address	Register
0x00	Rx Channel 0 Global Configuration Register A
0x04 – 0x1F	Reserved
0x20	Rx Channel 1 Global Configuration Register A
0x24 – 0x3F	Reserved
	:
	:
0x00 + N × 32	Rx Channel N Global Configuration Register A
0x04 + N × 32	Reserved

14.2.1.3 Infrastructure PKDMA

The infrastructure PKDMA is similar to the peripheral PKDMAs except that there is no peripheral engine to process data, and the streaming interface is connected in a loopback mode as shown in Figure 14.8. This is very useful for core-to-core data transfer and other peripheral-to-core transfers. It is also worth noting at this stage that data transmitted by the Tx DMA are received by the Rx DMA at the same clock cycle.

14.2.2 Descriptors

Descriptors are simple data structures that hold information as to how data are manipulated. These descriptors are used at runtime to pass argument information and are very handy when using message queues for passing messages (data) where the queue manager keeps a linked list as a message queue.

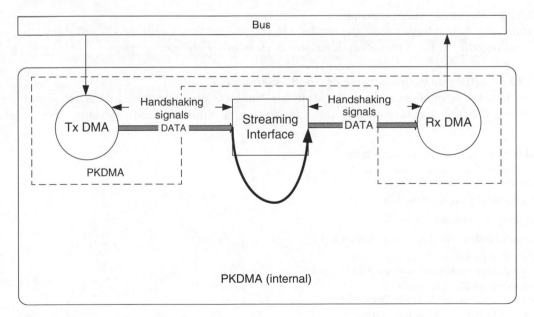

Figure 14.8 Infrastructure PKDMA.

These descriptors are stored into memory regions that are pushed to queues managed by the queue manager. There are two types of descriptors: the host packet descriptors and the monolithic descriptors.

14.2.2.1 Host packet descriptors

A host packet descriptor layout is shown in Table 14.3. The host packet descriptors hold the following information.

- Indicator that identifies the descriptor as a host packet descriptor
- Source and destination tags
- Packet type
- Packet length
- Protocol-specific region size
- Protocol-specific control/status bits
- Pointer to the first valid byte in the start of packet (SOP) data buffer
- Length of the SOP data buffer
- Pointer to the next buffer descriptor in the packet
- Software-specific information.

For instance, Words 3, 4 and 5 shown in Table 14.3 represent the buffer length, the buffer pointer and the next descriptor pointer, respectively.

The complete description of each field in Table 14.3 can be found in Ref. [10], and as an example, a packet descriptor used in the laboratory section is shown in Figure 14.9.

The descriptors can be linked together as shown in Figure 14.10. Note the Middle of Packets (MOP) and the End of Packets (EOP) are called *buffer descriptors*. The MOPs are useful when data to be transmitted are scattered and/or large. The structure of the host buffer descriptors is similar to that of the host packet descriptors. However, not all the fields are required; see Table 14.4 and Figure 14.11. Field descriptions can be found in Ref. [10].

14.2.2.2 Monolithic packet descriptor

Monolithic descriptors contain the header and the payload (see Figure 14.12) and cannot be linked together as the host packet descriptor does, but they are easy to use.

14.2.2.3 Setting up the memory regions for the descriptors

The linked descriptors must be of the same size, and these descriptors are located in memory regions. Different memory regions are available to hold different sized descriptors. For the

Table 14.3 Host packet descriptor layout

Packet info (3 words): (Word 0 to Word 2)

Buffer info (2 words): (Words 3 and 4)

Linking info (1 word): (Word 5)

Original buffer info (2 words): (Words 6 and 7)

Extended packet info block (optional):
Includes timestamp and software data (4 words)

Protocol-specific data (optional):
0 to M bytes, where M is a multiple of 4

Other SW data (optional and user defined)

```
MNAV_HostPacketDescriptor* host_pkt;
/*****************************************************************/
typedef struct
{
/* word 0 */
      uint32_t packet_length : 22;
      uint32_t ps_reg_loc : 1;
      uint32_t reserved_w0 : 2;
      uint32_t packet_type : 5;
      uint32_t type_id : 2;
/* word 1 */
      uint32_t dest_tag_lo : 8;
      uint32_t dest_tag_hi : 8;
      uint32_t src_tag_lo : 8;
      uint32_t src_tag_hi : 8;
/* word 2 */
      uint32_t pkt_return_qnum : 12;
      uint32_t pkt_return_qmgr : 2;
      uint32_t ret_push_policy : 1;
      uint32_t return_policy : 1;

      uint32_t ps_flags : 4;
      uint32_t err_flags : 4;
      uint32_t psv_word_count : 6;
      uint32_t reserved_w2 : 1;
      uint32_t epib : 1;
/* word 3 */
      uint32_t buffer_len : 22;
      uint32_t reserved_w3 : 10;
/* word 4 */
      uint32_t buffer_ptr;
/* word 5 */
      uint32_t next_desc_ptr;
/* word 6 */
      uint32_t orig_buff0_len : 22;
      uint32_t orig_buff0_refc : 6;
      uint32_t orig_buff0_pool : 4;
/* word 7 */
      uint32_t orig_buff0_ptr;
} MNAV_HostPacketDescriptor;
```

Figure 14.9 Host packet descriptor structure: example.

KeyStone I there are 20 regions, and for the KeyStone II there are 64 regions. These regions are configured by the three memory region registers as shown in Table 14.5. All registers are written by the host. The first register is written in order to set the base address of the memory region, and the second register is written to configure the index of the first descriptor in the memory region; for instance, if we have chosen to have two memory regions because we have two different descriptors, we need to define the first index for each region (Index 0 and Index 16) as illustrated in Figure 14.13. Finally, the last register is for specifying the descriptor size and the memory region size [10].

This can be implemented by the code shown in Figure 14.14.

The base address of the QMSS configuration registers can be found in the appropriate chip user's manuals, as shown in Figure 14.15 and Figure 14.16.

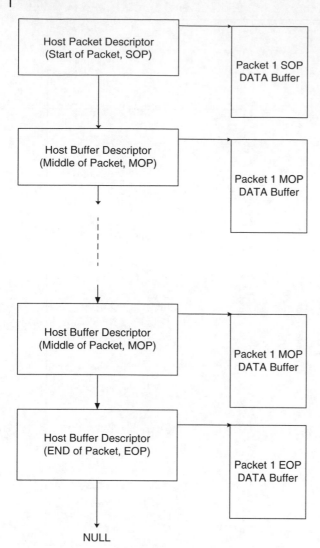

Figure 14.10 Host and buffer packets linked.

Table 14.4 Host buffer descriptor layout

Reserved (10 bytes)
Buffer reclamation info (2 bytes)
Buffer info (8 bytes)
Linking info (4 bytes)
Original buffer info (8 bytes)

14.2.3 Queue Manager Subsystem

The QMSS is composed of the following blocks (see Figure 14.18 and Figure 14.19):

1) Queue Manager (QM)
2) Infrastructure PKDMA (see Section 14.2.1)

```
typedef struct
{
/* word 0 */
    uint32_t reserved_w0;
/* word 1 */
    uint32_t reserved_w1;
/* word 2 */
    uint32_t pkt_return_qnum : 12;
    uint32_t pkt_return_qmgr : 2;
    uint32_t ret_push_policy : 1; //0=return to queue tail, 1=queue head
    uint32_t reserved_w2 : 17;
/* word 3 */
    uint32_t buffer_len : 22;
    uint32_t reserved_w3 : 10;
/* word 4 */
    uint32_t buffer_ptr;
/* word 5 */
    uint32_t next_desc_ptr;
/* word 6 */
    uint32_t orig_buff0_len : 22;
    uint32_t orig_buff0_refc : 6;
    uint32_t orig_buff0_pool : 4;
/* word 7 */
    uint32_t orig_buff0_ptr;
} MNAV_HostBufferDescriptor;
```

Figure 14.11 Host buffer descriptor structure: example.

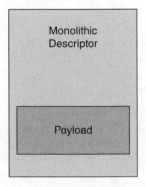

Figure 14.12 Monolithic descriptor.

Table 14.5 Description memory setup region registers [10]

Offset	Name	Description
0x00000000 + 16 × R	Memory Region R Base Address Register (0...19, or 0...63 for KeyStone II)	The Memory Region R Base Address Register is written by the host to set the base address of memory region R. This memory region will store a number of descriptors of a particular size as determined by the Memory Region R Control Register.
0x00000004 + 16 × R	Memory Region R Start Index Register (0...19, or 0...63 for KeyStone II)	The Memory Region R Start Index Register is written by the host to configure index of the first descriptor in this memory region.
0x00000008 + 16 × R	Memory Region R Descriptor Setup Register (0...19, or 0...63 for KeyStone II)	The Memory Region R Descriptor Setup Register is written by the host to configure various descriptor related parameters of this memory region.

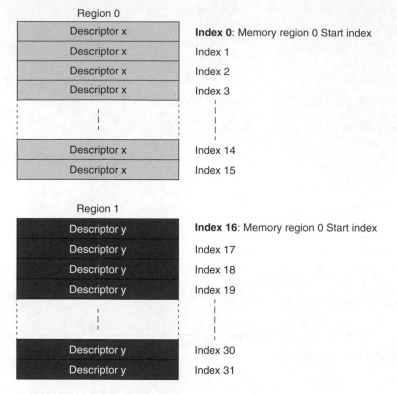

Figure 14.13 Memory region indexing.

```
/* Descriptor Memory Region */
#define QM_REG_MEM_REGION_BASE 0x000
#define QM_REG_MEM_REGION_INDEX 0x004
#define QM_REG_MEM_REGION_SETUP 0x008

#define QMSS_CFG_BASE (0x02a00000u)

#define QM_DESC_REGION (QMSS_CFG_BASE + 0x0006a000u)


reg = (uint32_t *)(QM_DESC_REGION + QM_REG_MEM_REGION_BASE + (regn * 16));
      *reg = addr;

reg = (uint32_t *)(QM_DESC_REGION + QM_REG_MEM_REGION_INDEX + (regn * 16));
      *reg = indx;

reg = (uint32_t *)(QM_DESC_REGION + QM_REG_MEM_REGION_SETUP + (regn * 16));
      *reg = setup;
```

Figure 14.14 Configuring the memory region registers.

02921000	029FFFFF	0 02921000	0 029FFFFF	1M-132K	Reserved
02A00000	02BFFFFF	0 02A00000	0 02BFFFFF	2M	Queue manager subsystem configuration
02C00000	07FFFFFF	0 02C00000	0 07FFFFFF	84M	Reserved

Figure 14.15 The base address of the QMSS configuration registers for the KeyStone I [1].

00 02A4 0000	00 02A7 FFFF	768K	Reserved		Reserved		Reserved	
00 02A0 0000	00 02AF FFFF	1M	Navigator configuration		Navigator configuration		Navigator configuration	
00 02B0 0000	00 02BF FFFF	1M	Navigator linking RAM		Navigator linking RAM		Navigator linking RAM	
00 02C0 0000	00 02C0 FFFF	64K	Reserved		Reserved		Reserved	

Figure 14.16 The base address of the QMSS configuration registers for the KeyStone II [2].

3) Accumulator packet data structure processors (APDSPs) (see Section 14.2.5)
4) Interrupt distributor (INTD) module.

14.2.4 Queue Manager

The QM manages the hardware queues which hold pointers to packets which are represented by descriptors. The KeyStone II has two queue managers, Q1 and Q2.

There are four management queue registers with fields that can be read, written to or both. These registers are used to push descriptors into the queues, pop descriptors from the queues or simply read information from the queues. These registers are:

a) Queue N Register A. This is a read-only register that specifies how many descriptors are in the current queue.
b) Queue N Register B. This is also a read-only register; it specifies the total number of bytes (for all packets) that are in the queue.
c) Queue N Register C. This register has two fields: one field is to specify if the descriptor will be pushed into the tail or the head of the queue (the default is the tail of the queue), and the other field is the packet size.
d) Queue N Register D. This register is written to in order to push a descriptor into the queue, and read to pop a descriptor from the queue. This register has two fields: one is used to specify the 16-byte-aligned address that points to the descriptor, and the other field is to specify the descriptor's size.

14.2.4.1 Queue peek registers

In general, peek operations on queues are the same as pops and pushes but without modifying the data. Therefore, it will be handy for the application to use these registers. In the QMSS, these peek registers provide the number of packets that are queued on a specific queue, the total number of bytes in all packed in the specific queue and the size of the packet in the head of a specific queue. There is also another register that allows the setup of a threshold value that the queue size is compared against and a bit field that is specified if the size is greater than, equal to, or less than the threshold before allowing a bit in the status register to be set; see Table 14.6.

As an example, the host can get the info of a queue **qn** and set a threshold as shown in the code in Figure 14.17.

Table 14.6 Peek registers

Offset	Name of the register	Description
0x00000000 + 16 × N	Queue N Status and Configuration Register A	Number of packets in the specific queue
0x00000004 + 16 × N	Queue N Status and Configuration Register B	Total number of bytes in all packets queued in the specific queue
0x00000008 + 16 × N	Queue N Status and Configuration Register C	Packet size of the head packet in the specific queue
0x0000000C + 16 × N	Queue N Status and Configuration Register D	Threshold value and when to set a bit in the status register

```
uint32_t get_byte_count(uint16_t qn)
{
      uint32_t *reg;
      uint32_t count;

      reg = (uint32_t *)(QM_PEEK_REGION + QM_REG_QUE_REG_B + (qn * 16));
      count = *reg;
      return(count);
}

uint32_t get_descriptor_count(uint16_t qn)
{
      uint32_t *reg;
      uint32_t count;

      reg = (uint32_t *)(QM_PEEK_REGION + QM_REG_QUE_REG_A + (qn * 16));
      count = *reg;
      return(count);
}

void set_queue_threshold(uint16_t qn, uint32_t value)
{
      uint32_t *reg;

      reg = (uint32_t *)(QM_PEEK_REGION + QM_REG_QUE_STATUS_REG_D + (qn * 16));
      *reg = value;
}
```

Figure 14.17 Code for reading the number of bytes and the number of descriptors in the queue.

14.2.4.2 Link RAM

The link RAM holds descriptors for the Keystone I and KeyStone II (see Figure 14.18 and Figure 14.19, respectively). The link RAM can hold 16K descriptors for the KeyStone I and 32K descriptors for the KeyStone II. If more descriptors are required, an external link RAM can be created into DDR or L2 memories.

14.2.5 Accumulator packet data structure processors

The uRISC cores (APDSPs) allow queue monitoring and management with programmable interrupts. Within the QMSS, there are two APDSPs in the KeyStone I and eight PDSPs in the KeyStone II. These APDSPs can be configured to perform some of the following tasks:

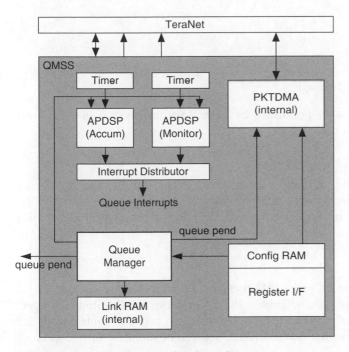

Figure 14.18 QMSS architecture (KeyStone I).

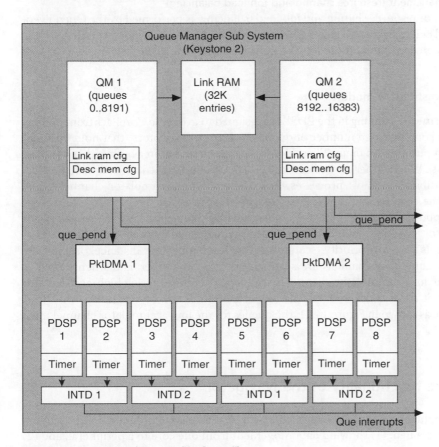

Figure 14.19 QMSS architecture (KeyStone II).

a) Accumulation
b) Quality of service (QoS)
c) Event management (job load balancing).

The implementation of these tasks is performed by Texas Instruments–provided firmware images that need to be loaded into the APDSPs.

14.2.5.1 Accumulation

This task allows the APDSPs to pull a number of selected queues and pop the appropriate descriptors to a host-provided buffer. When a certain number or a timeout is reached, an interrupt is generated to the host which in turn reads this buffer and takes action. The firmware takes care of recycling the descriptors.

14.2.5.2 Quality of service

Queues that can be used for QoS are fed to the PDSPs and are referred to as the *QoS queues*; see the highlighted queue maps in Table 14.7 and Table 14.8 for KeyStone I and KeyStone II, respectively. These PDSPs need to be loaded with the QoS firmware that provides control of packet flow to the host and peripherals. Further details on the QoS can be found in Refs. [10,11].

14.2.5.3 Event management (resource sharing and job load balancing)

Event management, or resource sharing and job load balancing, is performed by the Open Event Machine (OpenEM). The event management is performed by the scheduler running on the PDSPs and the dispatchers running on the CorePac; see Figure 14.20.

14.2.6 Interrupt distributor module

The accumulator firmware running in the PDSPs is designed to avoid the CorePacs from polling. The firmware polls some selected number of queues to extract descriptors (that hold information) that have been pushed into them. These descriptors are then put in a buffer (that has been provided by the host). When either the buffer is full or a timeout is reached, interrupts to the host are generated and, finally, the host processes the corresponding accumulator channel as the mapping between the channel and the host is fixed; see Figure 14.21.

The firmware supports 32, 16 or 48 (32 + 16) accumulator channels. These are known as the 32 high-accumulator channels and have high priority (scanned faster), and the 16 low-accumulator channels have low priority (scanned at low frequency). For instance, on the KeyStone I, the high-accumulator channels are scanned one after the other; and when a cycle is completed, then one low-accumulator channel is scanned. Therefore, if each high-accumulator channel is scanned at an N cycles interval, the low accumulator will be scanned at a $32N + 1$ cycles interval. Also, as shown in Figure 14.21, the queues for the low-accumulator channel must be contiguous [10].

14.3 Complete functionality of the Navigator

Once the QMSS is configured, the application code will be fairly easy to program. In this section, two scenarios are explained: (1) showing data movement from one core to a peripheral, and (2) showing data movement between two cores.

Table 14.7 Queue map for the KeyStone I

TCI6616 queues	TCI660x / C667x queues	TCI6618 / C6670 queues	TCI6614 queues	C665x queues	Purpose
0–511 (512)	Same	Same	Same	Same	Normally used by low-priority accumulation. The low-priority accumulator uses up to 512 queues divided into 16 channels, each channel being 32 continuous queues. Each channel triggers one broadcast interrupt. These queues can also be used as general-purpose queues.
512–639 (128)	Same	Same	Same		AIF2 Tx queues. Each queue has a dedicated queue pending signal that drives a Tx DMA channel.
640–648 (9)	Same	Same	Same		NetCP Tx queues. Each queue has a dedicated queue pending signal that drives a Tx DMA channel.
			650–657 (8)		ARM queue pend queues. These queues have dedicated queue pending signals wired directly to the ARM.
662–671 (10)	652–671 (20)	662–671 (10)	662–671 (10)		INTC0/INTC1 queue pend queues. These queues have dedicated queue pending signals wired directly into the chip-level INTC0 and/ or INTC1. Note that the event mapping can differ for each device.
			670–671 (2)		ARM queue pend queues. These queues have dedicated queue pending signals wired directly to the ARM. Note that these are also routed to INTC0.
672–687 (16)	Same	Same	Same	Same	SRIO Tx queues. Each queue has a dedicated queue pending signal that drives a Tx DMA channel.
688–695 (8)	Same	Same	Same		FFTC_A and _B Tx queues. Each queue has a dedicated queue pending signal that drives a Tx DMA channel.
704–735 (32)	Same	Same	Same	Same	Normally used by high-priority accumulation. The high-priority accumulator uses up to 32 queues, one per channel. Each channel triggers a core-specific interrupt. These queues can also be used as general-purpose queues.
736–799 (64)	Same	Same	Same	Same	Queues with starvation counters readable by the host. Starvation counters increment each time a pop is performed on an empty queue, and reset when the starvation count is read.
800–831 (32)	Same	Same	Same	Same	QMSS Tx queues. Used for infrastructure (core-to-core) DMA copies and notification.
832–863 (32)	Same	Same	Same	Same	General-purpose queues, or may be configured for use by QoS traffic-shaping firmware.
		864–867 (4)			FFTC_C Tx queues. Each queue has a dedicated queue pending signal that drives a Tx DMA channel.

(Continued)

Table 14.7 (Continued)

TCl6616 queues	TCl660x / C667x queues	TCl6618 / C6670 queues	TCl6614 queues	C665x queues	Purpose
864–895 (32)	Same	Same	Same	Same	HyperLink queue pend queues. These queues have dedicated queue pending signals wired directly into HyperLink. On some devices, these overlap. They cannot be used simultaneously for both IPs (i.e. use queue 864 for either FFTC_C or Hyperlink).
		868–875 (8)	864–871 (8)		BCP Tx queues. Each queue has a dedicated queue pending signal that drives a Tx DMA channel. Also routed to HyperLink.
896–8191	Same	Same	Same	Same	General purpose. Due to the mapping of logical to physical queues in the PKDMA interfaces, the use of 0xFFF in PKDMA *qnum* fields is reserved to specify non-override conditions.

Scenario 1. Consider Figure 14.22, where the QMMS is configured and the appropriate queues are selected and initialised. From this point, the following sequence is followed:

1) Core 0 reads a free descriptor from the Transmit Free Descriptor Queue (Tx Free Desc Queue). This descriptor will point to a free buffer that the application sets aside.
2) The application running in Core 0 will fill this buffer with the data to be transmitted and push the descriptor to the Transmit Queue (Tx Queue).
3) The queue manager will then generate a queue pend signal to the Transmit PKDMA (Tx PKDMA).
4) The Tx PKDMA automatically pops a descriptor from the Tx Queue, loads the data from the buffer, passes it to the streaming interface and recycles the descriptor.
5) The streaming interface then passes the processed data to the Receive PKDMA (Rx PKDMA).

Scenario 2. This scenario is similar to Scenario 1. Consider Figure 14.23, where the QMMS is configured and the appropriate queues are selected and initialised. From this point, the following sequence is followed:

1) Core 0 reads a free descriptor from the Tx Free Desc Queue. This descriptor will point to a free buffer that the application sets aside.
2) The application running in Core 0 will fill this buffer (Buffer 1) with the data to be transmitted and push the descriptor to the Tx Queue.
3) The queue manager will then generate a queue pend signal to the Tx PKDMA.
4) The Tx PKDMA automatically pops a descriptor from the Tx Queue, loads the data from the buffer and passes it to the streaming interface.
5) The streaming interface activates the RX PKDMA.
6) The Rx PKDMA then pops a descriptor from the Rx free descriptor queue. This descriptor will point to a buffer (Buffer 2), where the data sent through the streaming interface has to be written to.

Table 14.8 Queue map for the KeyStone II

K2K queues	K2H queues	K2L queues	K2E queues	Purpose
0–511 (512)	Same	Same	Same	Normally used by low-priority accumulation. The low-priority accumulator uses up to 512 queues divided into 16 channels, each channel being 32 continuous queues. Each channel triggers one broadcast interrupt. These queues can also be used as general-purpose queues.
512–639 (128)	Same			AIF2 Tx queues. Each queue has a dedicated queue pending signal that drives a Tx DMA channel.
		560–569 (10)	Same	EDMA0 queue pend queues
640–648 (9)	Same	896–1023 (128)	Same (896–1023)	NetCP Tx queues. Each queue has a dedicated queue pending signal that drives a Tx DMA channel.
528–559 (32), 652–671 (20)	Same	570–687 (118)	652–691 (40)	Broadcast CICx/SOC queue pend queues. These queues have dedicated queue pending signals wired directly into the chip-level interrupt controllers.
672–687 (16)	Same			SRIO Tx queues. Each queue has a dedicated queue pending signal that drives a Tx DMA channel.
688–695 (8)	Same	Same		FFTC_A and _B Tx queues. Each queue has a dedicated queue pending signal that drives a Tx DMA channel.
704–735 (32)	Same	Same	Same	Normally used by high-priority accumulation. The high-priority accumulator uses up to 32 queues, one per channel. Each channel triggers a core-specific interrupt. These queues can also be used as general-purpose queues.
736–799 (64)	Same	Same	Same	Queues with starvation counters readable by the host. Starvation counters increment each time a pop is performed on an empty queue, and reset when the starvation count is read.
800–831 (32)	Same	Same	Same	QMSS Tx queues for PKDMA1. Used for infrastructure (core-to-core) DMA copies and notification.
832–863 (32)	Same			General-purpose queues, or may be configured for use by QoS traffic-shaping firmware
	Same	832–879 (48)		IQN2 Tx queues
864–871 (8)	Same	696–703 (8)		BCP Tx queues. Each queue has a dedicated queue pending signal that drives a Tx DMA channel.
872–887 (16)	Same			FFTC_C, _D, _E and _F Tx queues (four per FFTC). Each queue has a dedicated queue pending signal that drives a Tx DMA channel.
8192–8703 (512)	Same			Normally used by low-priority accumulation for QM2. The low-priority accumulator uses up to 512 queues divided into 16 channels, each channel being

(Continued)

Table 14.8 (Continued)

K2K queues	K2H queues	K2L queues	K2E queues	Purpose
				32 continuous queues. Each channel triggers one broadcast interrupt. These queues can also be used as general-purpose queues.
8704–8735 (32)	Same	528–559 (32)	Same (528–559)	ARM interrupt controller queue pend queues
		589, 590	570–580 (11)	EDMA1 queue pend queues
		591–602 (12)	581–588 (8)	EDMA2 queue pend queues
		603, 604	589–604 (16)	EDMA3 queue pend queues
8736–8743 (8)	Same		605-612 (8)	EDMA4 queue pend queues
8744–8751 (8)	Same			HyperLink broadcast queue pend queues
8752–8759 (8)	Same		692–699 (8)	XGE queue pend queues
8796–8811 (16)	Same		613–636 (24)	HyperLink 0 queue pend queues
8812–8843 (32)	Same			DXB queue pend queues
8844–8863 (20)	Same			INTC0/C1/C2 queue pend queues. These queues have dedicated queue pending signals wired directly into the chip-level interrupt controllers.
8864–8879 (16)	Same			HyperLink 1 queue pend queues
8896–8927 (32)	Same			Normally used by high-priority accumulation for QM2. The high-priority accumulator uses up to 32 queues, one per channel. Each channel triggers a core-specific interrupt. These queues can also be used as general-purpose queues.
8928–8991 (64)	Same			Queues with starvation counters readable by the host. Starvation counters increment each time a pop is performed on an empty queue, and reset when the starvation count is read.
8992–9023 (32)	Same			QMSS Tx queues for PKDMA2. Used for infrastructure (core-to-core) DMA copies and notification.
9024–16383	Same			General-purpose queues

7) Rx PKDMA will fill Buffer 2.
8) The descriptor is then pushed into the Rx queue by the Rx PKDMA.
9) Core 1 is interrupted.
10) Core 1 reads Buffer 2.

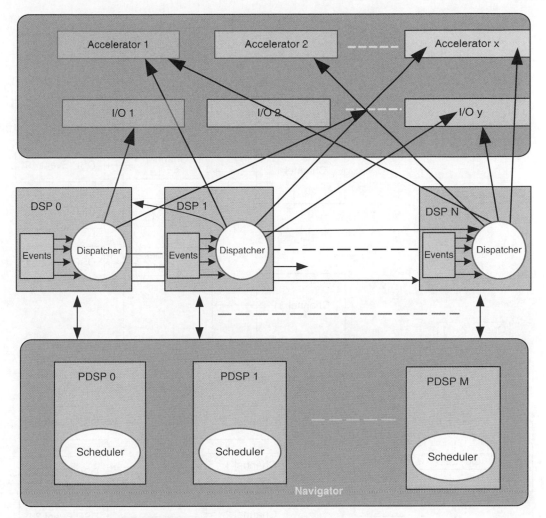

Figure 14.20 Event management with the Navigator.

14.4 Laboratory experiment

A complete implementation of a core-to-core data transmission shown in Figure 14.23 using the Navigator has been implemented on the KeyStone I.

Project location:

\Chapter_14_Code\Lab1

Step 1. Load the project and build it. No errors should happen.
Step 2. Group Core 0 and Core 1.
Step 3. Run the group.
Step 4. Explore the files.
Step 5. Referring to Figure 14.23 and Section 14.3 (Scenario 2), identify sequences 1 to 10 in the code.

For more details, refer to video tutorials and blogs on the Navigator [12–15]. The performance of the multicore Navigator is also reported in Ref. [16].

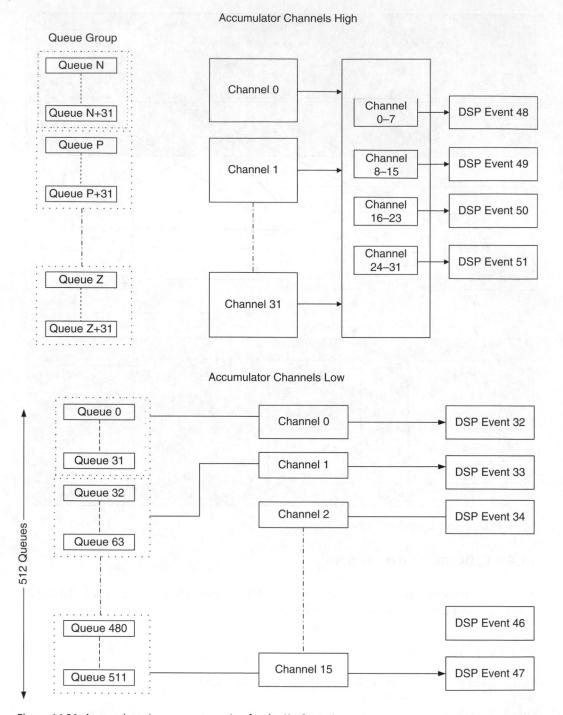

Figure 14.21 Accumulator interrupt generation for the KeyStone I.

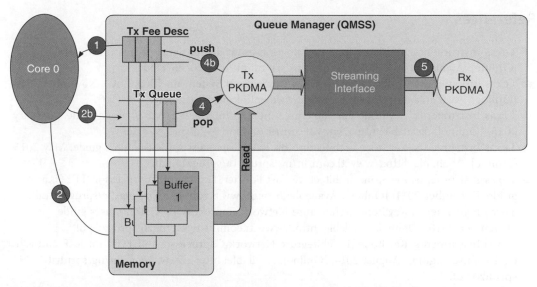

Figure 14.22 Core-to-peripheral movement.

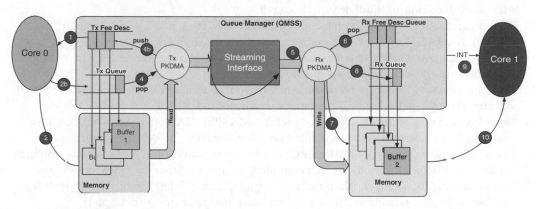

Figure 14.23 Core-to-core data movement using the Navigator.

14.5 Conclusion

The Navigator is an infrastructure (hardware, software and network) that can be used for inter-core communication and synchronisation, CorePacs to accelerator/peripheral data movements and core-to-core data movements. Once the Navigator is configured and initialised, the application running in a CorePac only pushes a descriptor (holding data and information) into a specific queue, and the PKDMA takes care of passing the data and notifying the receiver (i.e. another CorePac, an accelerator or an I/O device), leaving the CorePac to deal with other tasks, reducing software and enabling lock-free programming models. The Navigator functionality is also enhanced by the PDSPs which monitor queues and manage programmable interrupts to CorePacs. The Navigator can also perform resource sharing and job management to further increase the performance of the SoC.

References

1 Texas Instruments, Multicore fixed and floating-point digital signal processor, March 2014. [Online]. Available: http://www.ti.com/lit/ds/symlink/tms320c6678.pdf.

2 Texas Instruments, Multicore DSP + ARM KeyStone II System-on-Chip (SoC), November 2013. [Online]. Available: http://www.ti.com/lit/ds/symlink/66ak2h12.pdf.

3 Texas Instruments, KeyStone I architecture Antenna Interface 2 (AIF2), user's guide, February 2015. [Online]. Available: http://www.ti.com/lit/ug/sprugv7e/sprugv7e.pdf.

4 Texas Instruments, KeyStone architecture Bit Rate Coprocessor (BCP), user's guide, May 2015. [Online]. Available: http://www.ti.com/lit/ug/sprugz1a/sprugz1a.pdf.

5 Texas Instruments, KeyStone architecture Fast Fourier Transform Coprocessor (FFTC), user guide, December 2011. [Online]. Available: http://www.ti.com/lit/ug/sprugs2c/sprugs2c.pdf.

6 Texas Instruments, KeyStone architecture Network Coprocessor (NETCP), user guide, November 2010. [Online]. Available: http://www.ti.com/lit/ug/sprugz6/sprugz6.pdf.

7 Texas Instruments, KeyStone II architecture Network Coprocessor (NETCP) for K2E and K2L devices, user's guide, August 2014. [Online]. Available: http://www.ti.com/lit/ug/spruhz0/spruhz0.pdf.

8 Texas Instruments, KeyStone architecture Serial Rapid IO (SRIO), user guide, November 2012. [Online]. Available: http://www.ti.com/lit/ug/sprugw1b/sprugw1b.pdf.

9 Texas Instruments, KeyStone II architecture IQNet2, user's guide, May 2015. [Online]. Available: http://www.ti.com/lit/ug/spruho6a/spruho6a.pdf.

10 Texas Instruments, KeyStone architecture Multicore Navigator, user's guide, April 2015. [Online]. Available: http://www.ti.com/lit/ug/sprugr9h/sprugr9h.pdf.

11 Texas Instruments, Quality of service on KeyStone™ II architecture, May 2015. [Online]. Available: http://www.ti.com/lit/wp/spry287/spry287.pdf.

12 Texas Instruments, Multicore Navigator overview (online training), [Online]. Available: http://software-dl.ti.com/public/hpmp/KeyStone/04_Navigator/index.html.

13 Texas Instruments, Multicore Navigator packet DMA (PKDMA) (online training), [Online]. Available: http://software-dl.ti.com/public/hpmp/KeyStone/06_PKDMA/index.html.

14 Texas Instruments, Multicore Navigator Queue Manager Subsystem (QMSS) (online training), [Online]. Available: http://software-dl.ti.com/public/hpmp/KeyStone/05_QMSS/index.html.

15 Texas Instruments, Multicore Navigator tips 'n tricks, 2016. [Online]. Available: https://e2e.ti.com/blogs_/b/process/archive/2012/09/20/multicore-navigator-tips-n-tricks-pt-1.

15

FIR filter implementation

15.1 Introduction

Amongst all the obvious advantages that digital filters offer, the finite impulse response (FIR) filter can also guarantee linear phase characteristics that neither analogue nor infinite impulse response (IIR) filters can achieve. There are many commercially available software packages for filter design. However, without basic theoretical knowledge of the FIR filter, it will be difficult to use them.

The main purpose of this chapter is twofold. First is to show how to design an FIR filter and implement it on the TMS320C66x processor, and second to show how to interface a PC with the TMS320C6678 EVM using the UART interface.

Multicore DSP: From Algorithms to Real-time Implementation on the TMS320C66x SoC, First Edition. Naim Dahnoun.
© 2018 John Wiley & Sons Ltd. Published 2018 by John Wiley & Sons Ltd.
Companion website: www.wiley.com/go/dahnoun/multicoredsp

15.2 Properties of an FIR filter

15.2.1 Filter coefficients

An FIR filter can be defined by the following expression:

$$y[n] = \sum_{k=0}^{N} b_k \cdot x[n-k] \tag{15.1}$$

where:

$x[n]$ represents the filter input.
b_k represents the filter coefficients.
$y[n]$ represents the filter output.

From Equation 15.1, it is clear that any FIR filter can be fully characterised if its coefficients are known. However, there is also a hidden parameter, the sampling rate of the input signal. In the next few sections, the sampling rate will be taken into account. So let us look closer at the coefficients and see what they actually represent. If the signal $x[n]$ is replaced by an impulse $\delta[n]$, Equation 15.1 will lead to:

$$y[n] = \sum_{k=0}^{N} b_k \delta[n-k] \tag{15.2}$$

$$= h[n]$$

where $h[n]$ is the impulse response of the filter. Since:

$$\delta[n-k] = \begin{cases} 0 \text{ for } n \neq k \\ 1 \text{ for } n = k \end{cases}$$

Equation 15.2 gives:

$$b_0 = h[0]$$
$$b_1 = h[1]$$
$$\vdots \tag{15.3}$$
$$b_k = h[k] \; .$$

We can say that the coefficients of a filter are the same as the impulse response samples of the filter.

15.2.2 Frequency response of an FIR filter

By taking the z-transform of $h[n](H(z))$:

$$H(z) = \sum_{n=0}^{\infty} h[n]z^{-n} \; . \tag{15.4}$$

To find the frequency response, the parameter z is replaced by $e^{-j\omega}$. Therefore:

$$H(z)|_{z=e^{-j\omega}} = H(\omega) = \sum_{n=0}^{\infty} h[n]e^{-jn\omega} \; . \tag{15.5}$$

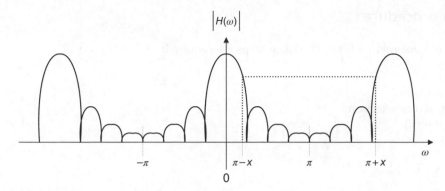

Figure 15.1 An arbitrary frequency response of an FIR filter showing the periodicity.

Since $e^{-j2\pi n} = 1$, then:

$$H(\omega + 2\pi) = \sum_{n=0}^{\infty} h[n] e^{-jn(\omega + 2\pi)} = \sum_{n=0}^{\infty} h[n] e^{-jn\omega} e^{-j2\pi n}$$

$$H(\omega + 2\pi) = \sum_{n=0}^{\infty} h[n] e^{-jn\omega} = H(\omega).$$

It has therefore been shown that the response of an FIR filter is periodic, and its period is 2π (see Figure 15.1).

This property has many practical consequences, one of which is the aliasing effect. For instance, if we choose a signal of frequency $\pi + x$ as shown in Figure 15.1, it will appear at the output of the filter as a signal of frequency $\pi - x$. To avoid aliasing, an analogue filter can be used to remove frequencies beyond the region of interest. Again, in a practical situation, this can be very challenging since analogue filters introduce non-linear phase distortion.

15.2.3 Phase linearity of an FIR filter

One of the most important characteristics of an FIR filter is its phase response. The phase response of an FIR filter can be made linear; this is quite important in communications technology.

It can be shown in Refs. [1–3] that for a causal FIR filter whose impulse response is symmetrical (i.e. $h[n] = h[N-1-n]$ for $n = 0, 1, ..., N-1$), its phase is guaranteed to be linear. This is summarised in Table 15.1.

Table 15.1 Conditions for linear phase

Condition	Phase $\left(k = -\frac{N-1}{2}\right)$	Phase property
$h[n] = h[N-n-1]$	$k\omega$	Linear phase
$h[n] = -h[N-n-1]$	$\frac{\pi}{2} + k\omega$	Linear phase and 90° phase shift

15.3   Design procedure

To fully design and implement a filter, five main steps are required:

1) Filter specification
2) Coefficients calculation
3) Appropriate structure selection
4) Simulation (optional)
5) Implementation.

15.3.1   Specifications

A system can be fully specified by its transfer function, which is magnitude $|H(\omega)|$ and phase $\theta(\omega)$. In general, $\theta(\omega)$ can be simply specified to be either linear or non-linear. However, the magnitude response is specified by various parameters as shown in Figure 15.2.

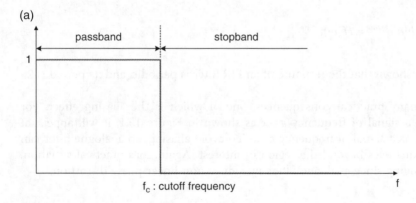

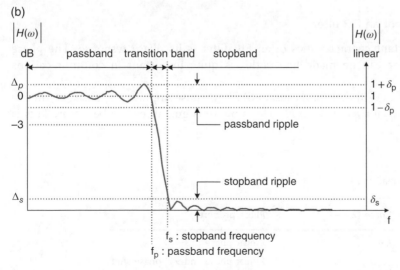

Figure 15.2 Filter specifications: (a) ideal, (b) practical.

The relation between the linear and logarithmic scale is:

$$\Delta p = -20\log_{10}(1 + \delta p) \text{ or } \delta p = 10^{-\Delta p/20} - 1$$
$$\Delta s = -20\log_{10}(\delta s) \text{ or } \delta s = 10^{-\Delta s/20}.$$

15.3.2 Coefficients calculation

The ability to achieve a frequency response 'identical' to or within the limits specified will depend mainly on the method used to calculate the coefficients. The most common methods are the window, frequency sampling or Park–McClellan (also known as *optimal*, *equiripple* or *optimal equiripple*) methods. For more details on the design methods, the reader is referred to Refs. [2–7]. In this chapter, the window method has been selected.

15.3.2.1 Window method

To extract the coefficients with this method, there are four steps required, each of which is described here.

Step 1. Specify the frequency response (see Figure 15.3).

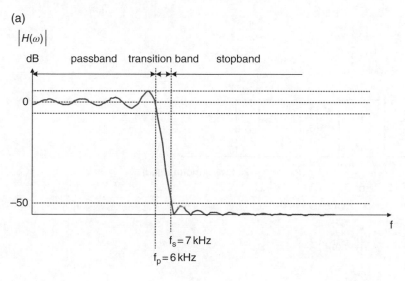

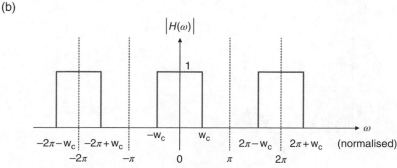

Figure 15.3 Frequency response: (a) desired, (b) ideal.

Step 2. Calculate the coefficients of the impulse response of the ideal filter (see Figure 15.4).

$$h_d(n) = \frac{1}{2\pi} \int_{-\pi}^{\pi} H_d(\omega) e^{-j\omega n} d\omega$$

$$= \frac{1}{2\pi} \int_{-\omega_c}^{\omega_c} 1 \cdot e^{j\omega n} d\omega$$

$$= \begin{cases} \dfrac{2f_c \sin(n\omega_c)}{n\omega_c}; & \text{for } n \neq 0 \\ 2f_c; & \text{for } n = 0. \end{cases}$$

As $\dfrac{\sin(\theta)}{\theta} = \dfrac{\sin(-\theta)}{-\theta}$, $h_d(n)$ is therefore symmetrical, and as a result the phase of the filter is linear. To achieve a magnitude of a transfer function identical to the desired one, the length of the filter should be infinite, which means that in order to implement this filter we need a large

(a)

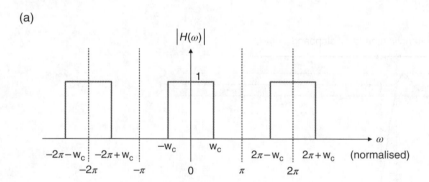

(b)

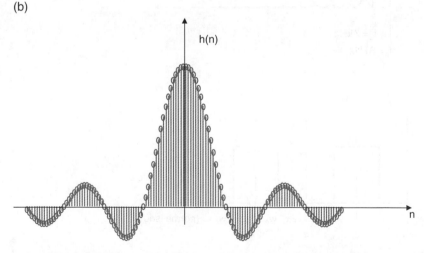

Figure 15.4 Ideal frequency response of a low-pass filter and (b) its impulse response.

amount of time to calculate each output. To get around this problem, the impulse response of the filter should be truncated. Consequently, the effect of truncation introduces overshoots and ripples. This is known as the *Gibbs phenomenon*. In order to reduce the overshoot and ripple, many window functions have been investigated. Figure 15.5 shows the frequently used windows, and Table 15.2 shows the features of each window.

Step 3. Determine the filter length, N.

For a Hamming window, we have:

$$N = \frac{3.3}{\Delta f}$$

where Δf is the normalised transition width, that is $\frac{\delta f}{f_s}$, where f_s is the sampling frequency and δf ($\delta f = f_s - f_p$) is the transition width (see Figure 15.2). Therefore,

$$N = \frac{3.3}{\delta f} f_s = \frac{3.3}{1\ kHz} 40\ kHz = 132.$$

Window's Name	Window Function $w(n)$, $n = 0, 1, ..., N-1$	Window Sequence
Rectangular	1	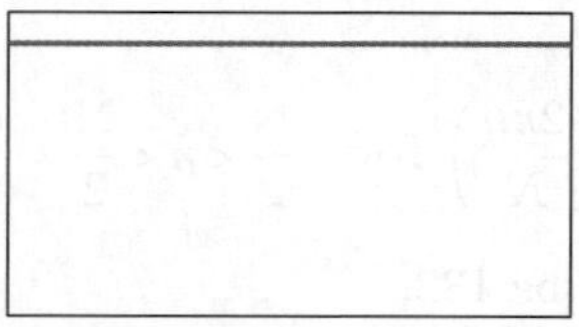
Hanning	$0.5\left[1 - \cos\left(\frac{2\pi n}{N-1}\right)\right]$	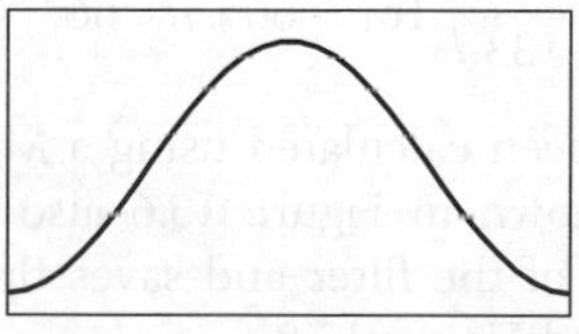
Hamming	$0.54 - 0.46\cos\left(\frac{2\pi n}{N-1}\right)$	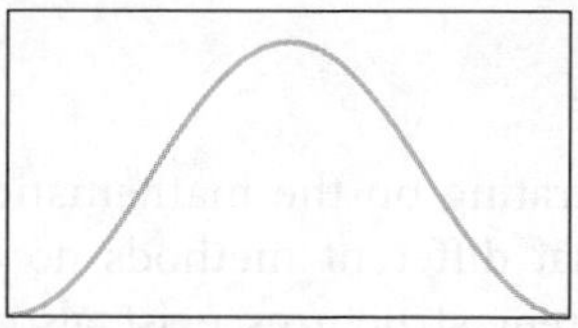
Blackman	$0.42 - 0.5\cos\left(\frac{2\pi n}{N-1}\right)$ $+ 0.08\cos\left(\frac{4\pi n}{N-1}\right)$	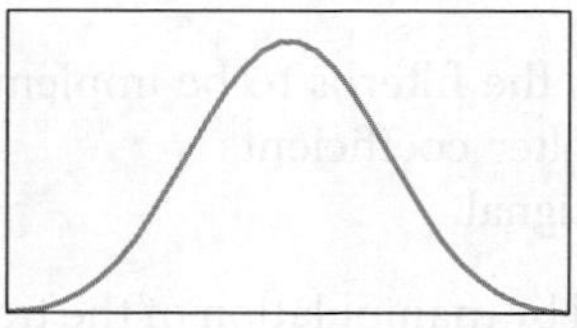

Figure 15.5 Frequently used windows.

Table 15.2 Window features

Window type	Normalised transition width ($\Delta f(Hz)$)	Passband ripple (dB)	Stopband attenuation (dB)
Rectangular	$\dfrac{0.9\pi}{N}$	0.7416	21
Hanning	$\dfrac{3.4\pi}{N}$	0.0546	44
Hamming	$\dfrac{3.3\pi}{N}$	0.0194	53
Blackman	$\dfrac{5.5\pi}{N}$	0.0017	74

It is worth observing that the larger the transition width, the smaller the length of the filter. Let us choose N equal to 133 in order to have odd symmetry for the purpose stated in Section 15.3.3.2.

Step 4. Calculate the set of truncated impulse response coefficients, $\{h(n)\}$.

$$h(n) = h_d(n) \cdot W(n) \text{ for } -\frac{N}{2} < n < \frac{N}{2}$$

or

$$W(n) = 0.54 + 0.46\cos\left(\frac{2\pi n}{N}\right) \text{ for } -\frac{N}{2} < n < \frac{N}{2}.$$

Since N has been chosen to be 133,

$$W(n) = 0.54 + 0.46\cos\left(\frac{2\pi n}{133}\right) \text{ for } -66 < n < 66.$$

The $h(n)$ coefficients have been calculated using a MATLAB program (see Figure 15.6 and Table 15.3). The program shown in Figure 15.6 also generates the impulse response $h(n)$, plots the frequency response of the filter and saves the coefficients in Q15 format to a file. See Table 15.3, Figure 15.7 and Figure 15.8.

15.3.3 Realisation structure

So far, we have been concentrating on the mathematical side in order to calculate the filter coefficients, and we found that different methods do exist. To either realise, implement or program a digital filter, different structures exist also. Which structure to use depends on factors such as:

1) DSP architecture on which the filter is to be implemented
2) Sensitivity to error in the filter coefficients
3) Sensitivity to error in the signal.

The structure can be derived by manipulation of the transfer function, as shown further in this section. For more details on the structures and factors influencing the choice of one structure over another, the reader is referred to Ref. [3].

```
fc= 6000/40000;
N=127
%N=133 -1
n=-(N/2):(N/2);
n=n+(n==0)*eps   ; %avoiding division by zero

[h]=sin(n*2*pi*fc)./(n*pi);
[w]= .54 +0.46*cos(2*pi*n/N);
d=h.*w;
d2 =d.*2^15
for i=1:8:N,
fprintf('c:\C66\chapter16\fir_coef.txt',' %8.0f, %8.0f, %8.0f, %8.0f,
%8.0f, %8.0f, %8.0f ,%8.0f\n',d2(i:i+7));
end;

figure(1)
[g,f]=freqz(d,1,512,40000);
plot(f,20*log10(abs(g)));
[g,f]=freqz(d,1,512,40000);
axis([0  2*10^4 -100 1]);
figure(2);
plot(n,d);
figure(3)
axis([0  2 -100 0]);
freqz(d,1,512,40000);
end
```

Figure 15.6 MATLAB program for generating the impulse response coefficients.

Table 15.3 FIR coefficients converted into Q15 format

−2, 10, 14, 7, −7, −17, −13, 3
19, 21, 4, −21, −32, −16, 18, 43
34, −8, −51, −56, −11, 53, 81, 41
−44, −104, −81, 19, 119, 129, 24, −119
−178, −88, 95, 222, 171, −41, −248, −266
−50, 244, 366, 181, −195, −457, −353, 85
522, 568, 109, −540, −831, −424, 474, 1163
953, −245, −1661, −2042, −463, 2940, 6859, 9469
9469, 6859, 2940, −463, −2042, −1661, −245, 953
1163, 474, −424, −831, −540, 109, 568, 522
85, −353, −457, −195, 181, 366, 244, −50
−266, −248, −41, 171, 222, 95, −88, −178
−119, 24, 129, 119, 19, −81, −104, −44
41, 81, 53, −11, −56, −51, −8, 34
43, 18, −16, −32, −21, 4, 21, 19
3, −13, −17, −7, 7, 14, 10, −2

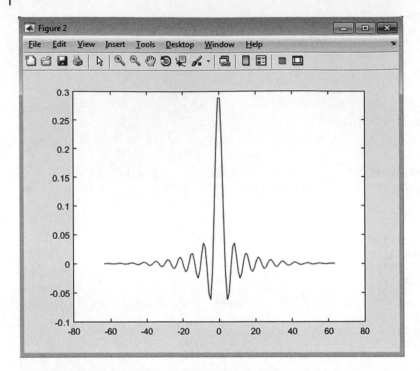

Figure 15.7 Plot of the filter coefficients *h*(*n*).

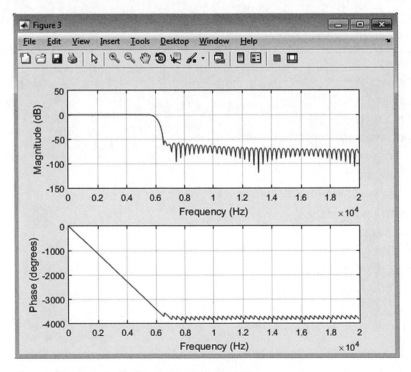

Figure 15.8 Transfer function of the designed filter.

15.3.3.1 Direct structure

We have seen previously that the transfer function of an FIR filter can be written as follows:

$$H(z) = \sum_{k=0}^{N-1} b_k z^{-k} \tag{15.6}$$

The transfer function can also be expressed as:

$$Y(z) = H(z) \cdot X(z) \tag{15.7}$$

where $X(z)$ and $Y(z)$ represent the z-transform of the input and the output sequences.

From Equation 15.6 and Equation 15.7, the difference equation can be derived as:

$$y(n) = b_0 x(n) + b_1 x(n-1) + \;....\; + b_{N-1} x(n-N+1) \tag{15.8}$$

A 'direct' implementation of Equation 15.8 is illustrated in Figure 15.9. This structure is known as the *direct form* or *transversal*.

15.3.3.2 Linear phase structures

By examining Figure 15.9, we can see that in order to implement an FIR filter of length N ($N-1$ taps), N multiplications and $N-1$ additions are required. We have seen previously that in order to have a linear phase, the coefficients must be symmetrical, that is, $b_0 = b_{N-1}$, $b_1 = b_{N-2}$ and so on. Since we are grouping the coefficients in pairs, we need to decide if the length of the filter is an even or odd number (see Figure 15.10).

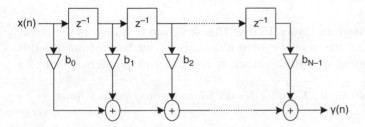

Figure 15.9 Direct form structure for an FIR filter.

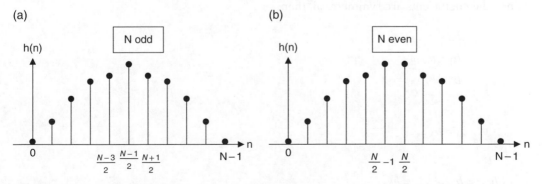

Figure 15.10 Impulse response, $h(n)$, for (a) N odd and (b) N even.

If N is even (Figure 15.10b), then:

$$H(z) = \sum_{k=0}^{N-1} b_k z^{-k} = \sum_{k=0}^{N/2-1} b_k z^{-k} + \sum_{k=N/2}^{N-1} b_k z^{-k}.$$

Since the coefficients are symmetrical, then:

$$b_0 = b_{N-1}$$

$$b_1 = b_{N-2}$$

$$b_2 = b_{N-3}$$

$$\vdots$$

$$b_k = b_{N-k-1}$$

$$\vdots$$

$$b_{\frac{N}{2}-1} = b_{N-\left(\frac{N}{2}-1\right)-1} = b_{\frac{N}{2}}$$

Therefore:

$$H(z) = \sum_{k=0}^{\frac{N}{2}-1} b_k \left(z^{-k} + z^{N-k-1} \right).$$

This leads to the structure shown in Figure 15.11a. This structure is known as the *linear phase structure*, and it reduces the number of multiplications by half. However, this structure introduces programming complexity since it requires two pointers moving in opposite directions.

If N is odd, then by referring to Figure 15.10 for N odd, we can write $H(z)$ as follows:

$$H(z) = \sum_{k=0}^{N-1} b_k z^{-k} = \sum_{k=0}^{\frac{N-3}{2}} b_k z^{-k} + b_{\frac{N-1}{2}} z^{-\frac{N-1}{2}} + \sum_{k=\frac{N+1}{2}}^{N-1} b_k z^{-k}.$$

Since the coefficients are symmetrical, then:

$$b_0 = b_{N-1}$$

$$b_1 = b_{N-2}$$

$$b_2 = b_{N-3}$$

$$\vdots$$

$$b_k = b_{N-k-1}$$

$$\vdots$$

$$b_{\frac{N-3}{2}} = b_{N-\left(\frac{N-3}{2}\right)-1} = b_{\frac{N+1}{2}}.$$

(a)

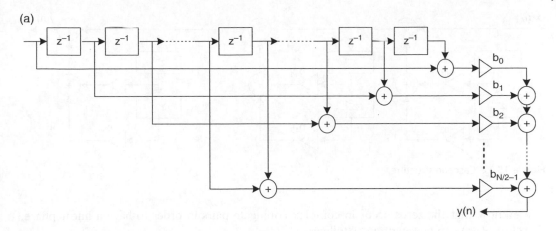

(b)

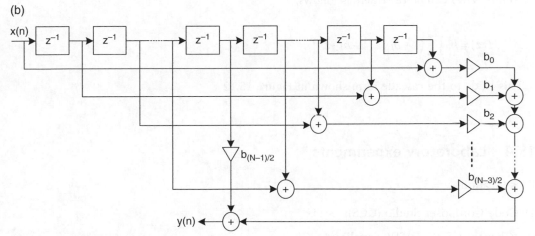

Figure 15.11 Linear phase structure for (a) *N* even and (b) *N* odd.

Therefore:

$$H(z) = \sum_{k=0}^{\frac{N-1}{2}} b_k \left(z^{-k} + z^{N-k-1} \right) + b_{\frac{N-1}{2}} z^{-\frac{N-1}{2}}.$$

This leads to the structure shown in Figure 15.11b and is similar to the structure for *N* even except for the term $b_{\frac{N-1}{2}}$.

15.3.3.3 Cascade structures

The cascade structure converts the transfer function into a product of second-order functions, as follows:

$$H(z) = \sum_{k=0}^{N-1} b_k z^{-k} = b_0 + b_1 z^{-1} + b_2 z^{-2} + \ldots + b_{N-1} z^{-(N-1)}$$

$$= b_0 \left[1 + \frac{b_1}{b_0} z^{-1} + \frac{b_2}{b_0} z^{-2} + \ldots + \frac{b_{N-1}}{b_0} z^{-(N-1)} \right].$$

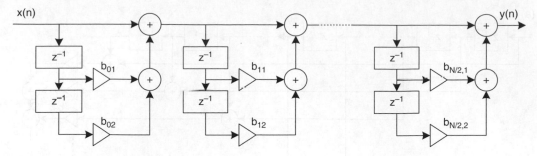

Figure 15.12 Cascade structure.

We know that the zeros occur in complex conjugate pairs in order to have a linear phase; therefore, $H(z)$ can be rewritten as follows:

$$H(z) = b_0 \prod_{k=1}^{\frac{N}{2}} \left(1 + b_{k,1}z^{-1} + b_{k,2}z^{-2}\right).$$

This leads to the cascade form shown in Figure 15.12.

15.4 Laboratory experiments

Files location:

1) Code Composer Studio (CCS):

   \Chapter_15_Code\FIR_SerialPort_CCS

2) Java code:

   \Chapter_15_Code\FIR_Eclipse_WS\AudioSerial2 (source code)
   \Chapter_15_Code\ audioSerial.jar (executable code).

In this laboratory experiment, a signal is generated by a PC and sent to the EVM via a serial port interface as shown in Figure 15.13 and Figure 15.14.

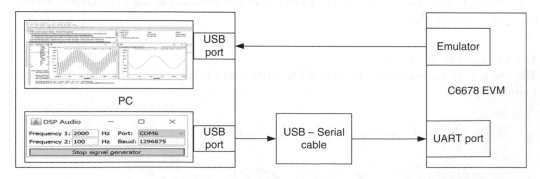

Figure 15.13 PC-to-DSP connections.

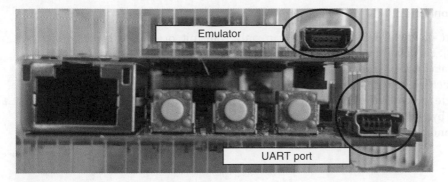

Figure 15.14 USB and UART ports of the EVM.

On the PC side, an application (DSP Audio: **audioSerial.jar**) is written in order to generate a signal composed of two sinewaves with user-defined specific frequencies (with an 8 KHz sampling rate) and sends the 16-bit sign signal as two bytes: the LSB first and the MSB last. On the DSP side, it will receive bytes and echo them so that the PC can determine if the LSB and MSB bytes are received in the correct order.

The DSP Audio application is written in Java, and the source code is provided.

15.4.1 Filter implementation

Once the filter coefficients have been calculated and a structure has been selected, the filter can be implemented. To make the implementation simple, let us first implement the filter designed in Section 15.3 using the direct form structure. The difference equation of the filter has been shown to be:

$$y(n) = b_0 x_0 + b_1 x_1 + \ldots + b_{N-1} x_{N-1}.$$

This equation can be converted into C code as shown in Figure 15.15.

```
R_in[0] = sample << 3;      // Read a sample and amplify it

acc = 0;                    // Zero the accumulator.

for (ii=0; ii<128; ii++)    // 128 taps
{
prod = (h[ii]*R_in[ii]);    // Perform Q.15 multiplication

acc = acc + prod;           // Update 32-bit accumulator, catering
}                           // for temporary overflow.

output = (acc>>15);         // Cast output to 16-bits.

R_out[0] = (short) output;


for (ii=127; ii>0; ii--)    // Shift delay samples.
R_in[ii]=R_in[ii-1];
```

Figure 15.15 C implementation of an FIR filter.

15.4.2 Synchronisation

The UART port on the KeyStone devices can only transmit or receive bytes (8-bit) (see Figure 15.16) [8]. However, in this application, the data sent from the PC needs to be 16-bit. Therefore, the data must be truncated and sent over two bytes. This is likely to cause a synchronisation issue when a byte is dropped. To synchronise data between the PC and the DSP, the PC sends the first byte and when the DSP receives this byte it will echo the character '1'. The PC then tests the receive byte to see if it is a character '1'; if yes, then it sends the second byte. The DSP receives the second byte and echoes the character '2'. The PC checks if it is character '2'. If true,

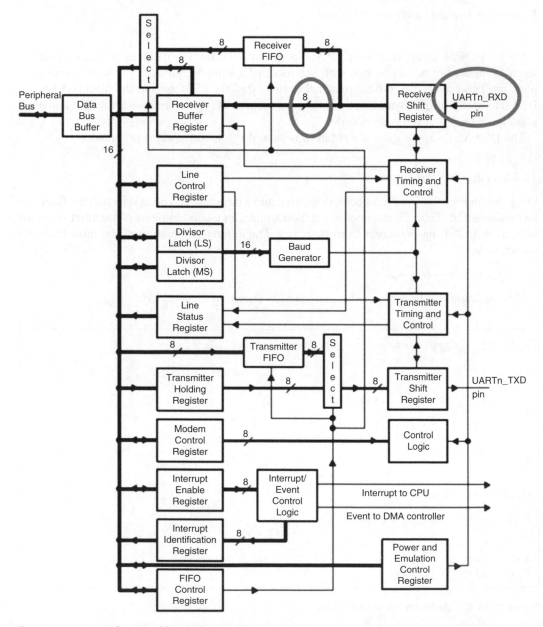

Figure 15.16 UART functional block diagram [8].

the PC waits for an acknowledgement character 'a' from the DSP when it finishes the processing of the FIR filter and starts the process again. However, if a byte is dropped, the PC receives the wrong byte. In this case, if it is sent byte 1 and received character '2', the PC will not send byte 2 but just wait for the acknowledgement character 'a' and start the process again. If the PC sends byte 2 and receives the character '1', it will send a dummy value. This is illustrated in Figure 15.17 Step 1 to Step 12.

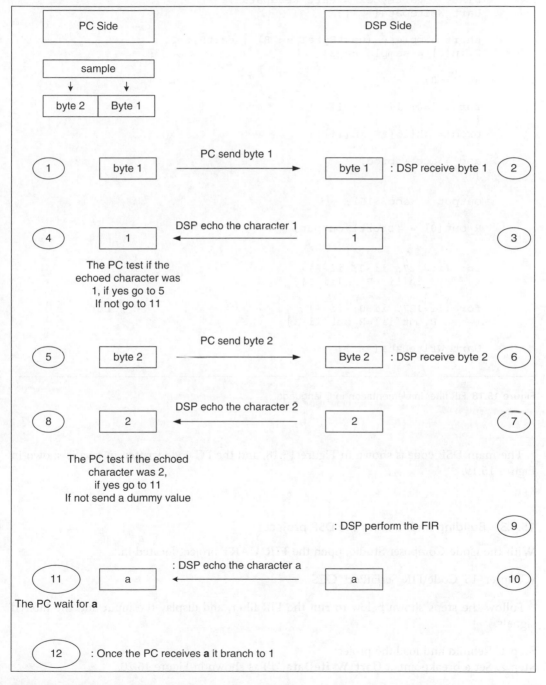

Figure 15.17 Illustration of the synchronisation mechanism.

```
while(1) {

    while(!Uart_IsDataReady());
    int8_t uartByte1 = Uart_ReadData();
    Uart_WriteData('1');
    while(!Uart_IsDataReady());
    uint8_t uartByte2 = Uart_ReadData();
    Uart_WriteData('2');

    short sample = (uartByte1 << 8) | uartByte2;
    R_in[0] = sample << 3;          // Read a sample << 3

    acc = 0;                         // Zero accumulator.

    for (ii=0; ii<128; ii++)         // 128 taps
    {
    prod = (h[ii]*R_in[ii]);    // Perform Q.15 multiplication

    acc = acc + prod;          // Update 32-bit accumulator, catering
    }                                // for temporary overflow.

    output = (acc>>15);           // Cast output to 16-bits.

    R_out[0] = (short) output;

    for (ii=127; ii>0; ii--)       // Shift delay samples.
        R_in[ii]=R_in[ii-1];

    for (ii=127; ii>0; ii--)       // Shift delay samples.
        R_out[ii]=R_out[ii-1];

    Uart_WriteData('a');
}
```

Figure 15.18 FIR filter implementation in C language.

The main DSP code is shown in Figure 15.18, and the PC code written in Java is shown in Figure 15.19.

15.4.3   Building and running the DSP project

With the Code Composer Studio, open the **FIR_UART** project located in:

\Chapter_15_Code\FIR_SerialPort_CCS

Follow the steps shown below to run the FIR filter, and display the input and the filtered signals.

Step 1. Rebuild and load the project.
Step 2. Set a breakpoint at **Uart_WriteData('a')** as shown in Figure 15.20.

```
for (int  i = 0; i < buffer.length - 1; i += 2) {

      // send byte 1, DSP should reply with '1'
      while(true) {
            try {
                  serial.writeByte(buffer[i]);
                  byte[] b = serial.readBytes(1, 500);

                  if((char) b[0] == '1') {
                        // DSP acknowledged correct byte position
                        break;
                  } else if((char) b[0] == '2') {
System.out.println("DSP out of sync, sent byte 1 but received ack 2. Resynching");
                        serial.readBytes(1, 500); // should be a final ack
                        continue;
                  } else {
System.out.println("Received acknowledge (b1). Next byte will be 1");
                        continue;
                  }
            } catch(SerialPortTimeoutException f) {
                  continue;
            }
      }

      while(true) {
            try {
                  serial.writeByte(buffer[i + 1]);
                  byte[] b = serial.readBytes(1, 500);

                  if((char) b[0] == '2') {
                        // DSP acknowledged correct byte position
                        break;
                  } else if((char) b[0] == '1') {
System.out.println("DSP out of sync, sent byte 2 but received ack 1. Sending dummy
value");
                        continue;
                  } else {
System.out.println("Received acknowledge (b). Next byte will be 1");
                        skipLoop = true;
                        break;
                  }
            } catch(SerialPortTimeoutException f) {
                  continue;
            }
      }

      if(skipLoop) continue;

      Thread.sleep(0, (int) targetSleep);

      try {
            serial.readBytes(1, 350);
      } catch(SerialPortTimeoutException f) {
            continue;
      }
}
```

Figure 15.19 Java synchronisation code.

```
106            Uart_WriteData('a');
107
```

Figure 15.20 Setting a breakpoint before echoing the character 'a'.

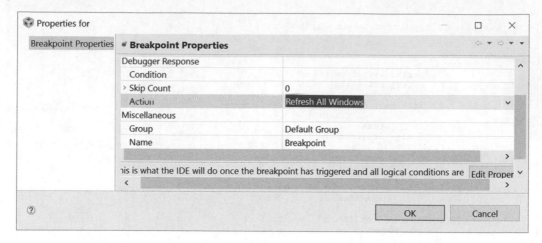

Figure 15.21 Setting the breakpoint property to **Refresh All Windows**.

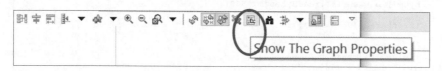

Figure 15.22 Icon used for displaying the graph properties.

Step 3. Point to the breakpoint, right-click, select **Breakpoint Properties** and change the **Action** to **Refresh All Windows** as shown in Figure 15.21, so that the display will be updated every time a breakpoint is reached.

Step 4. Open two graph display windows for displaying data in **R_in** and **R_out** arrays which hold the input and the filtered signals, respectively. In each graph, right-click and deselect the **Auto Scale** and import the configuration provided by selecting **Show the Graph Properties** as shown in Figure 15.22. The graph properties imported are shown in Figure 15.23 and Figure 15.24 by using the provided functions **R_in.graphProp** and **R_out.graphProp** located in the project directory.

Step 5. Do not run the code as yet. Move to Section 15.4.4.

15.4.4   Building and running the PC project

The PC application is written in Java running under the Eclipse Integrated Development Environment (IDE).

Step 1. Install Eclipse if not installed yet. Eclipse can be downloaded from Ref. [9]; see Figure 15.25.

Step 2. Open Eclipse and import the Eclipse project **AudioSerial2** located in:

\Chapter_15_Code\FIR_Eclipse_WS\AudioSerial2

Figure 15.26 to Figure 15.29 show how to import an Eclipse project and open it.

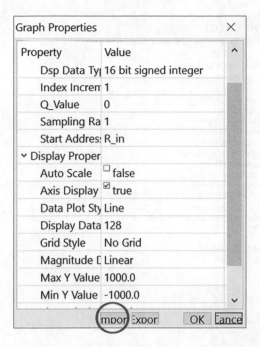

Figure 15.23 Graph priority for the input signal (using **R_in.graphProp**).

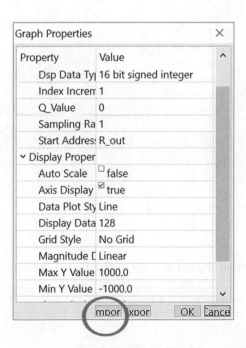

Figure 15.24 Graph priority for the filtered signal (using **R_out.graphProp**).

Java SE Runtime Environment 8u101

You must accept the Oracle Binary Code License Agreement for Java SE to download this software.

◯ Accept License Agreement ◉ Decline License Agreement

Product / File Description	File Size	Download
Linux x86	54.79 MB	jre-8u101-linux-i586.rpm
Linux x86	70.58 MB	jre-8u101-linux-i586.tar.gz
Linux x64	52.68 MB	jre-8u101-linux-x64.rpm
Linux x64	68.49 MB	jre-8u101-linux-x64.tar.gz
Mac OS X	55.99 MB	jre-8u101-macosx-x64.tar.gz
Mac OS X	64.32 MB	jre-8u101-macosx-x64.dmg
Solaris SPARC 64-bit	52 MB	jre-8u101-solaris-sparcv9.tar.gz
Solaris x64	49.85 MB	jre-8u101-solaris-x64.tar.gz
Windows x86 Online	0.71 MB	jre-8u101-windows-i586-iftw.exe
Windows x86 Offline	52.63 MB	jre-8u101-windows-i586.exe
Windows x86	59.42 MB	jre-8u101-windows-i586.tar.gz
Windows x64 Offline	59.17 MB	jre-8u101-windows-x64.exe
Windows x64	62.77 MB	jre-8u101-windows-x64.tar.gz

Figure 15.25 Java download location (using Windows x64 Offline) [9].

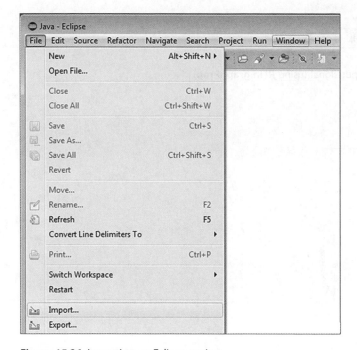

Figure 15.26 Importing an Eclipse project.

Step 3. Explore the **SendAudio.java** file, and check that the correct sampling rate is selected:

```
private float sampleRate = 44100; // sample rate 44100
```

Step 4. Run the project. This will open the default window shown in Figure 15.30.

Step 5. Modify the input variables as shown in Figure 15.30. Make sure the COM port is selected correctly. When completed, press **Start signal generator** as shown in Figure 15.31. To stop sending data to the DSP, press **Stop signal generator** as shown in Figure 15.32.

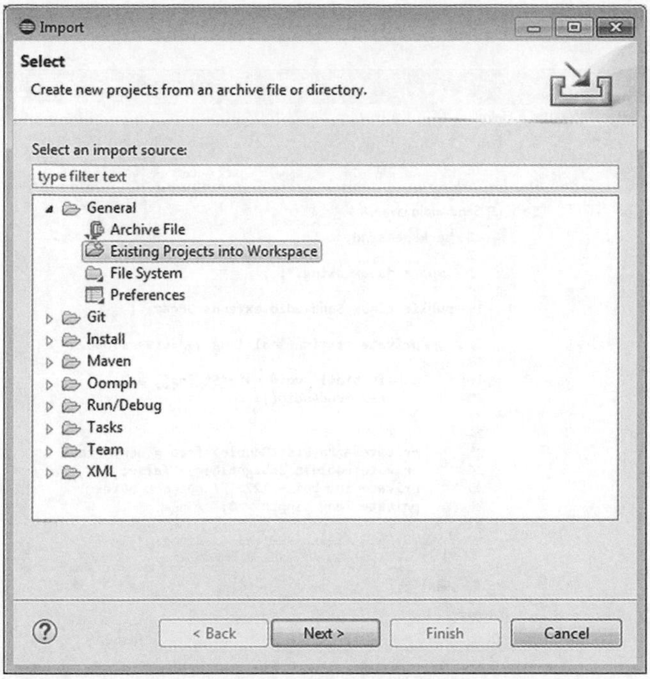

Figure 15.27 Opening an existing Eclipse project.

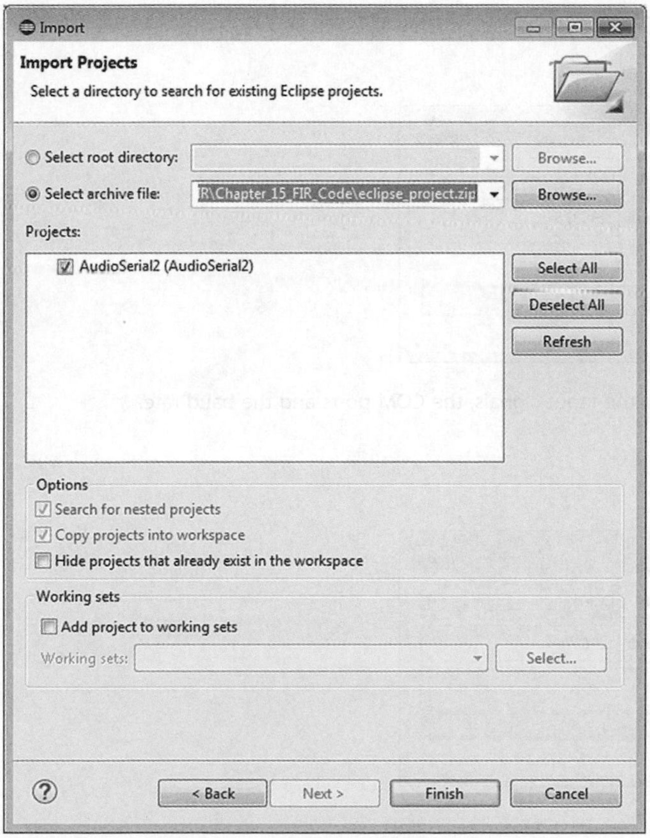

Figure 15.28 Importing a compressed Eclipse project.

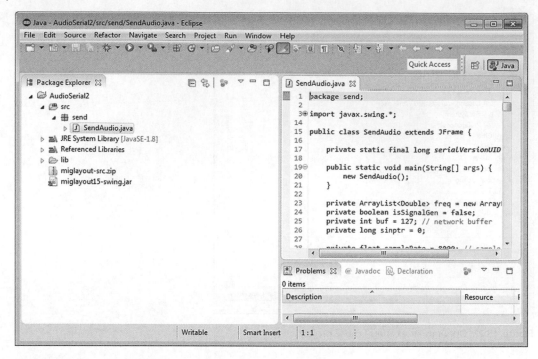

Figure 15.29 Opening the Java source code.

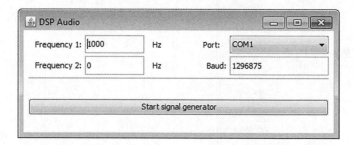

Figure 15.30 Java window for controlling the input signals, the COM ports and the baud rate.

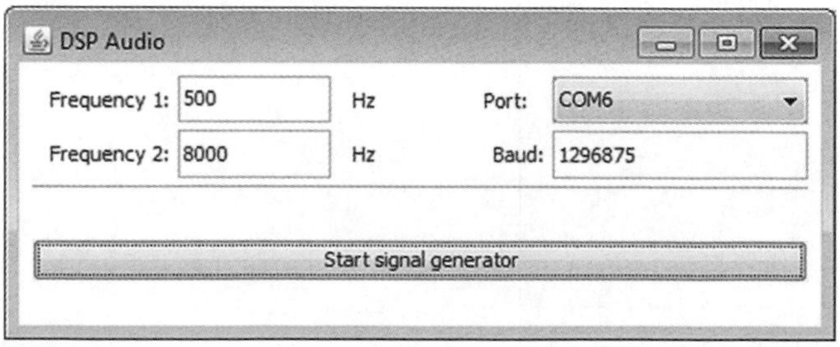

Figure 15.31 Default Java application.

Step 6. Go back to the CCS and run the code. If everything was set correctly, the CCS's console will display the message shown in Figure 15.33 and the Java IDE will show the message shown in Figure 15.34. However, if a message is lost, the Java application will generate a message as shown in Figure 15.35.

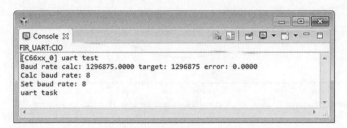

Figure 15.32 Java application sending data to the UART.

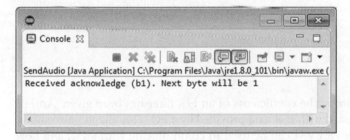

Figure 15.33 Code Composer Studio console output.

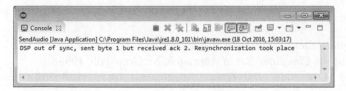

Figure 15.34 Java window message when no data are lost.

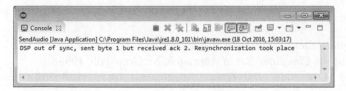

Figure 15.35 Java output when data are lost.

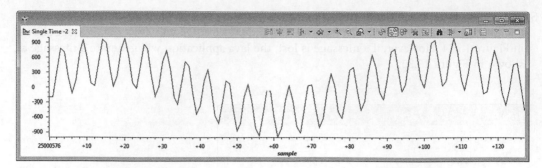

Figure 15.36 Input signal display.

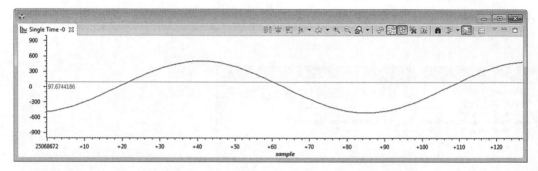

Figure 15.37 Filtered signal display.

Finally, Figure 15.36 and Figure 15.37 show the input and filtered signal, respectively, displayed with the CCS graph utility.

15.5 Conclusion

In this chapter, the theory of obtaining the coefficients of an FIR filter has been given. An FIR filter has been implemented and tested with real data provided by a PC. This chapter also shows how the UART of the TMS320C6678 EVM can be used to communicate with a PC, and how data can be synchronised.

References

1 T. W. Parks and C. S. Burrus, *Digital Filter Design*, John Wiley, 1987.
2 S. K. Mitra, *Digital Signal Processing: A Computer-Based Approach*, McGraw-Hill, 1998.
3 J. H. McClellan, R. W. Schafer and M. A. Yoder, *DSP First: A Multimedia Approach*, Prentice Hall, 1998.
4 L. C. Ludeman, *Fundamentals of Digital Signal Processing*, John Wiley, 1987.

5 R. D. Strum and D. E. Kirk, *First Principles of Discrete Systems and Digital Signal Processing*, Addison-Wesley, 1998.

6 J. Proakis, C. M. Rader, F. Ling and C. L. Nikias, *Advanced Digital Signal Processing*, Macmillan, 1992.

7 E. Ifeachor and B. Jervis, *Digital Signal Processing: A Practical Approach*, Addison-Wesley, 1993.

8 Texas Instruments, KeyStone Architecture Universal Asynchronous Receiver/Transmitter (UART) user guide, November 2010. [Online]. Available: http://www.ti.com/lit/ug/sprugp1/sprugp1.pdf.

9 Oracle, Java SE Runtime Environment 8 downloads, [Online]. Available: http://www.oracle.com/technetwork/java/javase/downloads/jre8-downloads-2133155.html.

16

IIR filter implementation

16.1 Introduction

Infinite impulse response (IIR) filters are the first choice when speed is paramount and the phase non-linearity characteristic is acceptable. IIR filters are computationally more efficient than finite impulse response (FIR) filters since they require fewer coefficients due to the fact that they use feedback or poles. However, this feedback can result in the filter being unstable if the coefficients deviate from their true values. This can happen during coefficient scaling or quantisation. In this chapter, it will be shown that the design and implementation of IIR filters are different than those for FIR filters.

The general equations of an IIR filter can be expressed as follows:

$$H(z) = \frac{b_0 + b_1 z^{-1} + \ldots + b_N z^{-N}}{1 + a_1 z^{-1} + \ldots + a_M z^{-M}}$$

$$= \frac{\displaystyle\sum_{k=0}^{N} b_k z^{-k}}{1 + \displaystyle\sum_{k=1}^{M} a_k z^{-k}}$$

where a_k and b_k are the filter coefficients.

Multicore DSP: From Algorithms to Real-time Implementation on the TMS320C66x SoC, First Edition. Naim Dahnoun.
© 2018 John Wiley & Sons Ltd. Published 2018 by John Wiley & Sons Ltd.
Companion website: www.wiley.com/go/dahnoun/multicoredsp

This transfer function can be factorised to give

$$H(z) = k\frac{(z-z_1)(z-z_2)...(z-z_3)}{(z-p_1)(z-p_2)...(z-p_3)} = \frac{Y(z)}{X(z)}$$

where:

$z_1, z_2, ..., z_N$ are the zeros.
$p_1, p_2, ..., p_N$ are the poles.

In terms of the difference equation that is useful for implementation, the transfer function leads to the following equation:

$$y(n) = \sum_{k=0}^{\infty} h(k)x(n-k)$$

$$= \sum_{k=0}^{N} b_k x(n-k) - \sum_{k=1}^{M} a_k y(n-k).$$

16.2 Design procedure

Similar to FIR design, there are five main steps to follow for designing an IIR filter:

1) Filter specification (refer to Chapter 15)
2) Coefficients calculation
3) Appropriate structure selection
4) Simulation (optional)
5) Implementation.

16.3 Coefficients calculation

There are two main approaches for deriving the z-transfer function of an IIR filter. The first approach is based on direct placement of poles and zeros, and the second is based on analogue filter design. Both approaches are described in this section.

16.3.1 Pole–zero placement approach

This is the easiest method for designing simple filters. All that is required is the knowledge that by placing a zero near or on the unit circle in the z-plane, the transfer function will be minimised at these points, whereas by placing a pole near or on the unit circle in the z-plane, the transfer function will be maximised at these points.

To obtain real coefficients, the poles and zeros must either be real or occur in complex conjugate pairs.

16.3.2 Analogue-to-digital filter design

This is the most popular method for calculating the filter coefficients. The popularity of this method comes from the rich analysis of the well-established analogue filters. There are two principle methods, as presented in this book; these are the bilinear transform (Section 16.3.3) and the impulse invariant (Section 16.3.4) methods.

16.3.3 Bilinear transform (BZT) method

The BZT method is perhaps the most common and effective method for deriving a digital filter from its counterpart analogue filter. This method is relatively simple and consists of mapping the s-plane to the z-plane.

Since we know that $z = e^{j\omega T_S}$ and $s = j\omega$, we can establish the following relationship between the s and z domains:

$$z = e^{sT_s} \tag{16.1}$$

or:

$$s = \frac{1}{T_S}\ln z. \tag{16.2}$$

Substituting Equation (16.2) for s into a transfer function $H(s)$ will lead to an equation in $ln(z)$ which is not practical. To eliminate the logarithmic term, an approximation to $ln(z)$ is used:

$$\ln(z) = 2\left[\frac{z-1}{z+1} + \frac{1}{3}\left(\frac{z-1}{z+1}\right)^3 + \cdots + \frac{1}{n+1}\left(\frac{z-1}{z+1}\right)^{n+1} + \cdots\right] \tag{16.3}$$

By considering only the first element in Equation (16.3), a new relationship between the s and z domains can be obtained, namely:

$$s = \frac{1}{T_S}\ln(z) = \frac{2}{T_S}\frac{z-1}{z+1} \tag{16.4}$$

As shown previously, Equation (16.4), which is known as the bilinear transform, is an approximation and therefore introduces distortion. However, Equation (16.4) offers a good approximation and is widely used. The bilinear z-transform mapping from the s to the z-plane is shown in Figure 16.1.

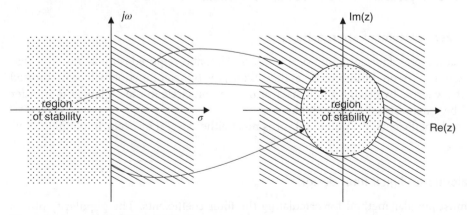

Figure 16.1 s- to z-plane mapping.

The mapping from the s to the z-plane introduces non-linearity between the analogue and digital frequencies as shown here. By using:

$$z = e^{j\omega_p T_s},$$

$$s = j\omega,$$

and

$$s = \frac{2}{T_s}\frac{z-1}{z+1},$$

we can deduce the relationship between ω and ω_p:

$$\omega = \frac{2}{T_s}\tan\left(\frac{\omega_p T_s}{2}\right) \tag{16.5}$$

Equation (16.5) is sketched in Figure 16.2. Equation (16.5) shows that when choosing a digital frequency, for instance the cut-off frequency, it has to be converted before being used as the analogue frequency. Figure 16.3 illustrates the difference between the analogue filter and its counterpart digital filter. Note that aliasing does not occur since all frequencies from zero to infinity are mapped between zero and π/T_s.

When deriving a digital filter with the BZT, the following logical design flow can be used.

1) Specify the analogue filter prototype.
2) Determine the cut-off frequency, ω_p, of the digital filter and find its equivalent analogue cut-off frequency, ω_c, using Equation (16.5). This is known as *pre-warping*.
3) De-normalise the analogue filter by ω_c. This can be done by replacing s with s/ω_c.
4) Finally, apply the BZT to the filter obtained in Step 3 by replacing s with $k\dfrac{z-1}{z+1}$; $\left(k = \dfrac{2}{T_s}\right)$.

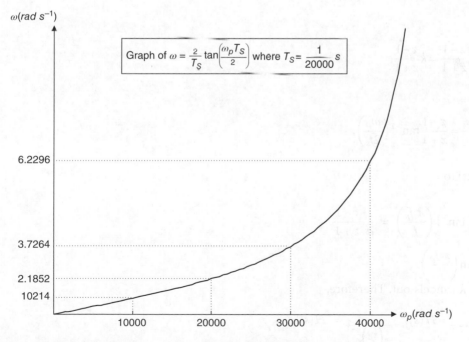

Figure 16.2 Relationship between the analogue and digital frequencies.

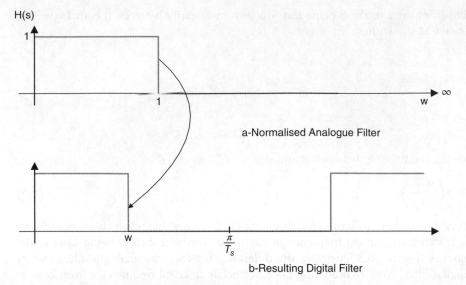

a-Normalised Analogue Filter

b-Resulting Digital Filter

Figure 16.3 Relationship between the analogue and digital frequency responses when using the bilinear transform (BZT).

To group Steps 2–4 into a single step, let us express s in terms of z whilst taking them into account.

$$\omega_c = 2f_s \tan\left(\frac{\omega_p}{2f_s}\right) = k \tan\left(\frac{\omega_p}{2f_s}\right)$$

and

$$s = 2f_s \frac{z-1}{z+1} = k \frac{z-1}{z+1},$$

hence,

$$\frac{s}{\omega_c} = \frac{z-1}{z+1} \tan^{-1}\left(\frac{\omega_p}{2f_s}\right),$$

which is equal to

$$\frac{z-1}{z+1} \tan^{-1}\left(\frac{\pi f_p}{f_s}\right) = \frac{1}{a} \frac{z-1}{z+1}.$$

where $a = \tan\left(\frac{\pi f_p}{f_s}\right)$.

Note that k cancels out. Therefore,

$$H(s)\big|_{s=\frac{s}{\omega_c}} = H(s)\big|_{s=\frac{1}{\tan\left(\frac{\pi f_p}{f_s}\right)}\frac{z-1}{z+1}} = H(s)\big|_{s=\frac{sz-1}{az+1}}.$$

Finally, we have two steps for converting an analogue filter to a digital filter using the BZT. These steps are:

1) Specify the analogue filter prototype.
2) Find $H(z)$ using $H(z) = H(s)|_{s = \frac{1}{\tan\left(\frac{\pi f_p}{f_s}\right)} \frac{z-1}{z+1}}$.

16.3.3.1 Practical example of the bilinear transform method

In this section, the design of a digital filter to approximate a second-order low-pass analogue filter described by the following transfer function is considered:

$$H(s) = \frac{1}{s^2 + \sqrt{2}s + 1}.$$

The digital filter should have a cut-off frequency of 6 kHz and a sampling frequency of 20 kHz.

16.3.3.2 Coefficients calculation

Now that the specifications have been given, we can proceed to Step 2 to determine the $H(z)$ function.

$$H(z) = H(s)|_{s = \frac{1}{\tan\left(\frac{\pi f_p}{f_s}\right)} \frac{z-1}{z+1}};$$

$$H(z) = \frac{1}{s^2 + \sqrt{2}s + 1}\bigg|_{s = \frac{1}{a}\frac{z-1}{z+1}}$$

$$= \frac{1}{\frac{1}{a^2}\left(\frac{z-1}{z+1}\right)^2 + \frac{\sqrt{2}}{a}\left(\frac{z-1}{z+1}\right) + 1} = \frac{a^2(z^2 + 2z + 1)}{(z^2 - 2z + 1) + \sqrt{2}a(z^2 - 1) + a^2(z^2 + 2z + 1)}$$

$$= a^2 \frac{z^2 + 2z + 1}{z^2\left(1 + \sqrt{2}a + a^2\right) + z(2a^2 - 2) + \left(1 + a^2 - \sqrt{2}a\right)};$$

$$H(z) = \frac{1 + 2z^{-1} + z^{-2}}{\left(\frac{1 + \sqrt{2}a + a^2}{a^2}\right) + 2\left(\frac{a^2 - 1}{a^2}\right)z^{-1} + \left(\frac{1 + a^2 - \sqrt{2}a}{a^2}\right)z^{-2}}$$

$$= \frac{b_0 + b_1 z^{-1} + b_2 z^{-2}}{1 + a_1 z^{-1} + a_2 z^{-2}},$$

where:

$$a = \tan\left(\frac{\pi f_p}{f_s}\right) = \tan\left(\frac{\pi.6000}{20000}\right) = 1.376382$$

$$b_0 = \frac{a^2}{1 + \sqrt{2}a + a^2}$$

$$b_1 = 2b_0$$

$$b_2 = b_0$$

$$a_1 = 2\frac{a^2 - 1}{\left(1 + \sqrt{2}a + a^2\right)}$$

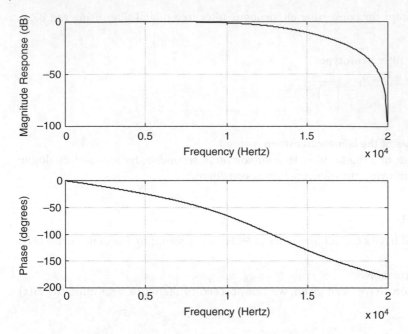

Figure 16.4 Transfer function of an IIR filter designed with the bilinear transform method.

$$a_2 = \frac{1 + a^2 - \sqrt{2}a}{1 + a^2 + \sqrt{2}a}.$$

Figure 16.4 shows the transfer function of $H(z)$ using MATLAB.

16.3.3.3 Realisation structures

As stated in Chapter 15, the choice of a structure depends on a few factors and can be obtained by manipulation of the transfer functions. In this chapter, the two frequently used structures are described, that is, the direct and cascade structures.

Direct form structure   As shown in this chapter, the transfer function of an IIR filter can be expressed as shown in Equation (16.6):

$$H(z) = \frac{Y(z)}{X(z)} = \frac{\displaystyle\sum_{n=0}^{N} b_n z^{-n}}{\displaystyle\sum_{n=0}^{N} a_n z^{-n}} = \frac{b_0 + b_1 z^{-1} + \ldots + b_N z^{-N}}{1 + a_1 z^{-1} + \ldots + a_N z^{-N}} \tag{16.6}$$

The difference equation (Equation (16.7)) can be derived from Equation (16.6) and can be written as:

$$y(n) = \sum_{k=0}^{N} b_k x(n-k) + \sum_{k=1}^{M} a_k y(n-k) \tag{16.7}$$

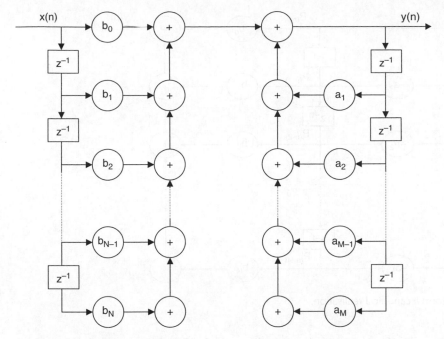

Figure 16.5 Direct form I structure.

Equation (16.7) leads to the direct structure shown in Figure 16.5 and is known as the *direct form I structure*. By using this structure, $N + M$ elements are required. This also means that $N + M$ data memory moves are required; to reduce this number, the transfer function can be manipulated as follows:

$$H(z) = H_1(z) \cdot H_2(z) = \frac{1}{1 + \displaystyle\sum_{k=1}^{N} a_k z^{-k}} \sum_{k=0}^{N} b_k z^{-k}; \quad (\text{for } N - M)$$

$$= \frac{P(z)}{X(z)} \cdot \frac{Y(z)}{P(z)},$$

where $\dfrac{P(z)}{X(z)} = \dfrac{1}{1 + \displaystyle\sum_{k=1}^{N} a_k z^{-k}}$ implies that $P(z) = \dfrac{X(z)}{1 + \displaystyle\sum_{k=1}^{N} a_k z^{-k}}$,

and $\dfrac{Y(z)}{P(z)} = \displaystyle\sum_{k=0}^{N} b_k z^{-k}$ implies that $Y(z) = P(z) \displaystyle\sum_{k=0}^{N} b_k z^{-k}$.

Taking the inverse z-transform of $P(z)$ and $Y(z)$ leads to Equation (16.8) and Equation (16.9).

$$p(n) = x(n) - \sum_{k=1}^{N} a_k p(n-k) \tag{16.8}$$

$$y(n) = \sum_{k=0}^{N} b_k p(n-k) \tag{16.9}$$

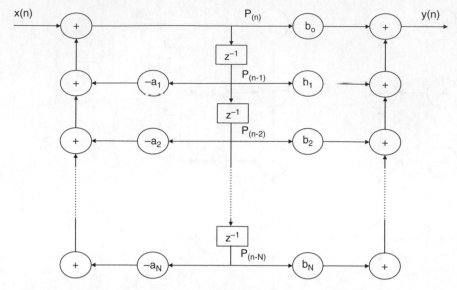

Figure 16.6 Direct form II canonical realisation.

These lead to the realisation structure shown in Figure 16.6, which is known as the *direct form II canonical realisation*. Note that with this structure, the number of delays has been reduced to N; this will save storage space and cycles during implementation.

Combining Equation (16.8) and Equation (16.9) leads to:

$$y(n) = b_0 p(n) + \sum_{k=1}^{N} b_k p(n-k)$$

$$= b_0 \left[x(n) - \sum_{k=1}^{N} a_k p(n-k) \right] + \sum_{k=1}^{N} b_k p(n-k)$$

$$y(n) = b_0 x(n) + \sum_{k=1}^{N} (b_k - b_0 a_k) p(n-k) \tag{16.10}$$

Equation (16.10) leads to an alternative to the direct form II canonical realisation as shown in Figure 16.7. This structure has the advantage of scaling down the amplitude and therefore improves the performance of the filter.

Cascade realisation   To obtain the cascade realisation, the transfer function $H(z)$ is expressed as a product of transfer functions $H_1(z)$, $H_2(z)$,... and $H_N(z)$:

$$H(z) = \frac{Y(z)}{X(z)} = H_1(z) \cdot H_2(z) \cdot ... \cdot H_N(z). \tag{16.11}$$

$H_n(z)$ are often called *bi-quadratic expressions*, which can be determined by one of the methods shown here, and take the following form:

$$H_n(z) = \frac{b_{0n} + b_{1n} z^{-1} + b_{2n} z^{-2}}{1 + a_{1n} z^{-1} + a_{2n} z^{-2}}, \quad n = 1, 2, ..., N \tag{16.12}$$

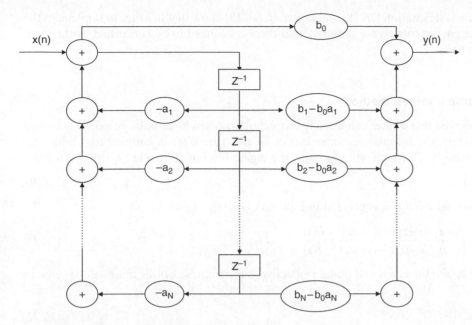

Figure 16.7 Alternative to the direct form II realisation.

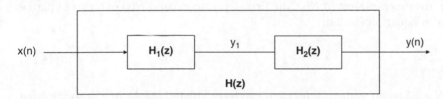

Figure 16.8 Cascade realisation using direct form II.

$H_n(z)$ can be implemented with the direct form I or II as shown here. However, the advantage of the direct form II structure makes it the first choice. Let us take a two-stage cascade realisation using the direct form II structure. The overall representation of $H(z)$ is shown in Figure 16.8. Equation (16.8) and Equation (16.9) are rewritten to take into account the cascade realisation. This is expressed in Equation (16.13) to Equation (16.19).

Equations for $H_1(z)$:

$$p_1(n) = x(n) - a_{11}p_1(n-1) - a_{21}p_2(n-2) \tag{16.13}$$

$$y_1(n) = b_{01}p_1(n) + b_{11}p_1(n-1) + b_{21}p_1(n-2). \tag{16.14}$$

Equations for $H_2(z)$:

$$p_2(n) = y_1(n) - a_{12}p_2(n-1) - a_{22}p_2(n-2) \tag{16.15}$$

$$y(n) = b_{02}p_2(n) + b_{12}p_2(n-1) + b_{22}p_2(n-2). \tag{16.16}$$

Substituting $y_1(n)$ from Equation (16.14) into Equation (16.15) leads to:

$$p_1(n) = x(n) - a_{11}p_1(n-1) - a_{21}p_2(n-2) \tag{16.17}$$

$$p_2(n) = b_{01}p_1(n) + b_{11}p_1(n-1) + b_{21}p_1(n-2) - a_{12}p_2(n-1) - a_{22}p_2(n-2) \tag{16.18}$$

$$y(n) = b_{02}p_2(n) + b_{12}p_2(n-1) + b_{22}p_2(n-2). \tag{16.19}$$

Equation (16.17), Equation (16.18) and Equation (16.19) show that in order to implement this filter, only the coefficients p_n are required, and there is no need to be concerned by the intermediate outputs y_n.

16.3.4 Impulse invariant method

Chapter 15 showed that a filter can be fully characterised when its impulse response is known. With this method, the impulse response $h(t)$ of an analogue filter is sampled to produce the impulse response $h(nT)$ which will represent the digital filter as shown in Equation (16.20):

$$h(n) = h(t)|_{t=nT} = h(nT) \tag{16.20}$$

Given an analogue filter as expressed in Equation (16.21),

$$H(s) = \frac{(s-z_1)(s-z_2)\cdots(s-z_n)}{(s-p_1)(s-p_2)\cdots(s-p_n)} = \frac{b(s)}{a(s)} \tag{16.21}$$

where z_n and p_n are the zeros and poles, respectively, the impulse of this filter will be given by Equation (16.22):

$$h(t) = L^{-1}[H(s)] \tag{16.22}$$

where L^{-1} is the inverse Laplace transform.

To calculate the inverse transform of $H(s)$, the first step is to expand $H(s)$ in terms of partial fractions as shown in Equation (16.23):

$$H(s) = \sum_{i=1}^{N} \frac{r_i}{s - p_i}. \tag{16.23}$$

The r_i, $i = 1,...,N$ are known as the *residues* and can be calculated one by one by multiplying the following equation by the appropriate term as shown here:

$$H(s) = \frac{b(s)}{a(s)} = \frac{r_1}{s - p_1} + \frac{r_2}{s - p_2} + \cdots + \frac{r_n}{s - p_n} + k \tag{16.24}$$

For calculation of r_1, Equation (16.23) and Equation (16.24) lead to:

$$\frac{b(s)(s-p_1)}{a(s)} = r_1 + \frac{r_2(s-p_1)}{s-p_2} + \frac{r_3(s-p_1)}{s-p_3} + \cdots + \frac{r_n(s-p_1)}{s-p_3} + k(s-p_1)$$

$$= r_1 + (s-p_1)\left[\frac{r_2}{s-p_2} + \frac{r_3}{s-p_3} + \cdots + \frac{r_n}{s-p_n} + k\right].$$

Therefore,

$$\frac{b(s)\,(s-p_1)}{a(s)}\bigg|_{s=p_1} = \frac{b(s)}{(s-p_2)(s-p_3)\cdots(s-p_n)}\bigg|_{s=p_1} = r_1$$

Hence,

$$r_1 = \frac{b(s)}{(s-p_2)(s-p_3)\cdots(s-p_n)}\bigg|_{s=p_1}$$

The other residues can be calculated in a similar way.

Now that we have all the residues, we can easily find the inverse Laplace transform of *H(s)* by using the following:

$$L^{-1}\left(\frac{r_i}{s+p_i}\right) = r_i e^{p_i t}.$$

Therefore:

$$L^{-1}(H(s)) = h(t) = \sum_{i=1}^{N} r_i e^{p_i t}.$$

By sampling the impulse response, we obtain:

$$h[nT] = \sum_{i=1}^{N} r_i e^{p_i nT} = \sum_{i=1}^{N} r_i \left(e^{p_i T}\right)^n.$$

Therefore,

$$H(z) = \sum_{n=0}^{\infty} h(nT) z^{-n} = \sum_{n=0}^{\infty} \left[\sum_{i=1}^{N} r_i \left[e^{p_i T}\right]^n\right] z^{-n}.$$

From *z*-transform tables, we know that:

$$\sum_{n=0}^{\infty} e^{pnT} z^{-n} = \frac{z}{z - e^{pT}},$$

and so

$$H(z) = \sum_{i=1}^{N} \frac{r_i z}{z - e^{p_i T}}.$$

In conclusion, in order to design a digital filter by the impulse invariant method, the following steps are required:

1) Specify the analogue prototype.
2) De-normalise the filter by replacing *s* with *s*/ω_c, where ω_c is the cut-off frequency.
3) Expand *H(s)* in terms of partial fractions.

16.3.4.1 Practical example of the impulse invariant method

Let us design a low-pass digital filter which has a cut-off frequency of 2 kHz and a sampling frequency of 20 kHz, and is based on an analogue filter with the following transfer function, *H(s)*:

$$H(s) = \frac{1}{s^2 + \sqrt{2}s + 1}$$

The solution is:

$$H(s) = H(s)|_{s=\frac{s}{\omega_c}} = \frac{\omega_c^2}{s^2 + \sqrt{2}\omega_c s + \omega_c^2}$$

$$= \frac{\omega_c^2}{(s - p_1)(s - p_2)} = \frac{r_1}{s - p_1} + \frac{r_2}{s - p_2},$$

where:

$$p_{1,2} = \frac{-\sqrt{2}\omega_c \pm \sqrt{-2\omega_c^2}}{2}$$

$$p_1 = \frac{-\sqrt{2}\omega_c - j\sqrt{2}\omega_c}{2} = \frac{-\omega_c\sqrt{2}(1+j)}{2}$$

and

$$p_2 = \frac{-\omega_c\sqrt{2}(1-j)}{2} = p_1^*.$$

Let us now find r_1 and r_2:

$$\frac{\omega_c^2}{(s-p_1)(s-p_2)}(s-p_1) = r_1 + \frac{r_2}{s-p_2}(s-p_1)$$

Hence,

$$r_1 = \frac{\omega_c^2}{(s-p_2)}\bigg|_{s=p_1} = \frac{\omega_c^2}{p_1-p_2}$$

$$= \frac{\omega_c^2}{p_1-p_1^*} = -j\frac{\omega_c}{\sqrt{2}}.$$

Similarly,

$$r_2 = \frac{\omega_c^2}{(s-p_1)}\bigg|_{s=p_2} = \frac{\omega_c^2}{p_2-p_1}$$

$$= \frac{\omega_c^2}{p_2-p_2^*} = -j\frac{\omega_c}{\sqrt{2}}$$

$$= j\frac{\omega_c}{\sqrt{2}} = r_1^*.$$

Therefore,

$$\begin{aligned}
H(z) &= \frac{r_1 z}{z - e^{p_1 T}} + \frac{r_2 z}{z - e^{p_2 T}} \\
&= \frac{r_1}{1 - e^{p_1 T}z^{-1}} + \frac{r_2}{1 - e^{p_2 T}z^{-1}} \\
&= \frac{r_1(1 - e^{p_2 T}z^{-1}) + r_2(1 - e^{p_1 T}z^{-1})}{(1 - e^{p_1 T}z^{-1})(1 - e^{p_2 T}z^{-1})} \\
&= \frac{r_1 + r_2 - (r_1 e^{p_1 T} + r_2 e^{p_2 T})z^{-1}}{1 - (e^{p_1 T} + e^{p_2 T})z^{-1} + e^{(p_1 + p_2)T}z^{-2}}.
\end{aligned} \tag{16.25}$$

Knowing that:

$$r_1 = r_2^* = r_r + jr_i,$$

$$p_1 = p_2^* = p_r + jp_i,$$

it follows that:

$$e^{p_1 T} + e^{p_2 T} = e^{(p_r + jp_i)T} + e^{(p_r - jp_i)T}$$
$$= 2e^{p_r T} \cos(p_i T), \tag{16.26}$$
$$e^{(p_1 + p_2)T} = e^{2p_r T} \tag{16.27}$$

and

$$
\begin{aligned}
r_1 e^{p_1 T} + r_2 e^{p_2 T} &= (r_r + jr_i)\left(e^{p_r T + jp_i T}\right) + (r_r - jr_i)\left(e^{p_r T - jp_i T}\right) \\
&= (r_r + jr_i)\left(e^{+p_r T}\right)\left(e^{+jp_i T}\right) + (r_r - jr_i)\left(e^{+p_r T}\right)\left(e^{-jp_i T}\right) \\
&= r_r\left(e^{+p_r T}\left(e^{+jp_i T} + e^{-jp_i T}\right)\right) + jr_i\left(e^{+p_r T}\left(e^{+jp_i T} - e^{-jp_i T}\right)\right) \\
&= r_r e^{+p_r T}(2\cos(p_i T)) + jr_i e^{+p_r T}(-2j\sin(p_i T)) \\
&= [2r_r \cos(p_i T) + 2r_i \sin(p_i T)]\, e^{+p_r T}. \tag{16.28}
\end{aligned}
$$

By substituting Equation (16.26), Equation (16.27) and Equation (16.28) into Equation (16.25), we obtain:

$$H(z) = \frac{2r_r - 2[r_r \cos(p_i T) + r_i \sin(p_i T)]e^{+p_r T}z^{-1}}{1 - 2e^{p_r T}\cos(p_i T)z^{-1} + e^{2p_r T}z^{-2}}$$

$$= \frac{b_0 + b_1 z^{-1}}{1 + a_1 z^{-1} + a_2 z^{-2}}$$

where:

$$r_r = 0,$$

$$r_i = -\frac{\omega_c}{\sqrt{2}}$$

$$p_i = p_r = -\frac{\omega_c \sqrt{2}}{2}$$

$$b_0 = 2r_r = 0$$

$$b_1 = 2[r_r \cos(p_i T) + r_i \sin(p_i T)]\, e^{p_r T}$$

$$\quad = 2[r_i \sin(p_i T)]\, e^{p_r T} = 2\left[j\frac{\omega_c}{2}\sin\left(-\frac{\omega_c \sqrt{2}}{2} T\right)\right] e^{-\left(\frac{\omega_c \sqrt{2}}{2} T\right)}$$

$$a_1 = 2e^{p_r T_s}\cos(p_i T) = 2\left[j\frac{\omega_c}{2}\cos\left(-\frac{\omega_c \sqrt{2}}{2} T\right)\right] e^{-\left(\frac{\omega_c \sqrt{2}}{2} T\right)}$$

$$a_2 = e^{2p_r T} = e^{-2\left(\frac{\omega_c \sqrt{2}}{2} T\right)}.$$

When the sampling frequency is 20 kHz (1/T) and the cut-off frequency is 2 kHz, the coefficients are as follows:

$$b_0 = 0$$

$$b_1 = 4.8984e + 003$$

$$a_1 = -1.1580$$

$$a_2 = 0.4112.$$

By using MATLAB, one can verify these results by simply using the following commands:

```
T=1/20000;
Wc=2*pi*2000;
Rr=0;
ri=-wc/sqrt(2);
pii=-wc/sqrt(2);
pr=pii;
b0=2*rr
b1=2*(ri*sin(pii*T))*exp(pr*T)
a1=-2*exp(pr*T)*cos(pii*T)
a2=exp(2*pr*T)
end
```

The output of the above program is shown in Figure 16.9. The laborious task of finding the coefficients can be overcome by using the **impinvar** MATLAB command.

16.4 IIR filter implementation

Now that the transfer function has been determined, by choosing a realisation structure, the digital filter can finally be implemented. In this example, the direct form II is chosen (using the coefficients calculated in Section 16.3.3.2); this is shown in Figure 16.10, and Equation (16.8) and Equation (16.9) can be reduced to Equation (16.29) and Equation (16.30).

$$p(n) = x(n) - a_1 p(n-1) - a_2 p(n-2) \tag{16.29}$$
$$y(n) = b_0 p(n) + b_1 p(n-1) + b_2 p(n-2). \tag{16.30}$$

The program to implement the above IIR filter is shown in Figure 16.11. Note that a cascade of three second-order filters was implemented. The code shown in Figure 16.11 can be optimised

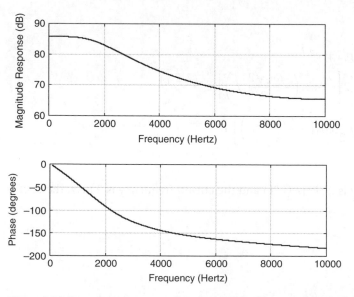

Figure 16.9 Frequency response of a second-order filter using the impulse invariant method.

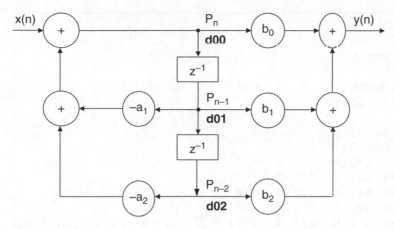

Figure 16.10 Direct form II structure.

by using the SIMD instruction available for the TMS320C66x devices. For instance, the following instructions:

```
prod1 = _mpy(d02,a2)>>15;
prod2 = _mpy(d01,a1)>>15;
d00 = xn + (short)(prod1 + prod2);
```

can be replaced by:

```
sum = (short) (_dotp2(d02d01,a2a1) >> 15);
d00T00 = (xn + sum) << 16; //result in d00
```

and the following code:

```
prod3 = _mpy(d01,b1);
prod4 = _mpy(d02,b2);
prod5 = _mpy(d00,b0);
y0 = (short)((prod3+prod4+prod5)>>15);
d02 = d01;
d01 = d00;
```

can be replaced by the following code:

```
d02d01d00T00= _dmv (d02d01, d00T00);
y0 = (short) ((_dotp4h(b2b1b0T0,d02d01d00T00)) >> 15);
```

In this case, only three multiplications are required and dummy values T0 and T00 are used to make use of the **dotp4h** instruction as illustrated in Figure 16.12. To use the two 32-bit registers **d0201** and **d00T00** as a pair register, the instruction **dmv** is used to concatenate them as shown above. Notice that **y0** is kept as 32-bit for the next stage to use.

```
#include "c6x.h"

IIR_C_MPY (short xn)
{
      short a1 = 0xd0b4;    // Negative value of a1
      short a2 = 0xc6f0,    // Negative value of a2

      short b0 = 0x3217;
      short b1 = 0x642e;
      short b2  = 0x3217;
      static short d01=0, d02=0, d00;
      static short d11=0, d12=0, d10;
      static short d21=0, d22=0, d20;
      static short d31=0, d32=0, d30;
      short y0, y1, y2, y3;
      int prod1, prod2, prod3, prod4, prod5;


      //============== stage 0 ==============
      prod1 = _mpy(d02,a2)>>15;
      prod2 = _mpy(d01,a1)>>15;
      d00 = xn + (short)(prod1 + prod2);
      prod3 = _mpy(d01,b1);
      prod4 = _mpy(d02,b2);
      prod5 = _mpy(d00,b0);
      y0 = (short)((prod3+prod4+prod5)>>15);
      d02 = d01;
      d01 = d00;
      //============== stage 1 ==============
      prod1 = _mpy(d12,a2)>>15;
      prod2 = _mpy(d11,a1)>>15;
      d10 = y0 + (short)(prod1 + prod2);
      prod3 = _mpy(d11,b1);
      prod4 = _mpy(d12,b2);
      prod5 = _mpy(d10,b0);
      y1 = (short)((prod3+prod4+prod5)>>15);
      d12 = d11;
      d11 = d10;
      //============== stage 2 ==============
      prod1 = _mpy(d22,a2)>>15;
      prod2 = _mpy(d21,a1)>>15;
      d20 = y1 + (short)(prod1 + prod2);
      prod3 = _mpy(d21,b1);
      prod4 = _mpy(d22,b2);
      prod5 = _mpy(d20,b0);
      y2 = (short)((prod3+prod4+prod5)>>15);
      d22 = d21;
      d21 = d20;
```

Figure 16.11 C code for the implementation of an IIR filter.

```
        //=============== stage 3 ===============
        prod1 = _mpy(d32,a2)>>15;
        prod2 = _mpy(d31,a1)>>15;
        d30 = y2 + (short)(prod1 + prod2);
        prod3 = _mpy(d31,b1);
        prod4 = _mpy(d32,b2);
        prod5 = _mpy(d30,b0);
        y3 = (short)((prod3+prod4+prod5)>>15);
        d32 = d31;
        d31 = d30;

        return(y3);
}
```

Figure 16.11 (Continued)

Figure 16.12 dotp4 with dummy T0 and T00 values.

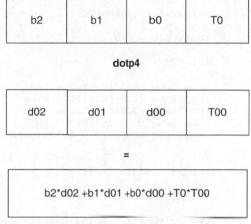

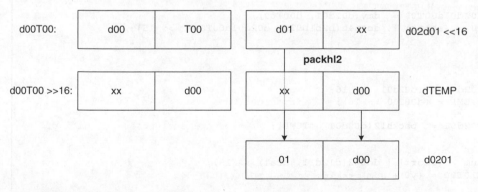

Figure 16.13 Updating **d02** and **d01**.

Finally, **d02** has to be replaced by **d01**, and **d01** has to be replaced by **d00**. To do so, the register **d02d01** is shifted to the left by 16 bits and the **packhl2** instruction is used to move **d00** to **d01** without affecting **d02**, as illustrated in Figure 16.13. The code is:

```
d02d01 = (d02d01)<<16; // move d01 to d02
dTEMP= (d00T00>>16);// format d00T00 for the next instruction
d02d01 =_packhl2 (d02d01, dTEMP); //leave d02 and update d01 with d00
```

The complete code for the four stages is shown in Figure 16.14.

```
#include "c6x.h"

volatile int d00T00 = 0x0, d02d01 = 0x0;
volatile int d10T00 = 0x0, d12d11 = 0x0;
volatile int d20T00 = 0x0, d22d21 = 0x0;
volatile int d30T00 = 0x0, d32d31 = 0x0;

IIR_C(short xn16) {
      int sum = 0;

      int a2a1 = 0xE6F0D0B4;

      short d00 = 0x0;

      int dTEMP = 0;

      long long d02d01d00T00 = 0x0;
      long long d12d11d10T00 = 0x0;
      long long d22d21d20T00 = 0x0;
      long long d32d31d30T00 = 0x0;
      long long b2b1b0T0 = 0x3217642E32170000;

      int y0, y1, y2, y3;
      int xn = (int) (xn16);

      //stage 0

      sum = (short) (_dotp2(d02d01, a2a1) >> 15);
      d00T00 = (xn + sum) << 16; //result in d00
      //  feedforward
      d02d01d00T00 = _dmv(d02d01, d00T00);
      y0 = (short) ((_dotp4h(b2b1b0T0, d02d01d00T00)) >> 15);
       //Multiply two sets of four signed 16-bit values and return the

      //32-bit sum

      d02d01 = (d02d01) << 16; // move d01 to d02
      dTEMP = (d00T00 >> 16); // format d00T00 for the next instruction

      d02d01 = _packhl2(d02d01, dTEMP); //leave d02 and update d01 with d00

      // stage 1
      sum = (short) (_dotp2(d12d11, a2a1) >> 15);
      d10T00 = (y0 + sum) << 16; //result in d10
      //  feedforward
      d12d11d10T00 = _dmv(d12d11, d10T00);
      y1 = (short) ((_dotp4h(b2b1b0T0, d12d11d10T00)) >> 15);
       //Multiply two sets of four signed 16-bit values and return the
      //32-bit sum
```

Figure 16.14 C code for the implementation of an IIR filter using SIMD instructions.

```
      d12d11 = d12d11 << 16; // move d11 to d12
      dTEMP = d10T00 >> 16; // format d10T00 for the next instrction
      d12d11 = _packhl2(d12d11, dTEMP); //leave d12 and update d11 with d10

      // stage 2
      sum = (short) (_dotp2(d22d21, a2a1) >> 15);
      d20T00 = (y1 + sum) << 16; //result in d20
      // feedforward
      d22d21d20T00 = _dmv(d22d21, d20T00);
      y2 = (short) ((_dotp4h(b2b1b0T0, d22d21d20T00)) >> 15);
      //Multiply two sets of four signed 16-bit values and return the
      //32-bit sum
      d22d21 = d22d21 << 16; // move d21 to d22
      dTEMP = d20T00 >> 16; // format d20T00 for the next instrction
      d22d21 = _packhl2(d22d21, dTEMP); //leave d22 and update d21 with d20

      // stage 3
      sum = (short) (_dotp2(d32d31, a2a1) >> 15);
      d30T00 = (y2 + sum) << 16; //result in d20
      // feedforward
      d32d31d30T00 = _dmv(d32d31, d30T00);
      y3 = (short) ((_dotp4h(b2b1b0T0, d32d31d30T00)) >> 15);
      //Multiply two sets of four  signed 16-bit values and return the
      //32-bit sum
      d32d31 = d32d31 << 16;  // move d31 to d32
      dTEMP = d30T00 >> 16;   // format d30T00 for the next instrction
      d32d31 = _packhl2(d32d31, dTEMP);  //leave d32 and update d31 with d30

      return ((short) y3);
}
```

Figure 16.14 (Continued)

16.5 Laboratory experiment

File location:

\Chapter_16_Code\IIR_filter

Open the project **IIR_filter** and explore the **main.c** function which contains the sinewave generator [1] and calls to two functions, **IIR_C()** and **IIR_C_MPY()**. Both functions are implemented with intrinsics. The **IIR_C_MPY()** function uses the **mpy** instruction, whereas the **IIR_C()** function uses **dotp2** and **dotp4h**.

Notice from Figure 16.15 that **IIR_C_MPY()** performed better than the **IIR_C()** function, which was more time-consuming to develop. The input and the filtered signals are shown in Figure 16.17 when built with the options shown in Figure 16.16.

16.6 Conclusion

This chapter introduces the IIR filters and describes two popular design methods, that is, the bilinear and the impulse invariant methods. Step by step, this chapter shows the procedures necessary to implement typical IIR filters specified by their transfer functions. Finally, this chapter provides complete implementation of an IIR filter in C language using intrinsics.

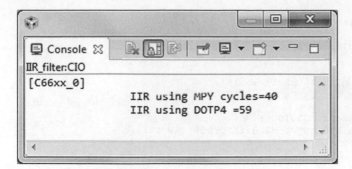

Figure 16.15 Performance comparison.

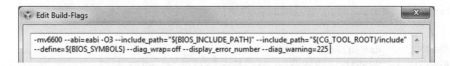

Figure 16.16 Build options.

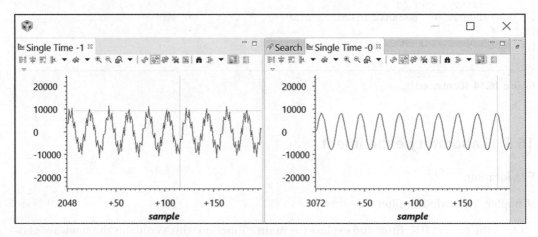

Figure 16.17 Input signal (Single time −1) and the filtered signal (Single time −0).

It has been shown again in this chapter that the compiler can perform better than hand-optimised code unless more effort is put in.

Reference

1 F. Kua, *Generation of a sine wave using a TMS320C54x digital signal processor*, Texas Instruments, July 2004. [Online]. Available: http://www.ti.com/lit/an/spra819/spra819.pdf. [Accessed December 2016].

17

Adaptive filter implementation

CHAPTER MENU

17.1 Introduction

Adaptive filters differ from other filters, such as finite impulse response (FIR) and infinite impulse response (IIR) filters, in the sense that the filter coefficients are not fixed and are determined by some desired specifications. In adaptive filters, the specifications are not known and change with time. Applications of adaptive filters are numerous and include process control, medical instrumentation, speech processing, echo and noise cancellation and channel equalisation.

The general procedure for constructing an adaptive filter is firstly to choose a FIR or IIR filter, and secondly to have a mechanism or algorithm to optimally adjust the FIR or IIR coefficients (see Figure 17.1) [1–3].

The real challenge for designing an adaptive filter resides with the adaptive algorithm. The latter needs to be practical to implement, adapt the filter coefficients quickly and provide the desired performance.

How is the performance measured? The main criterion that provides a good measure of the performance is based on the mean square error (MSE). This chapter will show how to calculate the filter coefficients using the MSE criterion before presenting the least mean square (LMS) algorithm. Finally, it will show how the LMS algorithm can be implemented in both C and linear assembly.

Multicore DSP: From Algorithms to Real-time Implementation on the TMS320C66x SoC, First Edition. Naim Dahnoun.
© 2018 John Wiley & Sons Ltd. Published 2018 by John Wiley & Sons Ltd.
Companion website: www.wiley.com/go/dahnoun/multicoredsp

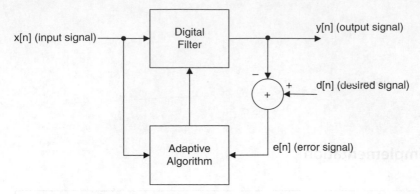

Figure 17.1 Basic block diagram of an adaptive filter.

17.2 Mean square error

Referring to Figure 17.1, we can write the following equations:

$$y(n) = \sum_{k=0}^{N-1} h(k)x(n-k) \tag{17.1}$$

$$e(n) = d(n) - y(n) \;=\; d(n) - \sum_{k=0}^{N-1} h(k)x(n-k) \tag{17.2}$$

Let $E[.]$ be the expectation operator.
Taking the expectation of $e^2(n)$:

$$E\left[e^2(n)\right] = E\left[(d(n)-y(n))^2\right]$$
$$= E\left[d^2(n)\right] - 2E[d(n)y(n)] + E\left[y^2(n)\right] .$$

Using the following vector notation:

$$y(n) = \sum_{i=0}^{N-1} h_i x(n-i)$$

can be written as:

$$y(n) = H^T X(n)$$

The expectation of $e^2(n)$ is now:

$$E\left[e^2(n)\right] = E\left[d^2(n)\right] - 2E\left[d(n)X^T(n)\right]H + E\left[H^T X(n)X^T(n)H\right] .$$

Finally, the above notation can be written as:

$$E\left[e^2(n)\right] = P_d - 2R_{dx}^T H + H^T R_{XX} H$$

where:

$$R_{dx} = E\left[d(n)X^T(n)\right]$$

$$R_{XX} = E\left[X(n)X^T(n)\right]$$

$$P_d = E[d^2(n)]$$

The vector R_{dx} is the cross-correlation between the desired signal $d(n)$ and the input signal $X(n)$. The R_{XX} matrix is the auto-correlation matrix of the input signal. To minimise $E[e^2(n)]$, the following equation should hold:

$$\frac{\partial E[e^2(n)]}{\partial h_i} = 0, \quad i = 0, 1, \ldots, N-1.$$

Therefore,

$$\frac{\partial E[e^2(n)]}{\partial h_i} = \frac{\partial P_d}{\partial h_i} - 2\frac{\partial R_{dx}^T H}{\partial h_i} + \frac{\partial H^T R_{XX} H}{\partial h_i}, \quad i = 0, 1, \ldots, N-1$$

$$= -2R_{dx} + 2R_{XX}H$$

$$= 0.$$

Finally, the optimum coefficient vector is:

$$H_{opt} = \frac{R_{dx}}{R} = R_{dx}R^{-1}. \tag{17.3}$$

Equation (17.3) is known as the Wiener–Hopf equation. It is evident that for real-time applications, the calculation of H_{opt} is time-consuming and impractical as it involves a matrix inversion.

17.3 Least mean square

The basic premise of the LMS algorithm is the use of the steepest descent algorithm shown here:

$$h_n(k) = h_{n-1}(k) + \beta\Delta_{n,k}.$$

where β is a positive value known as the *step size parameter*, and $\Delta_{n,k}$ is a gradient vector that makes $H(n)$ approach the optimal value H_{opt}. It has been shown [1] that:

$$\Delta_{n,k} = e(n)x(n-k).$$

Finally,

$$h_n(k) = h_{n-1}(k) + \beta e(n)x(n-k).$$

17.4 Implementation of an adaptive filter using the LMS algorithm

In this section, the adaptive filter shown in Figure 17.1 is considered by using a FIR digital filter and the LMS algorithm. Figure 17.2 shows the procedure for implementation. Note that the initialisation is done only once. The complete subroutine is shown in Figure 17.3.

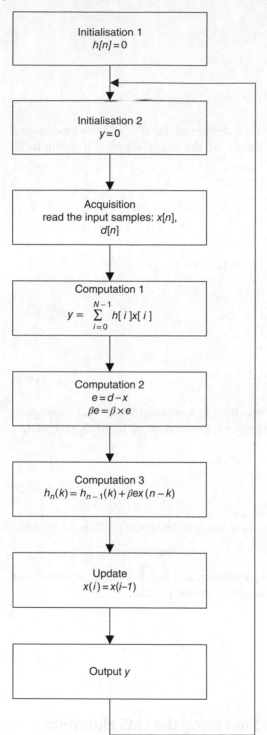

Figure 17.2 Steps for implementing an LMS adaptive filter.

```
for(i=0;i<N;i++)
      Y = Y + ((_mpy(h[i],X[i])) << 1) ;

E = D -(short) (Y>>16);
BETA_E =(short)((_mpy(beta,E)) >>15);

for(i=N-1;i>=0;i--)
{
      h[i] = h[i] +((_mpy(BETA_E,X[i])) >> 15);
      X[i]=X[i-1];
}
```

Figure 17.3 LMS algorithm in C language.

17.5   Implementation using linear assembly

To implement the algorithm in linear assembly language is more convenient as it is easier to code and maintain than assembly language, as discussed in Chapter 5. Let's follow the steps in Figure 17.2.

1) Computation 1. To implement $y(n) = \sum_{k=0}^{N-1} h(k)x(n-k)$, two pointers **hpt** and **xpt** are cre-

   ated, and the data and coefficients are stored as shown in Figure 17.4 so they can be accessed with LDDW instructions. The code to calculate $y(n)$ is shown here:

```
;Load h and x and calculate Y
      zero Y
      lddw  *hpt, h32:h10
      lddw  *xpt, x32:x10
      lddw  *hpt[1], h76:h54
      lddw  *xpt[1], x76:x54

      ddotp4h  h76:h54:h32:h10, x76:x54:x32:x10, acc3:acc4

      add Y,acc3,Y
      add Y,acc4,Y
      shr Y,15,Y
```

2) Computation 2. The code to implement $\mathbf{e} = \mathbf{d} - \mathbf{x}$ and $\boldsymbol{\beta}^*\mathbf{e}$ is shown here. Note that $\boldsymbol{\beta}$ was given as 0x174. However, to eliminate the shift (<<1), $\boldsymbol{\beta}$ is multiplied by 2 in advance.

```
;Calculate BETA

      mvkl   0x2E8, beta; beta=beta<<1= 0x174 x 2 = 0x2E8
      sub    D, Y, E
      mpy    beta, E, BETA_E
```

3) Computation 3. To update the 16-bit coefficients, one can use the **DSMPY2** instruction to update four coefficients in the same cycle, as illustrated in Figure 17.5. To update all the coefficients, the code used is shown here:

```
;make 8 16-bit with the same value BETA_E
    packh2 BETA_E, BETA_E, BETA_E
    MV     BETA_E, BETA_E2

    DSMPY2  x32:x10, BETA_E:BETA_E2, r3:r2:r1:r0
    DSMPY2  x76:x54, BETA_E:BETA_E2, r7:r6:r5:r4

    packh2 r1, r0, r1
    packh2 r3, r2, r3

    add2   h32, r3, h32
    add2   h10, r1, h10
    stdw   h32:h10, *hpt

    packh2 r7, r6, r7
    packh2 r5, r4, r5
    add2   h76, r7, h76
    add2   h54, r5, h54
    stdw   h76:h54, *hpt[1]
```

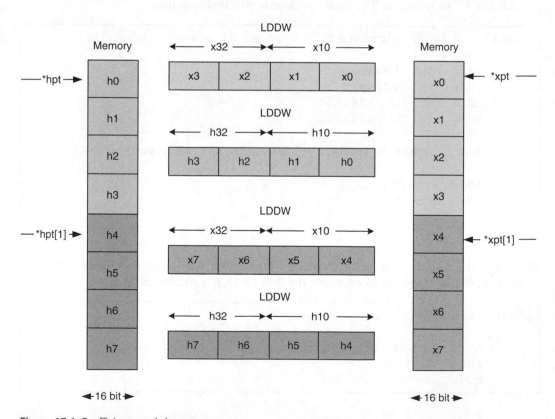

Figure 17.4 Coefficients and data storage.

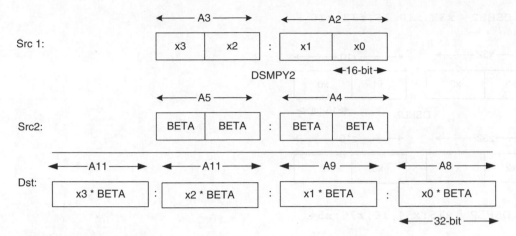

Figure 17.5 DSMPY2 operation.

Note that the instruction **packh2** illustrated in Figure 17.6 is convenient for converting and moving two 32-bit fixed-point values to two 16-bit values in one register.

4) Updating the input data. The sample data need to be updated, as shown in Figure 17.7. To do so, one can use the 4-way shift-left instruction (**DSHL2**) and the instruction **packlh2** illustrated in Figure 17.8. The code is:

```
DSHL2       x32:x10,16,z32:z10
DSHL2       x76:x54,16,z76:z54
packlh2     x54,x32,z54
```

Figure 17.6 Illustration of the **packh2** instruction.

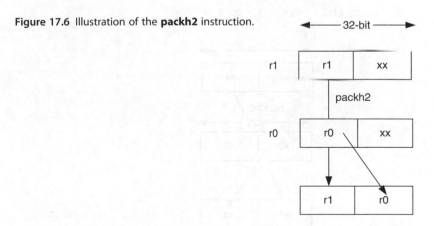

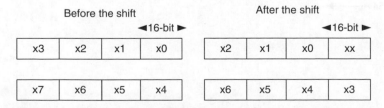

Figure 17.7 Data location before and after the shift.

1. DSHL2   **x32**:x10,16,**z32**:z10

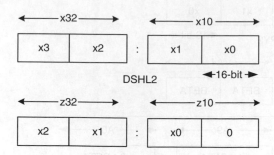

2. DSHL2   **x76**:x54,16,**z76:z54**

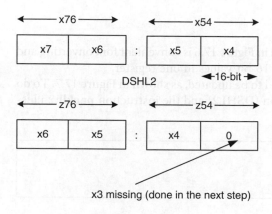

x3 missing (done in the next step)

3. packlh2   x54,x32,z54

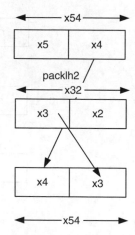

Figure 17.8 Updating data.

```
#define N 8
#define beta 0x00000174

short hC[8]={0,0,0,0,0,0,0,0};
short XC[8]={0,0,0,0,0,0,0,0};
int YC=0;
int EC=0;

#pragma DATA_ALIGN(hC,8)
#pragma DATA_ALIGN(XC,8)

lms_c (short newvalue)
{

    _nassert((int)(hC) % 8 == 0);
    _nassert((int)(XC) % 8 == 0);
    int i;
    short temp_out;
    short BETA_E,D;

    XC[0] = newvalue;
    D = XC[0];
    YC=0;

#pragma MUST_ITERATE(8,,8)
#pragma UNROLL(8)
    for(i=0;i<N;i++)
    {
        YC = YC + ((_mpy(hC[i],XC[i]))) ;
    }

    EC = D -(short) (YC>>15);
    BETA_E =(short)((_mpy(beta,EC)) >>15);

#pragma MUST_ITERATE(8,,8)
#pragma UNROLL(8)
    for(i=N-1;i>=0;i--)
    {
        hC[i] = hC[i] +((_mpy(BETA_E,XC[i])) >> 15);
        XC[i]=XC[i-1];
    }

    temp_out =  (short)(YC>>16);

    return(temp_out);

}
```

Figure 17.9 Using the compiler switches to increase the performance.

17.6 Implementation in C language with compiler switches

The performance of the code shown in Figure 17.3 can be improved by using the appropriate compiler switches learnt in Chapter 5. The code is shown in Figure 17.9.

17.7 Laboratory experiment

Files location:
  \Chapter_17_Code\Adaptive_Filter
  Open the project **Adaptive_filter** and explore the **main.c** function which contains the sine-wave generator [4] and calls to two functions, **lms_c()** and **lms_sa()**. The **lms_()** is implemented in C language, and the **lms_sa()** is implemented with linear assembly.
  Run the project and observe the output. Notice from Figure 17.10 that the code implemented in C language performed better than the code implemented in linear assembly. Open the graph display and set the properties as shown in Figure 17.11, and observe the outputs as shown in Figure 17.12.

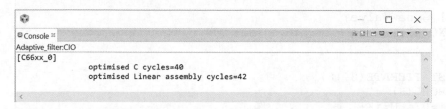

Figure 17.10 Cycles consumed.

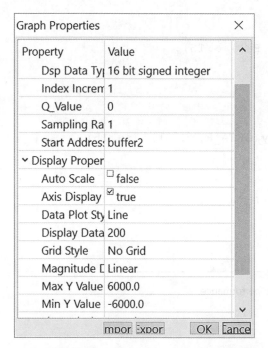

Figure 17.11 Graph properties settings.

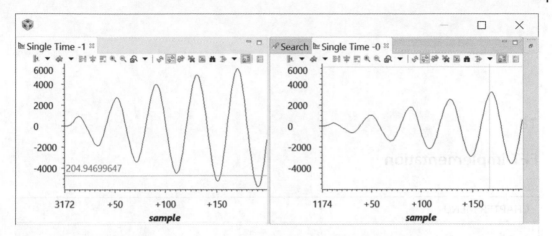

Figure 17.12 Output signal for the C code (*Single Time -1*) and (*Single Time -0*) for the linear assembly.

17.8 Conclusion

This chapter introduces the need for an adaptive filter in communications. It then shows how to calculate the filter coefficients using the mean square error (MSE) criterion, discusses the least mean square (LMS) algorithm and finally shows how the LMS algorithm is implemented in C language and linear assembly.

It has been shown in this chapter that the compiler can perform better than hand-optimised code unless more effort is put in.

References

1 B. Widrow and S. D. Stearns, *Adaptive Signal Processing*, New York: Pearson, 1985.
2 S. M. Bozic, *Digital and Kalman Filtering*, New York: Halsted Press, 1994.
3 S. S. Haykin, *Adaptive Filter Theory*, Upper Saddle River, NJ: Prentice Hall, 1996.
4 F. Kua, Texas Instruments, 2004, [Online]. Available: http://www.ti.com/lit/an/spra819/spra819.pdf. [Accessed December 2016].

18

FFT implementation

18.1   Introduction

The Fourier transform and its reverse transform convert a signal from a time or space domain to a frequency domain and from the frequency domain to a time or space domain, respectively. These transforms are very important in electrical engineering, communication, geology, medicine and optics, and the list is endless. However, for speed, these transforms are optimised to run fast. This has led to many algorithms, amongst these being the fast Fourier transform (FFT) Cooley–Tukey algorithm (1965) [1], the prime-factor FFT algorithm [2] and the split-radix FFT algorithm [3].

This chapter will introduce a derivation of an FFT algorithm and show its implementation.

18.2   FFT algorithm

18.2.1   Fourier series

Any periodic signal represented by a function $f(x)$ can be expressed by an infinite series of sines and cosines, as shown in Equation (18.1).

Multicore DSP: From Algorithms to Real-time Implementation on the TMS320C66x SoC, First Edition. Naim Dahnoun.
© 2018 John Wiley & Sons Ltd. Published 2018 by John Wiley & Sons Ltd.
Companion website: www.wiley.com/go/dahnoun/multicoredsp

$$f(x) = a_0 + \sum_{n-1}^{\infty} (a_n \, \cos(nx) + b_n \, \sin(nx)) \tag{18.1}$$

where:

$$a_0 = \int_{-\pi}^{\pi} f(x)dx$$

$$a_n = \int_{-\pi}^{\pi} f(x)\cos(nx)dx$$

$$b_n = \int_{-\pi}^{\pi} f(x)\sin(nx)dx$$

$$n = 1, 2, 3, \ldots$$

18.2.2 Fourier transform

The Fourier transform of the function $f(x)$ can be expressed as shown in Equation (18.2), and the inverse Fourier transform is shown in Equation (18.3).

$$F(k) = \int_{-\infty}^{+\infty} f(x)e^{-j2\Pi kx}dt \tag{18.2}$$

$$f(x) = \int_{-\infty}^{+\infty} F(k)e^{-j2\Pi kx}dk \tag{18.3}$$

where:

$$k = 0, 1, 2\ldots.N-1$$
$$x(nT) = x[n]$$
$$e^{jx} = \cos(x) + j\sin(x).$$

18.2.3 Discrete Fourier transform

The discrete Fourier transform (DFT) of a discrete-time signal $x(nT)$ is given by Equation (18.4).

$$X(k) = \sum_{n=0}^{N-1} x[n]e^{-j\frac{2\pi}{N}nk} \tag{18.4}$$

where:

$$k = 0, 1, \ldots N-1$$
and $x[n] = x(nT)$.

If we let $e^{-j\frac{2\pi}{N}} = W_N$, then Equation (18.4) can be rewritten as Equation (18.5).

$$X(k) = \sum_{n=0}^{N-1} x[n]W_N^{nk} \tag{18.5}$$

Table 18.1 DFT calculation for every frequency bin

$$X(0) = x[0]\,W_N^{\,0} + x[1]\,W_N^{\,0*1} + \dots + x[N-1]\,W_N^{\,0*(N-1)}$$

$$X(1) = x[0]\,W_N^{\,0} + x[1]\,W_N^{\,1*1} + \dots + x[N-1]\,W_N^{\,1*(N-1)}$$

$$\vdots$$

$$X(k) = x[0]\,W_N^{\,0} + x[1]\,W_N^{\,k*1} + \dots + x[N-1]\,W_N^{\,k*(N-1)}$$

$$\vdots$$

$$X(N-1) = x[0]\,W_N^{\,0} + x[1]\,W_N^{\,(N-1)*1} + \dots + x[N-1]\,W_N^{\,(N-1)(N-1)}$$

where:

$x[n]$ = input
$X[k]$ = frequency bins
W = twiddle factors.

Therefore, for N samples of x, we have N frequencies (frequency bins) representing the signal, and each frequency bin can be calculated, as shown in Equation (18.5).

From Table 18.1, it is clear that the DFT requires $N*N$ complex multiplications and $(N-1)*N$ complex additions.

18.2.4 Fast Fourier transform

A large amount of work has been devoted to reducing the computation time of a DFT. This has led to efficient algorithms known as *fast Fourier transforms*. In this section, the algorithm known as Radix 2 FFT is developed.

18.2.4.1 Splitting the DFT into two DFTs

By manipulating the twiddle factor W, as shown here and in Figure 18.1, Equation (18.5) can be rewritten as shown in Equation (18.6).

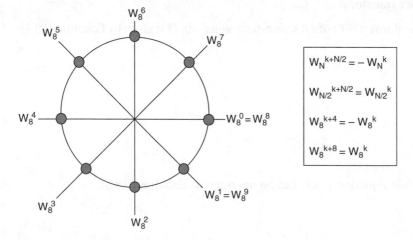

$$W_N^{\,k+N/2} = -W_N^{\,k}$$

$$W_{N/2}^{\,k+N/2} = W_{N/2}^{\,k}$$

$$W_8^{\,k+4} = -W_8^{\,k}$$

$$W_8^{\,k+8} = W_8^{\,k}$$

Figure 18.1 Twiddle factors for $N = 8$.

$$W_N^{2nk} = e^{-j\frac{2\pi}{N}2nk} = e^{-j\frac{2\pi}{N/2}nk} = W_{\frac{N}{2}}^{nk}$$

$$W_N^{(2n+1)k} = W_N^k \cdot W_{\frac{N}{2}}^{nk}$$

$$X(k) = \sum_{n=0}^{\frac{N}{2}-1} x[2n] W_N^{2nk} + \sum_{n=0}^{\frac{N}{2}-1} x[2n+1] W_N^{(2n+1)k} \tag{18.6}$$

Replacing $W_N^{(2n+1)k}$ *by* $W_N^k \cdot W_{\frac{N}{2}}^{nk}$, $x[2n]$ *by* $x_1[n]$ and $x[2n+1]$ *by* $x_2[n]$, Equation (18.6) can be rewritten as shown in Equation (18.7).

$$X(k) = \sum_{n=0}^{\frac{N}{2}-1} x_1[n] W_{\frac{N}{2}}^{nk} + W_N^k \sum_{n=0}^{\frac{N}{2}-1} x_2[n] W_{\frac{N}{2}}^{nk}$$
$$= Y(k) + W_N^k Z(k) \tag{18.7}$$

This shows that an N-point DFT can be divided into two $\frac{N}{2}$-point DFTs, $Y(k)$ and $Z(k)$, operating on the even and odd samples, respectively.

18.2.4.2 Exploiting the periodicity and symmetry of the twiddle factors

By exploiting the symmetry and periodicity of the twiddle factors, as shown here and illustrated in Figure 18.1, pairs of DFTs $X(k)$ *and* $X\left(k + \frac{N}{2}\right)$ can be grouped together, as they share the same twiddle factors and therefore will reduce the number of calculations; see Equations (18.7) and (18.8).

Symmetry: $W_N^{k+\frac{N}{2}} = e^{-j\frac{2\pi}{N}k} e^{-j\frac{2\pi}{N}\frac{N}{2}} = e^{-j\frac{2\pi}{N}k} e^{-j\pi} = -e^{-j\frac{2\pi}{N}k} = -W_N^k$

Periodicity: $W_{\frac{N}{2}}^{k+\frac{N}{2}} = e^{-j\frac{2\pi}{N/2}k} e^{-j\frac{2\pi}{N/2}\frac{N}{2}} = e^{-j\frac{2\pi}{N/2}k} = W_{\frac{N}{2}}^k$.

$$X(k) = \sum_{n=0}^{\frac{N}{2}-1} x_1[n] W_{\frac{N}{2}}^{nk} + W_N^k \sum_{n=0}^{\frac{N}{2}-1} x_2[n] W_{\frac{N}{2}}^{nk} \tag{18.7}$$

$$\vdots$$

$$X\left(k + \frac{N}{2}\right) = \sum_{n=0}^{\frac{N}{2}-1} x_1[n] W_{\frac{N}{2}}^{n\left(k+\frac{N}{2}\right)} + W_N^{k+\frac{N}{2}} \sum_{n=0}^{\frac{N}{2}-1} x_2[n] W_{\frac{N}{2}}^{n\left(k+\frac{N}{2}\right)} \tag{18.8}$$

Using the periodicity, the pair of DFTs can be rewritten as shown in Equations (18.9) and (18.10):

$$X(k) = \sum_{n=0}^{\frac{N}{2}-1} x_1[n] W_{\frac{N}{2}}^{nk} + W_N^k \sum_{n=0}^{\frac{N}{2}-1} x_2[n] W_{\frac{N}{2}}^{nk} \tag{18.9}$$

$$X\left(k + \frac{N}{2}\right) = \sum_{n=0}^{\frac{N}{2}-1} x_1[n] W_{\frac{N}{2}}^{nk} - W_N^k \sum_{n=0}^{\frac{N}{2}-1} x_2[n] W_{\frac{N}{2}}^{nk} \tag{18.10}$$

Equations (18.9) and (18.10) can be rewritten as Equations (18.11) and (18.12).

$$X(k) = Y(k) + W_N^k Z(k); \quad k = 0,\ldots\left(\frac{N}{2}-1\right) \tag{18.11}$$

$$X\left(k + \frac{N}{2}\right) = Y(k) - W_N^k Z(k); \; k = 0, \ldots \left(\frac{N}{2} - 1\right) \tag{18.12}$$

where $Y(k)$ *and* $Z(k)$ can be considered as two $\frac{N}{4}$ DFTs. By using the same process as shown in this section, $Y(k)$ *and* $Z(k)$ can be expressed as shown in Equations (18.13) to (18.16).

$$Y(k) = U(k) + W_{\frac{N}{2}}^k V(k) \tag{18.13}$$

$$Y\left(k + \frac{N}{4}\right) = U(k) - W_{\frac{N}{2}}^k V(k) \tag{18.14}$$

$$Z(k) = P(k) + W_{\frac{N}{2}}^k Q(k) \tag{18.15}$$

$$Z\left(k + \frac{N}{4}\right) = P(k) - W_{\frac{N}{2}}^k Q(k) \tag{18.16}$$

The process continues until 2-point DFTs are reached. This is known as the *decimation in time* (DIT) *Radix 2 FFT*, as illustrated in Figure 18.2.

Note that input data need to be out of order (e.g. 0, 4, 2, 6, 1, 5, 3, 7, for an 8-point FFT), and therefore the original data have to be reordered. This can be done by using a bit reversal since, when the binary representation of an index is reversed, it will lead to the correct value; see Figure 18.3. Reordering the input data of Figure 18.2 will lead to the decimation in frequency (DIF). However, the reordering should be performed at the end of the FFT; see Figure 18.4.

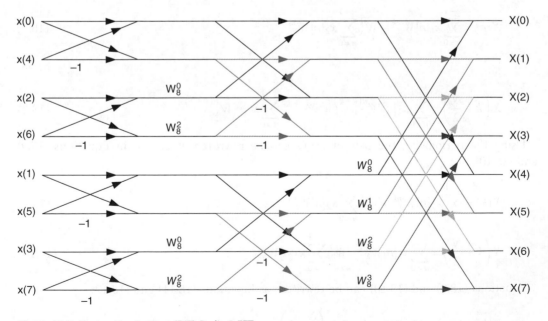

Figure 18.2 Decimation in time (DIT) Radix 2 FFT.

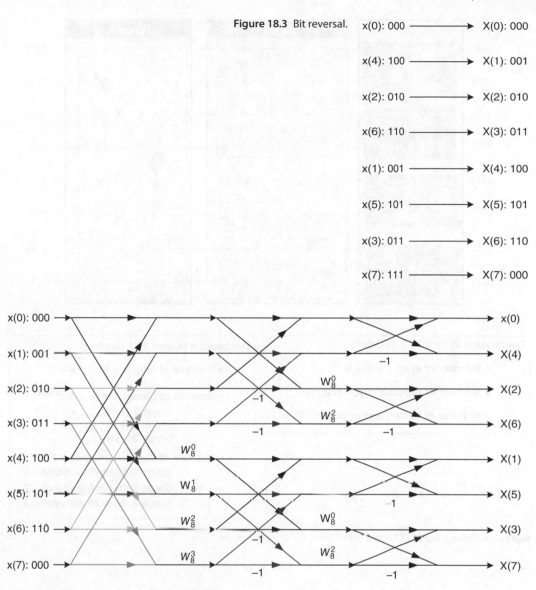

Figure 18.3 Bit reversal.

x(0): 000 ⟶ X(0): 000

x(4): 100 ⟶ X(1): 001

x(2): 010 ⟶ X(2): 010

x(6): 110 ⟶ X(3): 011

x(1): 001 ⟶ X(4): 100

x(5): 101 ⟶ X(5): 101

x(3): 011 ⟶ X(6): 110

x(7): 111 ⟶ X(7): 000

Figure 18.4 Decimation in frequency (DIF) Radix 2 FFT.

18.3 FFT implementation

To implement the FFT as illustrated in Figure 18.2, a few observations should be made:

1) For an N-point FFT, there are $\log_2 N$ stages. For $N = 8$, there will be three stages; see Figure 18.5.
2) There are $N/2^{\text{stage}}$ blocks per stage.
3) There are $2^{\text{stage}-1}$ butterflies per block.
4) The difference in indices for each butterfly is the same in each block within a stage.
5) The difference in indices for each butterfly in a group is twice that of the previous stage. Within Stage 1, there is a difference in indices of 1.

Decimation in time FFT (N-point):

- Number of stages = $\log_2 N$
- Number of blocks/stage = $N/2^{stage}$
- Number of butterflies/block = $2^{stage-1}$

Decimation in time FFT (8-point):

- Number of stages = $\log_2 N = 3$
- Number of blocks/stage = $N/2^{stage}$
  - Stage 1: 1: $N_{blocks} = 4$
  - Stage 2: 1: $N_{blocks} = 2$
  - Stage 3: 1: $N_{blocks} = 1$
- Number of butterflies/block = $2^{stage-1}$
  - Stage 1: butterflies/block = 1
  - Stage 2: butterflies/block = 2
  - Stage 3: butterflies/block = 4

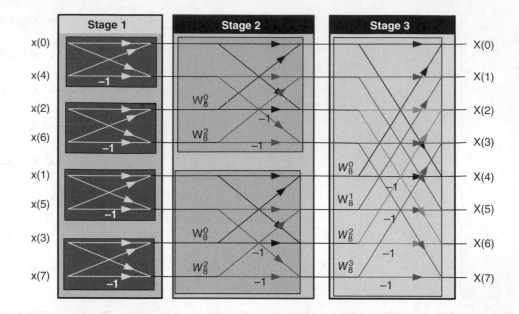

Figure 18.5 Diagram of DIT Radix 2 FFT used for the implementation.

To implement the inner loop (butterfly), let's calculate the outputs A and B of the butterfly, as shown in Figure 18.6, and derive the C code.

$$A = U + L = [U_R + L_R] + j[U_I + L_I]$$
$$B = (U - L) * W_N^k = ([U_R - L_R] + j[U_I - L_I]) * W_N^k$$

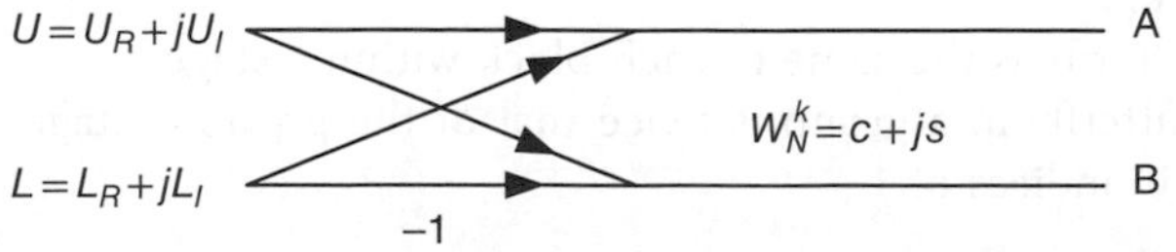

Figure 18.6 Flow graph of a butterfly.

Let's simplify Equations A and B by assuming:

$temp\, 2R = [U_R + L_R]$

$temp\, 2I = [U_I + L_I]$

$temp\, 1R = [U_R - L_I]$

$temp\, 1I = [U_I - L_I]$

$W_N^k = c + js$

These will lead to:

$A = temp\, 2R + j\, temp\, 2I$

and

$B = temp\, 1R + jtemp\, 1I)(c + js)$

$\quad = (temp\, 1R{*}c - temp\, 1I{*}s) + j(temp\, 1R{*}s + temp\, 1I{*}c)$

This can be implemented in C language by the C code shown in Figure 18.7. The complete implementation will require the calculations of each butterfly in each block starting from Stage 1 to the last stage. These can be performed by three loops, as shown in Figure 18.8.

```c
for(upperIdx = j; upperIdx < N; upperIdx+=k) //Do for a Number of butterflies
        {
                lowerIdx=upperIdx+BLdiff;

                temp1R = (Y[upperIdx].real - Y[lowerIdx].real)>>1;
                temp2R = (Y[upperIdx].real + Y[lowerIdx].real)>>1;
                temp1I = (Y[upperIdx].imag - Y[lowerIdx].imag)>>1;
                temp2I = (Y[upperIdx].imag + Y[lowerIdx].imag)>>1;

                Y[upperIdx].real   =   (short) temp2R;
                Y[upperIdx].imag   =   (short) temp2I;

                temp2R    = (c*temp1R - s*temp1I)>>15;
                temp2I    = (c*temp1I + s*temp1R)>>15;
                Y[lowerIdx].real   =   (short) temp2R;
                Y[lowerIdx].imag   =   (short) temp2I;
        }
```

Figure 18.7 Implementation of the butterfly.

```c
for(k = N; k > 1; k = (k>>1))    // Do log(base 2)(N) Stages
     {

  for(j = 0; j < BLdiff; j++) // Do for a Number of blocks  */
     {

    for(upperIdx = j; upperIdx < N; upperIdx+=k) //Do for a Number of butterfly

       {
       }


     }
```

Figure 18.8 Main loops for implementing a Radix 2 FFT.

18.4 Laboratory experiment

18.4.1 Part 1: Implementation of DIF FFT

Files location:

\Chapter_18_Code\FFT_in_C
Project: FFT_in_C.

In this laboratory experiment, a signal composed of two sinewaves is generated (this can be modified to add more sinewaves) and an FFT is performed on this signal, as shown in Figure 18.9.

Step 1. Open the **FFT_in_C** project and check the tools used for this project, as shown in Figures 18.10 and 18.11. Rebuild the project, compile, load and run the code.
Step 2. Display the signal and its FFT.

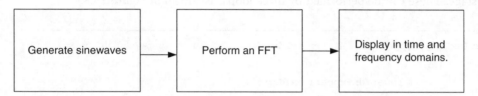

Figure 18.9 Tasks to perform.

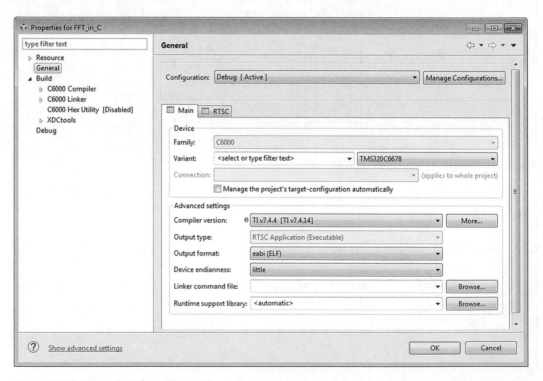

Figure 18.10 General configuration used.

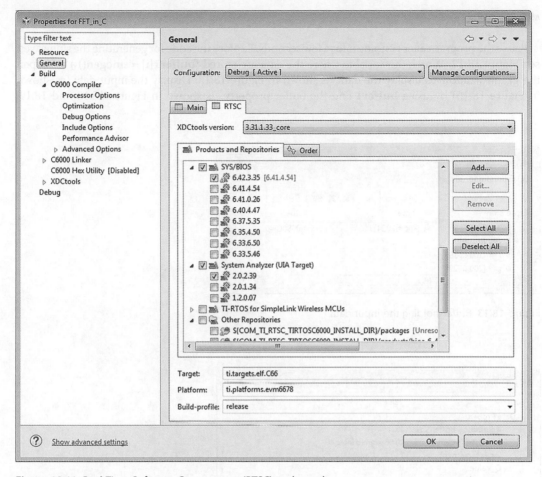

Figure 18.11 Real-Time Software Components (RTSC) tools used.

```
short sinegen()
{
    short temp;

    y[0] = ((((int)y[1]*(int)A))>>14) - y[2];
    y[2] = y[1];
    y[1] = y[0];

    z[0] = ((((int)z[1]*(int)B))>>14) - z[2];
    z[2] = z[1];
    z[1] = z[0];

    temp = (y[0] + z[0]/2);
    //           temp = (y[0]<<1);

    return(temp);
}
```

Figure 18.12 Code for generating sinewaves.

Pause the program after running it for a few seconds. Explore the code for generating the sinewaves; see Figure 18.12. Watch the variable where the data are stored (**buffer1[i] = sinegen()**) and open the graph window to display the data, as shown in Figure 18.13. Display the input data (as shown in Figure 18.15) by using **buffer1** and the buffer property, as shown in Figures 18.13 and 18.14.

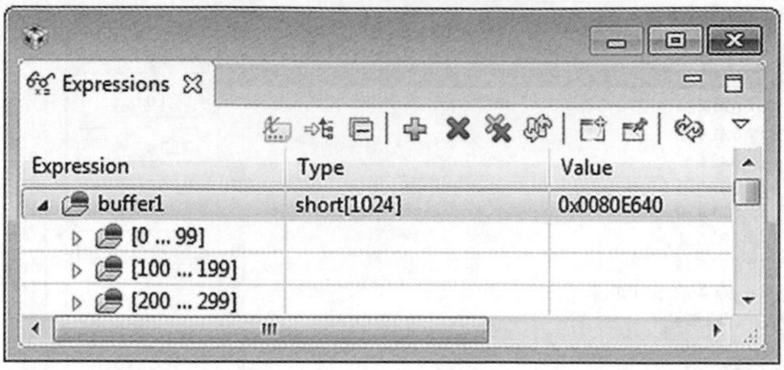

Figure 18.13 Buffer holding the input data.

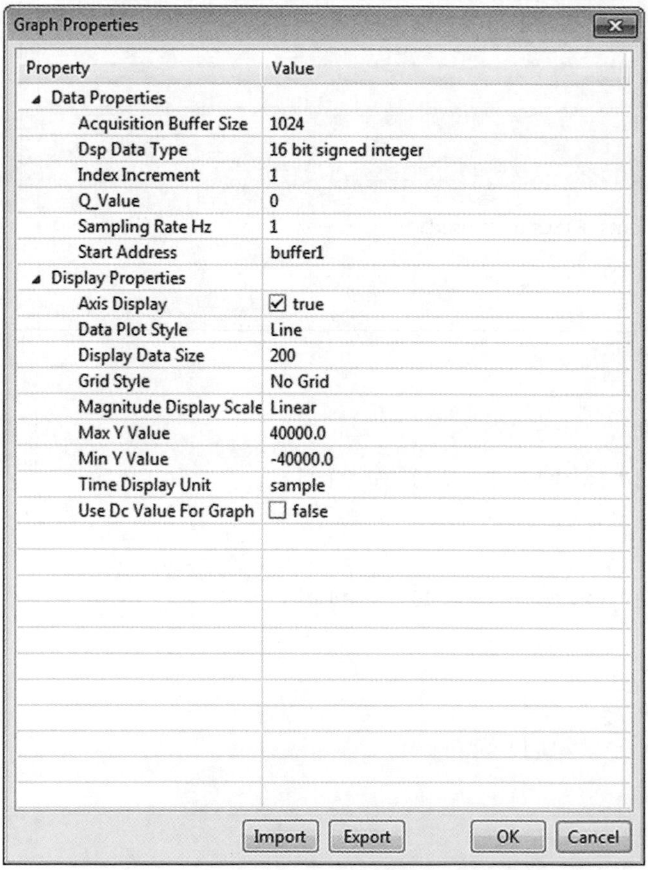

Figure 18.14 Graph properties used (**Display_Time.graphProp**).

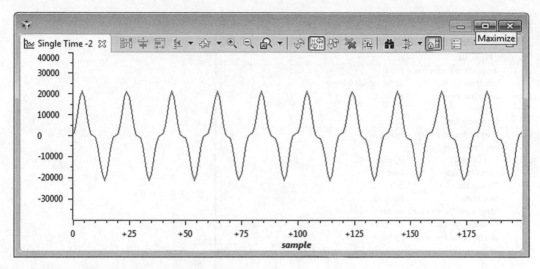

Figure 18.15 Display of the input data.

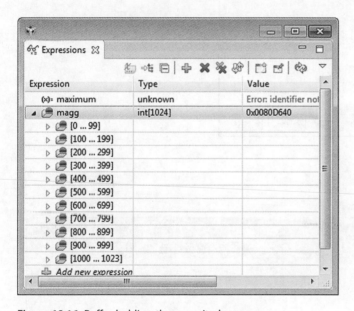

Figure 18.16 Buffer holding the magnitude **magg**.

Finally, display the FFT of the signal (as shown in Figure 18.18) by using the **magg** buffer and the buffer proprieties shown in Figures 18.16 and 18.17, respectively.

18.4.2 Part 2: Using ping-pong EDMA

Files location:

\Chapter_18_Code\FFT_in_C_EDMA.

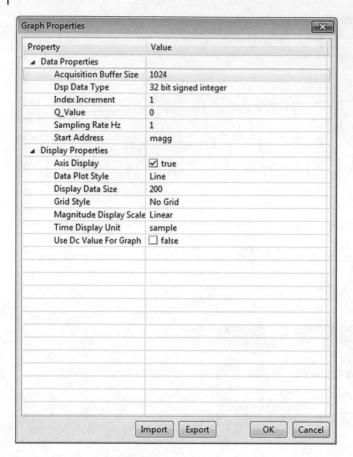

Figure 18.17 Graph properties.

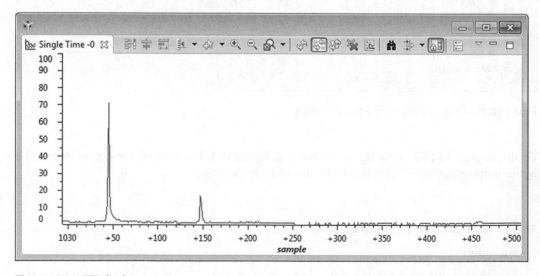

Figure 18.18 FFT display.

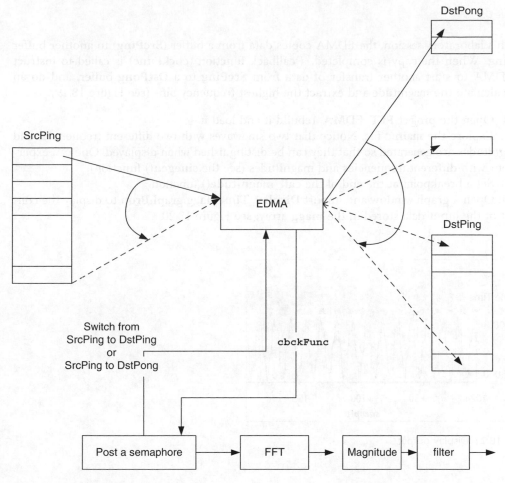

Figure 18.19 EDMA ping-pong.

Figure 18.20 Properties for displaying the input data in the array **magg**.

In this laboratory session, the EDMA copies data from a buffer (**SrcPing**) to another buffer **DstPing**. When the copy is completed, a callback function (**cbckFunc**) is called to instruct the EDMA to start another transfer of data from **SrcPing** to a **DstPong** buffer, and do an FFT, calculate the magnitude and extract the highest frequency bins (see Figure 18.19).

Step 1. Open the project **FFT_EDMA**, rebuild it and load it.
Step 2. Explore the **main.c** file. Notice that two sinewaves with two different frequencies and magnitudes are generated so that they can be distinguished when displayed. One can experiment with different frequencies and magnitudes (see the **sinegen()** function).
Step 3. Set a breakpoint at the end of the **call_magnitude()** function.
Step 4. Open a graph window and import **Display_Timemagg.graphProp** to display the content of the input data stored in the **magg** array; see Figure 18.20.

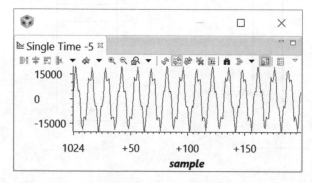

Figure 18.21 Display output.

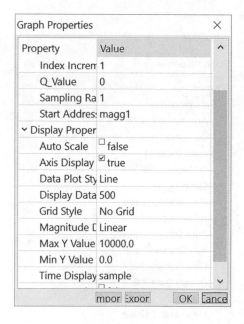

Figure 18.22 Properties for displaying the input data in the array **magg1**.

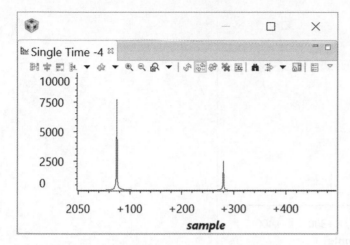

Figure 18.23 FFT magnitude display of data in **dstPong**.

Figure 18.24 Properties for displaying the input data in the array **magg2**.

Step 5. Run and observe the output graph; see Figure 18.21.
Step 6. Now that the data are correct, let's explore the magnitude of the FFT of the data stored in **magg1**. Open a graph window, and import **Display_FFTmagg1.graphProp** to display the content of the FFT magnitude stored in the **magg1** array (see Figure 18.22; for the FFT output, see Figure 18.23). The code is located in the main.c file.
Step 7. Repeat Step 6 to display the FFT of the data in **magg2** (see Figures 18.24 and 18.25). The graph properties for displaying **magg2** are located in **Display_FFTmagg2.graphProp**.

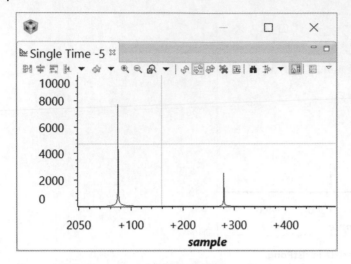

Figure 18.25 FFT magnitude display of data in **dstPing**.

18.5 Conclusion

In this chapter, the FFT Radix 2 has been studied, and a detailed implementation in C language has been provided. The implementation also includes using the EDMA to increase the performance.

References

1 J. W. Cooley and J. W. Tukey, An algorithm for the machine calculation of complex Fourier series, *Mathematics of Computation*, vol. **19**, pp. 297–301, 1965.
2 I. J. Good, The interaction algorithm and practical Fourier analysis, *Journal of the Royal Statistical Society*, vol. **20**, no. 2, pp. 361–372, 1958.
3 R. Yavne, An economical method for calculating the discrete Fourier transform, in *AFIPS Fall Joint Computer Conference, Part I*, San Francisco, CA, 1968.

19

Hough transform

19.1   Introduction

In this chapter, the implementation of the Hough transform (HT) is described. However, in order to fully understand the implementation, a review of the algorithm will be given first.

The HT is a popular technique for feature extraction in computer vision and digital image processing. It was first filed for a patent in 1960 by Paul V. C. Hough [1], and used later for the first time by Duda and Hart in 1972 to detect lines and curves in pictures [2]. With the HT, one can detect lines, circles or any structures with a known parametric equation. In this chapter, the implementation is limited to detect lines that will be used for lane detection in automotive applications.

19.2   Theory

Consider the Cartesian representation in Figure 19.1. P_i represents a point with the coordinates (x_i, y_i). There are an infinite number of lines passing by the point P_i. Each line can be fully represented by r (the normal to a line passing the point P_i) and the angle θ, as shown in Figure 19.1.

Let's express a line passing by the point P_i in terms of r and θ (see Equations (19.1), (19.2) and (19.3)) and then generalise this for each line.

$$y = ax + b \tag{19.1}$$
$$if\ y = 0;\ x = -\frac{b}{a}$$
$$if\ x = 0;\ y = b$$

Multicore DSP: From Algorithms to Real-time Implementation on the TMS320C66x SoC, First Edition. Naim Dahnoun.
© 2018 John Wiley & Sons Ltd. Published 2018 by John Wiley & Sons Ltd.
Companion website: www.wiley.com/go/dahnoun/multicoredsp

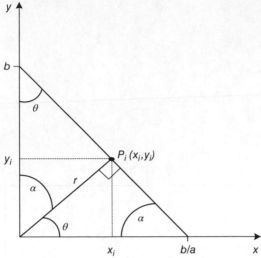

Figure 19.1 Cartesian representation of a line.

Also:

$$\frac{r}{b} = sin\theta \leftrightarrow \frac{1}{b} = \frac{sin\theta}{r} \tag{19.2}$$

$$\frac{r}{b}a = cos\theta \leftrightarrow \frac{a}{b} = \frac{cos\theta}{r} \tag{19.3}$$

Let's now express x and y by θ and r, since any point on the line $y = ax + b$ will have the same θ and r values.

Equation (19.1) can be rewritten as Equation (19.4) or (19.5):

$$\frac{y - ax}{b} = 1 \tag{19.4}$$

$$\frac{y}{b} - \frac{a}{b}x = 1 \tag{19.5}$$

Using Equations (19.2), (19.3) and (19.5), we can deduce Equations (19.6) and (19.7):

$$y\frac{sin\theta}{r} + x\frac{cos\theta}{r} = 1 \tag{19.6}$$

Finally, we can write:

$$x\,cos\theta + y\,sin\theta = r \tag{19.7}$$

From Equation (19.7), we can deduce that:

1) For every point (x_i, y_i), there is an infinite combination of (r, θ); see Figure 19.2.
2) For every combination of (r, θ), there is a number of finite points (x_i, y_i), and this number depends on how points are aligned. This constitutes the main idea behind the HT. This is illustrated in Figure 19.2b. Figure 19.2 shows that the ten aligned points will generated an (r, θ) point where all ten curves cross, and the number of crossings is referred to later as an *accumulator*.

As an example, Figure 19.3 shows some points and their corresponding polar representations. Figure 19.3d shows that there are three intersections and each intersection is between two

Figure 19.2 (a) Polar and (b) Cartesian coordinates.

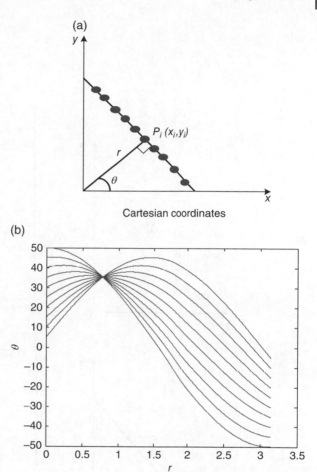

(a)

Cartesian coordinates

(b)

Polar coordinates

curves. That means that there are three straight lines and each line goes through two points. In Figure 19.3b, all the points are aligned and the curves (Figure 19.3e) intersect at one point. This means that there are three points belonging to the same straight line.

In Figure 19.3c, there are three points aligned on one line and three points aligned on another line. Figure 19.3c shows that there are five points, and Figure 19.3f shows many intersections and each intersection reveals how many points are aligned. For instance, there are two points with three intersections each; this means that there are two lines with three points on each (points (1,3), (2,2) and (3,1) and points (1,1) (2,2) and (3,3)).

19.3 Limits of *r* and θ

It is important to limit *r* and θ in order to reduce the number of calculations and the size of the accumulator and therefore the memory size, especially when the image size is large.

The limits of *r* and θ can be represented as shown in Equations (19.8) and (19.9), respectively:

$$\theta \in [\theta_{Min}, \theta_{Max}] \tag{19.8}$$

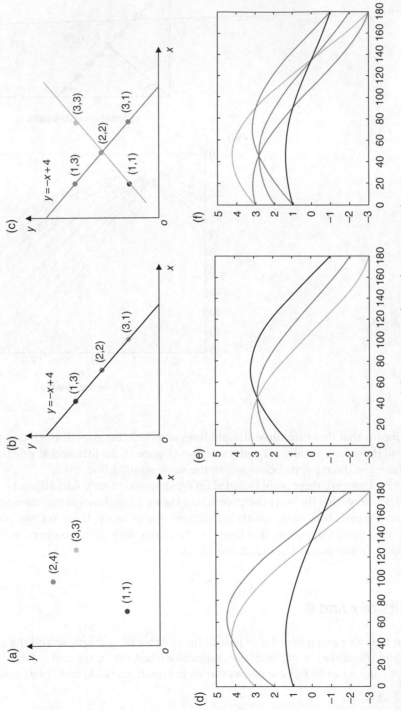

Figure 19.3 Example of points on a Cartesian representation and their corresponding polar representation.

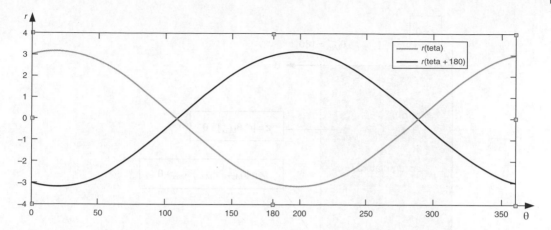

Figure 19.4 Plot of $r(\theta)$ *and* $r(\theta + 180)$.

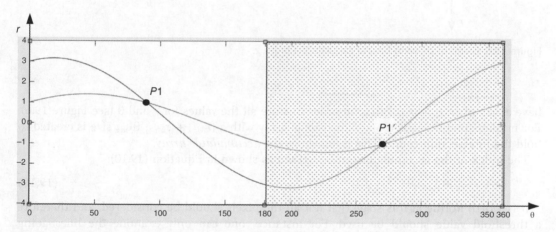

Figure 19.5 The r function is antisymmetric.

$$r \in [r_{Min}, r_{Max}] \tag{19.9}$$

where θ_{Min}, θ_{Max}, r_{Min} *and* r_{Max} need to be defined.

From Equation (19.7), one can see that the function is antisymmetric as $r(\theta) = -r(\theta + 180)$. This is illustrated in Figure 19.4. Therefore, all intersections of $r(\theta)$ will be antisymmetric and will occur twice (as illustrated in Figure 19.5). Therefore, the range could be limited from 0 to 180°.

The limits of r can also be obtained from Equation (19.7), where $r_{Max} = \sqrt{(x_M)^2 + (y_M)^2}$, x_M is the largest value of x and y_M is the largest value of y. However, $r_{Min} = -x_M$ since $x_M \gg y_M$ and $sin\theta > 0$ for all values of θ ($\theta = 0 : 180$).

19.4 Hough transform implementation

An array should be created in order to hold all values of r and θ. Section 19.3 showed that r spans from $-|r_{Min}|$ to r_{Max} and θ spans from 0 to 180°. Therefore, the accumulator should

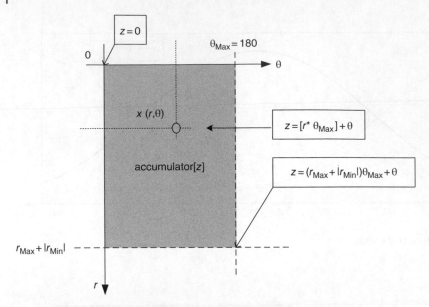

Figure 19.6 Diagram showing how to calculate the index *z*.

have a size of $(r_{Max} + |r_{Min}|)\theta_{Max}$ in order to store all the values for r and θ (see Figure 19.6). For implementation ease, a one-dimensional array with a $(r_{Max} + |r_{Min}|)\theta_{Max}$ size is created to hold the intersection count. This is known as the *accumulator array*.

The index of the accumulator can be written as shown in Equation (19.10):

$$(r + r_{Max})\theta_{Max} + \theta \tag{19.10}$$

It is worth noting at this stage that not all the pixels should be considered, and therefore a threshold value should be used. For instance, one can only examine the lines going through the brightest pixels and therefore set a threshold value of 255. Also, if one knows the range of the line as in lane detection for automotive applications, θ could be also restricted [3–5].

19.5 Laboratory experiment

In this laboratory experiment, an image is generated, an edge detection is applied to the image, the accumulator values are calculated and finally five maxima of the accumulator are extracted (extracting five lines); see Figure 19.7.

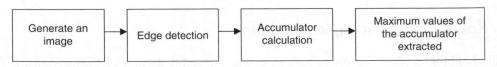

Figure 19.7 System to implement.

Files location: \Chapter_19_Code\Hough_Transform_1

Step 1. Generate an image. Using any drawing package, generate a picture with known lines to ease debugging as shown in Figure 19.8, and select the size of the image to be 1152 × 648 as shown in Figure 19.9.

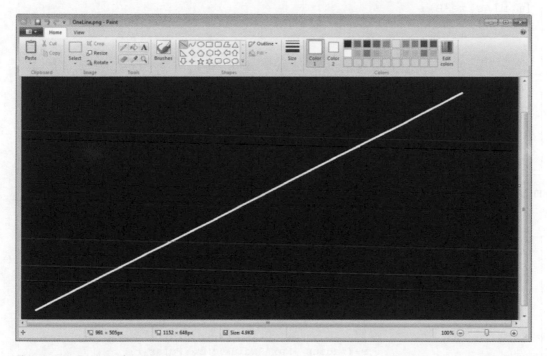

Figure 19.8 Generated test image.

Figure 19.9 Image property of 1152 × 648.

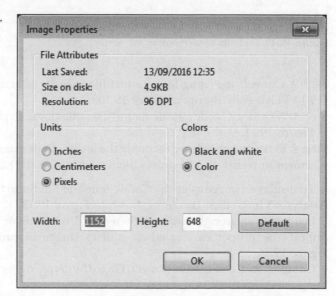

```
clear all
clc;
close all;
A=imread('H:\C66\Chapter_20_Hough_Transform\roadLine.png');
B=rgb2gray(A);
fid=fopen('Image2.txt','wt');
[m,n]=size(B);
for i=1:1:m
    for j=1:1:n
        if (j==n)
            fprintf(fid,'%g,\\ \n',B(i,j));
        else
            fprintf(fid,'%g,\t',B(i,j));
        end
    end
end
fclose(fid);
```

Figure 19.10 MATLAB code for generating the image header file.

```
for (y=0;y<HIGHT;y++)
            {
            for (x = 0; x < WIDTH; x++)
            {
             if (out[y*WIDTH+x] > 0){
              for (t = 0; t < Max_Polar; t++)
                {
                radius =(int) x*cos(t*DEG2RAD) + y*sin(t*DEG2RAD);
                z=(radius +Max_Radius)*Max_Polar+t;
                accumulator[z] = accumulator[z] +1;
                }
             }
            }
            }
```

Figure 19.11 Code for implementing the accumulator.

Step 2. Convert the *.**png** file to a text file in order to generate a header file. One can use the MATLAB code shown in Figure 19.10.

Step 3. Write the C code for calculating the radius for each pixel (x_i, y_i) and each angle; see code in Figure 19.11.

Step 4. Write the code to extract only the lines with the maximum numbers of points. The code shown in Figure 19.12 extracts the five lines with the maximum numbers of pixels (points).

To display an image using the Code Composer Studio, in the **Edit** mode, select **View, Other…**, **Analysis Views**, then **Image** as shown in Figure 19.13.

Inside the image, right click and then press **Properties** or **Import Properties**. For this project, import the properties located in project **Image_in_properties.txt**; see Figure 19.14 and Figure 19.15.

Compile, load and run the project. Go to the graph, right click and press **Refresh** to display the image; see Figure 19.16.

```
for (T = 0; T < 5; T++)
     {
     int MaximumVal = 0; // the maximum value can be estimated and used in order
                         //to reduce the number of unnecessary calculations.
          for (jj = 0; jj < Accumulator_Height; jj++)
               {
                    for (ii = 0 ; ii < Max_Polar ; ii++)

                    if (accumulator[jj*Max_Polar + ii] >= MaximumVal)
                         {
                    MaximumVal= accumulator[jj*Max_Polar + ii];
                    maximum[T][0] = MaximumVal;
                    maximum[T][1] = ii;
                    maximum[T][2] = jj - Max_Radius;
                         }
               }
     }
```

Figure 19.12 Code to extract the maxima of the accumulator.

Figure 19.13 Selecting the graphic display.

Figure 19.14 Changing, importing or exporting the properties of a graph.

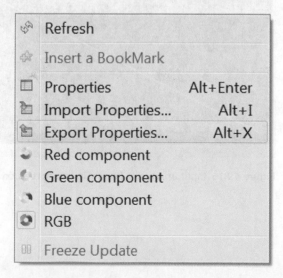

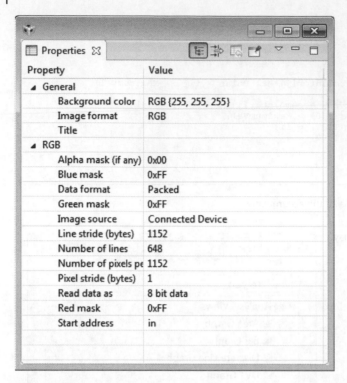

Figure 19.15 Properties for the input image (in): **Image_in_properties.txt**.

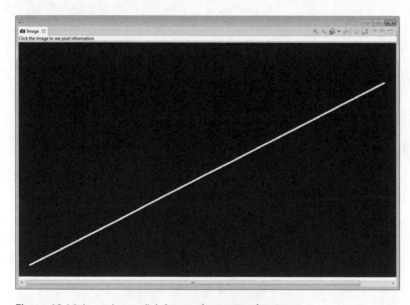

Figure 19.16 Input image (in): **Image_in_properties.txt**.

Now display the image after the edge detection is completed; see Figure 19.17 and Figure 19.18.
Now display the accumulator (see Figure 19.19 and Figure 19.20). To enhance the display of the accumulator, the accumulator has been incremented by 4 instead of 1, as shown here:

$$\text{accumulator}[z] = \text{accumulator}[z] + 4.$$

The indices of the accumulator with the five values are displayed in Figure 19.21.

Property	Value
◢ General	
Title	
Background color	RGB {255, 255, 255}
Image format	RGB
◢ RGB	
Number of pixels per line	1152
Number of lines	648
Data format	Packed
Pixel stride (bytes)	1
Red mask	0xFF
Green mask	0xFF
Blue mask	0xFF
Alpha mask (if any)	0x00
Line stride (bytes)	1152
Image source	Connected Device
Start address	out
Read data as	8 bit data

Figure 19.17 Properties for the image after the edge detection (out): **Image_out_properties.txt**.

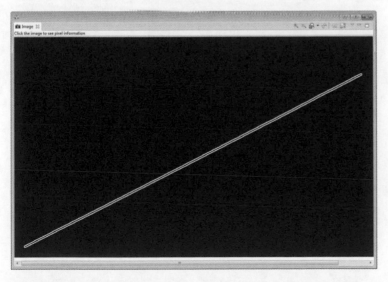

Figure 19.18 Image output after edge detection: **Image_out_properties.txt**.

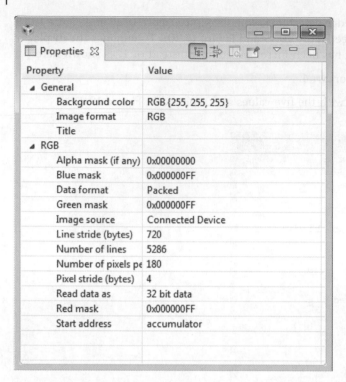

Figure 19.19 Image properties for the accumulator: **accumulator_properties.txt**.

Figure 19.20 Section of the accumulator output: **accumulator_properties.txt**.

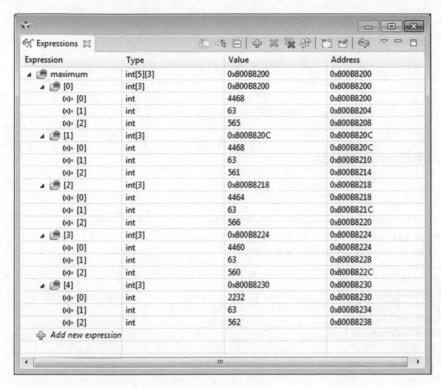

Figure 19.21 Values and coordinates of the five maxima.

19.6 Conclusion

This chapter shows the basic mathematics behind the Hough transform for detecting straight lines and how to implement the HT. This chapter also shows how to increase the performance by looking at the algorithm and minimising the number of operations required, and how to use the graphical display using the Code Composer Studio.

References

1 P. V. C. Hough, Method and means for recognizing complex patterns. USA Patent 3069654, December 1962.
2 R. O. Duda and P. E. Hart, Use of the Hough transformation to detect lines and curves in pictures, *Communications of the Association for Computing Machinery*, vol. **15**, no. 1, pp. 11–15, January 1972.
3 R. P. V. and D. N. Fan, Faster-than-real-time linear lane detection implementation using SoC DSP TMS320C6678, in *2016 IEEE International Conference on Imaging Systems and Techniques (IST)*, Chania, Crete, Greece, 2016.
4 U. Ozgunalp and N. Dahnoun, Lane detection based on improved feature map and efficient region of interest extraction, in *Global Conference on Signal and Information Processing (GlobalSIP)*, Orlando, Florida, USA, 2015.
5 P. Sun and H. Chen, Lane detection and tracking based on improved Hough transform and least squares method, in *Proceedings of International Symposium on Optoelectronic Technology and Application: Image Processing and Pattern Recognition*, Beijing, China, 2014.

20

Stereo vision implementation

20.1   Introduction

Moving from 2D to 3D representation of an object or a scene can offer an unparalleled advantage since it provides depth information. Stereo vision systems work in the same way as the human vision system which is composed of two eyes and a brain (two cameras and a processor). However, these systems cannot match the human vision system and therefore need to be limited to specific applications to improve the performance. Extracting depth information is a wide research topic that has been around for decades; the number of applications ranges from machine vision, robotics, autonomous vehicles and medicine, and now it is gaining more popularity in embedded and mobile environments due to the advance in low-cost miniature cameras and high-speed processes. An example is Advanced Driver Assistance Systems (ADAS) which have a big share of the ever-expanding automotive electronics market, and this has led to chip manufacturers like Texas Instruments developing the TDA2x SoC [1] and Freescale developing the S32V ADAS MCU that are designed to perform real-time video analytics, including stereo vision.

The principle behind stereo vision for depth calculation is fairly simple. Consider Figure 20.1, where an object P is located at an unknown distance Z from pinhole cameras with a focal length f. Therefore, by simple geometry one can extract the relationship between Z and the difference between the left and right cameras (L and R, respectively), as shown in Equation 20.1. This means that for specific cameras with the same focal length f and a baseline D, the difference in

Multicore DSP: From Algorithms to Real-time Implementation on the TMS320C66x SoC, First Edition. Naim Dahnoun.
© 2018 John Wiley & Sons Ltd. Published 2018 by John Wiley & Sons Ltd.
Companion website: www.wiley.com/go/dahnoun/multicoredsp

Figure 20.1 Computation of depth.

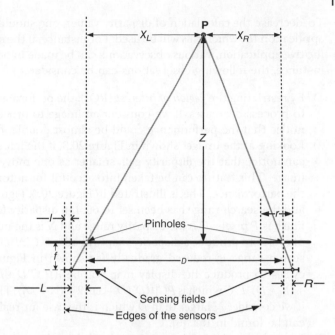

displacement between r and l is inversely proportional to the distance of an object. Therefore, the furthest the object P is from the cameras, the smallest is the difference between r and l. This difference is known as the *disparity*.

$$L + R = r - l = \frac{f\,D}{Z} \tag{20.1}$$

Camera calibration has to be performed before the depth information can be extracted. This is a big drawback of stereo vision as each set of cameras has to be calibrated separately.

There are five steps to consider for implementing a stereo vision system:

1) Decide which feature(s) to use. That could be intensity, colour or another structural feature like edges.
2) Decide to use either disparity between pixels or disparity between windows.
3) Decide how to process the whole image; see Section 20.2.
4) Decide on which cost function to use; see Section 20.3 [2, 3].
5) Select the method that uses the cost to determine the final value of the disparity [3].

20.2 Algorithm for performing depth calculation

There are different methods for evaluating the disparity values; a comprehensive review of stereo vision can be found in Refs. [4] and [5]. However, there is a compromise between precision of the disparity values (the quality of the disparity map) and the speed of performing the calculations.

To decrease the calculation of disparity values, one should consider the features in the specific application for which this will be used. For instance, if the stereo vision is to be used for an automotive application, various observations can be made in order to increase the performance. For instance, the following observations can be considered:

1) By restricting the *region of interest* (ROI), the performance increases as the number of pixels to process decreases. If we consider an image to process as shown in Figure 20.2, one can notice that the performance could be almost double if only the ROI is processed.
2) Looking at the images shown in Figure 20.3, if the left and right images are overlapped, one can notice that the disparity gets smaller as one moves from the bottom to the top of the image. This feature can be taken into account for automotive applications and can increase the performance. This is illustrated in Figure 20.4. Figure 20.4a shows an image of 100 lines, and the search range has been set to 60. If a disparity for each pixel takes C cycles and each line has P pixels, D is the disparity range and N is the number of lines, then the total number of cycles to produce a disparity map will take $P^*C^*N^*D$. However, for the same image, if the search range is reduced gradually (as shown in Figure 20.4b), then the total number of cycles to produce the display map will be $[(P^*C^*D + P^*C^*(D-1) + P^*C^*(D-2) + \ldots 2 + 1] + (N-D)^*P^*C$ which is $PC\{[(D^*(D+1)]/2 + (N-D)\}$. This shows an improvement of D times if we consider N larger than D which is the case for real applications. A practical application can be found in Ref. [6].
3) From the previous point, one can also reduce the search range but not gradually. Instead, the search range will be controlled by the surrounding pixels; see Figure 20.5 and Refs. [7] and [8].
4) The disparity calculations can be reduced without significantly affecting the quality by only considering the even lines and taking into account the previous frames [9].

20.3 Cost functions

It has been shown that the disparity calculation was based on the difference between the right and left of corresponding pixels or windows. However, the cost functions in stereo vision systems determine the similarity of two entities. These entities could be pixels (pixel-based) or a group of pixels (window-based). The common cost functions are shown in Table 20.1.

From Table 20.1, it can be seen that SAD is the easiest to implement and fastest to run. However, which cost function to use in a practical situation will depend on the application.

Figure 20.2 Reducing the ROI.

(a)

Left image

(b)

Right image

(c)

Left and right images overlapped, showing reduced disparity

Figure 20.3 (a) Left image, (b) right image and (c) left and right images merged showing the disparities reduced.

20.4 Implementation

The two main features of a stereoscopic system are the speed (real-time) and accuracy (quality). To calculate the disparity is to select a pixel on one image, find its corresponding pixel on the other image and measure the distance between these two pixels as illustrated in Figure 20.6. Figure 20.6 shows the left and right images with a pixel marked x. The pixel on the left image

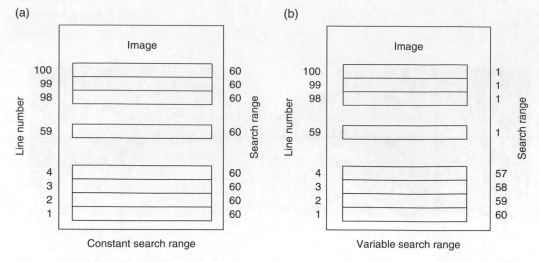

Figure 20.4 Reduced disparity, line by line.

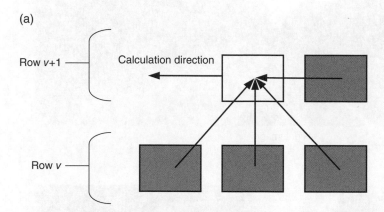

For the even-numbered row, the algorithm calculates the disparity values from right to left using support from bottom and right.

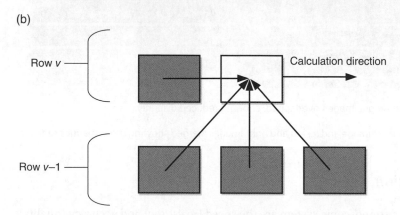

For the odd-numbered row, the algorithm calculates the disparity values from left to right using support from bottom and left.

Figure 20.5 Estimation of disparity range from neighbouring pixels [7].

Table 20.1 Most common cost functions

Sum of absolute differences (SAD)	$$SAD(i,j,k) = \sum_{x,y \in W} |I_L(x,y) - I_R(x,y-k)|$$
Zero-mean sum of absolute differences (ZSAD)	$$ZSAD(i,j,k) = \sum_{x,y \in W} |I_L(x,y) - M_L - I_R(x,y-k) + M_R|$$
Sum of squared differences (SSD)	$$SSD(i,j,k) = \sum_{x,y \in W} (I_L(x,y) - I_R(x,y-k))^2$$
Zero-mean sum of squared differences (ZSSD)	$$ZSSD(i,j,k) = \sum_{x,y \in W} (I_L(x,y) - M_L - I_R(x,y-k) + M_R)^2$$
Normalised cross correlation (NCC)	$$NCC(i,j,k) = \frac{\sum_{x,y \in W} (I_L(x,y) \times I_R(x,y-k))}{\sqrt{\sum_{x,y \in W} I_L(x,y)^2 \times \sum_{x,y \in W} I_R(x,y-k)^2}}$$
Zero-mean normalised cross correlation (ZNCC)	$$ZNCC(i,j,k) = \frac{\sum_{x,y \in W} [(I_L(x,y) - M_L) \times (I_R(x,y-k) - M_R)]}{\sqrt{\sum_{x,y \in W} (I_L(x,y) - M_L)^2 \times \sum_{x,y \in W} (I_R(x,y-k) - M_R)^2}}$$

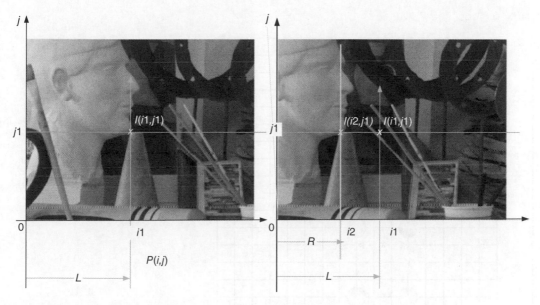

Figure 20.6 Corresponding pixels appear at different coordinates on the left and the right images.

with coordinates $p(i1, j1)$ appears on the right image at coordinates $p(i2,j1)$. The distance $(L - R)$ between the two pixels is known as the *disparity*.

In a practical situation, calculating the simple cost is very challenging due to issues like:

1) Alignment. The images should be aligned properly.
2) Illumination. The left and right images should be taken with the same lighting conditions.
3) Homogeneity. Neighbouring pixels can be identical, and this can lead to an ambiguous matching.

In addition to these points, any image transformation will cause inaccuracy. If the transformation is known, correction(s) can be made before calculating the disparity. However, this may be time-consuming.

The main idea for calculating the disparity is first to select the cost function that specifies how the pixels or windows are compared, then record the distance between the corresponding pixels or windows $(L - R)$.

20.4.1 Laboratory experiment

In this experiment, the SAD, NCC and ZNCC will be implemented in C language.

Files location:\Chapter_20_Code\Disparity_NCC_and_ZeroMean
Project's name: **Disparity_SAD_NCC**.

The implementations of the SAD, NCC and ZNCC cost functions are provided in this section.

20.4.1.1 SAD implementation
The main issue for basic implementation is with the calculations of the indices for each pixel. The images are two-dimensional but are represented with a one-dimensional array. Figure 20.7 illustrates how to calculate the indices for four pixels.

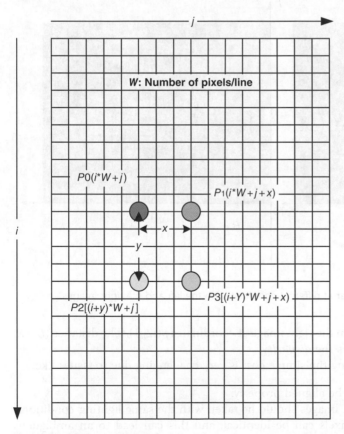

Figure 20.7 Indices calculation.

Step 1. Explore the SAD code, which is shown in Figure 20.9 and located in the **SAD.c** file.
Step 2. Open project **Disparity_SAD_NCC**, build and run.
Step 3. Display the input and the output images. Make sure the input images are displayed correctly before proceeding.

Open **Image**, as shown in Figure 20.8, then right click on the image and select **Import properties**. From the project directory, import the **Disparity_SAD.txt** file. Finally, right click on the image and press **Refresh**. The results are shown in Figure 20.10.

20.4.1.2   NCC implementation

Step 1. Explore the NCC code shown in Figure 20.11 and located in the **NCC.c** file.
Step 2. Open project **Disparity_SAD_NCC**, build and run.
Step 3. Display the results.

Open the image display, import the **Disparity_NCC.txt** file and refresh the display. The results are shown in Figure 20.12.

20.4.1.3   ZNCC implementation

Step 1. Explore the ZNCC code shown in Figure 20.13 and located in the **ZNCC.c** file.
Step 2. Open project **Disparity_NCC_Z.txt**, build and run.
Step 3. Display the results.

Open the image display and import the **Disparity_NCC_Z.txt** file, then refresh the display. The results are shown in Figure 20.14.

Figure 20.15 shows the time taken for each cost function. The results are obviously not acceptable for real-time applications. Figure 20.16 shows the time taken for each cost function when using the compiler optimiser with the **−O3** option. The results are still not good enough for

Figure 20.8 Selecting the image display feature.

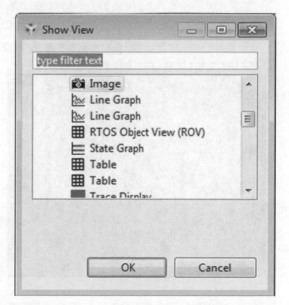

```
for (i=(H-1)-Radius;i>=0+Radius;i--)
    {
            for (j=(W-1)-Radius;j>=0+Radius;j--)
            {

            int Distance=0;
            int Minimize=100000;

            for (k=0;k<Search_Range;k++)
            {
                    Sum=0;
                    if (j-Radius-k>=0)
                    {
                            for (x=-Radius;x<=+Radius;x++){
                                    for (y=-Radius;y<=+Radius;y++)
                                    {

            Sum += abs(L[(i+x)*W+(j+y)]-R[(i+x)*W+(j-k+y)]);
                                    }
                            }

                            if (Sum<Minimize)
                            {
                                    Minimize=Sum;
                                    Distance=k;
                            }
                    }
            }
            Disparity_Map[i*W+j]= Distance<<1;
    }
    }
}
```

Figure 20.9 SAD implementation in C language.

Figure 20.10 SAD output image.

```
void NCC(unsigned char *Disparity, unsigned char *Left, unsigned char *Right)
{     int i,j, x, y, d;
float cost;
float up,ll,rr;
float aver_r,aver_l;
int N = 60;
int Radius = 4;
#define Height 223 // Height
#define Width 280  // Width
int Search_range = 40;

for (i = Height - 1 - Radius; i >=Radius; i--)
{
     // calculate the disparity
     for (j = Width - 1 - Radius; j >= Radius ; j--)
     {
          int Distance = 0;
          float max = 0;
          for (d = 0; d < Search_range; d++)
               if (j-Radius-d>=0)
               {
                    {
                         cost = 0;
                         up = ll = rr = 0;

                         for (x = -Radius ; x <=+ Radius; x++)
                         {
                              for (y = -Radius; y <=+ Radius; y++)
                              {
     ll += Left[(i + x)*Width + j + y] * Left[(i + x)*Width + j + y];
     up += Left[(i + x)*Width + j + y] * Right[(i + x)*Width + j + y - d];
     rr += Right[(i + x)*Width + j + y - d] * Right[(i + x)^Width + j + y - d];
                              }
                         }
                         cost = up / sqrt(ll*rr);

                         //extract the higher value (cost max =1)
                         if (cost >= max)
                         {
                              max = cost;
                              Distance = d;
                         }
                    }
                    // Store the disparity value in an array.
Disparity[i*Width + j] = Distance<<2 ; //multiply the Distance by 2 for enhancing
                                       //the display.
               }
     }
}
```

Figure 20.11 NCC implementation in C language.

real-time applications. However, by considering the points mentioned in Section 20.2 and using linear assembly, a better frame rate can be achieved.

20.5 Conclusion

This chapter shows the principle behind the stereo vision system and highlights different levels of optimisations for achieving real-time performance. Some techniques for reducing the processing time for calculating disparity values for automotive applications are also introduced.

Figure 20.12 NCC output image.

```c
void NCC_Mean(unsigned char *Disparity2, unsigned char *Left, unsigned char *Right)
{
N= (2*Radius +1)*(2*Radius+1);
    for (i = Height - 1 - Radius; i >=Radius; i--)
    {
        for (j = Width - 1 - Radius; j >= Radius; j--)
        {
            int Distance = 0;
            float max = 0;

        for (d = 0; d < Search_range; d++)
        {
            if (j-Radius-d>=0)
                                                                                {

            cost = 0;
            up = rr = ll = 0;
            aver_l = aver_r = 0;

            for (x = -Radius ; x <= Radius; x++)
            {
            for (y = -Radius; y < Radius; y++)
                {
                    aver_l += Left[(i + x)*Width + j + y];
                    aver_r += Right[(i + x)*Width + j + y - d];
                }
                }
                    aver_l = aver_l/N;
                    aver_r = aver_r/N;
            for (x = -Radius ; x <= Radius; x++)
            {
            for (y = -Radius; y < Radius; y++)
                {
ll += (Left[(i + x)*Width + j + y]-aver_l)*(Left[(i + x)*Width + j + y]-aver_l);
rr += (Right[(i + x)*Width + j + y - d]-aver_r)*(Right[(i + x)*Width + j + y - d]-aver_r);
up += (Left[(i + x)*Width + j + y]-aver_l)*(Right[(i + x)*Width + j + y - d]-aver_r);
                }
                }
                cost = up / sqrt(ll*rr);
                if (cost >= max)
                {
                    max = cost;
                    Distance = d;
                }
            }
            }
                Disparity2[i*Width + j] = Distance <<1;
                                                                                }
        }
    }
}
```

Figure 20.13 ZNCC implemented in C language.

Figure 20.14 ZNCC output image.

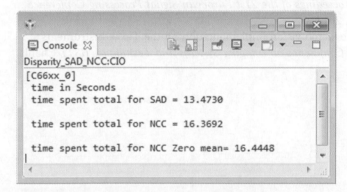

Figure 20.15 Time comparison between SAD, NCC and ZNCC with no optimisation.

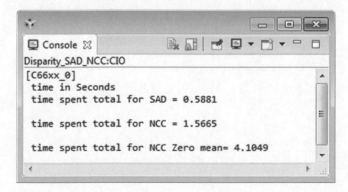

Figure 20.16 Time comparison between SAD, NCC and ZNCC with optimisation (−**O3**).

References

1 Texas Instruments, TDA2x ADAS System-on-Chip, [Online]. Available: http://www.ti.com/lit/ml/sprt681/sprt681.pdf [Accessed 5 July 2017].

2 S. Patil, J. S. Nadar, J. Gada, S. Motghare and S. S. Nair, Comparison of various stereo vision cost aggregation methods, *International Journal of Engineering and Innovative Technology (IJEIT)*, vol. **2**, no. 8, pp. 222–226, 2013.

3 H. Hirschmuller and D. Scharstein, Evaluation of cost functions for stereo matching, in *Conference on Computer Vision and Pattern Recognition (CVPR), 2007 (CVPR '07)*. IEEE, Minneapolis, Minnesota, USA, 2007.

4 N. Lazaros, G. C. Sirakoulis and A. Gasteratos, Review of stereo vision algorithms: from software to hardware, *International Journal of Optomechatronics*, vol. **2**, no. 4, p. 435–462, 2008.

5 B. Tippetts, D. J. Lee and J. Archibald, Review of stereo vision algorithms and their suitability for resource-limited systems, *Journal of Real-Time Image Processing*, vol. **11**, no. 1, p. 5–25, 2016.

6 Z. Zhang, Y. Wang, J. Brand and N. Dahnoun, Real-time obstacle detection based on stereo vision for automotive applications, in *5th European DSP Education and Research Conference (EDERC)*, Amsterdam, the Netherlands, 2012.

7 U. Ozgunalp, X. Ai, Z. Zhang, G. Koc and N. Dahnoun, Block-matching disparity map estimation using controlled search range, in *7th Computer Science and Electronic Engineering Conference (CEEC)*, Colchester, UK, 2015.

8 Z. Zhang, X. Ai and N. Dahnoun, Efficient disparity calculation based on stereo vision with ground obstacle assumption, in *Proceedings of the 21st European Signal Processing Conference (EUSIPCO)*, Marrakech, Morocco, 2013.

9 H. Cui and N. Dahnoun, Real-time stereo vision implementation on the Texas Instruments KeyStone II SoC, in *International Conference on Imaging Systems and Techniques (IST)*, Chania, Crete, Greece, 2016.

Index

Multicore DSP: From Algorithms to Real-time Implementation on the TMS320C66x SoC, First Edition. Naim Dahnoun.
© 2018 John Wiley & Sons Ltd. Published 2018 by John Wiley & Sons Ltd.
Companion website: www.wiley.com/go/dahnoun/multicoredsp